Mathematics 8
FOCUS ON UNDERSTANDING

Authors
Nancy Fournier
B.A., B.Ed.
Halifax Regional School Board

Daniel MacDonald
B.Sc., B.Ed.
Bridgewater, Nova Scotia

Jodie MacIlreith
B.A., B.Ed., M.Ed., M.Ed.
Halifax Regional School Board

David McKillop
B.Sc., B.Ed., M.Ed.
Truro, Nova Scotia

Jacob Speijer
B.Eng., M.Sc.Ed., P.Eng.
District School Board of Niagara

Sandy Szeto
B.Sc., B.Ed.
Toronto District School Board

Marilyn Price
Lower Sackville, Nova Scotia

Pedagogy Consultants
Beth Calabrese
Coxheath, Nova Scotia

Dan Gilfoy
Halifax Regional School Board

Therese Forsythe
Annapolis Valley Regional School Board

Tess Miller
Queen's University, Kingston, Ontario

Assessment Consultant
Tess Miller
Queen's University, Kingston, Ontario

Marilyn Price
Lower Sackville, Nova Scotia

Technology Consultant
Rona Chisholm-Cleary
Halifax Regional School Board

Literacy Consultant
Debby Vass
Halifax Regional School Board

The Nova Scotia Department of Education wishes to thank the following teachers and consultants for their thoughtful reviews and suggestions in the development of this textbook.

Sharon McCready
Provincial Mathematics Strategy Consultant

Anne Boyd
Retired from Strait Regional School Board

Donna Karsten
Provincial Mathematics Consultant

Peggy MacIntyre
Retired from Cape Breton-Victoria Regional School Board

Richard MacKinnon
MacMath Ed., Halifax

Ron MacLean
Retired Mathematics Consultant, Cape Breton-Victoria Regional School Board

Stephen McNeill
Annapolis Valley Regional School Board

LaJune Naud
Geometry Consultant, Halifax, Nova Scotia

Janet Porter
Provincial Literacy Support Consultant

The Nova Scotia Department of Education wishes to give a special thanks to **Nancy Chisholm**, IT Integration Consultant of the Department of Education, for her thoughtful reviews and advice on the use of technology for Grades 7 to 9.

Toronto Montréal Boston Burr Ridge, IL Dubuque, IA Madison, WI New York San Francisco St. Louis Bangkok Bogotá Caracas Kuala Lumpur Lisbon London Madrid Mexico City Milan New Delhi Santiago Seoul Singapore Sydney Taipei

The McGraw·Hill Companies

COPIES OF THIS BOOK MAY BE OBTAINED BY CONTACTING:

McGraw-Hill Ryerson Ltd.

WEB SITE:
http://www.mcgrawhill.ca

E-MAIL:
orders@mcgrawhill.ca

TOLL-FREE FAX:
1-800-463-5885

TOLL-FREE CALL:
1-800-565-5758

OR BY MAILING YOUR ORDER TO:
McGraw-Hill Ryerson
Order Department
300 Water Street
Whitby, ON L1N 9B6

Please quote the ISBN and title when placing your order.

McGraw-Hill Ryerson
Mathematics 8: Focus on Understanding

ISBN-13: 978-0-07-096588-1
ISBN-10: 0-07-096588-9

http://www.mcgrawhill.ca

1 2 3 4 5 6 7 8 9 10 TCP 1 0 9 8

Printed and bound in Canada

PUBLISHER: Linda Allison
PROJECT MANAGERS: Eileen Jung, Maggie Cheverie
DEVELOPMENTAL EDITORS: Bradley T. Smith, Susan Lishman, Ingrid D'Silva
MANAGER, EDITORIAL SERVICES: Crystal Shortt
COPY EDITORS: Laurel Sparrow, Loretta Johnson
PHOTO RESEARCH/PERMISSIONS: Robyn Craig Literary & Photo Research
PHOTO RESEARCH/SET-UP PHOTOGRAPHY: Roland W. Meisel, First Folio Resource Group, Inc., Dick Hemingway, Michelle Darling, Pippin Lee, Nick Beecham
EDITORIAL ASSISTANT: Erin Hartley
MANAGER, PRODUCTION SERVICES: Yolanda Pigden
PRODUCTION COORDINATOR: Jennifer Hall
COVER DESIGN: Dianna Little
ART DIRECTION: Tom Dart/First Folio Resource Group Inc.
ELECTRONIC PAGE MAKE-UP: Tom Dart, Kim Hutchinson, Adam Wood/ First Folio Resource Group, Inc.
COVER IMAGE: Courtesy of Dick Killam taken of Joggins, Western Shore, Nova Scotia

The authors, consultants, project managers, and publisher would like to thank Mahrie Boyle, Brett Bridges, Paul Ripley, and Adrienne Wong for their assistance in the development of this textbook.

Contents

CHAPTER 4

Ratio and Proportion 144

CHAPTER 5

Data Management and Probability 182

CHAPTER 6

Rational Numbers 240

CHAPTER 7

Algebraic Expressions and Solving Equations 290

CHAPTER 8

Patterns and Relations

CHAPTER 9

Geometry II

CHAPTER 10

Measurement

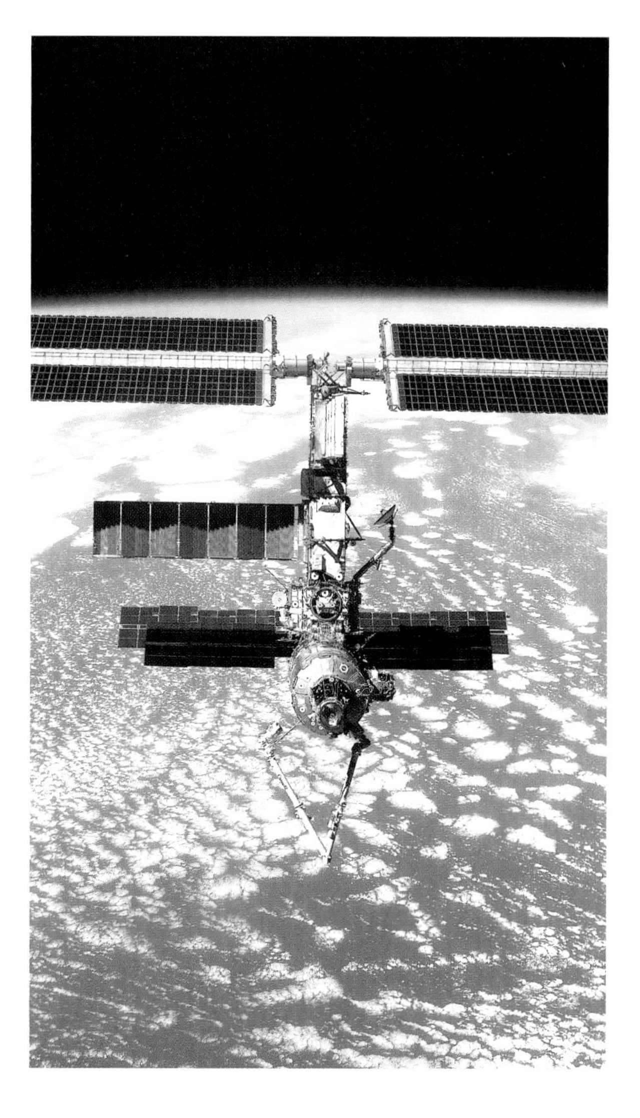

A Tour of Your Textbook

How is *Mathematics 8: Focus on Understanding* set up?

Each chapter starts off with a **Chapter Problem** that connects math and your world. You will be able to solve the problem using the math skills that you learn in the chapter.

CHAPTER 2

Fraction Operations

Each year Truro holds a tulip festival where beautiful flower gardens are open for public viewing. These gardens may contain many different colours of tulips. One colour may occupy only a fraction of the whole garden. What fractions can you use to describe this garden?

In this chapter, you will learn how to add, subtract, multiply, and divide fractions using materials, pictures, and symbols. You will apply the order of operations to solve problems involving operations with proper fractions, improper fractions, and mixed numbers.

Chapter Problem

Gardens often contain sections with different colours or types of flowers or vegetables. Think about a garden you have seen recently or a garden you would like to plant.

- Describe the different colours in the garden.
- What fraction of the garden is each colour?
- What fraction of the garden is each type of flower or vegetable?

47

You are asked to answer questions related to the problem throughout the chapter.

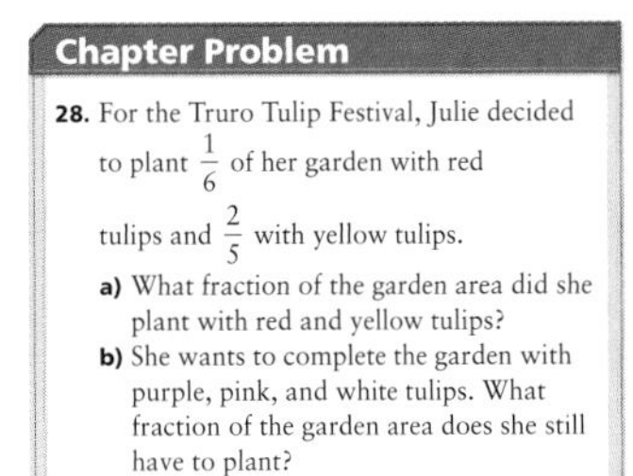

Chapter Problem

28. For the Truro Tulip Festival, Julie decided to plant $\frac{1}{6}$ of her garden with red tulips and $\frac{2}{5}$ with yellow tulips.

a) What fraction of the garden area did she plant with red and yellow tulips?

b) She wants to complete the garden with purple, pink, and white tulips. What fraction of the garden area does she still have to plant?

The **Chapter Problem Wrap-Up** is at the end of the chapter, on the second Practice Test page.

Chapter Problem Wrap-Up

1. The grade 8 classes are asked to suggest a design for the school flower garden. Your class is responsible for $\frac{1}{4}$ of the entire garden. This portion must contain four different colours of flowers.

a) Create a design for your class's part of the garden.

b) What fraction of the entire garden is each colour? Show how you know.

2. Create a design for your own garden. Use at least four different types of flowers or vegetables. Write as many fraction statements as you can about your garden. Use each of the four operations (+, −, ×, ÷) at least once in each statement. Use pictures, words, and symbols in your report.

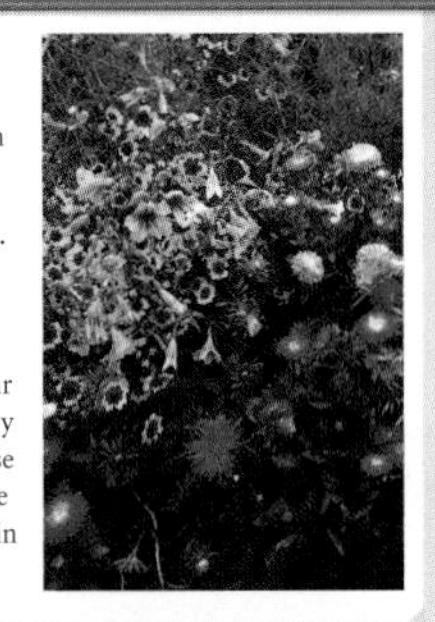

The **Get Ready** pages provide a brief review of skills from previous grades that are important for success with this chapter.

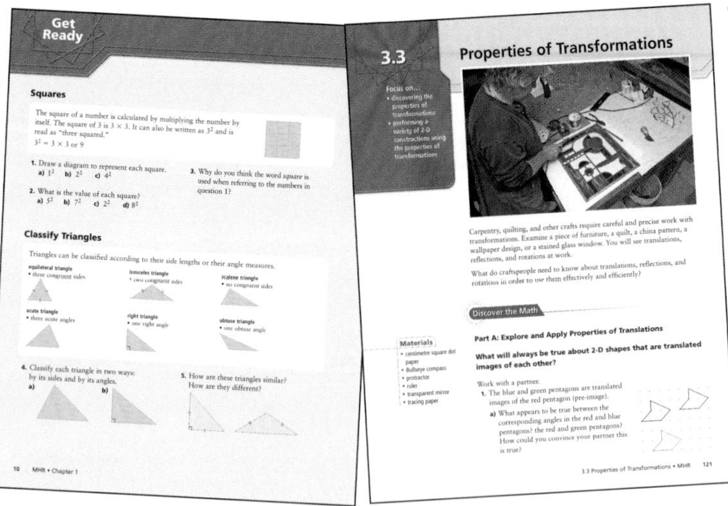

Get Ready

Squares

The square of a number is calculated by multiplying the number by itself. The square of 3 is 3×3. It can also be written as 3^2 and is read as "three squared."

$3^2 = 3 \times 3$ or 9

1. Draw a diagram to represent each square.
a) 1^2 **b)** 2^2 **c)** 4^2

2. What is the value of each square?
a) 5^2 **b)** 7^2 **c)** 2^2 **d)** 8^2

3. Why do you think the word *square* is used when referring to the numbers in question 1?

Classify Triangles

Triangles can be classified according to their side lengths or their angle measures.

equilateral triangle
• three congruent sides

isosceles triangle
• two congruent sides

scalene triangle
• no congruent sides

acute triangle
• three acute angles

right triangle
• one right angle

obtuse triangle
• one obtuse angle

4. Classify each triangle in two ways: by its sides and by its angles.
a) **b)**

5. How are these triangles similar? How are they different?

10 MHR • Chapter 1

3.3 **Properties of Transformations**

Focus on...
• discovering the properties of transformations
• performing a variety of 2-D constructions using the properties of transformations

Carpentry, quilting, and other crafts require careful and precise work with transformations. Examine a piece of furniture, a quilt, a china pattern, a wallpaper design, or a stained glass window. You will see translations, reflections, and rotations at work.

What do craftspeople need to know about translations, reflections, and rotations in order to use them effectively and efficiently?

Discover the Math

Part A: Explore and Apply Properties of Translations

What will always be true about 2-D shapes that are translated images of each other?

Materials
• centimetre square dot paper
• Bullseye compass
• protractor
• ruler
• transparent mirror
• tracing paper

Work with a partner.

1. The blue and green pentagons are translated images of the red pentagon (pre-image).

a) What appears to be true between the corresponding angles in the red and blue pentagons? the red and green pentagons? How could you convince your partner this is true?

3.3 Properties of Transformations • MHR 121

The **numbered sections** often start with a photo to connect the topic to a real setting. The purpose of this introduction is to help you make connections between the math in the section and the real world, or to make connections to previous knowledge.

A three-part lesson follows.

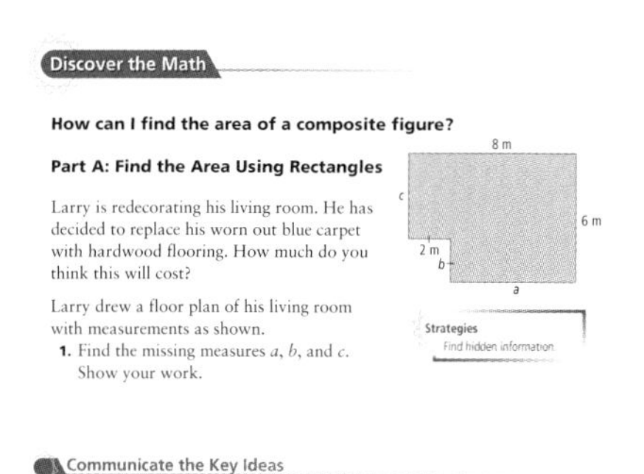
Discover the Math

How can I find the area of a composite figure?

Part A: Find the Area Using Rectangles

Larry is redecorating his living room. He has decided to replace his worn out blue carpet with hardwood flooring. How much do you think this will cost?

Larry drew a floor plan of his living room with measurements as shown.

1. Find the missing measures *a*, *b*, and *c*. Show your work.

The first part helps you find answers to the key question.

- An activity is designed to help you build your own understanding of the new concept and lead toward answers to the key question.
- **Examples** and **Solutions** demonstrate how to use the concept.

Communicate the Key Ideas

1. a) What is a composite figure? Draw an example.
   b) For the example you chose, identify the simple shapes that combine to form the composite figure.
2. a) Show or describe how each figure can be divided into two or more simple shapes.
   b) Show a different way to divide each figure in part a) into simple shapes.

- Questions in the **Communicate the Key Ideas** section let you talk or write about the concepts and assess whether or not you understand the ideas.

The **Check Your Understanding** section contains:

- Practice questions that are straightforward questions to check your knowledge and understanding of what you have learned.
- Application questions where you need to apply what you have learned to solve problems.
- Extension questions that may be a little more challenging and may make connections to other lessons.

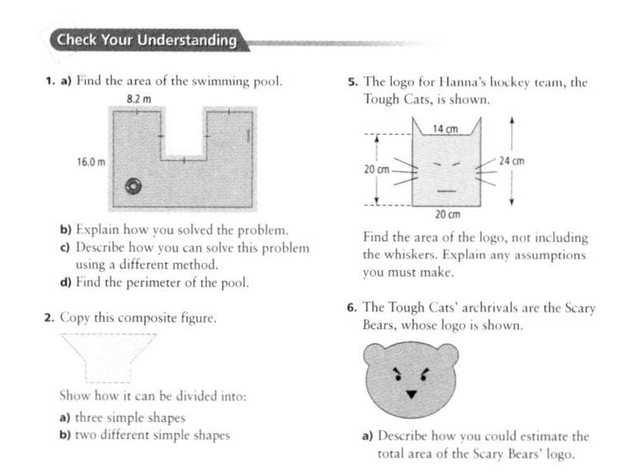
Check Your Understanding

1. a) Find the area of the swimming pool.
   b) Explain how you solved the problem.
   c) Describe how you can solve this problem using a different method.
   d) Find the perimeter of the pool.
2. Copy this composite figure.
   Show how it can be divided into:
   a) three simple shapes
   b) two different simple shapes
5. The logo for Hanna's hockey team, the Tough Cats, is shown.
   Find the area of the logo, not including the whiskers. Explain any assumptions you must make.
6. The Tough Cats' archrivals are the Scary Bears, whose logo is shown.
   a) Describe how you could estimate the total area of the Scary Bears' logo.

10. Determine if each triangle is unique. Explain how you know. Include diagrams.
    a) $\triangle ABC$ where AB = 4.2 cm, BC = 6.4 cm, and AC = 3.2 cm.
    b) $\triangle DEF$ where $\angle D = 60°$ and $\angle E = 110°$.
    c) $\triangle JKL$ where $\angle J = 90°$, JK = 4 cm, and KL = 5 m.
    d) $\triangle QRS$ where $\angle R = 45°$, $\angle Q = 74°$, and QS = 4.5 mm.
    e) $\triangle XYZ$ where $\angle X = 20°$, XY = 6.0 m, and YZ = 3.1 m.

The question(s) before the **Extend** questions is designed to assess your level of success with the section. Everyone should be able to respond to at least some part of each ? question.

Some numbered sections are followed by a **Use Technology** feature. This means some or part of the preceding section may be done using the technology shown. In this textbook, scientific calculators and *The Geometer's Sketchpad*® will be addressed.

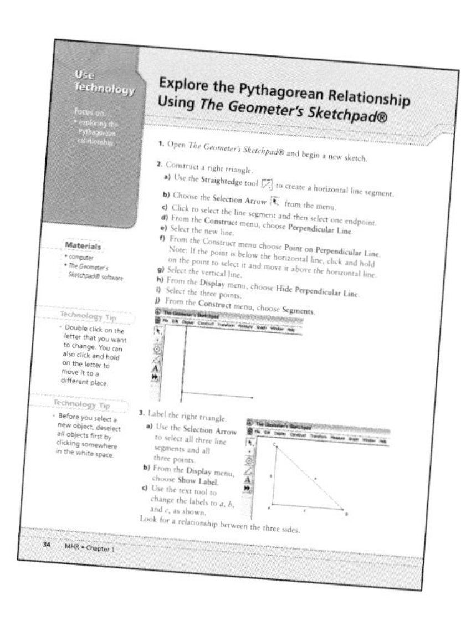
Use Technology

Focus on...
- exploring the Pythagorean relationship

Explore the Pythagorean Relationship Using *The Geometer's Sketchpad*®

Materials
- computer
- *The Geometer's Sketchpad*® software

1. Open *The Geometer's Sketchpad*® and begin a new sketch.
2. Construct a right triangle.
   a) Use the **Straightedge** tool to create a horizontal line segment.
   b) Choose the **Selection Arrow** from the menu.
   c) Click to select the line segment and then select one endpoint.
   d) From the **Construct** menu, choose **Perpendicular Line**.
   e) Select the new line.
   f) From the Construct menu choose **Point on Perpendicular Line**. Note: If the point is below the horizontal line, click and hold on the point to select it and move it above the horizontal line.
   g) Select the vertical line.
   h) From the **Display** menu, choose **Hide Perpendicular Line**.
   i) Select the three points.
   j) From the **Construct** menu, choose **Segments**.
3. Label the right triangle.
   a) Use the **Selection Arrow** to select all three line segments and all three points.
   b) From the **Display** menu, choose **Show Label**.
   c) Use the text tool to change the labels to *a*, *b*, and *c*, as shown.

Look for a relationship between the three sides.

Technology Tip
- Double click on the letter that you want to change. You can also click and hold on the letter to move it to a different place.

Technology Tip
- Before you select a new object, deselect all objects first by clicking somewhere in the white space.

34 MHR • Chapter 1

How does *Mathematics 8: Focus on Understanding* help you learn?

Understanding Vocabulary

Key words are listed on the Chapter Opener. Perhaps you already know the meaning of some of them. Great! If not, watch for these terms highlighted the first time they are used in the chapter. The meaning is given close by in the margin.

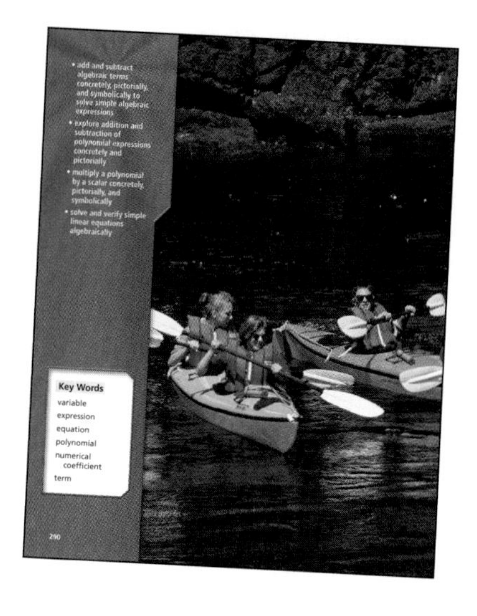

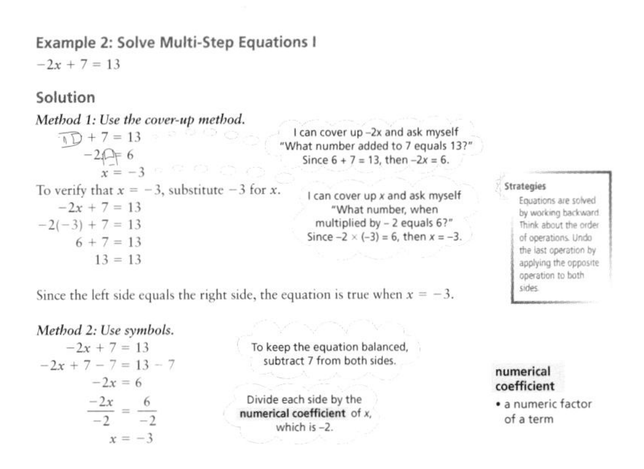

Example 2: Solve Multi-Step Equations I

$-2x + 7 = 13$

Solution

Method 1: Use the cover-up method.

$\square + 7 = 13$
$-2\square = 6$
$x = -3$

I can cover up $-2x$ and ask myself "What number added to 7 equals 13?" Since $6 + 7 = 13$, then $-2x = 6$.

I can cover up x and ask myself "What number, when multiplied by -2 equals 6?" Since $-2 \times (-3) = 6$, then $x = -3$.

To verify that $x = -3$, substitute -3 for x.

$-2x + 7 = 13$
$-2(-3) + 7 = 13$
$6 + 7 = 13$
$13 = 13$

Since the left side equals the right side, the equation is true when $x = -3$.

Method 2: Use symbols.

$-2x + 7 = 13$
$-2x + 7 - 7 = 13 - 7$
$-2x = 6$
$\frac{-2x}{-2} = \frac{6}{-2}$
$x = -3$

To keep the equation balanced, subtract 7 from both sides.

Divide each side by the **numerical coefficient** of x, which is -2.

Strategies
Equations are solved by working backward. Think about the order of operations. Undo the last operation by applying the opposite operation to both sides.

numerical coefficient
• a numeric factor of a term

Communicating Mathematically provide tips to help you read and interpret items in math. These tips will help you in other subjects as well.

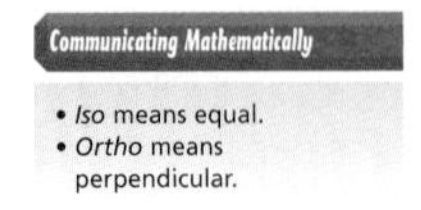

Understanding Concepts

The Discover the Math activity is designed to help you construct your own understanding of new concepts. The key question tells you what the activity is about. Short steps, with illustrations, lead you to be able to make some conclusions in the last step, the **Reflect** question.

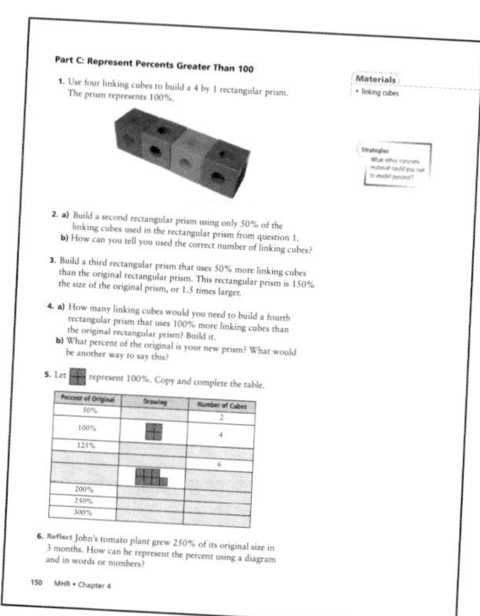

Part C: Represent Percents Greater Than 100

1. Use four linking cubes to build a 4 by 1 rectangular prism. The prism represents 100%.

2. a) Build a second rectangular prism using only 50% of the linking cubes used in the rectangular prism from question 1.
   b) How can you tell you used the correct number of linking cubes?

3. Build a third rectangular prism that uses 50% more linking cubes than the original rectangular prism. This rectangular prism is 150% the size of the original prism, or 1.5 times larger.

4. a) How many linking cubes would you need to build a fourth rectangular prism that uses 100% more linking cubes than the original rectangular prism? Build it.
   b) What percent of the original is your new prism? What would be another way to say this?

5. Let [square] represent 100%. Copy and complete the table.

Percent of Original	Drawing	Number of Cubes
50%		2
100%	[drawing]	4
125%		
		6
	[drawing]	
200%		
250%		
300%		

6. **Reflect** John's tomato plant grew 250% of its original size in 3 months. How can he represent the percent using a diagram and in words or numbers?

Materials
• linking cubes

Strategies
What other concrete material could you use to model percent?

150 MHR • Chapter 4

The **Examples** and their worked **Solutions** include several tools to help you understand the work.

- Notes in a thought or speech bubble help you think through the steps.
- Sometimes different methods of solving the same problem are shown. One way may make more sense to you than the other.
- **Problem Solving Strategies** are pointed out.

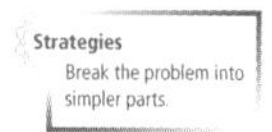

Example 1: Confirm That a 3-D Object is a Cone

Are these three-dimensional objects cones? Explain why or why not.

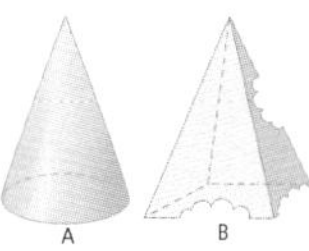

Solution

Object A is a cone because its base is flat, it has an apex, and all line segments that join points of the base to the apex along a surface are straight.

Object B is not a cone. Even though it has an apex, some line segments joining the base to the apex along a surface are not straight.

Communicate the Key Ideas

1. Who is correct? Explain where and why the other person went wrong in their thinking.

Justina, I drew a triangle that has side lengths 6 cm, 7 cm, and 8 cm. My triangle is unique.

Sorry, Ravi. I can draw the same triangle. So, your triangle is not unique.

2. Are these triangles unique? Explain.
   a) $\triangle DEF$ with $\angle D = 42°$, $\angle E = 75°$, and $DE = 12$ cm
   b) $\triangle LMN$ with $LM = 8$ cm, $MN = 9$ cm, and $\angle L = 37°$
   c) $\triangle ABC$ with $\angle A = 36°$, $\angle B = 47°$, and $\angle C = 97°$

The exercises begin with **Communicate the Key Ideas**. These short questions focus your thinking on the key ideas you learned in the section. By discussing these questions in a group, or doing the action called for, you can see whether you understand the main points and are ready to start the exercises.

The first few questions in **Check Your Understanding** can often be done by following one of the worked Examples.

Check Your Understanding

1. Carla drew this triangle. She says that it is unique. Is she correct? Explain.

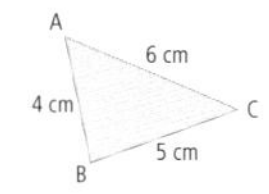

2. Draw $\triangle DEF$ with $DF = 8$ cm, $\angle D = 35°$, and $\angle F = 95°$. Is your triangle unique? Explain. Compare your triangle with several classmates to check.

3. When asked to construct $\triangle PQR$ with $PQ = 12$ cm, $QR = 15$ cm, and $\angle P = 27°$, Gina and Jackie drew congruent triangles. Are there any other possible triangles with these measurements that are not congruent to the ones that were drawn? Explain.

4. Show all the unique triangles you can make from $\triangle LMN$ by adding one more measurement. Explain why each triangle is unique.

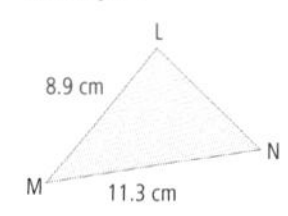

5. Explain how this could be true.

I have drawn two triangles. Both have a side length of 8.5 cm and two angles of 46° and 65°, but they look different.

What else will you find in *Mathematics 8: Focus on Understanding*?

Two special sections at the beginning of the book will help you to be successful with the grade 8 course.

Problem Solving

This is an overview of the four steps you can use to approach solving problems. Samples of 12 problem solving strategies are shown. You can refer back to this section if you need help choosing a strategy to solve a problem. You are also encouraged to use your own strategies.

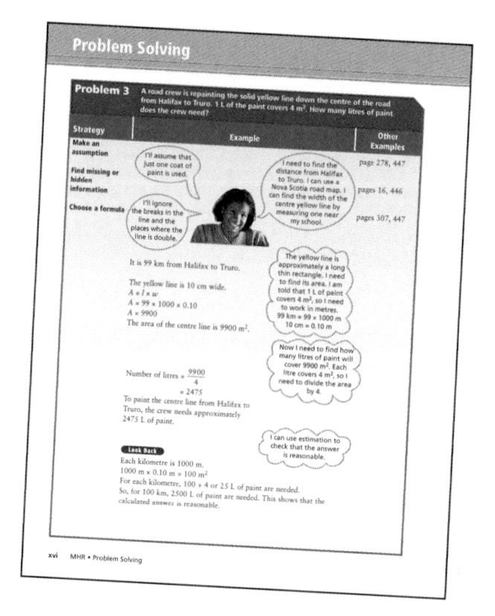

Problem Solving

Problem 3 A road crew is repainting the solid yellow line down the centre of the road from Halifax to Truro. 1 L of the paint covers 4 m². How many litres of paint does the crew need?

Strategy	Example	Other Examples
Make an assumption	I'll assume that just one coat of paint is used.	page 278, 447
Find missing or hidden information	I need to find the distance from Halifax to Truro. I can use a Nova Scotia road map. I can find the width of the centre yellow line by measuring one near my school.	pages 16, 446
Choose a formula	I'll ignore the breaks in the line and the places where the line is double.	pages 307, 447

It is 99 km from Halifax to Truro.

The yellow line is 10 cm wide.
$A = l \times w$
$A = 99 \times 1000 \times 0.10$
$A = 9900$
The area of the centre line is 9900 m².

The yellow line is approximately a long thin rectangle. I need to find its area. I am told that 1 L of paint covers 4 m², so I need to work in metres. 99 km = 99 × 1000 m 10 cm = 0.10 m

$$\text{Number of litres} = \frac{9900}{4} = 2475$$

To paint the centre line from Halifax to Truro, the crew needs approximately 2475 L of paint.

Now I need to find how many litres of paint will cover 9900 m². Each litre covers 4 m², so I need to divide the area by 4.

I can use estimation to check that the answer is reasonable.

Look Back
Each kilometre is 1000 m.
1000 m × 0.10 m = 100 m²
For each kilometre, 100 ÷ 4 or 25 L of paint are needed.
So, for 100 km, 2500 L of paint are needed. This shows that the calculated answer is reasonable.

xvi MHR • Problem Solving

Get Ready for Grade 8

These six pages present a brief review of basic concepts from earlier grades and ways of thinking about the concepts. Explain your thinking whenever possible.

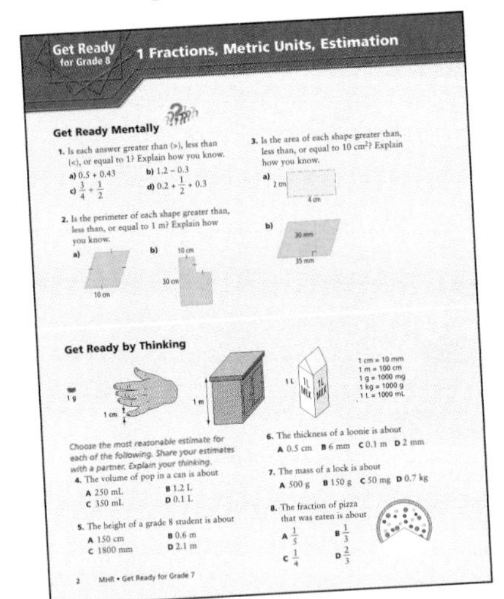

Get Ready for Grade 8 **1 Fractions, Metric Units, Estimation**

Get Ready Mentally

1. Is each answer greater than (>), less than (<), or equal to 1? Explain how you know.
   a) $0.5 + 0.43$ b) $1.2 - 0.3$
   c) $\frac{3}{4} + \frac{1}{2}$ d) $0.2 + \frac{1}{2} + 0.3$
2. Is the perimeter of each shape greater than, less than, or equal to 1 m? Explain how you know.
   a) 10 cm b) 10 cm, 30 cm
3. Is the area of each shape greater than, less than, or equal to 10 cm²? Explain how you know.
   a) 2 cm, 4 cm
   b) 30 mm, 35 mm

Get Ready by Thinking

1 g, 1 cm, 1 m, 1 L

1 cm = 10 mm
1 m = 100 cm
1 g = 1000 mg
1 kg = 1000 g
1 L = 1000 mL

Choose the most reasonable estimate for each of the following. Share your estimates with a partner. Explain your thinking.

4. The volume of pop in a can is about
   A 250 mL B 1.2 L
   C 350 mL D 0.1 L
5. The height of a grade 8 student is about
   A 150 cm B 0.6 m
   C 1800 mm D 2.1 m
6. The thickness of a loonie is about
   A 0.5 cm B 6 mm C 0.1 m D 2 mm
7. The mass of a lock is about
   A 500 g B 150 g C 50 mg D 0.7 kg
8. The fraction of pizza that was eaten is about
   A $\frac{1}{5}$ B $\frac{1}{3}$
   C $\frac{1}{4}$ D $\frac{2}{3}$

2 MHR • Get Ready for Grade 7

Other Special Features

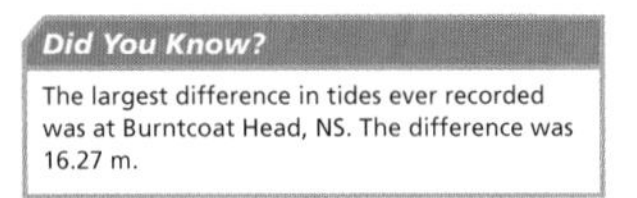

Did You Know?

The largest difference in tides ever recorded was at Burntcoat Head, NS. The difference was 16.27 m.

Did You Know?

These are interesting facts related to math topics you are learning.

Making Connections

These activities link the current topic to careers, games, another subject, a previous chapter, or lesson.

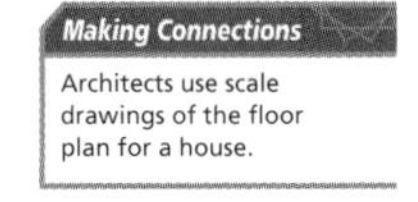

Making Connections

Architects use scale drawings of the floor plan for a house.

Internet Connect

Internet Connect To learn more about tides, go to **www.mcgrawhill.ca/links/math8NS** and follow the links.

Internet Connect You can find extra information related to some questions on the Internet. Log on to **www.mcgrawhill.ca/links/math8NS** and you will be able to link to recommended Web sites.

Puzzler

These are logic puzzles that are fun to do!

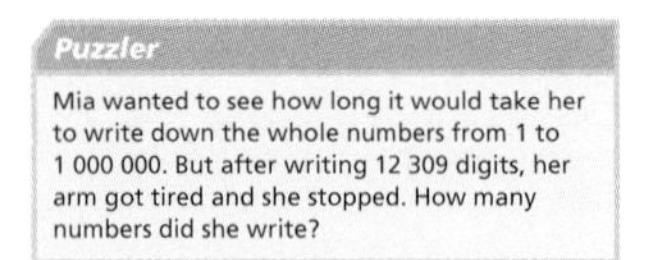

Puzzler

Mia wanted to see how long it would take her to write down the whole numbers from 1 to 1 000 000. But after writing 12 309 digits, her arm got tired and she stopped. How many numbers did she write?

Each chapter ends with a **Chapter Review** and a **Practice Test**. The chapter review starts with multiple choice questions. The test includes the different types of questions that you will find on provincial tests: multiple choice, short answer, and extended response.

Task

These projects follow several chapters. To provide a solution, you may need to combine skills from multiple chapters and your own creativity.

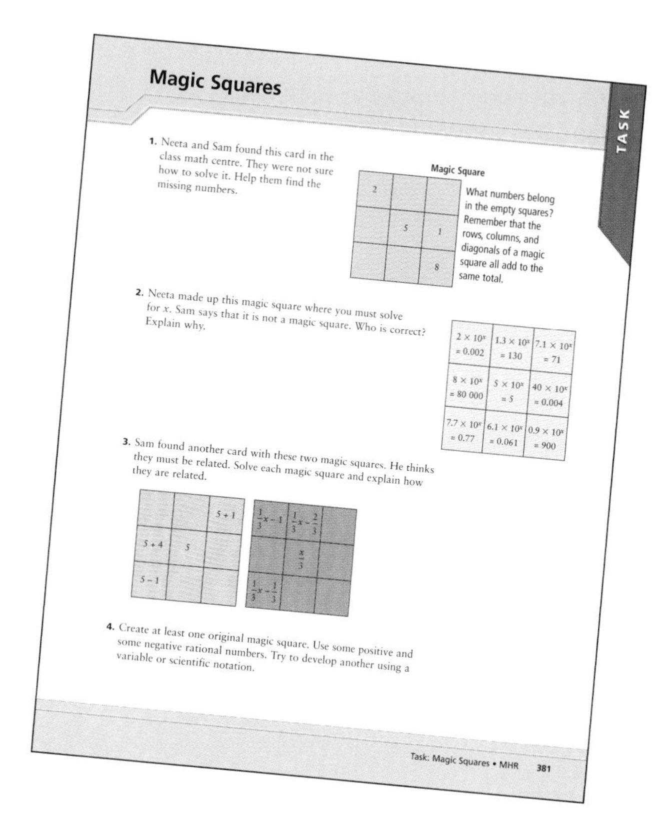

Magic Squares

TASK

1. Neeta and Sam found this card in the class math centre. They were not sure how to solve it. Help them find the missing numbers.

Magic Square

2		
	5	1
		8

What numbers belong in the empty squares? Remember that the rows, columns, and diagonals of a magic square all add to the same total.

2. Neeta made up this magic square where you must solve for x. Sam says that it is not a magic square. Who is correct? Explain why.

$2 \times 10^x = 0.002$	$1.3 \times 10^x = 130$	$7.1 \times 10^x = 71$
$8 \times 10^x = 80\,000$	$5 \times 10^x = 5$	$40 \times 10^x = 0.004$
$7.7 \times 10^x = 0.77$	$6.1 \times 10^x = 0.061$	$0.9 \times 10^x = 900$

3. Sam found another card with these two magic squares. He thinks they must be related. Solve each magic square and explain how they are related.

		$5 + 1$
$5 + 4$	5	
$5 - 1$		

$\frac{1}{3}x - 1$	$\frac{1}{3}x - \frac{2}{3}$	
	$\frac{x}{3}$	
$\frac{1}{3}x - \frac{1}{3}$		

4. Create at least one original magic square. Use some positive and some negative rational numbers. Try to develop another using a variable or scientific notation.

Task: Magic Squares • MHR 381

Reviews of the previous chapters can be found following Chapters 3, 6, and 9.

Glossary

Refer to the illustrated Glossary at the back of the text if you need to check the exact meaning of mathematical terms.

Problem Solving

How can you solve problems like the four below? Compare your ideas with the strategies that are shown on the following pages.

Problem 1

Dina's family owns and operates a small restaurant. They have many small square tables and folding chairs. What is the greatest number of people that can be seated when 10 tables are put together?

Problem 2

Amy's mother bought a basket of strawberries. Ben came in and ate half of them. Steve came home next and ate half of the remaining strawberries. Dora returned and ate half the number that remained. Amy came home last. She ate half of the remaining strawberries and left two whole strawberries for her mother. How many strawberries were orginally in the basket?

Problem 3

A road crew is repainting the solid yellow line down the centre of the road from Halifax to Truro. 1 L of the paint covers 4 m^2. How many litres of paint does the crew need?

Problem 4

Raj and his friend Matt live in a neighbourhood where the streets form a regular grid pattern. How many different routes are there from Raj's house to Matt's? Assume that Raj does not retrace his steps and always takes the shortest route.

People solve mathematical problems at home, at work, and at play. There are many different ways to solve problems. In *Mathematics 8: Focus on Understanding*, you are encouraged to try different methods and to use your own ideas. Your method may be different but it may also work.

A Problem Solving Model

Where do you begin with problem solving? It may help to use the following four-step process.

Understand

Read the problem three times.

- On the first reading, get a general sense of what the problem is about.
- On the second reading, determine what information is given and what you need to find.
  - What information do you have?
  - What additional information do you need?
  - What is the problem asking you to do?
- On the third reading, get the details straight.

Plan

Select a strategy or several strategies for solving the problem.

- Consider other problems you have solved. Is this problem similar? Can you use a similar strategy? Strategies that you might use include:
  - Make a model or diagram
  - Act it out
  - Brainstorm or make a web idea
  - Work backward
  - Look for a pattern
  - Make an assumption
  - Use systematic trial
  - Express a pattern using algebra
  - Set up an equation and solve
  - Choose an appropriate formula
  - Make an organized list, table, or chart
  - Solve a simpler problem
  - Choose a previously learned procedure
  - Find hidden or missing information
- Decide whether any of the following might help. Plan how to use them.
  - tools such as a ruler or a calculator
  - materials such as graph paper, a number line, or manipulatives

Do It!

Solve the problem by carrying out your plan.

- Use mental math to estimate a possible answer.
- Do the calculations.
- Record each step you are doing.
- Explain and justify your thinking.

Look Back

Examine your answer. Does it make sense?

- Is your answer close to your estimate?
- Does your answer fit the facts given in the problem?
- Is the answer reasonable? If not, make a new plan. Try a different strategy.
- Consider solving the problem a different way. Do you get the same answer? Compare your method with that of other students.

Problem Solving

Here are some different ways to solve the four problems on page xvi. Often you need to use more than one strategy to solve a problem. Your ideas on how to solve the problems might be different from any of these.

To see other examples of how to use these strategies, refer to the page references. These show where the strategy is used in other sections of *Mathematics 8: Focus of Understandng.*

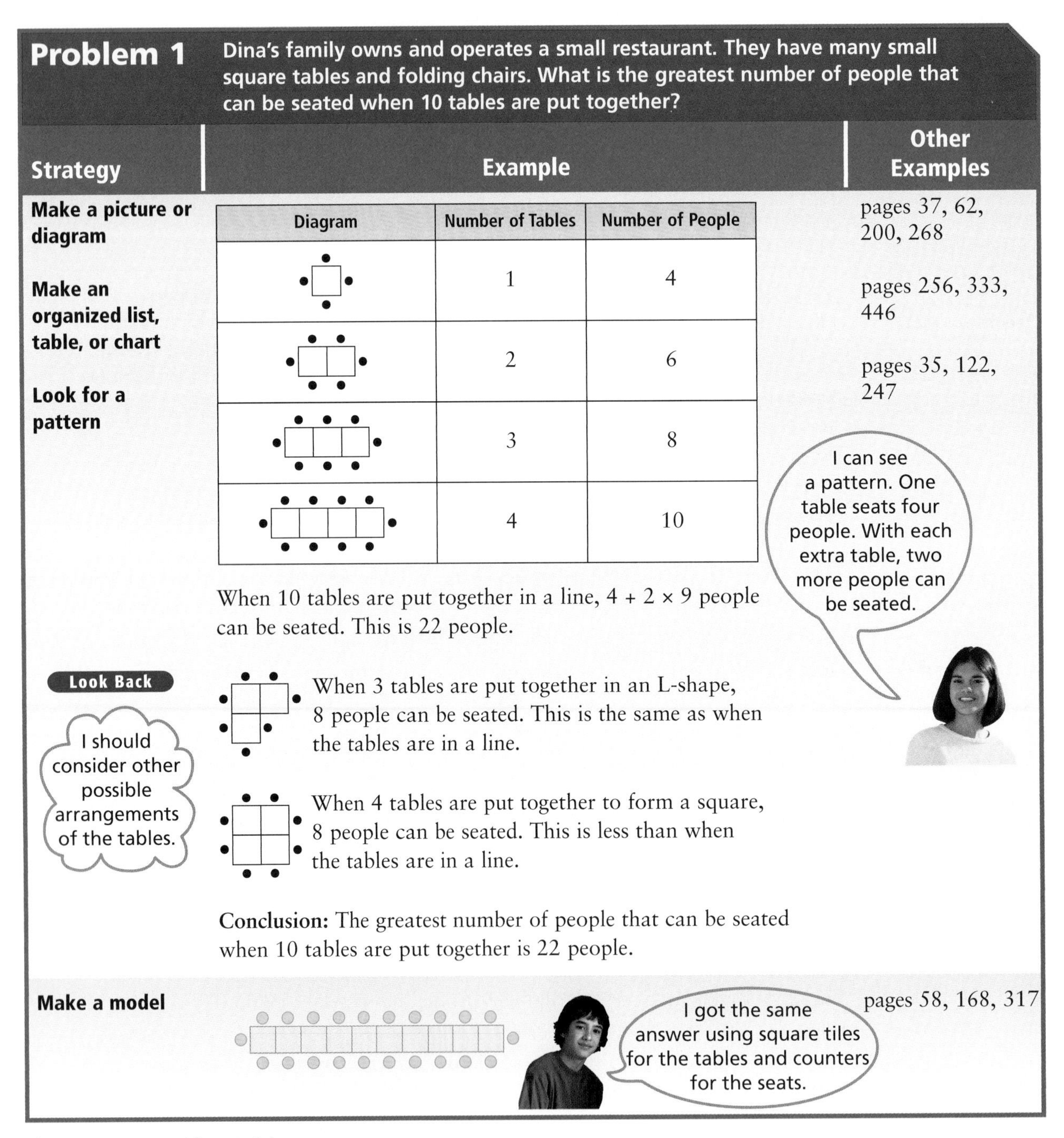

Problem 1

Dina's family owns and operates a small restaurant. They have many small square tables and folding chairs. What is the greatest number of people that can be seated when 10 tables are put together?

Strategy	Example	Other Examples
Make a picture or diagram		pages 37, 62, 200, 268
Make an organized list, table, or chart		pages 256, 333, 446
Look for a pattern		pages 35, 122, 247

Diagram	Number of Tables	Number of People
	1	4
	2	6
	3	8
	4	10

When 10 tables are put together in a line, $4 + 2 \times 9$ people can be seated. This is 22 people.

Look Back

When 3 tables are put together in an L-shape, 8 people can be seated. This is the same as when the tables are in a line.

When 4 tables are put together to form a square, 8 people can be seated. This is less than when the tables are in a line.

Conclusion: The greatest number of people that can be seated when 10 tables are put together is 22 people.

Make a model — pages 58, 168, 317

Problem 2

Amy's mother bought a basket of strawberries. Ben came in and ate half of them. Steve came home next and ate half of the remaining strawberries. Dora returned and ate half the number that remained. Amy came home last. She ate half of the remaining strawberries and left two whole strawberries for her mother. How many strawberries were orginally in the basket?

Strategy	Example	Other Examples

Work backward

Make a table or chart

Two strawberries were left for Amy's mother. The last person, Amy, must have eaten two. Half of 4 is 2. That's right. I will work backward from there.

Person	Number Eaten	Remaining	Total
Amy	2	2	4
Dora	4	4	8
Steve	8	8	16
Ben	16	16	32

Look Back Check that a basket containing 32 works.
Ben: half of 32 = 16
Steve: half of 16 = 8
Dora: half of 8 = 4
Amy: half of 4 = 2
Mother: 2
16 + 8 + 4 + 2 + 2 = 32 It checks.
There were 32 strawberries in the basket originally.

Other Examples: Work backward — pages 255, 311, 314; Make a table or chart — pages 14, 221, 256

Act it out

Work backward

Mother sees ● ●
Amy sees ● ● ● ●
Dora sees ● ● ● ● ● ● ● ●
Steve sees ● ● ● ● ● ● ● ●
● ● ● ● ● ● ● ●

So, Ben ate 16.

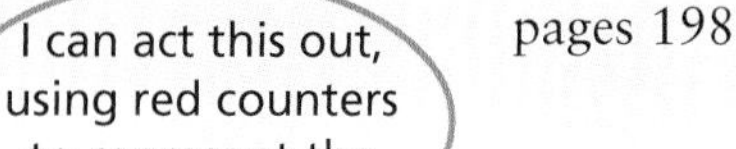

I can act this out, using red counters to represent the strawberries.

In the full basket, there were 32 strawberries.

Other Examples: pages 198

Use systematic trial

Ben: half of 40 = 20
Steve: half of 20 = 10
Dora: half of 10 = 5

I'm going to guess and test. I will try 40 first.

Look Back I need to change my starting number, because Amy would be eating $2\frac{1}{2}$ strawberries and leaving the same number for her mother. The problem says the mother got 2 whole berries.
Since 2 is less than $2\frac{1}{2}$, I will start with a smaller number.
When I repeat the steps with 32, I find it works perfectly.
There were 32 strawberries in the basket originally.

Other Examples: pages 316

Problem 3 A road crew is repainting the solid yellow line down the centre of the road from Halifax to Truro. 1 L of the paint covers 4 m^2. How many litres of paint does the crew need?

Strategy	Example	Other Examples
Make an assumption		page 278, 447
Find missing or hidden information		pages 16, 446
Choose a formula		pages 307, 447

I'll assume that just one coat of paint is used.

I'll ignore the breaks in the line and the places where the line is double.

I need to find the distance from Halifax to Truro. I can use a Nova Scotia road map. I can find the width of the centre yellow line by measuring one near my school.

It is 99 km from Halifax to Truro.

The yellow line is 10 cm wide.

$A = l \times w$

$A = 99 \times 1000 \times 0.10$

$A = 9900$

The area of the centre line is 9900 m^2.

The yellow line is approximately a long thin rectangle. I need to find its area. I am told that 1 L of paint covers 4 m^2, so I need to work in metres.
99 km = 99 × 1000 m
10 cm = 0.10 m

Now I need to find how many litres of paint will cover 9900 m^2. Each litre covers 4 m^2, so I need to divide the area by 4.

$$\text{Number of litres} = \frac{9900}{4}$$

$$= 2475$$

To paint the centre line from Halifax to Truro, the crew needs approximately 2475 L of paint.

I can use estimation to check that the answer is reasonable.

Look Back

Each kilometre is 1000 m.

$1000 \text{ m} \times 0.10 \text{ m} = 100 \text{ m}^2$

For each kilometre, 100 ÷ 4 or 25 L of paint are needed.

So, for 100 km, 2500 L of paint are needed. This shows that the calculated answer is reasonable.

Problem 4

Raj and his friend Matt live in a neighbourhood where the streets form a regular grid pattern. How many different routes are there from Raj's house to Matt's? Assume that Raj does not retrace his steps and always takes the shortest route.

Strategy	Example	Other Examples
Solve a simpler problem	What if they lived one block apart?	pages 20, 152, 278
Look for a pattern		pages 256, 363

What if they lived one block apart?

Raj
1
Matt
1 2

Record the number of routes to get to each corner.

There are 2 possible routes in this case, each 2 blocks long.

What if they lived two blocks apart?

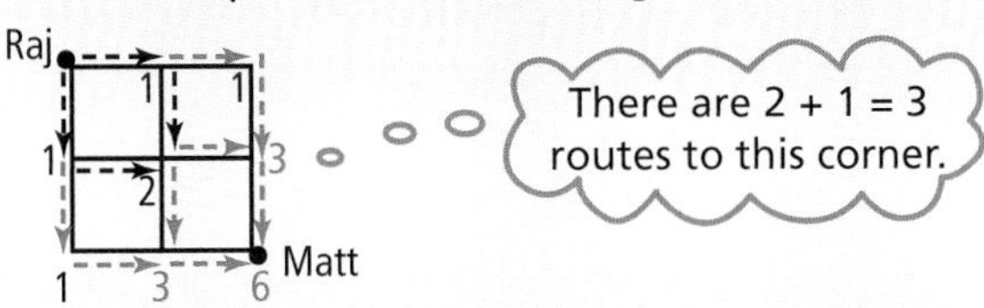
Raj
Matt
There are 2 + 1 = 3 routes to this corner.

There are 6 possible routes in this case, each 4 blocks long.

What if they lived three blocks apart?

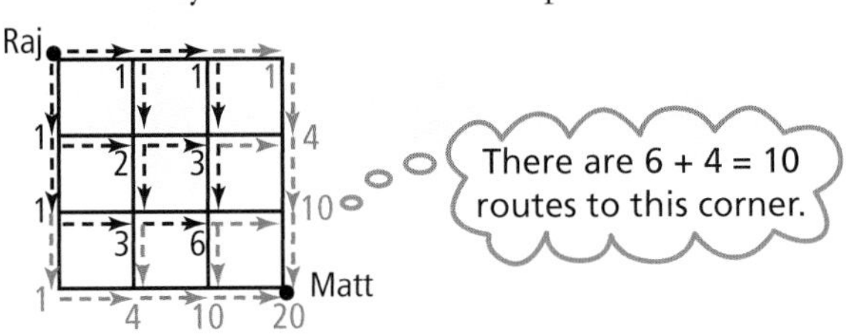
Raj
Matt
There are 6 + 4 = 10 routes to this corner.

There are 20 possible routes in this case, each 6 blocks long.

The last step: They actually live four blocks apart.

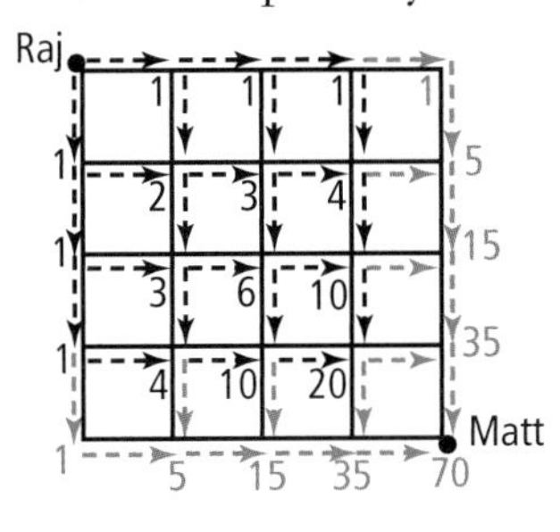
Raj
Matt

There are 70 possible routes in this case, each 8 blocks long.

Amazingly, there are 70 possible different routes from Raj's house to Matt's.

Get Ready for Grade 8

1 Fractions, Metric Units, Estimation

Get Ready Mentally

1. Is each answer greater than (>), less than (<), or equal to 1? Explain how you know.

a) $0.5 + 0.43$ **b)** $1.2 - 0.3$

c) $\frac{3}{4} + \frac{1}{2}$ **d)** $0.2 + \frac{1}{2} + 0.3$

2. Is the perimeter of each shape greater than, less than, or equal to 1 m? Explain how you know.

a)

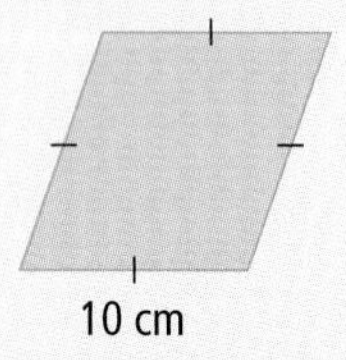

b)

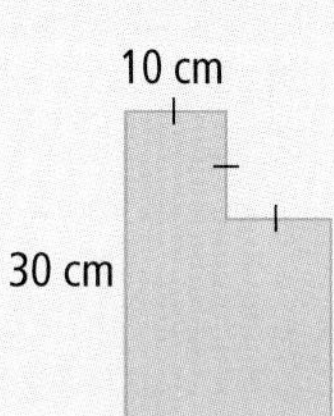

3. Is the area of each shape greater than, less than, or equal to 10 cm^2? Explain how you know.

a)

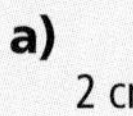

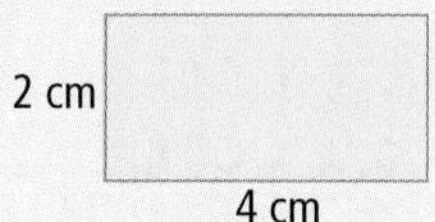

b)

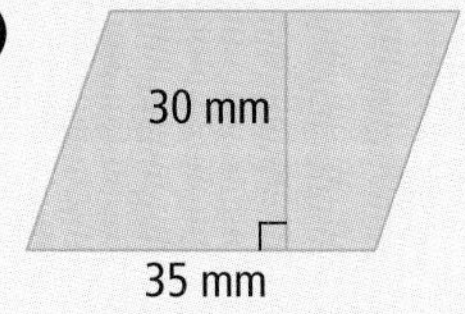

Get Ready by Thinking

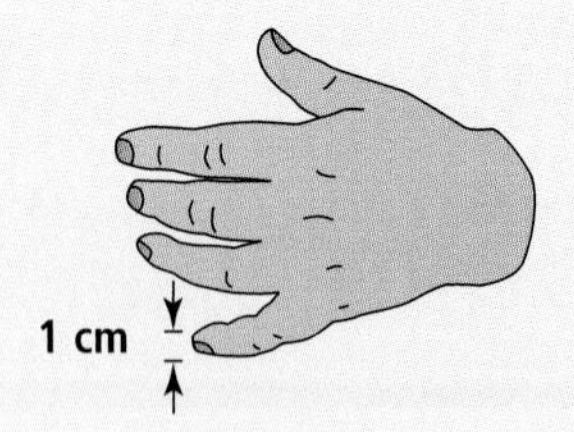

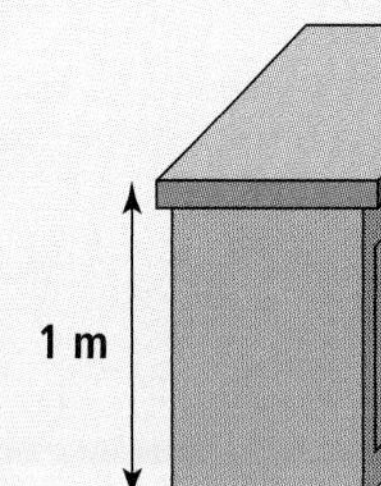

1 cm = 10 mm
1 m = 100 cm
1 g = 1000 mg
1 kg = 1000 g
1 L = 1000 mL

Choose the most reasonable estimate for each of the following. Share your estimates with a partner. Explain your thinking.

4. The volume of pop in a can is about

A 250 mL **B** 1.2 L
C 350 mL **D** 0.1 L

5. The height of a grade 8 student is about

A 150 cm **B** 0.6 m
C 1800 mm **D** 2.1 m

6. The thickness of a loonie is about

A 0.5 cm **B** 6 mm **C** 0.1 m **D** 2 mm

7. The mass of a lock is about

A 500 g **B** 150 g **C** 50 mg **D** 0.7 kg

8. The fraction of pizza that was eaten is about

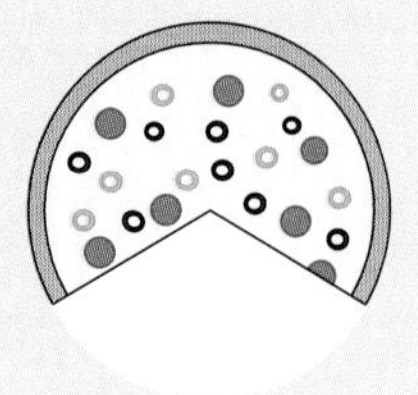

A $\frac{1}{5}$ **B** $\frac{1}{3}$

C $\frac{1}{4}$ **D** $\frac{2}{3}$

Get Ready by Exploring

Materials
- centimetre grid paper
- scissors
- BLM Get Ready Finger Puppets

9. Maya is designing some finger puppets for the drama club. She needs to cut two pieces of felt for each puppet.

a) Determine the area of felt needed for each puppet.

b) What do you notice about the areas?

c) What is the total area of felt used for these puppets?

10. For each puppet, Maya starts with a square piece of felt. She then cuts pieces from it and rearranges them. What is the area of the original square?

11. Look at the puppet designs on the worksheet. Show how the designs could be cut and rearranged back to squares.

Here are some hints to get you started on the first two puppets.
A: Cut and move one piece.
B: Cut and move one piece.

12. a) Draw a square with an area of 25 cm^2.

b) Cut the square into two or three pieces. Then rearrange the pieces to create your puppet. Decorate your puppet. Draw another square to show your solution.

Get Ready by Reflecting

13. How is area different from perimeter?

14. Explain why Maya might want to know the perimeter of her puppets. Would they all have the same perimeter?

Get Ready for Grade 8

2 Convert Fractions, Decimals, and Percents; Perfect Squares

Get Ready Mentally

1. Write each fraction as a percent.

a) $\frac{1}{2}$ **b)** $\frac{1}{4}$ **c)** $\frac{1}{5}$ **d)** $\frac{3}{5}$

2. Write each fraction as a decimal.

a) $\frac{1}{4}$ **b)** $\frac{1}{5}$ **c)** $\frac{1}{10}$ **d)** $\frac{3}{10}$

3. Write each decimal as a percent.

a) 0.25 **b)** 0.45 **c)** 0.9 **d)** 0.333

4. Write each percent as a decimal.

a) 50% **b)** 75% **c)** 95% **d)** 100%

5. Which numbers that are less than $\frac{1}{2}$?

a) 0.3 **b)** 0.45 **c)** 0.55 **d)** 0.6

Get Ready by Thinking

Origami is an ancient art. Square pieces of paper can be folded to form many different designs.

6. What is the area of a square piece of origami paper with each side length?

a) 4 cm **b)** 8 cm **c)** 100 mm

7. Ashton is organizing his origami paper. What is the side length of a square piece with each area?

a) 36 cm^2 **b)** 4 cm^2 **c)** 121 cm^2

8. What is the perimeter of a square piece of origami paper with each area?

a) 16 cm^2 **b)** 25 cm^2 **c)** 49 mm^2

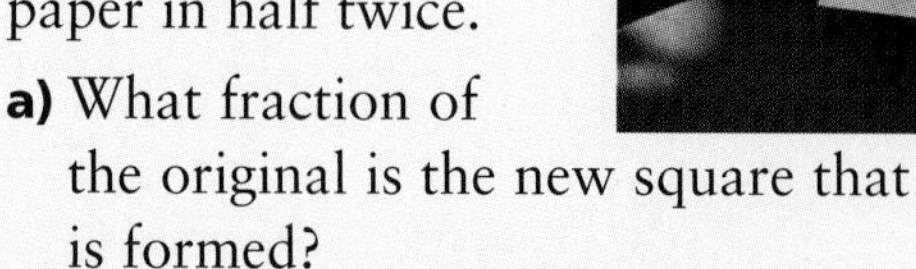

9. To make a paper crane, Kayle folds a piece of origami paper in half twice.

a) What fraction of the original is the new square that is formed?

b) Write your answer to part a) as a decimal and as a percent.

10. Shawana got 15 out of 20 marks for her design in an origami contest. Adwin got 19 out of 25 for precise folding. Who got the better mark? Explain.

Get Ready by Exploring

Elsa created a game called Math Spin for the Math Expo at school. To win, a player must spin the letters of the word MATH in the correct order.

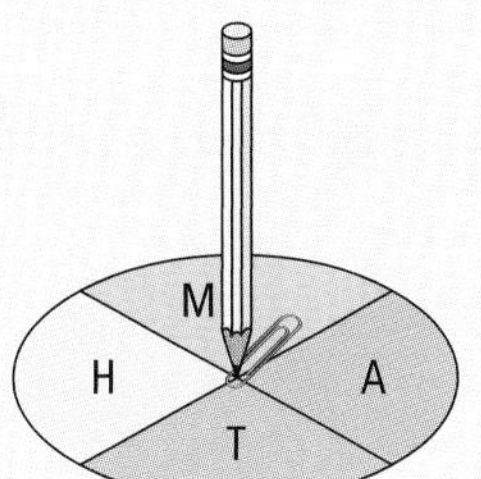

Materials

- BLM Get Ready Math Spinner

11. Express each answer as a fraction and as a percent. If you spin the spinner once, what is the probability that you will get

a) a consonant?

b) the letter T?

12. Aaliyah and Andre played the game. Their results are shown in the table.

Spin Number	Aaliyah	Andre
1	(M)	A
2	T	(M)
3	(A)	(A)
4	H	A
5	(T)	H
6	(H)	(T)

a) Who won the game? How many spins did it take?

b) What is the minimum number of spins needed to win a game of Math Spin?

13. a) Play Math Spin with a partner for 5 min. Record the results of each game, as well as the number of spins it took for someone to win, using a table.

b) Determine the mean, median, and mode of the number of spins to win the game. Show your calculations.

14. Gather the game results from the whole class. Use them to repeat question 13b) for the whole class.

I already have an M. I have a one in four chance of getting an H on my next spin.

Get Ready by Reflecting

15. Is Math Spin a fair game? Explain how you know.

16. Describe how you could change this game so that it could be won in fewer spins.

Get Ready for Grade 8

3 Patterns With Natural Numbers, Fractions, and Integers

Get Ready Mentally

1. List the next three numbers in each pattern.

a) 4, 9, 14, … **b)** –2, –5, –8, …
c) 1, 1, 2, 3, 5, … **d)** 4, 2, 0, …
e) 1, 4, 3, 6, 5, … **f)** –15, –11, –7, …

2. List the next three numbers in each pattern.

a) $\frac{1}{10}, \frac{1}{5}, \frac{3}{10}, \ldots$ **b)** $\frac{1}{4}, \frac{1}{2}, \frac{3}{4}, 1, 1\frac{1}{4}, \ldots$

c) $3\frac{1}{2}, 3, 2\frac{1}{2}, \ldots$ **d)** $\frac{1}{2}, -2, \frac{1}{3}, -3, \frac{1}{4}, \ldots$

Get Ready by Thinking

3. In each box, which number does not belong? Explain why.

a)

4	8	1
36	49	25

b)

2	11	13
5	3	6

c)

$\frac{2}{6}$	$\frac{4}{12}$	$\frac{3}{9}$
$\frac{6}{15}$	$\frac{10}{30}$	$\frac{7}{21}$

4. Each player at a chess tournament shakes hands once with every other player.

a) Copy and complete the table.

Number of Players	Number of Handshakes
1	0
2	1
3	3
4	
5	

b) Describe the pattern that relates the number of handshakes to the number of players.

c) How many handshakes would there be for 6 players? 10 players?

d) Describe the strategy you used to answer this question.

5. Squares can be divided into smaller squares using horizontal and vertical lines.

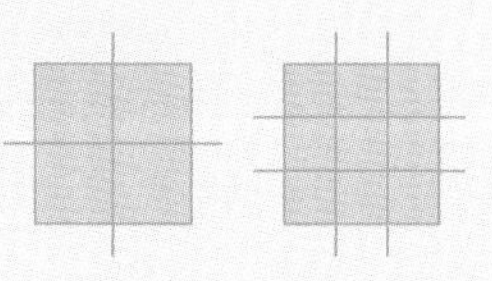

a) How many lines do you need to draw to get 36 smaller squares?

b) How many smaller squares will you get with 12 lines?

6. Describe what the number cruncher is doing to each set of numbers.

a)

In	Out
2	4
$\frac{1}{2}$	$2\frac{1}{2}$
–3	–1
7	9

b)

In	Out
3	5
6	11
$\frac{1}{2}$	0
10	19

Get Ready by Exploring

Ms. King asks each of her students to design a personal tile. The tiles will be joined together with clips at each vertex so that they can be displayed in the hallway.

Materials
- isometric dot paper

Ms. King arranged the tiles like this:

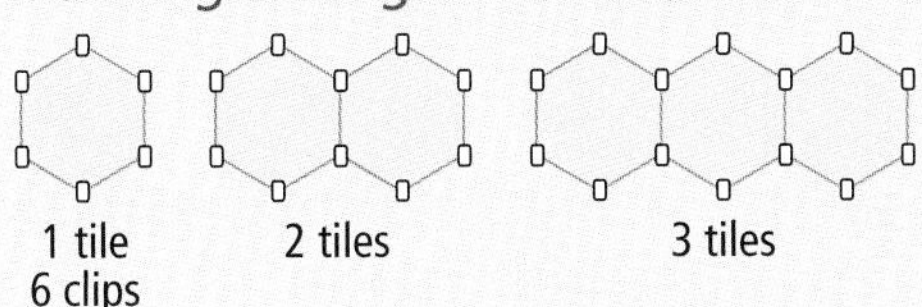

1 tile
6 clips

2 tiles

3 tiles

7. a) On isometric dot paper, sketch the arrangements for four to eight tiles, making sure to show the clips.

b) Copy and complete the table to show your results from part a).

Number of Tiles	Number of Clips
1	6
2	10
3	
4	
5	
6	
7	
8	

8. How many clips would Ms. King need for 12 students? Show two different ways to solve this.

9. a) Tiles cost \$0.65 each and clips cost \$0.07 each. How much would it cost to create a tile display for a class of 25 students?

b) Suggest another way to arrange the 25 tiles to save some money. Draw your arrangement on another piece of triangle dot paper.

10. a) What is the cost of your arrangement from question 9b)?

b) Describe some advantages and disadvantages of using your arrangement.

Get Ready by Reflecting

11. Describe a pattern that you have seen. How are the parts related? Explain using pictures, words, and numbers.

12. Write a paragraph about a real-life situation where you used a pattern to help solve a problem.

- model and link various representations of square roots of a number
- recognize perfect squares between 1 and 144 and apply patterns related to them
- distinguish between an exact square root and its decimal approximation
- find the square root of any number, using an appropriate method
- demonstrate understanding of the Pythagorean relationship, using models
- apply the Pythagorean relationship in problem situations

Key Words

Pythagorean relationship

hypotenuse

legs (of a right triangle)

perfect square

estimate

approximate

CHAPTER 1

Squares, Square Roots, and Pythagoras

Triangles are an important design feature. They are used in making sails. They are used in construction because they provide strength and stability. Right triangles are particularly important in building houses. Builders use a special property of the triangles' sides to ensure that walls meet at right angles.

In this chapter, you will explore how the sides of a right triangle are related. You will apply this relationship to solve problems.

Chapter Problem

You are given a wooden stick 100 cm in length. You can cut the stick once, to make two pieces. How can you use the two pieces of wood to build a kite frame?

Suppose you glue a sheet of kite paper to one side of your kite frame, as shown. What will be the perimeter of the kite paper? What will be its area?

In this chapter, you will explore different possible kites.

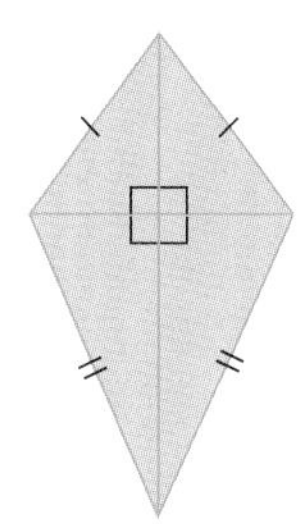

Get Ready

Squares

The square of a number is calculated by multiplying the number by itself. The square of 3 is 3×3. It can also be written as 3^2 and is read as "three squared."

$3^2 = 3 \times 3$ or 9

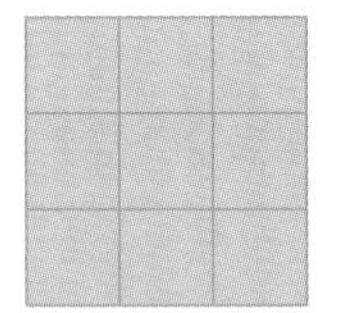

1. Draw a diagram to represent each square.
   a) 1^2 **b)** 2^2 **c)** 4^2

2. What is the value of each square?
   a) 5^2 **b)** 7^2 **c)** 2^2 **d)** 8^2

3. Why do you think the word *square* is used when referring to the numbers in question 1?

Classify Triangles

Triangles can be classified according to their side lengths or their angle measures.

equilateral triangle
- three congruent sides

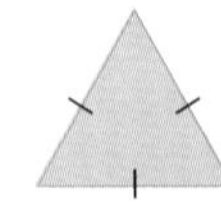

isosceles triangle
- two congruent sides

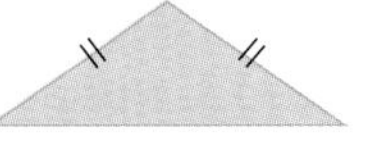

scalene triangle
- no congruent sides

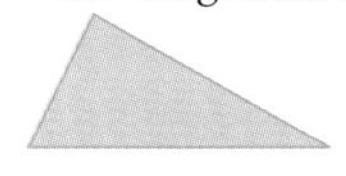

acute triangle
- three acute angles

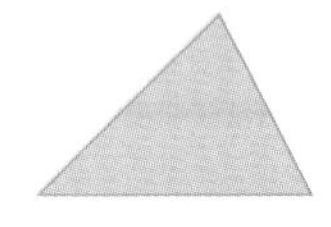

right triangle
- one right angle

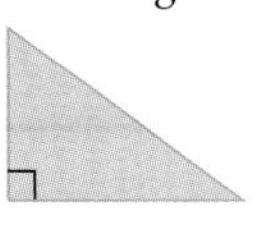

obtuse triangle
- one obtuse angle

4. Classify each triangle in two ways: by its sides and by its angles.
   a)

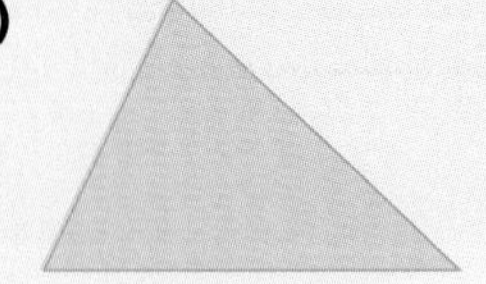

   b)

5. How are these triangles similar? How are they different?

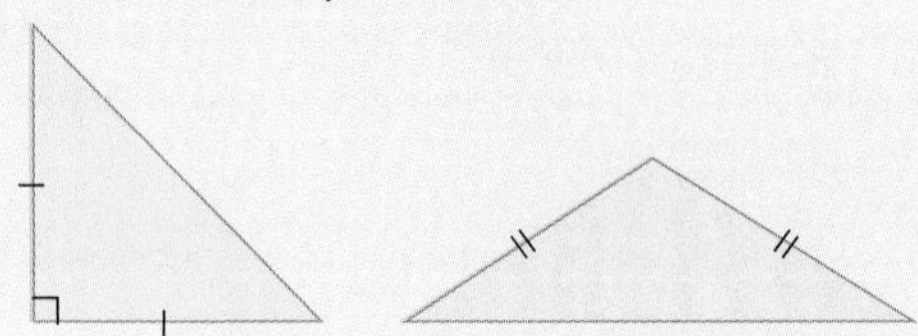

Perimeter and Area of Squares and Triangles

The distance around a shape is called its **perimeter**. Perimeter is measured in linear units, such as millimetres (mm), centimetres (cm), or metres (m).

The surface that a shape covers is called its **area**. Area is measured in square units, such as square millimetres (mm^2), square centimetres (cm^2), or square metres (m^2).

$P = 5 + 5 + 5 + 5$ or $P = 4 \times 5$
$P = 20$
The perimeter of the square is 20 m.

$A = l \times w$ or $A = s^2$
$A = 5 \times 5$ $\quad A = 5^2$
$A = 25$
The area of the square is 25 m^2.

$P = 5 + 12 + 13$
$P = 30$
The perimeter of the triangle is 30 m.

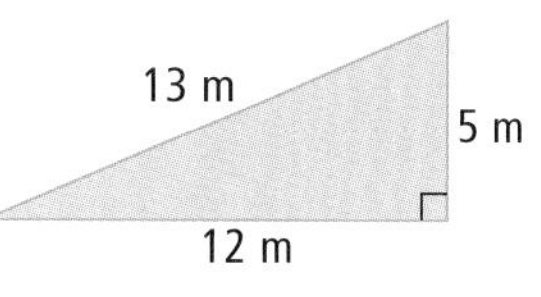

The triangle is half the area of a rectangle that is 12 m by 5 m.

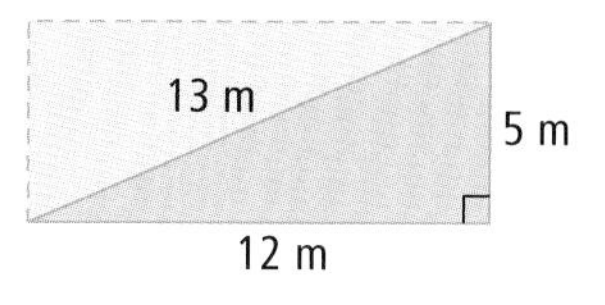

$$A = \frac{1}{2}(bh)$$

You can also express this as $A = (bh) \div 2$.

$$A = \frac{1}{2}(12 \times 5)$$

$$A = \frac{1}{2}(60)$$

$A = 30$
The area of the triangle is 30 m^2.

6. Find the perimeter and the area of each square.

a)

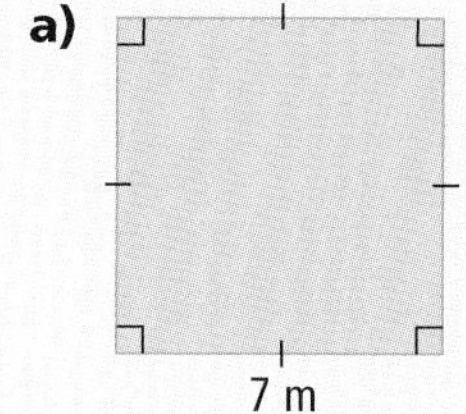

b)

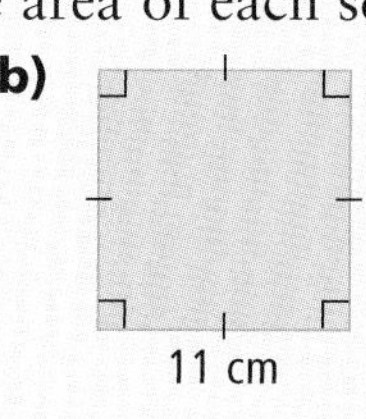

7. Find the perimeter and the area of each triangle.

a)

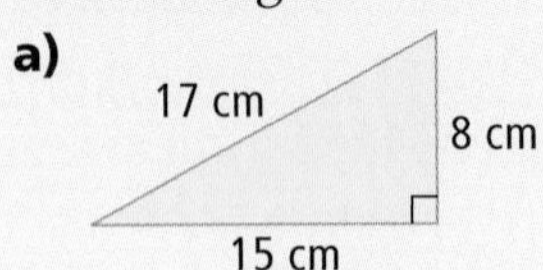

b)

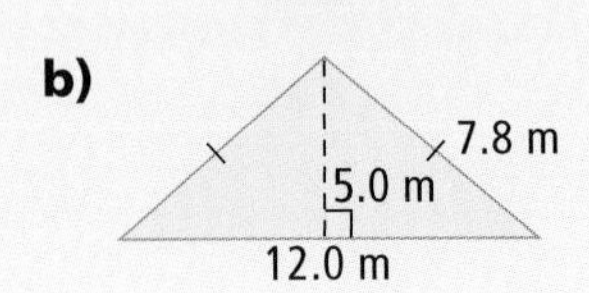

8. Use sectioning to find the area of each tilted square.

a) **b)**

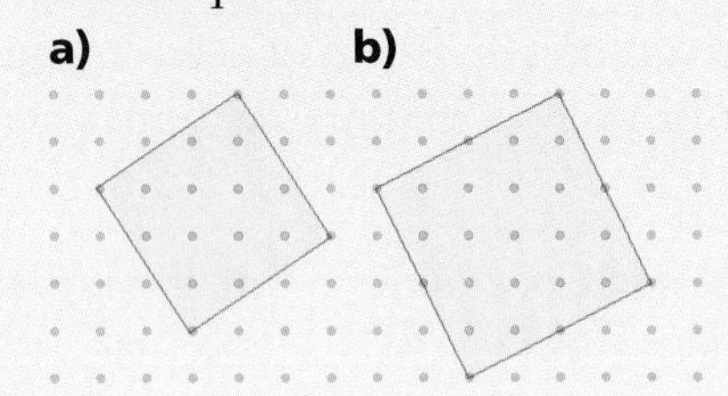

Solve Equations

Solve for x.

a) $x + 7 = 12$

I can see by inspection that x must be 5, because $5 + 7 = 12$.

$x = 5$

b) $25 = 16 + x$

Method 1: By inspection.

$x = 9$

I can see x must be 9, since $25 = 16 + 9$.

Method 2: Use an inverse operation.

$25 = 16 + x$, so

$25 - 16 = x$

$9 = x$

9. Solve for x.

a) $x + 10 = 18$

b) $x + 15 = 24$

c) $4 + x = 16$

d) $12 + x = 25$

10. Solve for x.

a) $20 = 9 + x$

b) $36 = x + 20$

c) $32 = x + 16$

d) $81 = x + 49$

Use Factor Trees to Find the Prime Factorization

Factor trees can be used to write a composite number as the product of its **prime factors** .

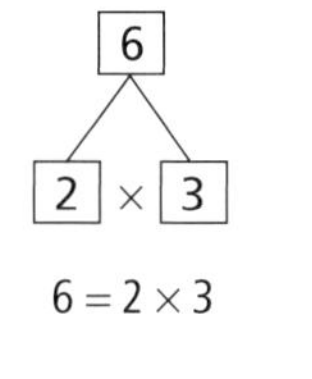

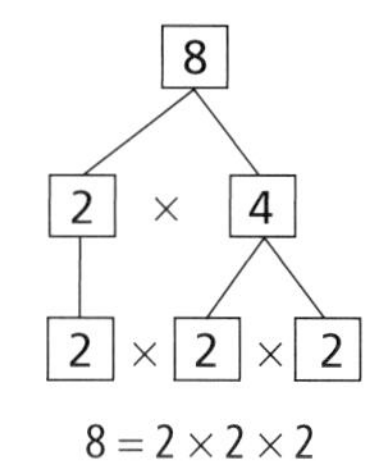

11. Find the prime factorization of each number.
a) 48 **b)** 72 **c)** 64 **d)** 150

12. Why is the number 1 not used when writing the prime factorization of a number?

1.1 Identify Perfect Squares and Related Patterns

Focus on...
- perfect squares
- patterns among perfect squares
- square roots of perfect squares

A SCRABBLE® board is made up of 225 small congruent squares. Colin wants to make a SCRABBLE® board for a woodworking project in technical education class. He decides to make the board from strips of small squares that are laminated together. How does Colin know how many squares to use for one row of his SCRABBLE® board?

Discover the Math

Materials
- 11 × 11 geoboard
- 1-cm square dot paper
- elastics

How does Colin know how many squares to put on one row in order to make a SCRABBLE® board? How does the area of a square relate to its side length?

1. a) On a geoboard, make many different squares. Some should be tilted on an angle. Here are three examples to get you started.

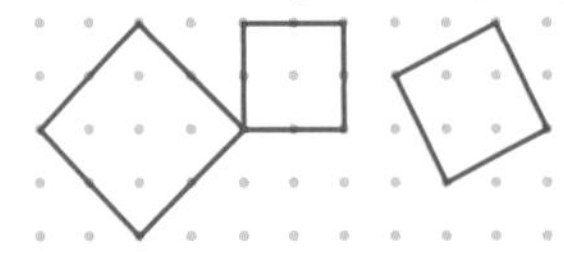

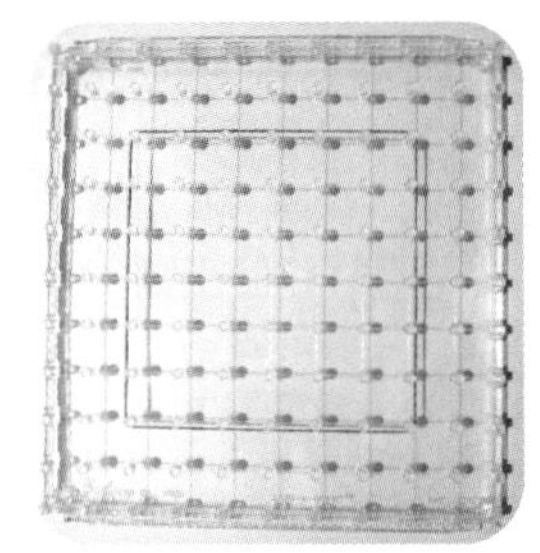

b) Draw each square on the dot paper.

Strategies
Use a table to organize your work.

c) Some of the squares you drew have side lengths and areas that can be found easily. Copy the table and record the areas and side lengths for the squares in the table.

Area of Square (unit^2)	1	4	9	16	25	36	49	64	81
Side Length of Square (unit)									

2. a) If a square has a side length of 10 units, what is its area?

b) If a square has an area of 121 square units, what is the length of one side?

c) If a square has an area of 144 square units, what is the length of one side?

3. Reflect In questions 1 and 2, you have just determined the square roots of the first 12 perfect squares. How is the area of each square in the table related to its side length? How is the side length of each square related to its area?

4. Reflect Examine one of the tilted squares that you drew. Find its area by sectioning it into squares or triangles. Measure its side length with a centimetre ruler. Do you think these tilted squares have the same relationship between side length and area as the squares in the table? Explain.

5. What do you think the term *perfect square* means? What do you think the term *square root* means? What do you think the phrase *a number that is not a perfect square* means?

Example 1: Find the Square Roots

Evaluate.

a) $\sqrt{49}$ **b)** $\sqrt{400}$ **c)** $\sqrt{121}$

d) $\sqrt{0.09}$ **e)** $\sqrt{1.44}$

Communicating Mathematically

- The symbol $\sqrt{\ }$ means the square root of.

Solution

a) $\sqrt{49} = 7$, because $7 \times 7 = 49$.

b) $\sqrt{400} = 20$, because $20 \times 20 = 400$.

c) $\sqrt{121} = 11$, because $11 \times 11 = 121$.

d) $\sqrt{0.09} = 0.3$, because $0.3 \times 0.3 = 0.09$.

e) $\sqrt{1.44} = 1.2$, because $1.2 \times 1.2 = 1.44$.

Example 2: Find the Perimeter of a Square

Find the perimeter of a square with an area of 9 cm^2.

> **Strategies**
> What is the hidden information you need to find before you can calculate the perimeter?

Solution

$\sqrt{9} = 3$, because $3 \times 3 = 9$.
That means that the side length of a square with an area of 9 cm^2 is 3 cm. The square has 4 sides, so its perimeter is 4×3 cm $= 12$ cm.

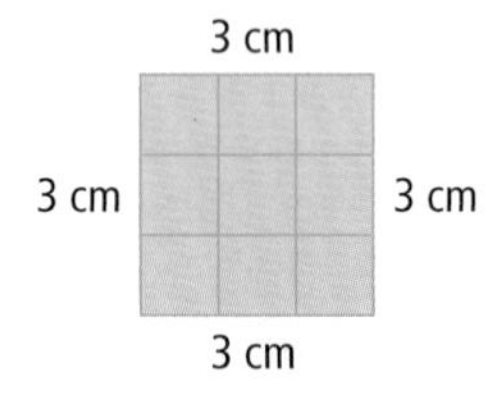

perfect square

- a number that can be expressed as the product of two equal integer factors

Example 3: True or False?

True or false? Show your work.

a) $\sqrt{25} + \sqrt{64} = \sqrt{25 + 64}$

b) $\sqrt{25} \times \sqrt{64} = \sqrt{25 \times 64}$

Solution

a) Left side $= \sqrt{25} + \sqrt{64}$
$= \sqrt{5 \times 5} + \sqrt{8 \times 8}$
$= 5 + 8$
$= 13$

Right side $= \sqrt{25 + 64}$
$= \sqrt{89}$
$\doteq 9.4$

> I used my calculator to evaluate $\sqrt{89}$.

Left side $\neq$ Right side
$\sqrt{25} + \sqrt{64} = \sqrt{25 + 64}$
This statement is false.

b) Left side $= \sqrt{25} \times \sqrt{64}$
$= \sqrt{5 \times 5} \times \sqrt{8 \times 8}$
$= 5 \times 8$
$= 40$

Right side $= \sqrt{25 \times 64}$
$= \sqrt{1600}$
$= 40$

Left side = Right side
$\sqrt{25} \times \sqrt{64} = \sqrt{25 \times 64}$
This statement is true.

Communicating Mathematically

- The symbol $\neq$ means not equal to.

Communicate the Key Ideas

1. a) Why is 100 a perfect square?
b) Why is 20 not a perfect square?

2. a) How does the first diagram show the square root of 16?
b) How does the second diagram show the square root of 8?

3. a) Why is 4^2 often read as "four squared"?
b) What is the relationship between the square and the square root of a number?

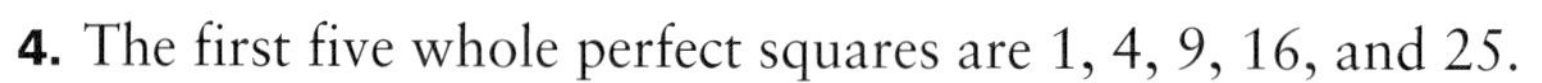

4. The first five whole perfect squares are 1, 4, 9, 16, and 25.
a) Subtract the first number from the second, the second from the third, and so on. The differences form a sequence.
b) Find the difference between pairs of consecutive numbers from part a).
c) Write the pattern in part b). Test this pattern on five more consecutive whole perfect squares of your choice.
d) Do you think that this pattern applies to all whole perfect squares?
e) Explain why finding the square root of 36 is the same as finding the side length of a square with an area of 36 square units.

5. Refer to the introduction on page 8. Colin needs 225 small congruent squares to make his SCRABBLE® board. What are the dimensions of his board? Explain.

Check Your Understanding

1. Evaluate. Give the value of each square root.

a) $\sqrt{121}$ **b)** $\sqrt{100}$
c) $\sqrt{169}$ **d)** $\sqrt{81}$

2. A square with an area of 16 square units has what side length?

3. Determine the perimeter of a square with an area of

a) 25 cm^2 **b)** 144 cm^2
c) 64 cm^2 **d)** 169 cm^2

4. If you know the area of a square, how can you find its perimeter?

5. True or false? Show your work.

a) $\sqrt{81} + \sqrt{121} = \sqrt{81 + 121}$
b) $\sqrt{81} \times \sqrt{121} = \sqrt{81 \times 121}$

6. Explore how to use the square root key on your calculator. Test it using the number 25. You know what answer to expect. Practise using your calculator by finding $\sqrt{36}$, $\sqrt{100}$, or the square root of other small perfect squares.

7. A marching band in the Canada Day Parade has 256 members. Marching bands change formations often. When the band changes formation to a square, how many musicians would be in each row? How many rows would there be?

8. Each face of a cube has an area of 36 cm^2. What is the volume of the cube?

9. A square has an area of 289 cm^2. What is the radius of the largest circle that can fit inside the square? Explain.

10. The sum of two perfect squares can sometimes equal another perfect square. For example, $9 + 16 = 25$. Find at least three more sums of two perfect squares that equal a perfect square.

11. **a)** The number 4 is two times the number 2. Does this mean that 4^2 is two times the value of 2^2?
    b) Consider 8 and 2. Is the value of 8^2 four times the value of 2^2? If not, what is the relationship?
    c) Consider 10 and 2. Is 10^2 five times 2^2? If not, what is the relationship?
    d) Look for a pattern in your results. Propose a relationship between the squares of two numbers, when one number is a multiple of the other. Test your relationship on at least three other pairs of numbers.

12. **a)** Write the next five lines in the following pattern.
    $2^2 = 3 \times 1 + 1 = 4$
    $3^2 = 4 \times 2 + 1 = 9$
    $4^2 = 5 \times 3 + 1 = 16$
    b) Describe the pattern in words.
    c) Use the pattern to determine each square.
    i) 29^2 **ii)** 41^2

13. A farmer wants to divide one of his fields into 80 congruent square plots to let people plant gardens. If his field is rectangular in shape and has an area of 2000 m^2, what are the dimensions of each garden?

14. **a)** Create each V figure using linking cubes.

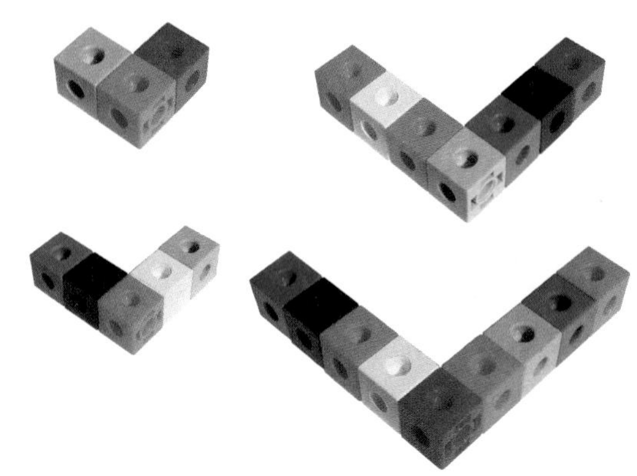

b) Add one cube to the V made with 3 cubes. What shape do you get? Record the total number of cubes as an addition statement: number of cubes added + number of cubes in the V = total number of cubes. Use a table like the one shown.

Number of Cubes in V Figure	Number of Cubes Added	Shape Description	Total Number of Cubes
3			

c) Add the figure from part b) to the V made with 5 cubes. What shape do you get? Record the information in the table.
d) Add the figure from part c) to the V with 7 cubes. Add the new figure to the V with 9 cubes. Record the information for each V in the table.
e) Examine the patterns in the table. Describe the patterns.

Extend

15. The square root of a perfect square number can be found by subtracting the consecutive odd whole numbers, beginning with 1, from the square until the result is 0.

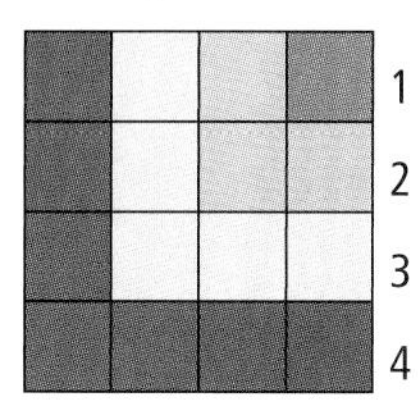

$16 - 1 = 15$
$15 - 3 = 12$
$12 - 5 = 7$
$7 - 7 = 0$

Four numbers were subtracted to reach zero, so

$\sqrt{16} = 4$.

a) Use this method to find the square root of each number.

i) 81

ii) 169

b) Why do you think this method works?

16. a) Choose two consecutive perfect squares that are less than 400. Calculate the difference between them.

b) Find the square root of each perfect square and add the square roots.

c) Choose two more consecutive perfect squares. Calculate the difference between them. Find the square root of each perfect square and add the square roots.

d) What pattern do you notice?

e) Use the pattern from part d) to find the next perfect square that is greater than 400.

17. Since $2^3 = 8$, 2 is known as the "cube root" of 8. Find the cube root of each of these numbers. Explain how you found your answer.

a) 27

b) 125

c) 1 000 000

18. Perfect cubes can be written as the sum of consecutive odd numbers.

For example, 8 is a perfect cube and can be written as the sum of two consecutive odd numbers, 3 + 5.

The next perfect cube is 27, which can be written as the sum of three consecutive odd numbers.

The next perfect cube is 64, which can be written as the sum of four consecutive odd numbers.

a) Copy and complete the table.

Perfect Cube Number	Sum of Consecutive Odd Numbers
8	3 + 5
27	7 + 9 + 11
64	
125	
216	
343	

b) What is the mean of the odd numbers added for each perfect cube? How does the mean relate to the number being cubed?

c) Use the pattern from part b) to find the cube root of each number.

i) 8 **ii)** 27

iii) 64 **iv)** 343

1.2 Find Square Roots

Focus on...
- finding square roots of any number
- distinguishing between an exact square root and a decimal approximation

Stained glass artisans use many rectangular patterns in their work. Sometimes they use square glass pieces with side lengths that are not in whole measurement units.

Discover the Math

Materials
- 1-cm square dot paper
- base-10 materials
- coloured pencil
- ruler in millimetres

Optional:
- centimetre grid paper
- BLM Base 10 Blocks (all)*

* See Mathematics Blackline Masters Grades P to 9

Do numbers that are not perfect squares have square roots? How can you find approximate square roots?

Recall from Section 1.1 the tilted squares (squares that are not perfect) in the Discover the Math activity.

1. On a sheet of square dot paper, draw as many different-sized tilted squares as you can.

2. Section each tilted square into a smaller square and four right triangles to find the area as shown in the diagrams. Then, measure the side length of the tilted square, to the nearest tenth of a centimetre. Write the area in the centre of the tilted square and record the side length on one side.

Strategies
Break the problem into simpler parts.

3. Square the side length of each tilted square. Notice how close your answers are to the actual areas. You have **approximated** the square roots of some squares that are not perfect.

approximate

- to find a number that is close enough to another number to make useful calculations
- For example, 3.14 may be used to approximate π.

4. How can you determine the square root of 5, using base-10 blocks?
   a) Start with five flats. Let one flat represent 1. Model the perfect square that is just less than 5. How many flats did you use?
   b) Trade the remaining flats for rods (tenths). Distribute the rods evenly on two sides, keeping the square shape. Try to make the square that approximates 5. To complete your square, trade in the remaining rods for unit cubes (hundredths). Add unit cubes to complete the square that represents 5. What is the side length of the square?
   c) Is your answer exact or approximate? Why?
   d) Find the value of $\sqrt{5}$ using your calculator. Explain the difference between your answer and the calculator answer.

5. **Reflect** What type of number has a whole number square root? What type of number has only an **estimated** or approximate square root? Give examples of each.

estimate

- to calculate approximately the value, amount, or size of something

Example 1: Estimate the Square Roots of Squares That Are Not Perfect

Estimate the value of $\sqrt{86}$, to one decimal place.

Solution

I can use estimation. 81 and 100 are the two perfect squares closest to 86. I know that $9^2 = 81$ and $10^2 = 100$. So, $\sqrt{86}$ is between 9 and 10.

I can use a number line to help me decide which perfect square is closest to 86. 86 is about one-quarter of the way between 81 and 100.

$\sqrt{86}$

$\sqrt{81}$ $\sqrt{100}$

8 9 10 11

On my calculator, the answer is 9.2736184954957, but I know that this number is an approximation.

An estimate for $\sqrt{86}$ is 9.3.

Example 2: Find the Square Root of Any Number

Find the value of $\sqrt{3600}$.

Solution

Method 1: Use patterns.

At first I thought that $\sqrt{3600}$ might be 600, but then I realized that $600^2 = 360\ 000$, which is too big. I then tried 60. This works because $60^2 = 3600$.

Method 2: Use factors.

3600 is equal to 36 × 100, which are both perfect squares.

$$\begin{aligned}\sqrt{3600} &= \sqrt{36 \times 100}\\ &= \sqrt{36} \times \sqrt{100}\\ &= \sqrt{6 \times 6} \times \sqrt{10 \times 10}\\ &= 6 \times 10\\ &= 60\end{aligned}$$

Example 3: Find the Square Roots of a Perfect Square by Prime Factorization

Find the value of $\sqrt{784}$. Use prime factorization.

Solution

$$\begin{aligned}\sqrt{784} &= \sqrt{4 \times 196}\\ &= \sqrt{2 \times 2 \times 14 \times 14}\\ &= \sqrt{2 \times 2 \times 2 \times 7 \times 2 \times 7}\\ &= \sqrt{(2 \times 2 \times 7) \times (2 \times 2 \times 7)}\\ &= 2 \times 2 \times 7\\ &= 28\end{aligned}$$

I can find the prime factors.

Communicate the Key Ideas

1. Which of the numbers below does not have a whole number square root? Explain why.

   36, 16, 144, 49, 10, 225

2. Rachel was trying to estimate $\sqrt{67}$ but did not know where to start. Suggest the steps that Rachel could use to estimate $\sqrt{67}$.

3. Morad said that $\sqrt{6}$ is exactly 2.449 489 742 because this is what his calculator displayed. Sabina said that this is an approximation. Who is correct? Justify your answer.

Check Your Understanding

1. Find the value of each square root. Use an estimate if the square root is not exact.

   a) $\sqrt{256}$
   b) $\sqrt{2500}$
   c) $\sqrt{225}$
   d) $\sqrt{26}$

2. Explain why $\sqrt{0.25}$ has an exact value, while $\sqrt{2.5}$ has only an approximate value.

3. Find the value of each square root. Use an estimate if the square root is not exact.

   a) $\sqrt{0.64}$
   b) $\sqrt{6400}$
   c) $\sqrt{6.4}$
   d) $\sqrt{640}$

4. Which of the following statements are true? Explain how you know. Correct the statements that are false.

   a) The value of $\sqrt{30}$ is between 5 and 6, and closer to 5.
   b) The value of $\sqrt{48}$ is between 6 and 7, and closer to 6.
   c) The value of $\sqrt{10}$ is between 4 and 5, and closer to 5.
   d) The value of $\sqrt{75}$ is between 8 and 9, and closer to 9.

5. Use perfect square numbers to explain how to make a reasonable estimate of $\sqrt{13}$, to one decimal place.

6. Tom wants to thank his art teacher by giving her one of his paintings. The painting is square with an area just slightly less than 900 cm^2. Tom wants the painting to fit snugly into a gift box. What estimates might Tom give for the dimensions of the box?

7. Estimate each square root to one decimal place. Explain your thinking.

   a) $\sqrt{10}$
   b) $\sqrt{40}$
   c) $\sqrt{72}$
   d) $\sqrt{55}$

8. Find each square root using patterns or factorization. Show your work.

   a) $\sqrt{4900}$
   b) $\sqrt{729}$
   c) $\sqrt{10\,000}$
   d) $\sqrt{484}$
   e) $\sqrt{0.04}$
   f) $\sqrt{0.000049}$

9. Is each statement below always true, sometimes true, or never true? Support your answer.

   a) The square of an odd number is odd.
   b) The square root of a number is greater than the square of the number.
   c) The square root of a perfect square is even.
   d) Only whole numbers have square roots.
   e) Perfect squares are found only between 1 and 10 000.

10. a) Write the numbers in increasing order.
    6, $\sqrt{30}$, $\sqrt{25}$, 5.8, $\sqrt{28}$
    b) Draw a number line and mark the approximate position of each value.

11. What are the dimensions of the square sticker?

area = 3.61 cm^2

12. A rectangular wire fence with dimensions 45 m by 20 m needs to be reshaped to form a square with the same area. What would be the length of each side of the fence? Would more or less wire be required, and if so, how much?

13. Which of the square roots are between 6 and 7? Explain how you know.

$\sqrt{52}$ $\sqrt{41}$ $\sqrt{35}$ $\sqrt{38}$ $\sqrt{45}$

14. The playing surface of a checkerboard has an area of 576 cm^2. What is the side length of each small square on the board? (Hint: The board is square and contains 64 congruent squares.)

15. A boat club is renovating its meeting room, which measures 15 m by 15 m. The directors decide to give 20% of the area of the meeting room to the junior club members for an activity room. If the activity room is also a square, estimate its dimensions, to one tenth of a metre.

16. Kayla wants to build a square pen for her dogs to play in. The dog pen she has now has dimensions of 36.00 m by 30.25 m.

a) If the pen is to have the same area what will its dimensions be, to one tenth of a metre?

b) If the fencing material costs $40/m, how much will she pay for the new fence?

17. Each face of the cube has an area of 2500 mm^2.

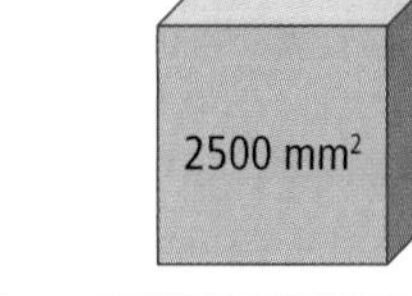

What is the volume of the cube in cubic millimetres (mm^3)? in cubic centimetres (cm^3)? (Hint: 1 cm = 10 mm)

Strategies

What hidden information do you need to find before you calculate the volume?

18. a) If a perfect square ends with four zeros, why will its square root end with two zeros?

b) If a square number has six decimal places, why will its square root have 3 decimal places?

Chapter Problem

19. Brian has a square piece of kite paper with an area of 1200 cm^2. If he uses the entire piece of paper without cutting it, what is the approximate side length of his kite, to the nearest centimetre?

Extend

20. Graph the area of a square versus its side length. Use whole side lengths from 0 to 10 units. Join the points using a smooth curve. Describe how you could use the completed graph to estimate roots of squares that are not perfect, such as $\sqrt{30}$.

Puzzler

Mia wanted to see how long it would take her to write down the whole numbers from 1 to 1 000 000. But after writing 12 309 digits, her arm got tired and she stopped. How many numbers did she write?

1.3 Discover the Pythagorean Relationship

Focus on...

- discovering the Pythagorean relationship

A fourth generation sail-maker from Lunenburg made the main sail for the Bluenose II in 2005. Her great-grandfather, who had crewed on the original Bluenose as a sail trimmer, passed down the art of sail-making. His method for cutting the original sailcloth was to start by laying the cloth on a frozen lake, marking the corners (one of which was a right angle), putting nails in the ice, tying string around the nails, and then marking the sailcloth to cut it. Seams and panels were then added to the sails.

Discover the Math

Materials

- ruler
- centimetre grid paper
- index card

How did the first sail maker for the *Bluenose* know the length of the longest edge of the sail?

Part A: Right Triangles

legs

- the two shorter sides of a right triangle
- legs meet at 90°

hypotenuse

- longest side of a right triangle
- opposite the right angle

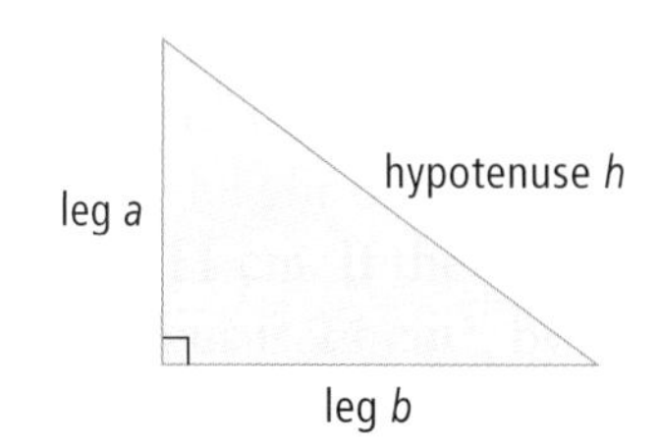

In this activity, you will learn a fascinating property of right triangles. The sides of right triangles have special names. The two sides that form the right angle are called the **legs** and the side that is opposite the right angle is called the **hypotenuse**. The hypotenuse is always the longest side in any right triangle. The legs have been labelled a and b, while the hypotenuse is labelled h. These letters stand for the length of each side.

1. On centimetre grid paper, draw a right triangle in which leg a is 1 cm and leg b is 1 cm. Draw the hypotenuse. Carefully construct squares on all three sides as shown in the figure on the left. Use a set square or the corner of an index card to draw the square on the hypotenuse.

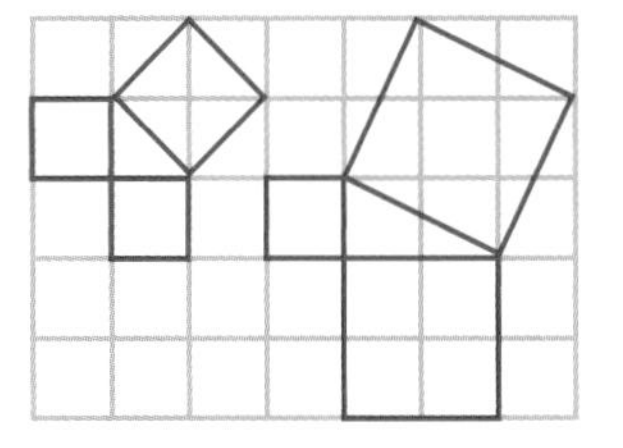

Strategies
Draw a diagram. Can you think of another way to construct these figures?

2. Find the areas of all three squares. To find the area of the square on the hypotenuse, section it into triangles.

3. Copy the table and complete the first row. Use the leg lengths given in the table to draw the remaining triangles. Construct squares on all three sides of each triangle, then find and record the areas of the squares.

Leg a (cm)	Leg b (cm)	Area of Square on Leg a, a^2 (cm²)	Area of Square on Leg b, b^2 (cm²)	Area of Square on Hypotenuse h, h^2 (cm²)
1	1			
1	2			
1	3			
2	3			
2	4			
3	4			

4. **Reflect**
   a) What pattern do you notice in the areas of the squares?
   b) Can you write a sentence to express the relationship you have discovered?

You have explored the **Pythagorean relationship**. *The sum of the areas of the squares on the two shorter sides of a right triangle is equal to the area of the square on the hypotenuse.*

The Pythagorean relationship can be expressed by the equation $h^2 = a^2 + b^2$, where h represents the length of the hypotenuse, and a and b represent the lengths of the legs.

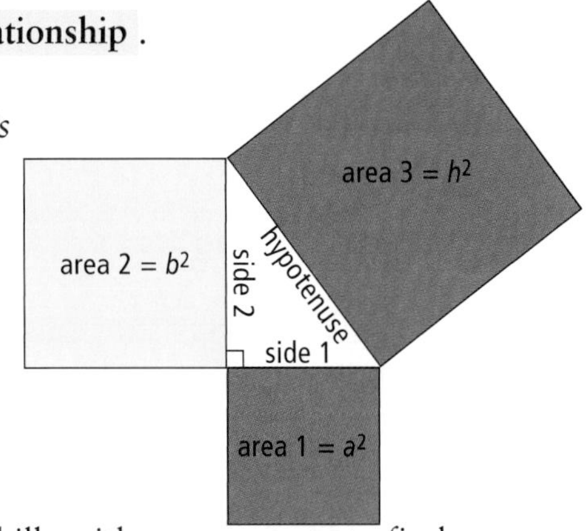

You can use this relationship and your skills with square roots to find the unknown length of a side of a right triangle if you know the lengths of the other two sides.

Pythagorean relationship
- the relationship between the lengths of the sides of a right triangle

Part B: Determine If a Triangle is a Right Triangle

Materials

- Bullseye
- centimetre grid paper
- scissors
- index card

Optional:

- ruler
- protractor

1. Draw each set of squares on centimetre grid paper and cut out each set. Write the area, in square centimetres, on each square as you draw it.

- 4 cm by 4 cm, 5 cm by 5 cm, and 8 cm by 8 cm
- 4 cm by 4 cm, 12 cm by 12 cm, and 14 cm by 14 cm
- 3 cm by 3 cm, 4 cm by 4 cm, and 5 cm by 5 cm
- 5 cm by 5 cm, 12 cm by 12 cm, and 13 cm by 13 cm
- 6 cm by 6 cm, 8 cm by 8 cm, and 10 cm by 10 cm

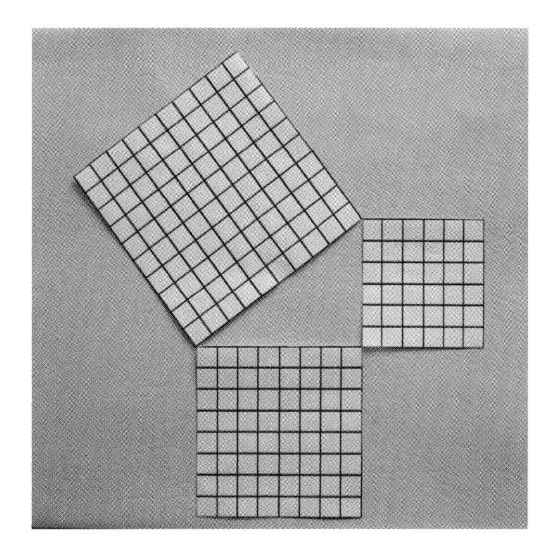

2. Make a triangle with each set of squares. Arrange the three squares so that one side of each square forms one side of the triangle. Tape each arrangement onto a piece of paper.

3. Use the corner of an index card to determine if any of the triangles contain a right angle.

4. Copy and complete the table, using your observations of the five triangles. The sides are identified as side 1 (a), side 2 (b), and side 3 (h).

Side 1, a (cm)	Side 2, b (cm)	Side 3, h (cm)	Square 1 Area, a^2 (cm^2)	Square 2, Area, b^2 (cm^2)	Square 3 Area, h^2 (cm^2)	Type of Triangle (acute, obtuse, or right)
4	5	8				
4	12	14				
3	4	5				
5	12	13				
6	8	10				

5. Reflect

a) Do you see any patterns in your results?

b) Consider the last three triangles. Compare the areas of the squares on the sides of each triangle. How are the areas of the squares related?

c) Examine the first two triangles. Does the relationship in the last three triangles apply to the first two triangles? Why or why not?

6. Which triangle is a right triangle? Explain how you know.

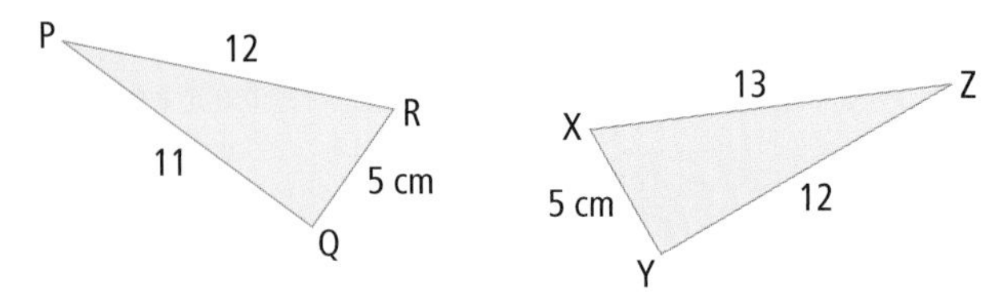

7. Use the Pythagorean relationship to determine which of the following sets of side lengths will form a right triangle. Verify two of your answers using a diagram.

Triangle	Side Lengths	Right Triangle?
ABC	AB = 7.0 cm BC = 8.0 cm AC = 9.0 cm	
WXY	WX = 6.0 cm XY = 8.0 cm WY = 10.0 cm	
JKL	JK = 5.4 cm KL = 3.0 cm JL = 3.6 cm	
RST	RS = 4.0 cm ST = 5.0 cm RT = 7.0 cm	
DEF	DE = 1.5 cm EF = 2.0 cm DF = 2.5 cm	

8. **Reflect** Write a statement that explains how you can tell if a triangle contains a right angle given the lengths of its sides.

Example 1: Find the Length of the Hypotenuse

A square is drawn on each side of $\triangle ABC$.

a) Find the area of the square on side AC.

b) Find the length of the hypotenuse, to the nearest tenth of a centimetre.

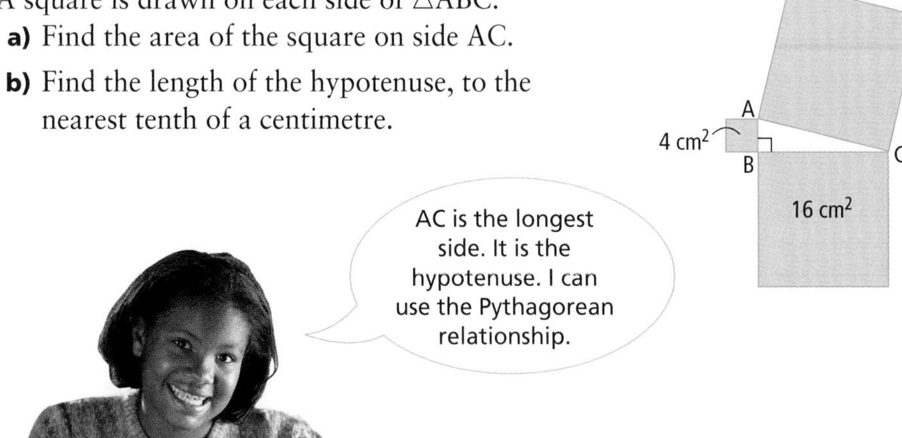

Solution

a) $AB^2 + BC^2 = AC^2$

$4 + 16 = AC^2$

$20 = AC^2$

The area of the square on side AC is 20 cm^2.

Strategies
Estimate.

b) $AC^2 = 20$

$AC = \sqrt{20}$

$AC \doteq 4.5$ C 2nd √x 20 =

The length of the hypotenuse is approximately 4.5 cm.

20 is between 16 and 25, so the length of AC must be between 4 cm and 5 cm.

Example 2: Find the Length of the Leg of a Right Triangle

A square is drawn on each side of △PQR.

a) Find the area of the square on side PQ.

b) Find the length of the side PQ, to the nearest tenth of a centimetre.

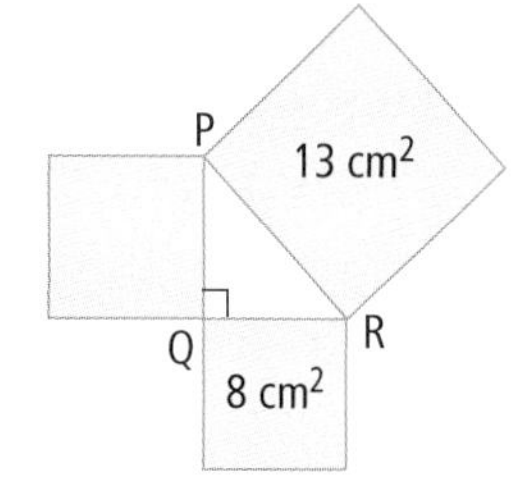

I know that the areas of the two smaller squares add up to the area of the square on the hypotenuse.

Solution

a) $QR^2 + PQ^2 = PR^2$ **Write the Pythagorean relationship.**

$8 + PQ^2 = 13$

$PQ^2 = 13 - 8$

$PQ^2 = 5$

The area of the square on side PQ is 5 cm^2.

Strategies
Set up and solve an algebraic equation.

b) $PQ^2 = 5$

$PQ = \sqrt{5}$

$PQ \doteq 2.2$ C 2nd √x 5 =

The length of PQ is approximately 2.2 cm.

5 is between 4 and 9, so the length of PQ must be between 2 cm and 3 cm.

Example 3: Right or Not?

A stained glass artist wants to include a right triangle in his panel. He designs the panel and measures the sides of the triangle. Is the triangle a right triangle? How do you know?

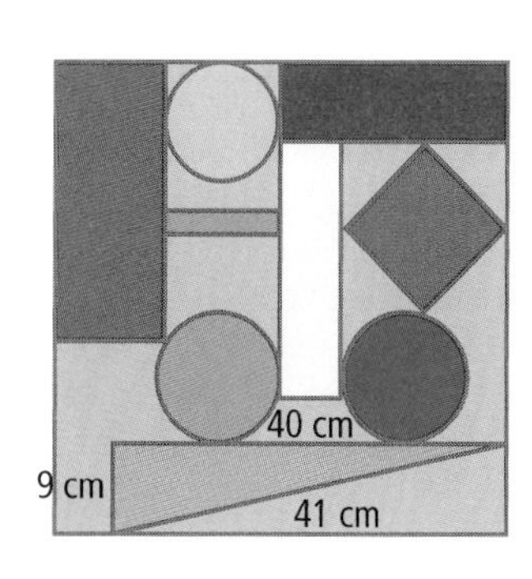

Solution

Draw and label a diagram.
Let h be the longest side.
Use the Pythagorean relationship.

$b = 40$ cm
$a = 9$ cm
$h = 41$ cm

$a^2 + b^2 = h^2$

$$\begin{aligned} \text{Left side} &= a^2 + b^2 \\ &= 9^2 + 40^2 \\ &= 81 + 1600 \\ &= 1681 \end{aligned} \qquad \begin{aligned} \text{Right side} &= h^2 \\ &= 41^2 \\ &= 1681 \end{aligned}$$

$$\text{Left side} = \text{Right side}$$

So, $9^2 + 40^2 = 41^2$.
The triangle is a right triangle.

Communicate the Key Ideas

1. What property must a triangle have before you can use the Pythagorean relationship to find a missing side measure? Explain how you know.

2. Three squares are placed together as shown.

Triangle ABC looks like a right triangle to me.

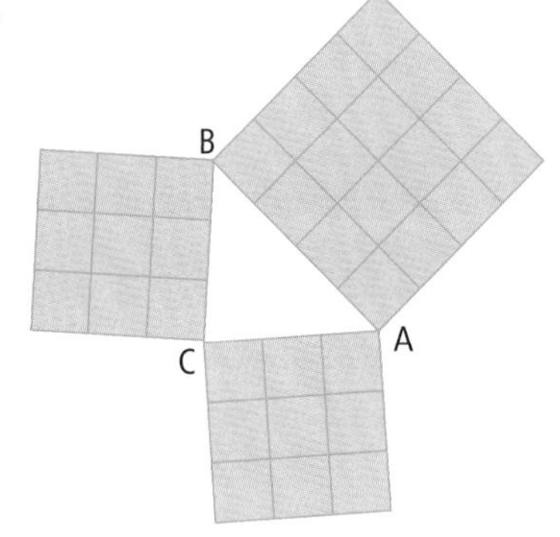

Is Linh correct? Explain how you know.

3. How are square roots used to find the missing side of a right triangle?

4. If you know the area of the square on the hypotenuse and the area of the square on one leg, how could you determine the area of the square on the other leg?

5. Who is correct? Why?

20.5 cm
4.5 cm
20.0 cm

Alene, the triangle that I drew is a right triangle.

I'm sorry Andre, but it isn't a right triangle because the side lengths are not multiples of the basic right triangle with sides 3, 4, and 5 units long.

Check Your Understanding

1. What is the area of the square on the hypotenuse of each triangle?

a)

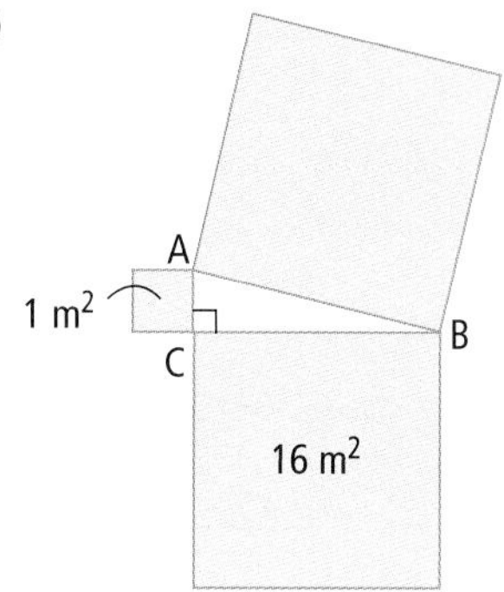

b)

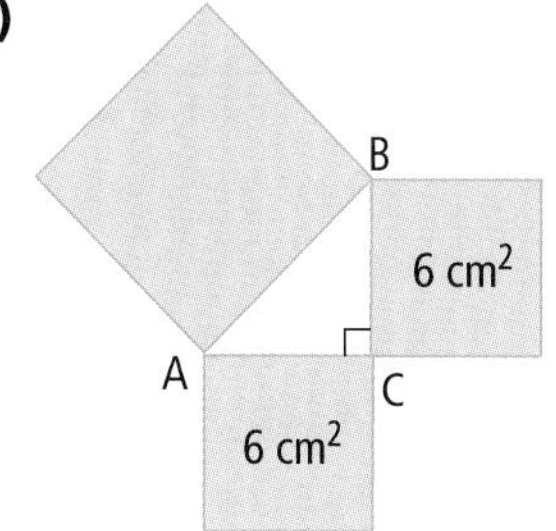

c)

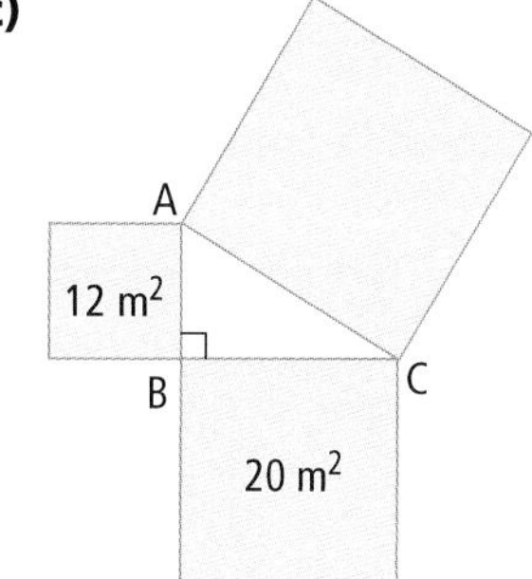

d)

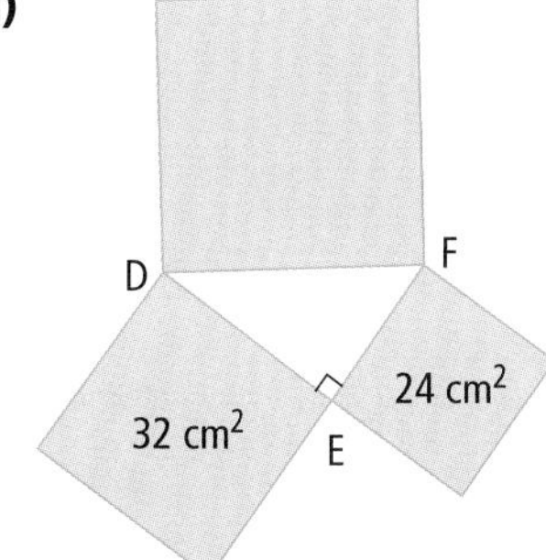

2. What is the area of the square on the leg of each triangle?

a)

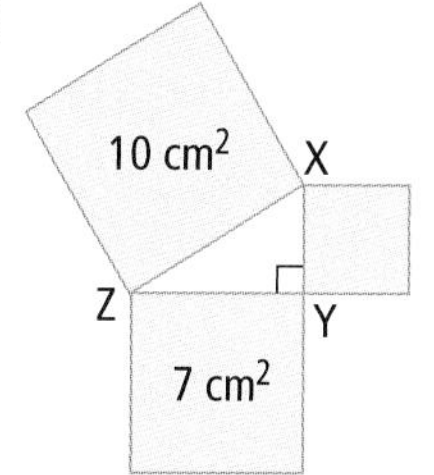

b)

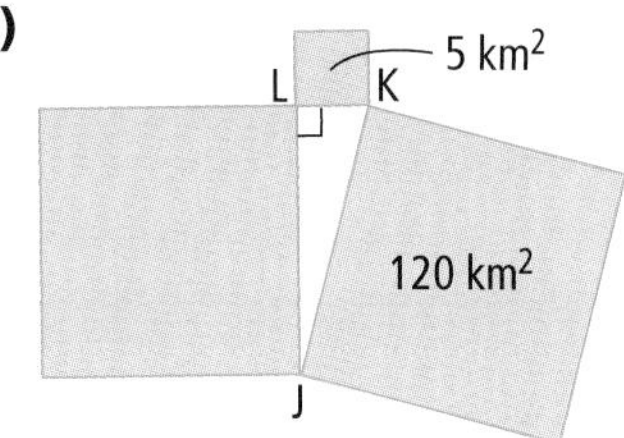

c)

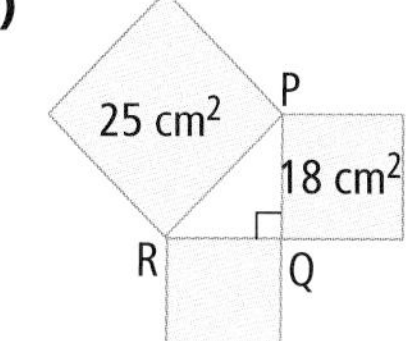

d)

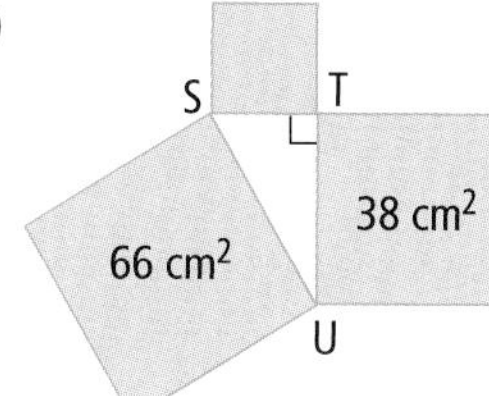

3. In each figure

- identify the hypotenuse
- use the Pythagorean relationship to find the missing area

a) **b)**

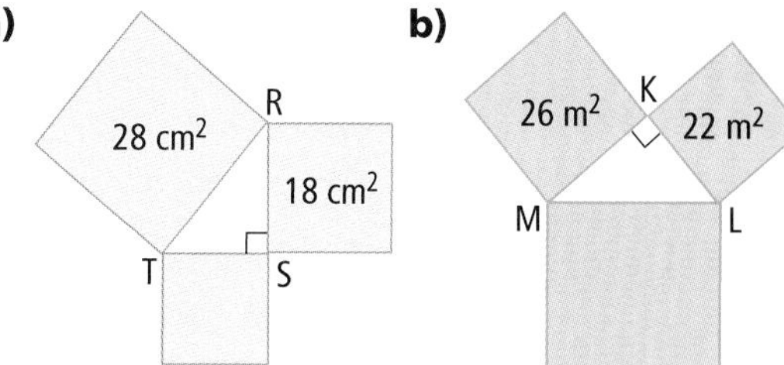

4. Is Armand correct? Explain how you know if his answer is reasonable. Do not work out the entire solution.

In a right triangle, if one leg is 3 cm and one leg is 4 cm, then the hypotenuse is 4 cm.

5. In a right triangle, what is the area of the square on the hypotenuse if the areas of the squares on the legs are the following?

a) 11 cm^2 and 15 cm^2
b) 10 m^2 and 4 m^2
c) 7 cm^2 and 21 cm^2
d) 26 km^2 and 13 km^2

Strategies
Diagrams would be helpful here.

6. Fizal found a geometric method of approximating some square roots. He drew a right triangle with a base of 3 cm and a height of 3 cm. He claims that by measuring the hypotenuse of the triangle he can find $\sqrt{18}$ to one decimal place.

a) On centimetre grid paper, draw a right triangle with leg lengths 3 cm and 3 cm.
b) Draw a square on the hypotenuse of the triangle.
c) How does Fizal know that the hypotenuse of the triangle measures $\sqrt{18}$?
d) Use your ruler to find an estimate for $\sqrt{18}$.

7. Do the following side lengths form right triangles? Show your work.

a) 9, 15, 12
b) 24, 7, 25
c) 4, 9, 11
d) 5, 5, $\sqrt{50}$

8. Draw a square with an area of 4 cm^2 on centimetre dot or grid paper. Then, draw another square with an area of 9 cm^2, as shown.

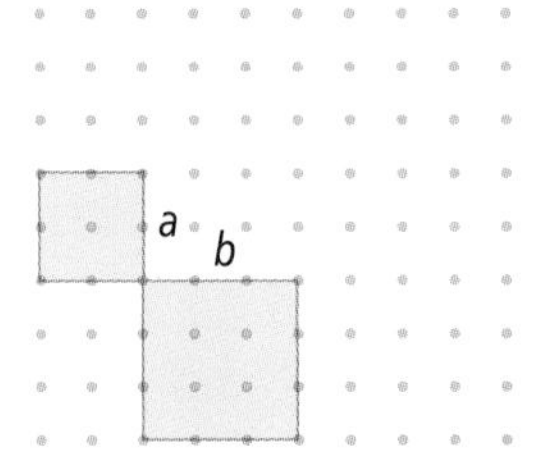

a) Express the length of each side of the two squares as a whole number and a square root.
b) Create a right triangle whose legs are a and b by connecting the top right vertices of both squares to form the hypotenuse.
c) What is the length of the hypotenuse of the triangle expressed as a square root?
d) Create a second right triangle by connecting the bottom left vertices of both squares. What is the length of the hypotenuse of the triangle expressed as a square root?

9. Construct a right isosceles triangle whose hypotenuse is $\sqrt{8}$ units, using a geoboard or square dot paper. What is the area of each square on the legs of the triangle?

10. For which of the sets of three squares does the Pythagorean relationship apply? Show your work.

a) 10 cm^2, 4 cm^2, 14 cm^2
b) 3 cm^2, 12 cm^2, 16 cm^2
c) 5 cm^2, 27 cm^2, 22 cm^2
d) 21 cm^2, 2 cm^2, 19 cm^2

11. The square on the hypotenuse of a right triangle has an area of 34 cm^2. The square on one of the legs has an area of 17 cm^2.

a) What is the area of the square on the remaining leg?
b) How else can you classify this right triangle?

12. Does the Pythagorean relationship apply to a right triangle with side lengths that are not all whole numbers? Explain your reasoning and provide examples.

13. Larry drew a triangle with side lengths of 5 cm, 7 cm, and 9 cm.

a) Find the area of the square on each side of the triangle.

b) Is the triangle a right triangle? Explain why or why not.

c) If the triangle is a right triangle, draw a diagram. Label the hypotenuse and the legs.

14. The areas of the squares on two sides of a right triangle are 26 cm^2 and 40 cm^2. What are the possible side lengths of the square on the third side? Is there more than one possible triangle? Draw sketches to support your answers.

15. Kara has several square stickers of four different sizes, as shown. Kara wonders if she can place any three stickers together to form a right triangle. Solve this problem for her. Use diagrams to show your answer.

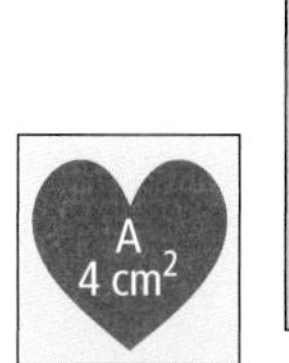

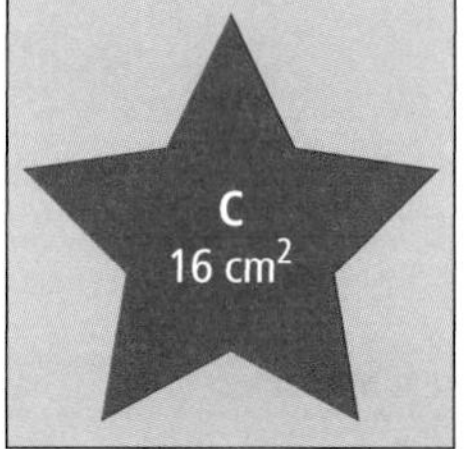

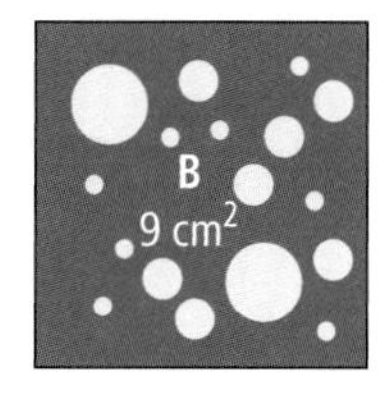

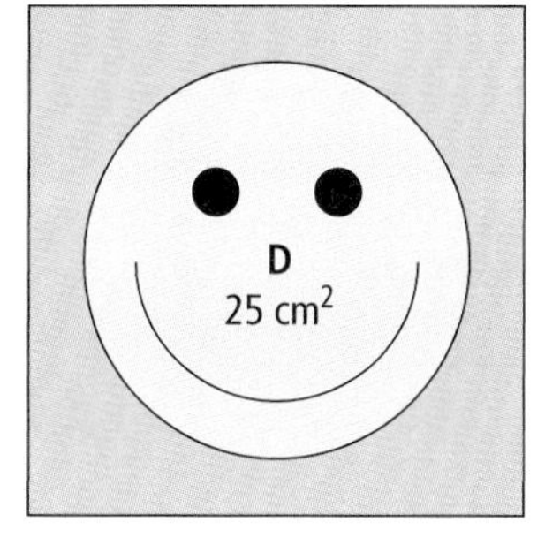

16. {3, 4, 5} and {5, 12, 13} are Pythagorean triples. Why do you think these sets of natural numbers are called Pythagorean triples? Can you find more Pythagorean triples?

Extend

17. Does the Pythagorean relationship apply if a figure other than a square is drawn on the sides of a right triangle? Experiment by replacing the square on each side of a right triangle with equilateral triangles. Find the area of each equilateral triangle. Is there a relationship between the areas of the equilateral triangles?

18. The Pythagorean relationship is named after the Greek mathematician Pythagoras, who lived in the sixth century B.C.E. However, the Babylonians and Chinese likely knew the relationship long before this. Research the history of the Pythagorean relationship. Why was Pythagoras given the credit for discovering it?

Internet Connect To learn more about Pythagoras and his work, go to **www.mcgrawhill.ca/links/math8NS** and follow the links. Make a poster of one of the proofs of the relationship and explain it to the class.

19. Is Justina correct? Use squares with side lengths of 1 unit and 4 units to help you explain.

17 is not a perfect square, so it is not possible to draw a square with an area of exactly 17 square units.

Use Technology

Explore the Pythagorean Relationship Using *The Geometer's Sketchpad®*

Focus on…

- exploring the Pythagorean relationship

Materials

- computer
- *The Geometer's Sketchpad®* software

1. Open *The Geometer's Sketchpad®* and begin a new sketch.

2. Construct a right triangle.
   - **a)** Use the **Straightedge** tool to create a horizontal line segment.
   - **b)** Choose the **Selection Arrow** from the menu.
   - **c)** Click to select the line segment and then select one endpoint.
   - **d)** From the **Construct** menu, choose **Perpendicular Line.**
   - **e)** Select the new line.
   - **f)** From the Construct menu choose **Point on Perpendicular Line.** Note: If the point is below the horizontal line, click and hold on the point to select it and move it above the horizontal line.
   - **g)** Select the vertical line.
   - **h)** From the **Display** menu, choose **Hide Perpendicular Line.**
   - **i)** Select the three points.
   - **j)** From the **Construct** menu, choose **Segments.**

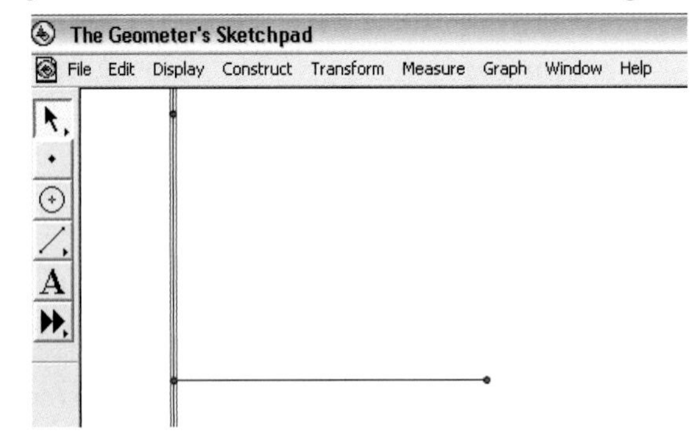

Technology Tip

- Double click on the letter that you want to change. You can also click and hold on the letter to move it to a different place.

Technology Tip

- Before you select a new object, deselect all objects first by clicking somewhere in the white space.

3. Label the right triangle.
   - **a)** Use the **Selection Arrow** to select all three line segments and all three points.
   - **b)** From the **Display** menu, choose **Show Label.**
   - **c)** Use the text tool to change the labels to a, b, and c, as shown.

   Look for a relationship between the three sides.

4. a) Use the **Selection Arrow** to select line segment *a*.

b) From the **Measure** menu, choose **Length**. The length of *a* is displayed on the screen.
Note: If your triangle appears on top of your measurements, drag a box around the triangle to select the entire triangle then click and hold to move the triangle over.

c) Follow the same steps to find the lengths of the sides *b* and *c*.

5. a) Use the **Selection Arrow** to select the measure of line segment *a*.

b) From the **Measure** menu, choose **Calculate**. A calculator window will appear.

c) Under **Values**, click on *a*.
- Click on the multiplication symbol, *.
- Click on *a* again.
- Then, click OK.

The computer will calculate and display the value for $a \times a$, or the square of the length of line segment *a*.

d) Follow the same steps to find the squares of the lengths of line segments *b* and *c*.

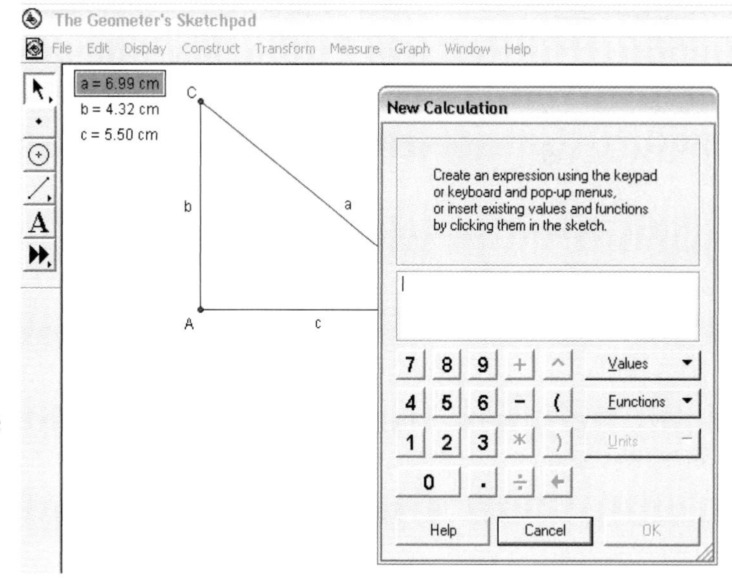

6. Compare the values for the squares of the three line segments. What relationship can you find?

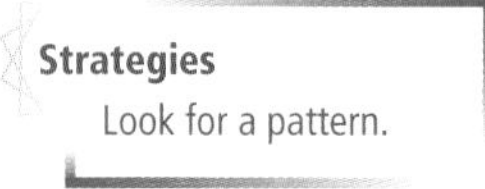

7. Use the **Selection Arrow** to select and drag the vertex of one of the acute angles to change the shape of the triangle.

a) Does the triangle remain a right triangle?

b) Notice how the measures and calculations displayed for the three line segments change as you drag the vertex. Does the relationship you discovered in question 6 still apply to the new triangle?

8. Reflect Write a statement about how the side lengths of a right triangle are related.

1.4 Apply the Pythagorean Relationship in Problem Solving

Focus on...

- applying the Pythagorean relationship in problem solving

The schooner *Amistad* visited Halifax in July 2006, stopping at the site of the historic black settlement of Africville. The original ship, *La Amistad*, was transporting 53 Africans (Sierra Leone), from the over 100 Africans who had been kidnapped from their homes in 1839. They were taken by force to Cuba.

The ship was famous in that the Africans, led by Cinque, revolted and took the ship to Connecticut, USA, where they were charged with murder and mutiny. With the help of abolitionist Theodore Joadson, attorney Roger S. Baldwin argued the case in favour of the Africans. When the decision was made to refuse their right to freedom, John Quincy Adams appealed the decision to the Supreme Court of the USA. Using the words of Cinque, he convinced the Supreme Court to free the 43 survivors and to return them to their homes in Sierra Leone.

If the ship sailed approximately 7515 km from Sierre Leone to Cuba and approximately 2261 km from Cuba to Connecticut, what was the distance of its return voyage? The path of the ship approximated a right triangle. How can you use the Pythagorean relationship to determine the distance?

To learn more about the *La Amistad*, go to **www.mcgrawhill.ca/links/math8NS** and follow the links.

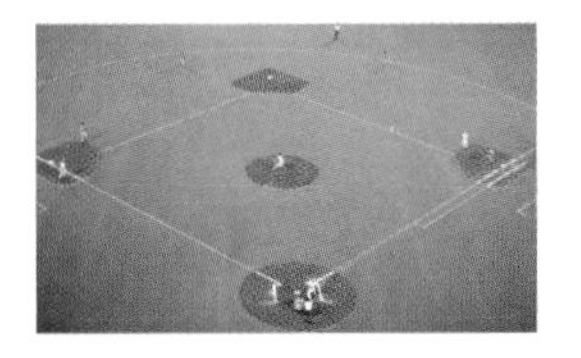

Example 1: Apply the Pythagorean Relationship

A baseball diamond is actually a square with sides 27.0 m in length. How far does the player at first base have to throw the ball to get a runner out at third base? Round your answer to the nearest tenth of a metre.

Solution

Draw a diagram of the baseball diamond so you can see what information is given and what you are being asked to find. The sides of the square are 27 m long. You are asked to find the distance from first to third base. This distance is the length of the diagonal.

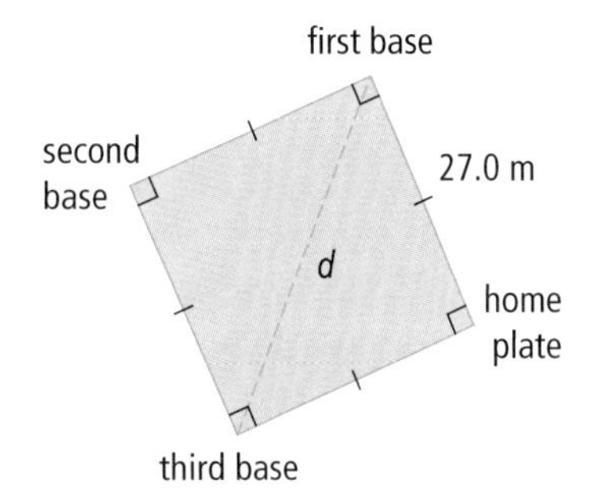

Strategies
Make a picture or diagram.

Plan

The distance from first base to third base is the hypotenuse of the right triangle. The lengths of the legs of the right triangle are both 27.0 m. Apply the Pythagorean relationship to find the hypotenuse.

Do It!

Let h represent the diagonal distance.

$$\text{Diagonal}^2 = (\text{Leg } 1)^2 + (\text{Leg } 2)^2$$
$$h^2 = 27^2 + 27^2$$
$$h^2 = 729 + 729$$
$$h^2 = 1458$$
$$h = \sqrt{1458}$$
$$h \doteq 38.2$$

Estimate: $\sqrt{1600} = 40$, therefore $\sqrt{1458}$ is a bit less than 40.

Look Back

The approximate answer is close to the estimate. The answer is reasonable for the given side lengths of the baseball diamond.

The player at first base must throw the ball about 38.2 m to get a runner out at third base.

Strategies
How many problem solving strategies can you find in this example?

Example 2: Find the Length of an Unknown Side

The gable end of a house is an isosceles triangle that has a height of 3 m. The sloping sides measure 7 m from the peak to the eaves. Find the width of the roof, to the nearest tenth of a metre.

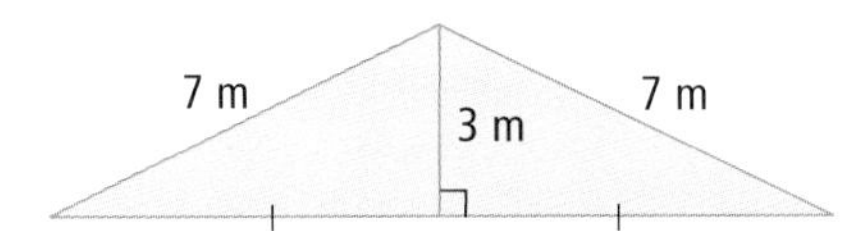

Solution

Let a represent the length of the unknown side.
Use the Pythagorean relationship.

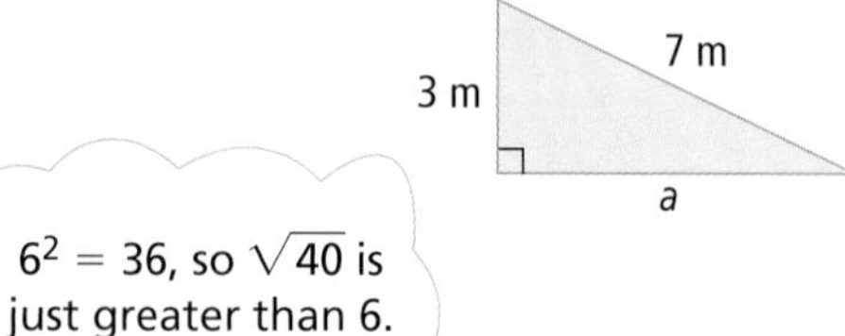

Half of the gable end forms a right triangle.

$$a^2 + 3^2 = 7^2$$
$$a^2 + 9 = 49$$
$$a^2 = 49 - 9$$
$$a^2 = 40$$
$$a = \sqrt{40}$$
$$a \doteq 6.3$$

$6^2 = 36$, so $\sqrt{40}$ is just greater than 6.

Half the width is approximately 6.3 m, so the whole width of the roof is twice the amount, or about 12.6 m.

Communicate the Key Ideas

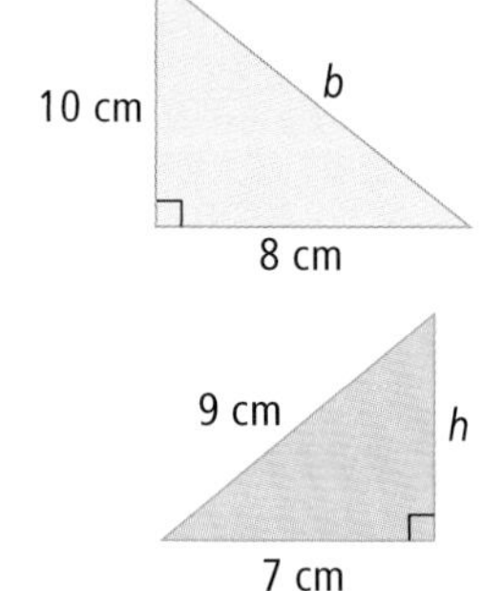

1. To find the length b in this diagram, Crystal wrote $b^2 = 10^2 + 8^2$. Is her method correct? If not, explain what she did wrong.

2. Darian wants to find the height, h, of this triangle. His method is to calculate $\sqrt{81 - 49}$. Will his answer be correct? Explain why or why not.

3. Describe the steps you would use to find the length of each hypotenuse. Which is greater, c or d, and by how much?

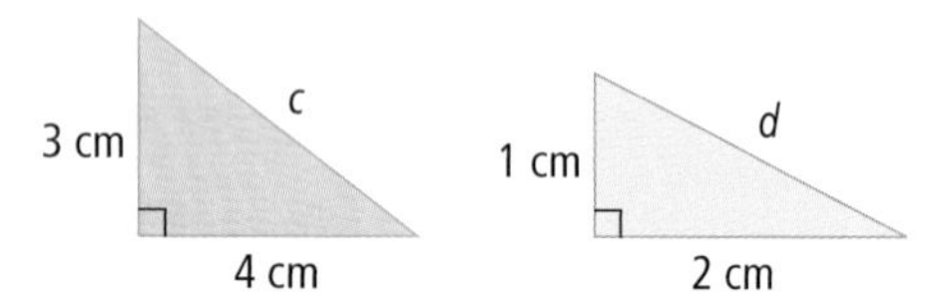

4. If you know the lengths of any two sides of a right triangle, you can use the Pythagorean relationship to find the length of the third side. Write a note explaining the steps to a student who missed this class.

Check Your Understanding

1. Find the length of the unknown side in each triangle.

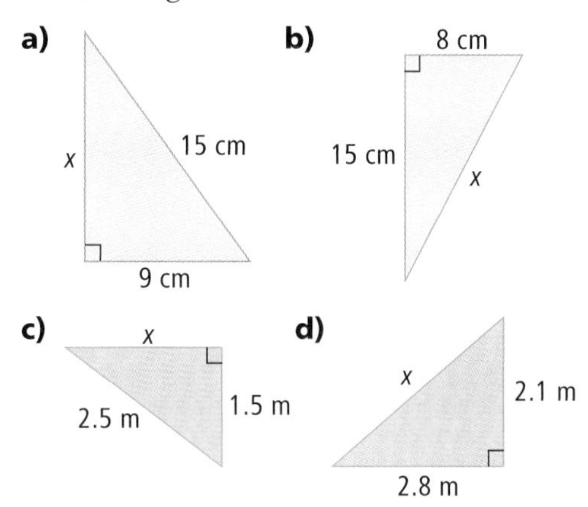

2. A wheelchair ramp is 9.6 m long. It spans a horizontal distance of 9.5 m. What is the ramp's vertical height? Round your answer to the nearest tenth of a metre.

3. Refer to the introduction.

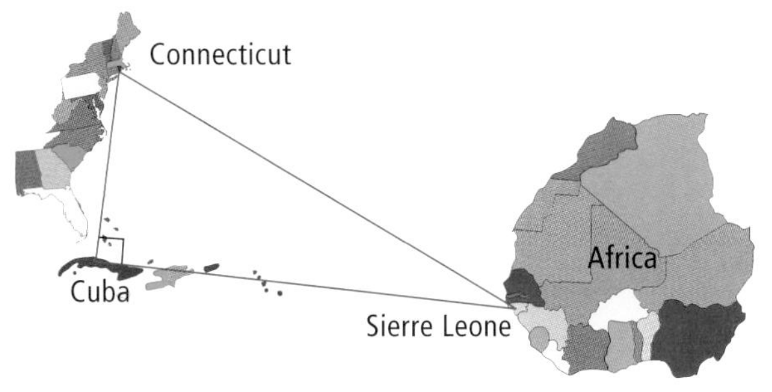

Find the distance from Connecticut to Sierra Leone, to the nearest tenth of a kilometre.

Did You Know?

The triangle created by the paths between the countries are called trade triangles. These were routes used for the trade of supplies.

4. Dev plans to build a toy boat for his younger brother. The side view of his plan is shown. What length of wood will he need to cut to make the top deck of the boat?

Strategies
Try breaking the figure into simpler parts.

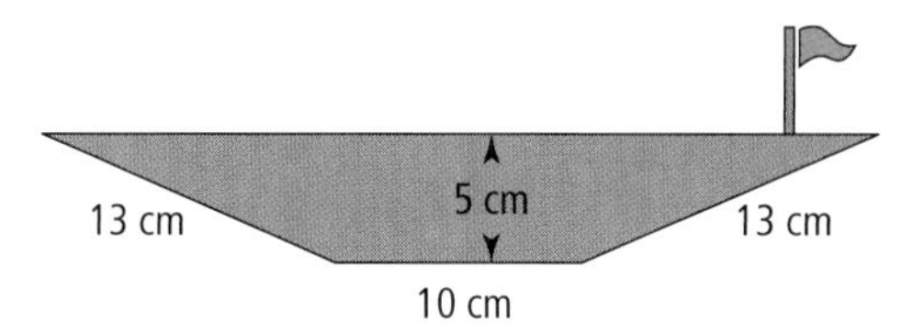

5. The Johnsons want to build a fence in their backyard. The fencing costs $9.95/m. What is the cost of the fencing?

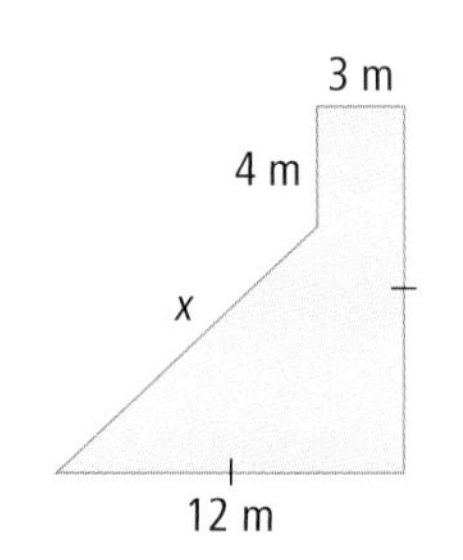

6. Find the missing dimensions using the Pythagorean relationship. Round your answers to one decimal place.

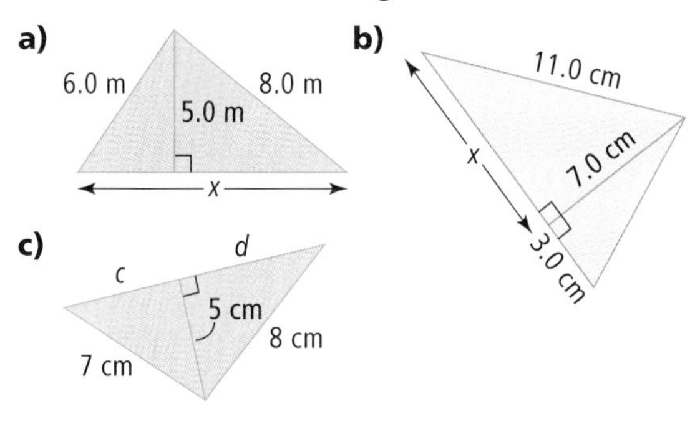

7. Find the area of this equilateral triangle.

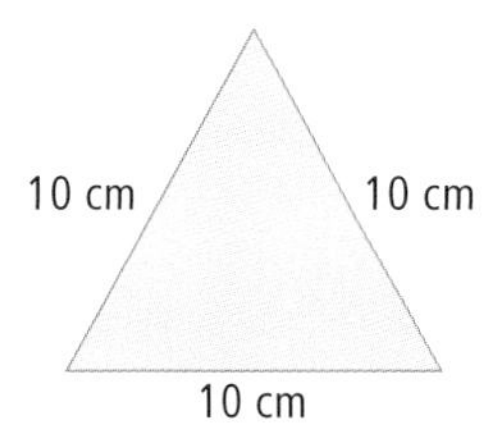

8. Find the height of the isosceles triangle using the Pythagorean relationship.

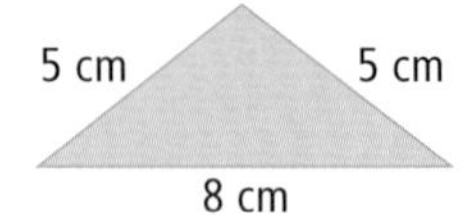

9. Ahmed starts from home and walks 120 m down his street. Then, he turns left at the corner and walks another 100 m to the library. Ahmed has a two-way radio with a range of 150 m. Can he call his brother, who is at home, from the library?

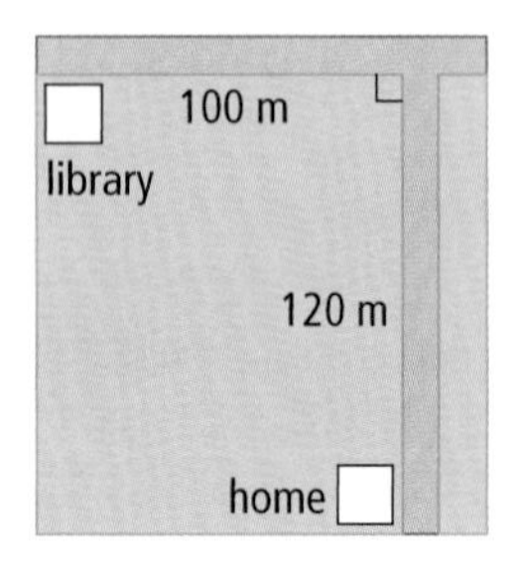

10. Tia is making a handrail for a staircase. The staircase rises 3.2 m over a horizontal distance of 3.6 m. How long should the handrail be, to the nearest tenth of a metre?

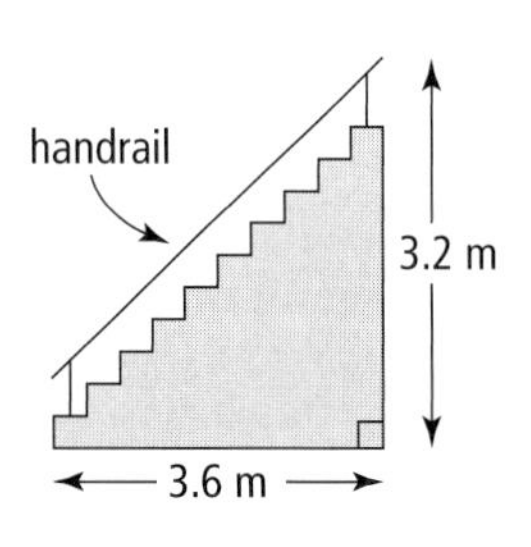

11. A square tablecloth has an area of 1 m^2.

a) What is the length of each side of the tablecloth, in centimetres?

b) What is the length of the diagonal of the tablecloth, to the nearest centimetre?

12. Tony must cross a rectangular parking lot. The lot measures 80 m by 85 m. How many fewer metres will Tony walk if he crosses the lot diagonally rather than walking the length and width? Round your answer to the nearest metre.

13. Each of Jessie's cartwheels covers a distance of 2.5 m. How many cartwheels can she perform along the diagonal of an 8 m by 8 m gymnasium mat?

14. What is the shortest distance from Anne's house to the bookstore?

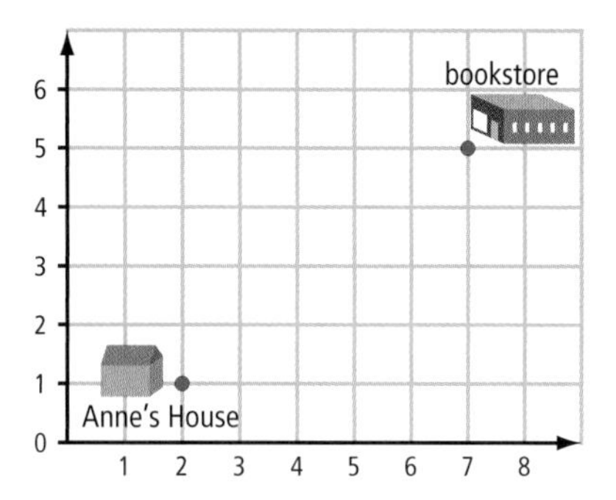

15. A fishing boat leaves St. John's, Newfoundland and Labrador, and travels due north at 7.2 km/h for 4 h. Then, the boat turns due east and continues its journey, at the same speed, for another $\frac{1}{2}$ h. How far is the boat from St. John's after $4\frac{1}{2}$ h at sea?

Chapter Problem

16. Katie is building a kite. She uses two wooden sticks, one measuring 30 cm and the other 60 cm, to form the frame. The centre point of the shorter stick is attached to a point one third of the way from the end of the longer stick.

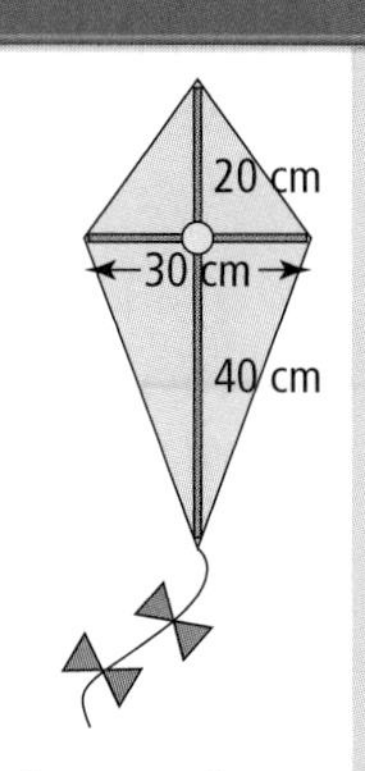

a) To the nearest tenth of a centimetre, how much ribbon will she need to trim the four outer edges of the kite paper?

b) What area of kite paper will she need to cover one side of the frame?

17. A doorway is 0.78 m wide and 2.00 m high. Will a round tabletop with a diameter of 2.50 m fit through the doorway?

a) Draw and label a diagram to model the problem.

b) Use numbers and words to justify your answer.

18. Find the distance, in grid units, of the most direct route between each pair of locations on the map. Do not use a ruler.

a) the ferry and the hotel

b) the art museum and the history museum

c) the mall and the amusement park

d) the amusement park and the boat tour

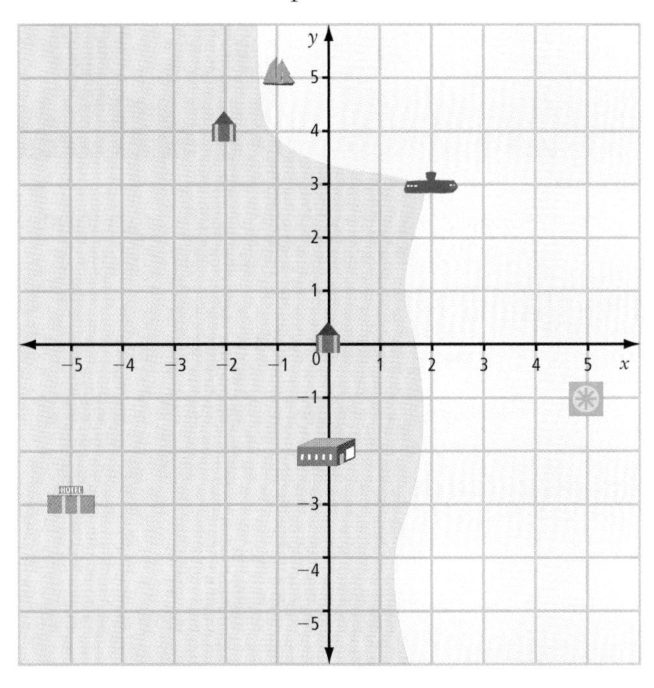

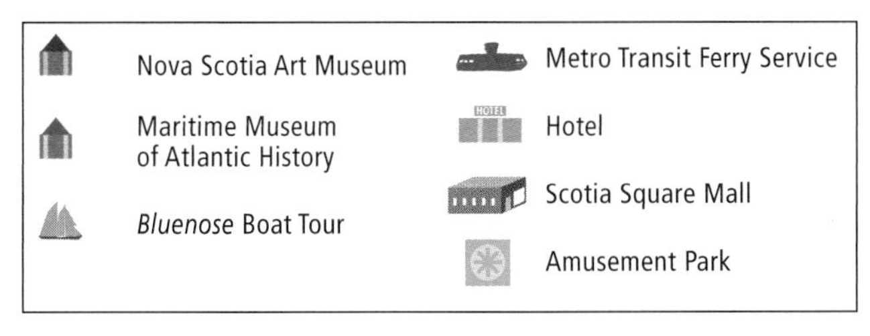

19. Many steel cables are used to support a suspension bridge. In the diagram, three of the cables are shown.

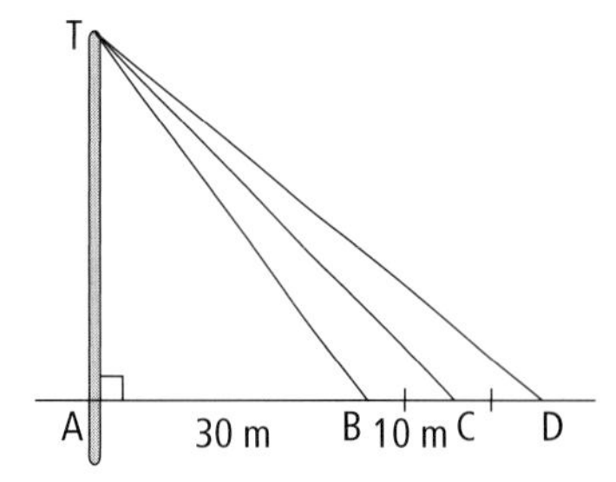

a) Cable TB is 50 m long. Find the height of the tower, TA.

b) Find the lengths of cables TC and TD.

c) How much longer is cable TC than cable TB? Can you add this amount to the length of TC to find the length of TD? Explain.

20. Find the shortest distance from A to B if the larger square has a perimeter of 36 cm and the smaller square has an area of 9 cm^2.

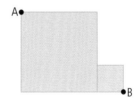

21. A ramp 3.5 m long is used to load furniture into a moving van. One end of the ramp rests on a porch, which is 0.5 m high. The other end of the ramp, leading into the back of the van, is 1.0 m above the ground. What is the horizontal distance from the back of the van to the edge of the porch, to the nearest tenth of a metre?

Extend

22. Refer to question 21. The movers want to load a metal pole with a length of 7 m into the van. The interior dimensions of the van are 2.5 m, 3 m, and 4 m. Will the pole fit inside the van? Explain.

CHAPTER 1 Review

1. Explain why 9 is a perfect square.

2. Evaluate without using a calculator.

   a) $\sqrt{16}$ b) $\sqrt{169}$ c) $\sqrt{6.25}$

3. A game board is made up of 100 small congruent squares. Each small square has an area of 9 cm^2. What is the perimeter of the game board?

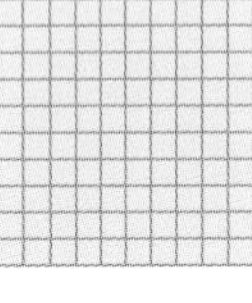

4. A square picture is glued to a square piece of construction paper. The paper has a side length of 30 cm and has four times the area of the square picture. What is the side length of the picture?

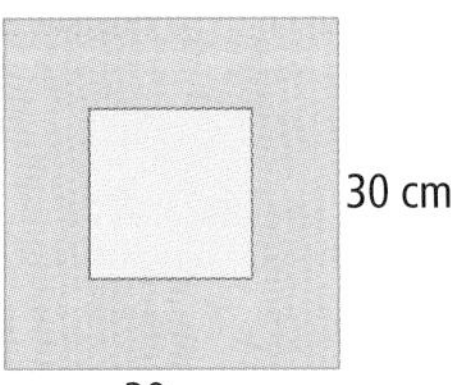

5. A right triangle is shown.

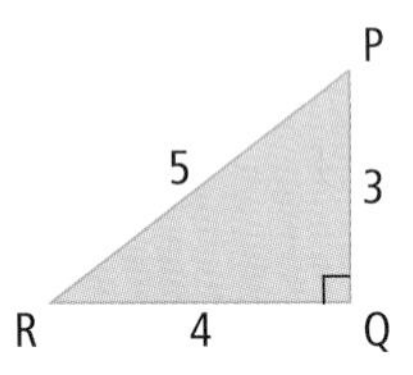

   a) Name the hypotenuse.
   b) Name the two legs of the triangle.
   c) How are the lengths of the sides related by the Pythagorean relationship?

6. The numbers 3 and 12 are not perfect squares. Explain why the product of 3 and 12 is a perfect square.

7. Draw a diagram of a right triangle and label the sides x, y, and z. Write an equation that represents the Pythagorean relationship for this triangle.

8. Which student, Alene or Ravi, is using mental math to estimate an answer? Which student is rounding to determine an approximate answer? Explain.

I think the value of $\sqrt{40}$ is about 6.4. The number 40 is between 36 and 49 and is a bit closer to 36.

According to my calculator, $\sqrt{40}$ is 6.3245552. I would give its value as 6.32.

9. Three squares are placed together to form a triangle. Determine if each triangle is a right triangle.

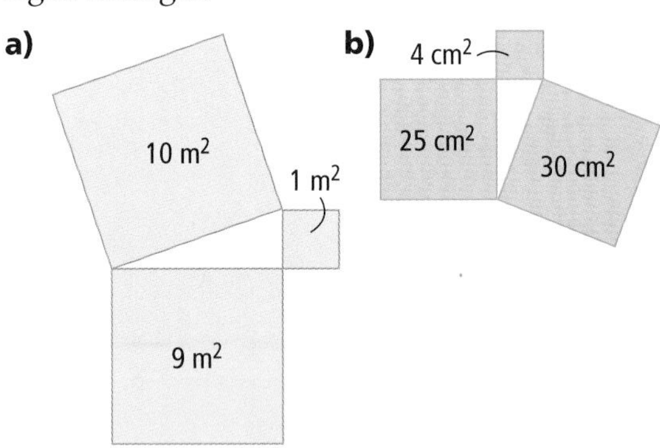

10. Find the length of the hypotenuse.

   a)

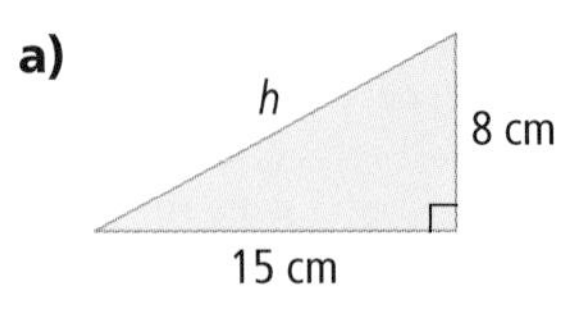

   b)

   7 km, h, 24 km

11. Find the unknown side lengths. Round your answers to one decimal place, if necessary.

a)

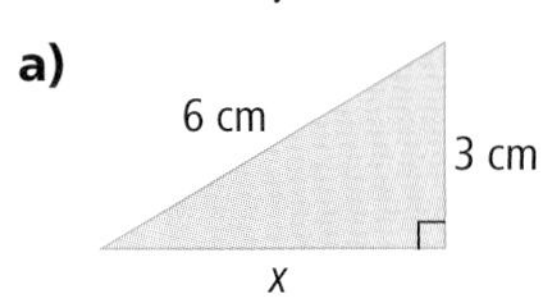

b)

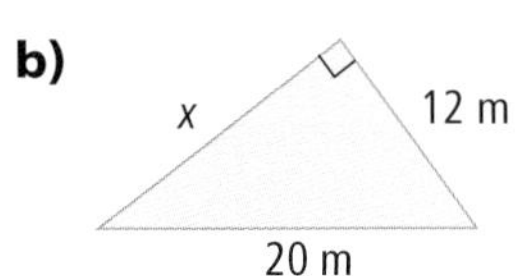

c)

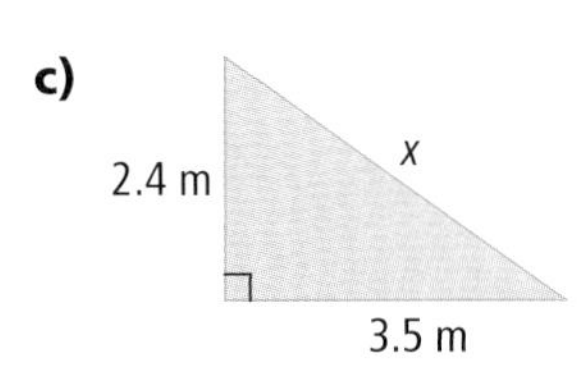

12. Avi uses a 3.6-m ladder to climb up to his tree house. For safety, the distance from the base of a ladder to the object it is resting against should be one third the length of the ladder.

a) How far from the tree should the base of the ladder be placed?

b) How far from the ground is the top of the ladder?

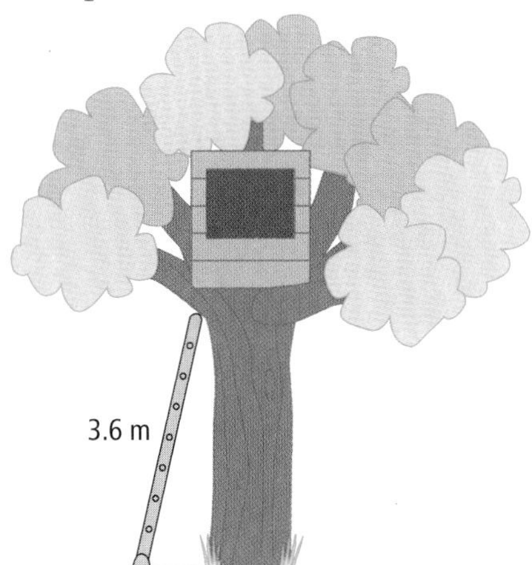

13. On a chessboard, each of the 64 small congruent squares has a side length of 3.2 cm.

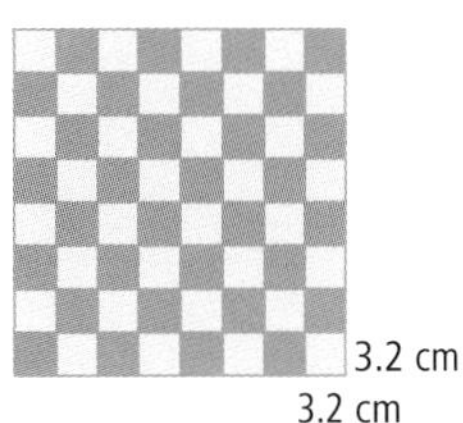

a) What is the length of the diagonal across the chessboard?

b) Solve part a) using a different method.

14. Use a factorization method to evaluate.

a) $\sqrt{324}$ **b)** $\sqrt{1225}$ **c)** $\sqrt{14\,400}$

15. Three solutions are presented for the following problem.
Find the length of the unknown side.

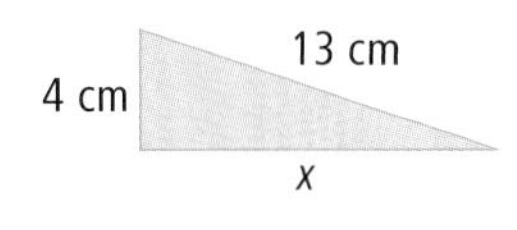

Solution 1

$$4 + x = 13$$
$$x = 13 - 4$$
$$x = 9$$
$$x = 9 \text{ cm}$$

Solution 2

$$13^2 + 4^2 = x^2$$
$$169 + 16 = x^2$$
$$185 = x^2$$
$$13.6 \doteq x$$
$$x \doteq 13.6 \text{ cm}$$

Solution 3

$$13^2 = 4^2 + x^2$$
$$169 = 16 + x^2$$
$$169 - 16 = x^2$$
$$153 = x^2$$
$$x = 153 \text{ cm}$$

a) For each solution
- Find the error, if it exists
- Correct the error

b) Find the correct value for x.

CHAPTER 1

Practice Test

Selected Response

Select the best answer.

1. A square has an area of 10 cm^2. Its side length is

A 100 cm
B between 3 cm and 4 cm
C between 1 cm and 2 cm
D between 2 cm and 5 cm

2. Which number is not a perfect square?

A 25 **B** 169
C 114 **D** 400

3. Which square root cannot be expressed as a whole number?

A $\sqrt{1}$ **B** $\sqrt{22}$
C $\sqrt{81}$ **D** $\sqrt{169}$

4. The first four numbers in a pattern are: 1, 4, 9, 16. The next number in the pattern is

A 20 **B** 32
C 25 **D** 144

5. The area of a square drawn on side YZ would be

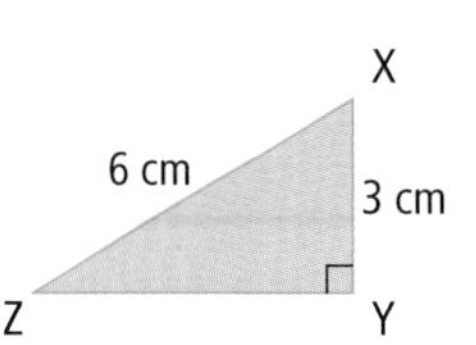

A 3 cm^2 **B** 9 cm^2
C 27 cm^2 **D** 45 cm^2

6. This diagram can be used to find the square root of

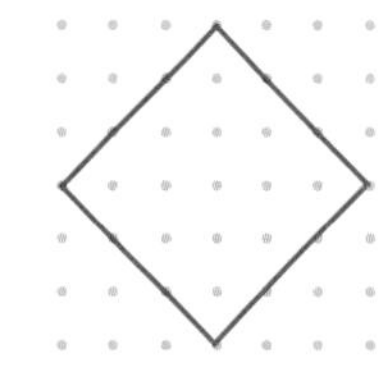

A 12 **B** 18
C 6 **D** 36

7. The value of $\sqrt{72}$ is closest to

A 8 **B** 8.1 **C** 8.5 **D** 9

8. The unknown side length is approximately

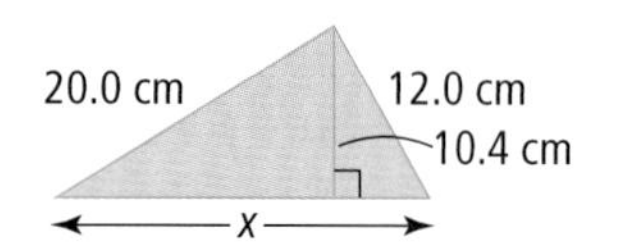

A 23.0 cm
B 6.0 cm
C 17.0 cm
D 20 cm

Short Response

Provide a complete solution.

9. Evaluate. Do not use a calculator.

a) $\sqrt{64}$ **b)** $\sqrt{400}$
c) $\sqrt{1.44}$ **d)** $\sqrt{2.25}$

10. Decide if each number is a perfect square. Show how you know.

a) 121 **b)** 47 **c)** 14 400

11. Does each set of side lengths form a right triangle? Show your work.

a) 6, 8, 10 **b)** 10, 24, 26
c) 7, 16, 17 **d)** 2, 2, $\sqrt{8}$

12. Find the missing dimensions using the Pythagorean relationship. Round your answers to one decimal place.

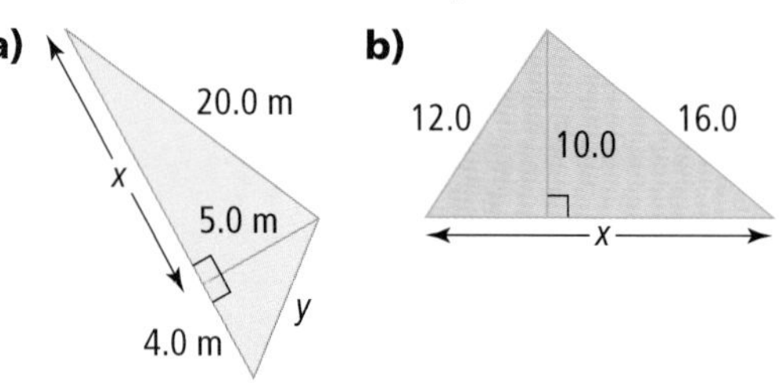

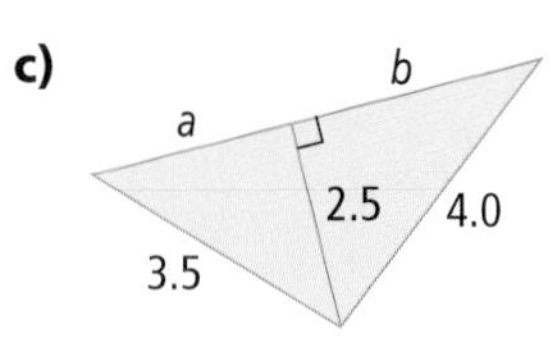

13. A square photograph has an area of 361 cm^2. Will it fit inside a 30 cm by 18 cm rectangular frame? Explain.

14. Calculate the perimeter and the area of the triangular plot of land.

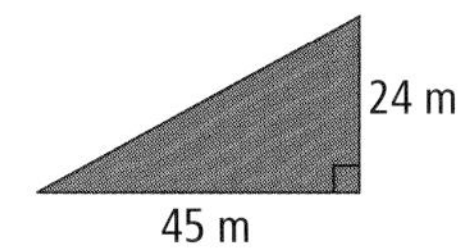

15. Explain why the following statement is incorrect. "The area of the square on the longest side of a triangle is equal to the sum of the areas of the squares on the two shorter sides." Use diagrams to support your reasoning.

Extended Response

Provide a complete solution.

16. You have two square sheets of construction paper with areas of 9 cm^2 and 36 cm^2. Describe how you can use these to find the approximate value of $\sqrt{45}$. What measurement tool will you need?

17. Three solutions are presented for the following problem.
Find the length of the unknown side.

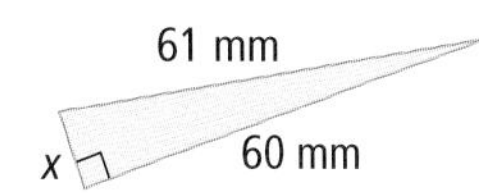

Solution 1

$$\begin{aligned} 61^2 &= 60^2 + x^2 \\ 3721 &= 3600 + x^2 \\ 3721 - 3600 &= x^2 \\ 121 &= x^2 \\ x &= 121 \text{ cm} \end{aligned}$$

Solution 2

$$\begin{aligned} 60 + x &= 61 \\ x &= 61 - 60 \\ x &= 1 \\ x &= 1 \text{ cm} \end{aligned}$$

Solution 3

$$\begin{aligned} 61^2 + 60^2 &= x^2 \\ 3721 + 3600 &= x^2 \\ 7321 &= x^2 \\ 85.6 &\doteq x \\ x &\doteq 85.6 \text{ cm} \end{aligned}$$

a) Correct any errors that you find.
b) Which solution is correct, if any?
c) Find the correct value for x.

Chapter Problem Wrap-Up

You are given a wooden stick 100 cm in length and are allowed to cut it once to make two pieces. With the two pieces of wood, you will build a kite frame.

1. Sketch two possible kite designs, labelling the lengths of the two pieces of wood.
2. What area of kite paper is needed for each design?
3. How much ribbon is needed to trim the perimeter of each kite?
4. How should the stick be cut to make a kite with the greatest area? Explain.

- add, subtract, multiply, and divide fractions using manipulatives, pictures, and symbols
- add, subtract, multiply, and divide fractions using estimation and mental math
- apply the order of operations to fractions using pencil and paper and the calculator
- model, solve, and create problems using fractions in meaningful contexts

Key Words

denominator

numerator

proper fraction

improper fraction

mixed number

lowest common denominator (LCD)

order of operations

CHAPTER 2

Fraction Operations

Each year Truro holds a tulip festival where beautiful flower gardens are open for public viewing. These gardens may contain many different colours of tulips. One colour may occupy only a fraction of the whole garden. What fractions can you use to describe this garden?

In this chapter, you will learn how to add, subtract, multiply, and divide fractions using materials, pictures, and symbols. You will apply the order of operations to solve problems involving operations with proper fractions, improper fractions, and mixed numbers.

Chapter Problem

Gardens often contain sections with different colours or types of flowers or vegetables. Think about a garden you have seen recently or a garden you would like to plant.

- Describe the different colours in the garden.
- What fraction of the garden is each colour?
- What fraction of the garden is each type of flower or vegetable?

Get Ready

The Meaning of Fractions

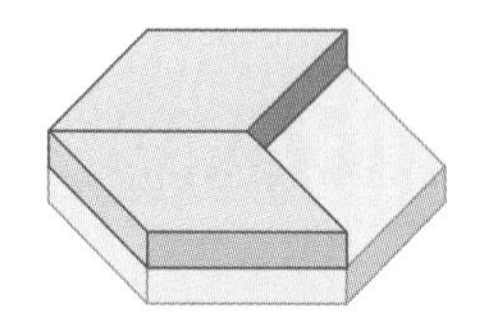

The hexagon represents a whole. It can be divided into 3 equal parts.

In this model, blue represents $\frac{2}{3}$. $\frac{1}{3}$ remains uncovered and is yellow.

The number $\frac{2}{3}$ is a **fraction**.

3 is called the **denominator**. It tells how many equal pieces the whole has been divided into.

2 is called the **numerator**. It tells how many of the equal pieces there are.

In a **proper fraction**, the numerator is less than the denominator, such as in $\frac{2}{3}$.

In an **improper fraction**, the numerator is greater than the denominator, such as in $\frac{4}{3}$.

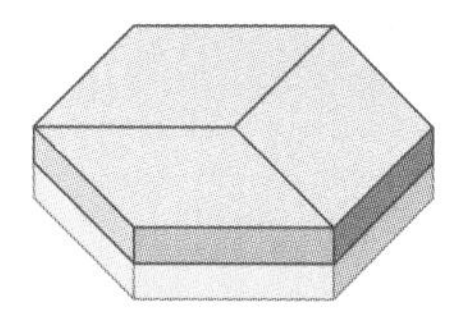

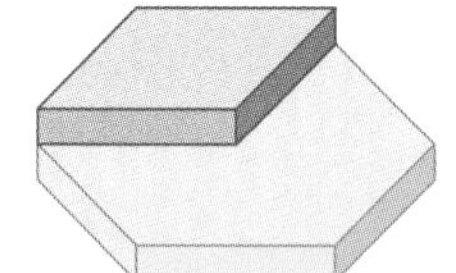

1. Write the fraction that each diagram represents.

a)

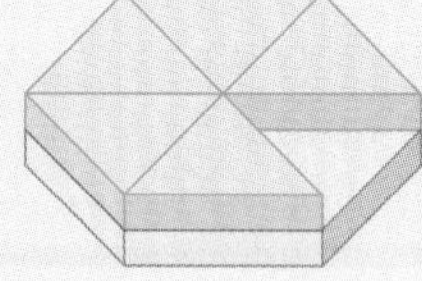

b)

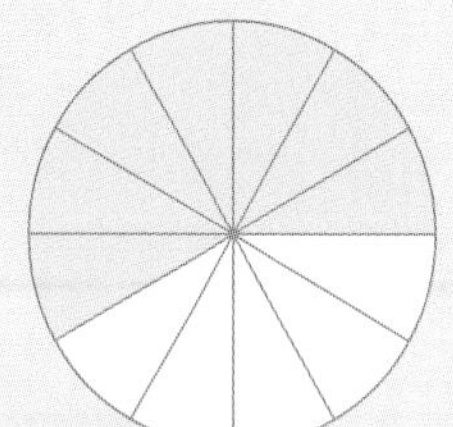

c)

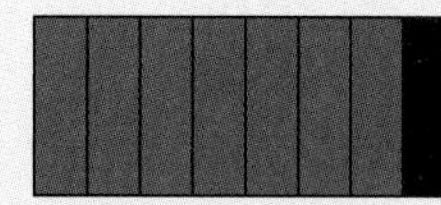

d)

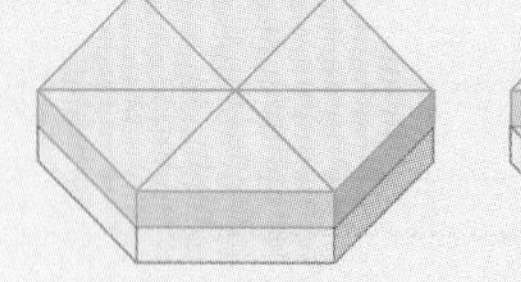

e)

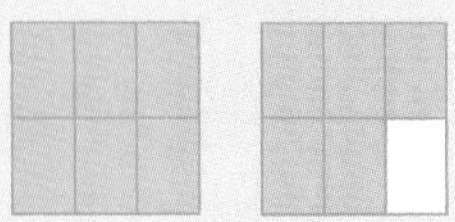

f)

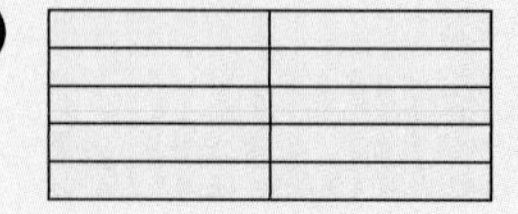

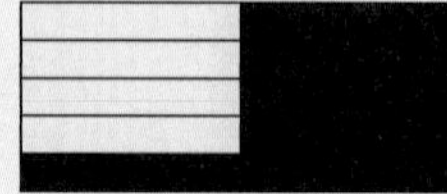

If one black rectangle represents 1, I could also represent the number with 2 black rectangles and 4 yellow Fraction Factory pieces.

Multiples and Factors

The first four **multiples** of 2 are 2, 4, 6, and 8.

Each of these multiples is the product of 2 and a **natural** or counting number.

$1 \times 2 = 2 \quad 2 \times 2 = 4 \quad 3 \times 2 = 6 \quad 4 \times 2 = 8$

The **lowest common multiple (LCM)** is the smallest number that is a multiple of several other numbers. The LCM of 3 and 5 is 15.

The **factors** of 8 are 1, 2, 4, and 8. Each of these factors divides evenly into 8 with no remainder. The **greatest common factor (GCF)** is the greatest number that is a factor of several numbers. The GCF of 8 and 12 is 4.

2. List the first four multiples of each number.
a) 6 **b)** 3 **c)** 5 **d)** 10

3. Write the LCM of each pair of numbers.
a) 3, 4 **b)** 2, 5 **c)** 5, 10 **d)** 6, 9

4. List the factors of each number, from least to greatest.
a) 12 **b)** 0 **c)** 36 **d)** 50

5. Write the GCF of each pair of numbers.
a) 2, 12 **b)** 10, 20 **c)** 8, 36 **d)** 3, 50

Equivalent Fractions

The piece of paper represents 1.
You can use paper folding and shading to show equivalent fractions.

$\frac{1}{2}$ of the paper is shaded blue.

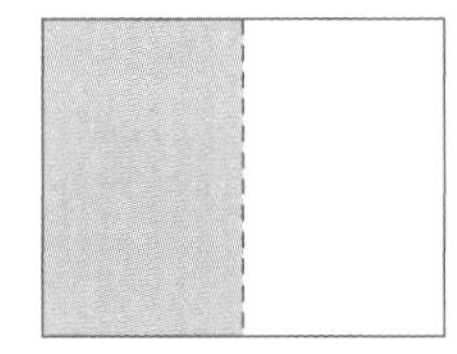

$\frac{2}{4}$ of the paper is blue.

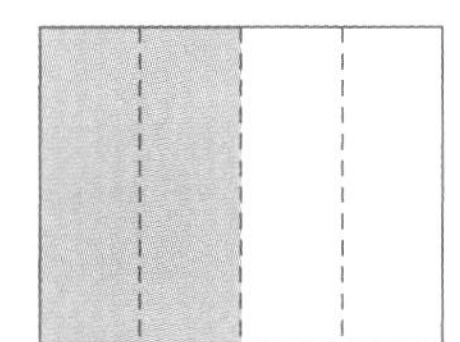

$\frac{3}{6}$ of the paper is blue.

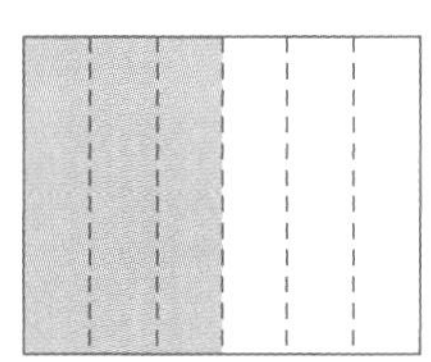

$\frac{1}{2}$, $\frac{2}{4}$, and $\frac{3}{6}$ each represent the same fraction of the whole paper that is shaded blue.

Get Ready

Since $\frac{1}{2} = \frac{2}{4} = \frac{3}{6}$, they are equivalent fractions.

They represent the same part of a whole that has been divided into different equal-sized parts.

$\frac{1 \times 2}{2 \times 2} = \frac{2}{4}$ $\quad \frac{1 \times 3}{2 \times 3} = \frac{3}{6}$

To write an equivalent fraction, you can multiply or divide the numerator and the denominator by the same number.

When you divide to write an equivalent fraction, you **simplify** the fraction.

To simplify $\frac{3}{6}$, divide the numerator and the denominator by 3: $\frac{3 \div 3}{6 \div 3} = \frac{1}{2}$

6. Use paper folding to find an equivalent fraction for each fraction. Record your models.

a) $\frac{2}{3}$ **b)** $\frac{4}{5}$ **c)** $\frac{1}{4}$

7. Simplify each fraction.

a) $\frac{8}{10}$ **b)** $\frac{6}{18}$ **c)** $\frac{2}{6}$

Compare and Order Fractions

You can compare fractions with a picture.

$\frac{1}{2}$ of this fraction block is orange.

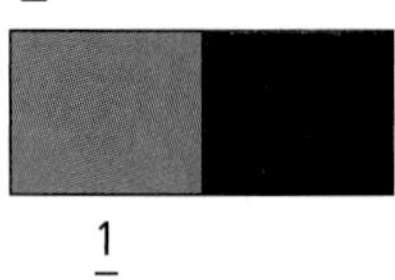

$\frac{1}{2}$

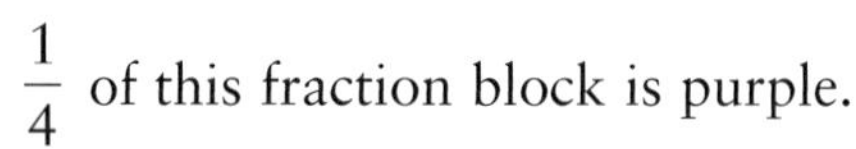

$\frac{1}{4}$ of this fraction block is purple.

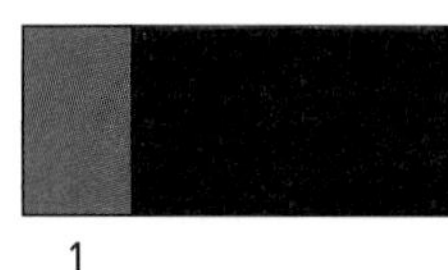

$\frac{1}{4}$

As you can see, the orange section is larger than the purple section, so $\frac{1}{2} > \frac{1}{4}$.

You can also compare fractions without a picture.

If two fractions have the same denominator, compare the fractions by comparing the numerators.

To determine which is greater, $\frac{2}{5}$ or $\frac{3}{5}$, you could ask yourself, "Would I get a bigger piece if I had $\frac{2}{5}$ of an apple or $\frac{3}{5}$ of the same apple?"

When the denominators are the same, the fraction with the greater numerator is larger: $\frac{3}{5} > \frac{2}{5}$.

If two fractions have different denominators, write equivalent fractions with a **common denominator**, and then compare the numerators.

An easy way to find a common denominator is to find the LCM of the denominators.

To compare $\frac{2}{3}$ and $\frac{1}{2}$, find the LCM of 3 and 2.
Multiples of 3: 3, (6), 9, 12, 15, 18, ...
Multiples of 2: 2, 4, (6), 8, 10, 12, ...
The LCM of 3 and 2 is 6.
Write each fraction with the **lowest common denominator (LCD)** of 6.

$$\frac{2}{3} = \frac{■}{6} \qquad \frac{1}{2} = \frac{■}{6}$$

$$\frac{2 \times 2}{3 \times 2} = \frac{4}{6} \qquad \frac{1 \times 3}{2 \times 3} = \frac{3}{6}$$

$\frac{4}{6} > \frac{3}{6}$ so $\frac{2}{3} > \frac{1}{2}$.

8. Write a fraction for each diagram. Which is larger?

a) 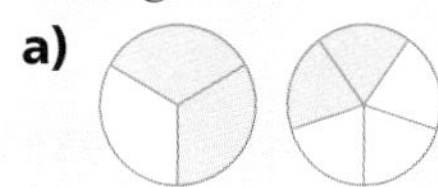

b) 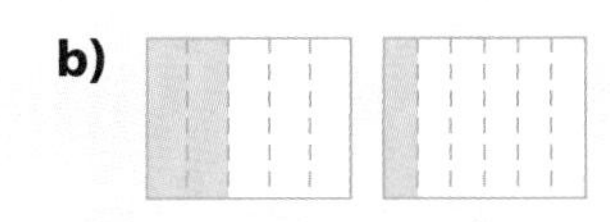

c) 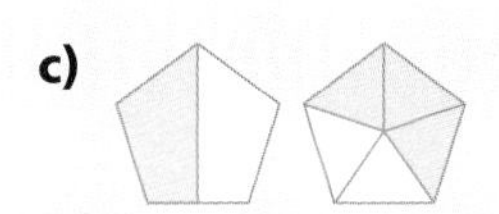

9. Find a common denominator for each pair of fractions.

a) $\frac{1}{2}, \frac{3}{4}$ **b)** $\frac{7}{8}, \frac{1}{4}$ **c)** $\frac{1}{3}, \frac{5}{6}$

10. Write each pair of fractions with a common denominator, and then compare them.

a) $\frac{3}{5}, \frac{2}{3}$ **b)** $\frac{5}{6}, \frac{7}{8}$ **c)** $\frac{3}{4}, \frac{5}{7}$

Get Ready

Mixed Numbers and Improper Fractions

Sometimes it is useful to write **mixed numbers** as improper fractions when working with fractions.

Since there are $\frac{4}{4}$ in whole, $2\frac{1}{4}$ can be written as $\frac{9}{4}$.

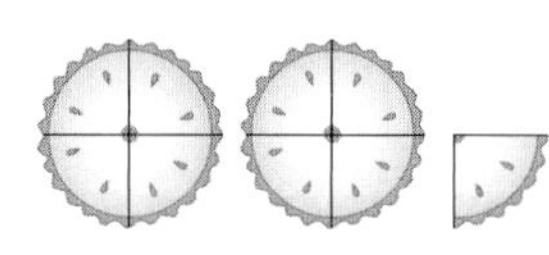

The thirds can be grouped to make 3 wholes with $\frac{2}{3}$ remaining.

Therefore, $\frac{11}{3}$ can be written as the mixed number $3\frac{2}{3}$.

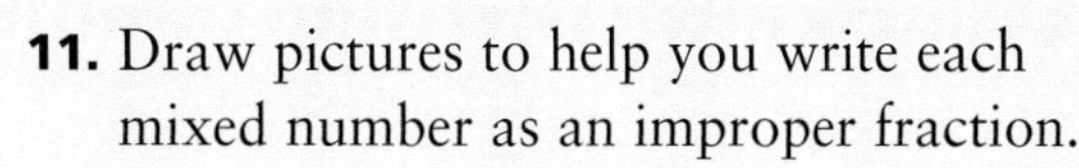

11. Draw pictures to help you write each mixed number as an improper fraction.

a) $4\frac{2}{3}$ **b)** $6\frac{1}{6}$ **c)** $3\frac{1}{2}$

12. Draw pictures to help you write each improper fraction as a mixed number.

a) $\frac{11}{4}$ **b)** $\frac{12}{5}$ **c)** $\frac{9}{2}$

Write Fractions in Simplest Form

A fraction is in **simplest form** when the numerator and denominator have no common factors other than 1. To write a fraction in simplest form, divide the numerator and the denominator by the greatest common factor (GCF). The result is an equivalent fraction, so its value is the same as the original fraction.

To write $\frac{12}{18}$ in simplest form, find the GCF of 12 and 18.

The factors of 12 are 1, 2, 3, 4, 6, and 12.
The factors of 18 are 1, 2, 3, 6, 9, and 18.

The greatest common factor is 6.

$$\frac{12}{18} = \frac{12 \div 6}{18 \div 6} = \frac{2}{3}$$

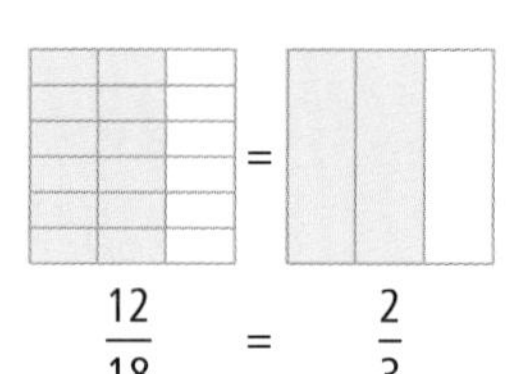

I could divide 12 and 18 by 2 to get 6 and 9. I could then divide 6 and 9 by 3 to get 2 and 3.

13. Write each fraction in simplest form.

a) $\frac{7}{49}$ **b)** $\frac{9}{39}$

c) $\frac{6}{20}$ **d)** $\frac{36}{48}$

e) $\frac{15}{20}$ **f)** $\frac{21}{30}$

g) $\frac{20}{36}$ **h)** $\frac{49}{56}$

14. Write each fraction in simplest form. Then, order the fractions from least to greatest.

$\frac{13}{15}, \frac{5}{20}, \frac{6}{12}, \frac{6}{18}$

Estimation

Use Benchmarks

It is often helpful to round a proper fraction to its nearest benchmark, 0, $\frac{1}{2}$, or 1, when estimating.

$3\frac{3}{5} + \frac{1}{10} + 1\frac{7}{8}$ can be approximated by $3\frac{1}{2} + 0 + 2$,

$5\frac{1}{2}$ is an estimate for $3\frac{3}{5} + \frac{1}{10} + 1\frac{7}{8}$.

Use Clustering

When fractions in an expression are all close to the same value, it is useful to round them to the same amount in order to estimate.

$\frac{1}{17} + \frac{1}{18} + \frac{1}{19}$ are all close to $\frac{1}{18}$ so $\frac{1}{18} + \frac{1}{18} + \frac{1}{18} = \frac{3}{18}$ or $\frac{1}{6}$.

$\frac{1}{6}$ is a good estimate for the sum.

$\frac{4}{15} + \frac{3}{11} + \frac{6}{25}$ are all close to $\frac{1}{4}$ so $\frac{1}{4} + \frac{1}{4} + \frac{1}{4} = \frac{3}{4}$.

$\frac{3}{4}$ is a good estimate for the sum.

15. Estimate. Show your steps.

a) $4\frac{2}{15} + 3\frac{19}{20}$ **b)** $9\frac{3}{20} + 2\frac{4}{7}$ **c)** $\frac{2}{7} + \frac{2}{8} + \frac{2}{9}$ **d)** $8\frac{1}{5} \times 4\frac{9}{10}$

Get Ready

Add and Subtract Fractions

This picture models $\frac{1}{6} + \frac{2}{6} = \frac{3}{6}$. This picture models $\frac{3}{6} - \frac{2}{6} = \frac{1}{6}$.

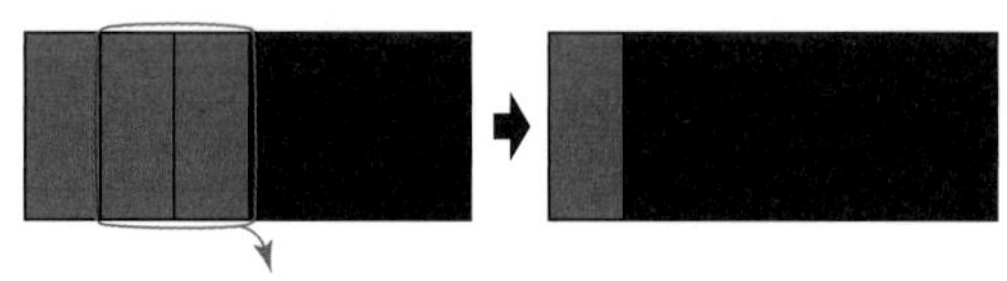

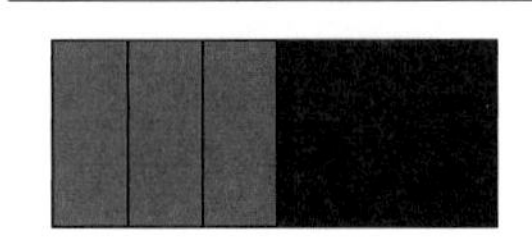

To add fractions with common denominators, add the numerators. Keep the denominator the same. Simplify the answer if necessary.

For example, $\frac{3}{4} + \frac{2}{4} = \frac{5}{4}$ or $1\frac{1}{4}$.

To subtract, take away one numerator from the other. Keep the denominator the same. Simplify the answer if necessary.

For example, $\frac{3}{4} - \frac{2}{4} = \frac{1}{4}$.

16. Write the number statement that represents each diagram and its result.

a)

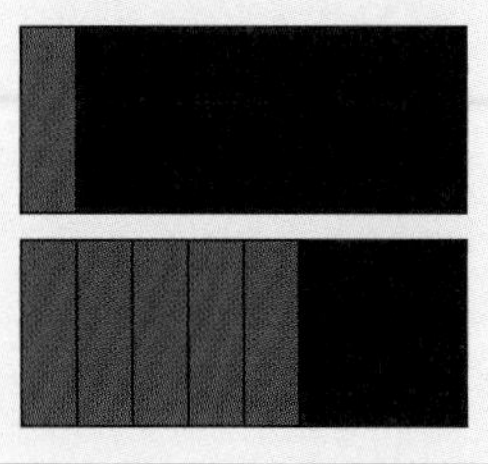

b)

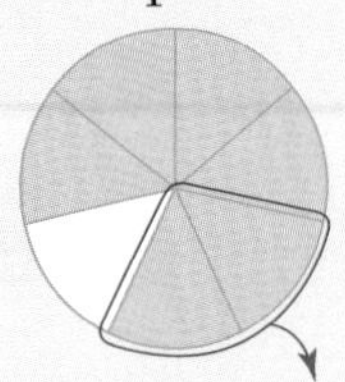

c)

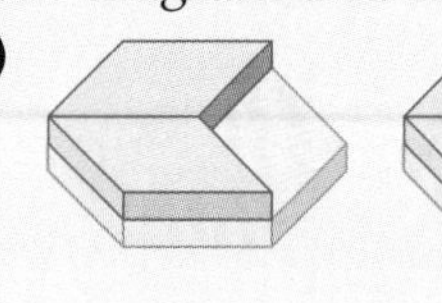

d)

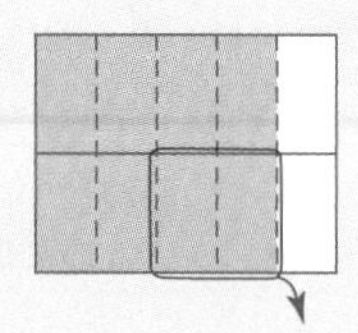

17. Evaluate.

a) $\frac{11}{12} - \frac{7}{12}$ **b)** $\frac{7}{10} + \frac{2}{10}$ **c)** $\frac{7}{8} - \frac{5}{8}$ **d)** $\frac{5}{6} + \frac{3}{6}$

Mentally Multiply a Fraction by a Whole Number and Vice Versa

You can often multiply a fraction by a whole number mentally.

$9 \times \frac{2}{3} = 6$

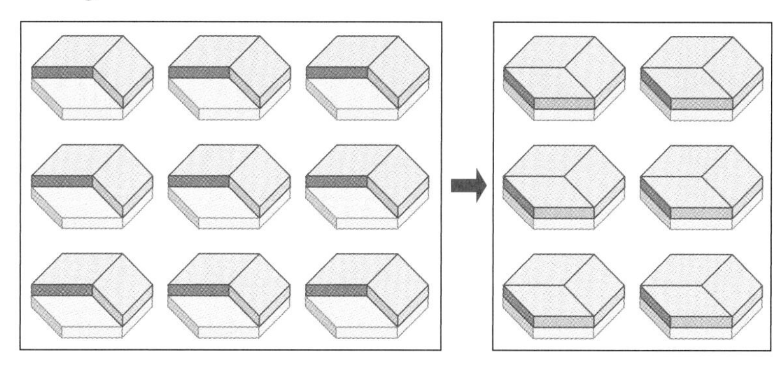

$9 \times \frac{2}{3}$ is 18 thirds or 6 wholes.

You can also multiply a whole number by a fraction mentally.

$\frac{1}{6} \times 12 = 2$

$\frac{1}{6} \times 12$ is 12 sixths or 2 wholes.

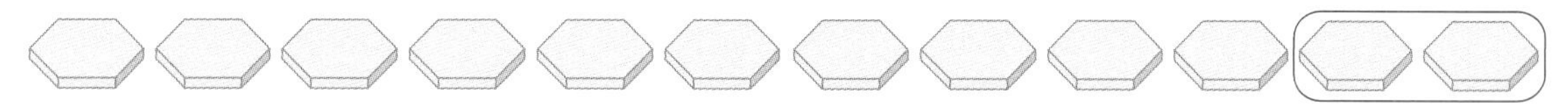

You can also multiply a fraction by a whole number using a number line.

$4 \times \frac{3}{4} = 3$

0 1 2 3

18. Multiply mentally. Write the multiplication sentence.

a) $6 \times \frac{1}{6}$ **b)** $\frac{1}{4} \times 12$ **c)** $\frac{3}{4} \times 12$

d) $2 \times \frac{1}{3}$ **e)** $\frac{1}{3} \times 18$ **f)** $\frac{3}{4} \times 20$

2.1 Add and Subtract Fractions

Focus on...
- adding and subtracting fractions using manipulatives, pictures, and symbols
- adding and subtracting fractions using mental math

Parties R Fun is a party supply store that is closed for inventory. Kelsey and her co-workers are counting and recording all of the supplies in the store.

Kelsey is taking inventory of the balloons. She finds two open boxes of balloons in the storeroom. The boxes are the same size. One is about $\frac{1}{2}$ full and the other is about $\frac{1}{3}$ full. Kelsey combines the contents into one box and records the sum of the contents.

Discover the Math

Materials
- Fraction Factory

Optional
- pattern blocks
- fraction blocks

Part A: How do you add fractions with different denominators?

Work with a partner.

1. Use Fraction Factory pieces to represent the contents of each box of balloons. Draw your model.

2. Use Fraction Factory pieces to help you determine the total contents of the combined box. Explain what you did and then provide diagrams. Write an addition sentence that models the sum of the contents.

3. Write another addition sentence below the first one that shows the trades you made to solve the problem.

4. Is the box full? If not, what fraction of the box is empty? Make and draw a model using the pieces you used in question 2 and write a subtraction sentence to answer this question.

5. **Reflect** How can you add the fractions you started with to find a solution? Why was it necessary to trade your original Fraction Factory pieces for different pieces?

Example 1: Add Fractions

Alan is taking inventory of the helium tanks. He estimates that one tank is $\frac{2}{3}$ full and the other is $\frac{1}{4}$ full. What single fraction represents the contents of the two tanks together?

Solution

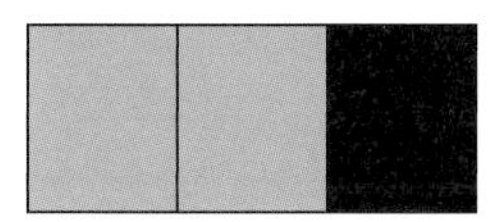

Use a picture to represent each fraction.

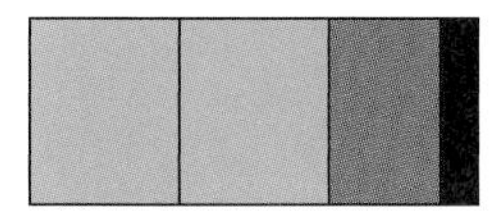

Estimate: $\frac{1}{4}$ is less than $\frac{1}{3}$, so $\frac{2}{3} + \frac{1}{4}$ will be less than $\frac{2}{3} + \frac{1}{3} = 1$.

Combine the pieces to represent $\frac{2}{3} + \frac{1}{4}$.

Is it possible to use one colour of Fraction Factory pieces to represent the combined amount?

$\frac{2}{3}$ and $\frac{8}{12}$ are equivalent fractions. $\frac{1}{4}$ and $\frac{3}{12}$ are equivalent fractions.

Replace the 2 green pieces with 8 beige pieces.

Replace the 1 purple piece with 3 beige pieces.

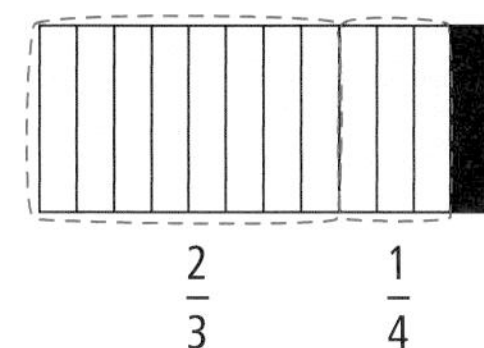

The combined amount is equal to $\frac{11}{12}$.

Write this addition as

$$\begin{aligned} &\frac{2}{3} + \frac{1}{4} \\ &= \frac{8}{12} + \frac{3}{12} \\ &= \frac{11}{12} \end{aligned}$$

12 is the LCD of $\frac{2}{3}$ and $\frac{1}{4}$ because it is the LCM of the denominators 3 and 4.

There is $\frac{11}{12}$ of a tank of helium altogether.

Discover the Math

Part B: How do you subtract fractions with different denominators?

Materials
- Fraction Factory

Cameron is taking inventory of the streamers. In the previous inventory, there was a box $\frac{2}{3}$ full of streamers. Then $\frac{1}{4}$ of that box was sold. He records how full the box of streamers should be in the inventory.

Work with a partner.

Strategies
Use a model.

1. Use Fraction Factory pieces to represent the contents in the original box and the amount that was sold. Draw your model.
2. Use Fraction Factory pieces to help you determine the fraction of the box that should be in stock. Explain what you did in words and with pictures of the pieces.
3. Write the subtraction sentence that the model represents. Write a second subtraction sentence below the first one that shows the trades you made to solve the problem.
4. What fraction of the box is empty? Make and draw a model using the pieces you used in question 2 on top of the whole and write a subtraction sentence and an addition sentence to answer this question.
5. **Reflect** How can you subtract the fractions you started with to find a solution? Why is it necessary to find the common denominators?

Example 2: Subtract Fractions

Simone rented a car. The gas tank was $\frac{3}{4}$ full when she rented it. When she returned the car, the tank was $\frac{3}{8}$ full. How much gas did Simone use?

Solution

Use Fraction Factory pieces to represent each fraction.

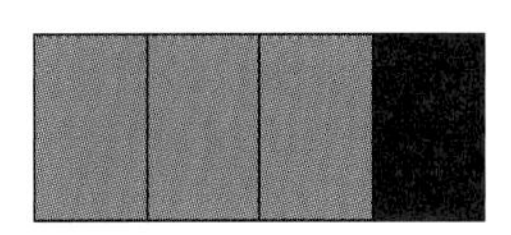

How much she started with.

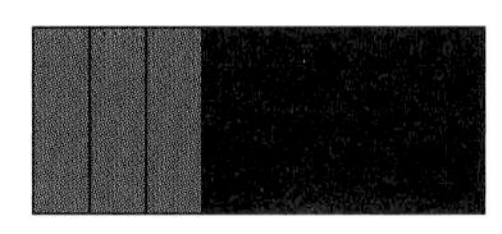

How much was left.

Since $\frac{3}{4} = \frac{6}{8}$, cover the purple tiles with brown tiles.

Compare the model for $\frac{3}{4}$ with the model for $\frac{3}{8}$.

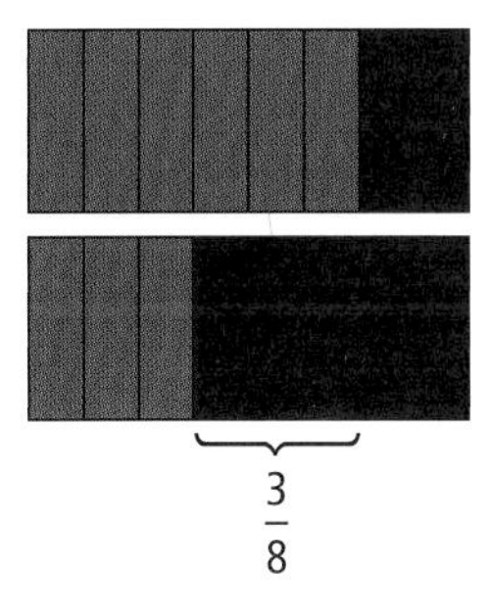

The difference is $\frac{3}{8}$.

Check:

Write this subtraction as
$$\begin{aligned} & \frac{3}{4} - \frac{3}{8} \\ &= \frac{6}{8} - \frac{3}{8} \\ &= \frac{3}{8} \end{aligned}$$

Simone used $\frac{3}{8}$ of a tank of gas.

Example 3: Subtract Mixed Numbers Using Equivalent Fractions

Hannah has $5\frac{1}{3}$ cups of whole-wheat flour in her pantry. She uses $3\frac{1}{2}$ cups for a bread recipe. How much flour is left?

Solution

Method 1: Use diagrams or pattern blocks.
Let a yellow hexagon represent 1.

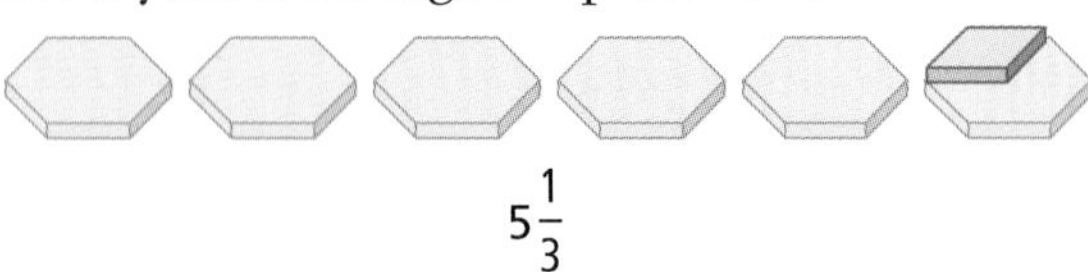

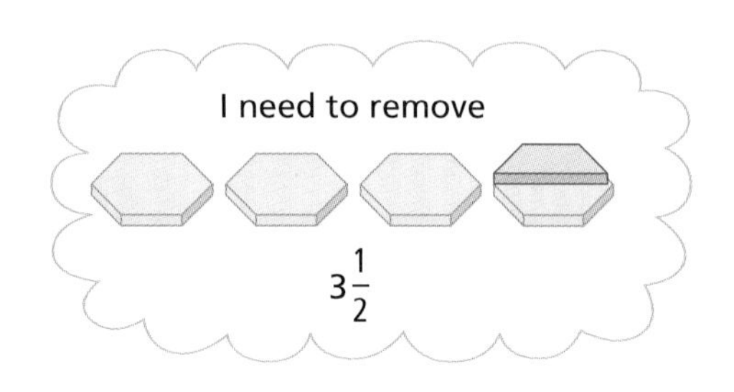

There are enough 1s to take away 3 from $5\frac{1}{3}$, but you cannot take $\frac{1}{2}$ from $\frac{1}{3}$ easily. So borrow 1 from $5\frac{1}{3}$. Change the borrowed 1 into six $\frac{1}{6}$ pieces.

I can represent $\frac{1}{3}$ and $\frac{1}{2}$ using $\frac{1}{6}$ pieces.
6 is the LCD of $\frac{1}{3}$ and $\frac{1}{2}$.

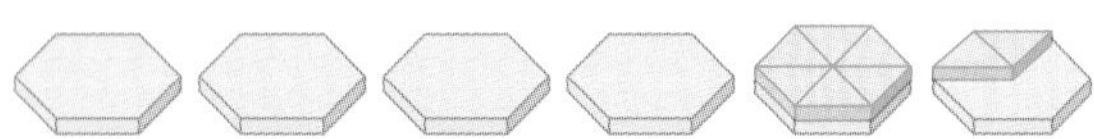

Now take away $3\frac{1}{2}$.

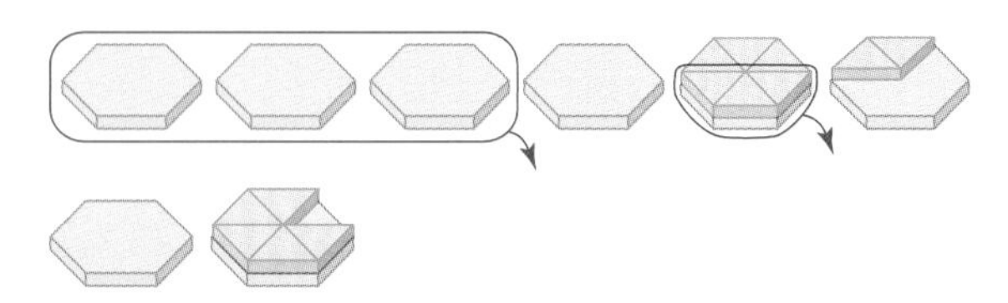

Hannah has $1\frac{5}{6}$ cups of flour left over.

Method 2: Regroup.
This subtraction can be done using regrouping.

$$5\frac{1}{3} - 3\frac{1}{2}$$

$$= 5\frac{2}{6} - 3\frac{3}{6}$$

$$= 4 + 1 + \frac{2}{6} - 3\frac{3}{6}$$

$$= 4 + \frac{6}{6} + \frac{2}{6} - 3\frac{3}{6}$$

$$= 4\frac{8}{6} - 3\frac{3}{6}$$

$$= 1\frac{5}{6}$$

Hannah has $1\frac{5}{6}$ cups of flour left over.

Method 3: Use improper fractions.

$$5\frac{1}{3} - 3\frac{1}{2} = \frac{16}{3} - \frac{7}{2}$$

Write the mixed numbers as improper fractions.

$$= \frac{32}{6} - \frac{21}{6}$$

Write equivalent fractions with the same denominator.
The LCD of $\frac{1}{3}$ and $\frac{1}{2}$ is 6.

$$= \frac{11}{6} \text{ or } 1\frac{5}{6}$$

Subtract the numerators. Write the improper fraction as a mixed number.

Hannah has $\frac{11}{6}$ or $1\frac{5}{6}$ cups of flour left over.

Example 4: Add and Subtract Fractions Using Pattern Blocks

Calculate.

a) $\frac{1}{6} + \frac{2}{3}$ **b)** $4 - \frac{2}{3}$

Solution

a) Let a yellow hexagon represent 1.

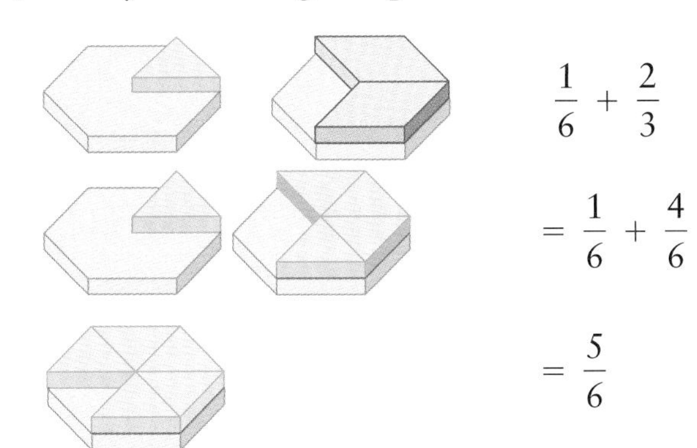

$$\frac{1}{6} + \frac{2}{3}$$

$$= \frac{1}{6} + \frac{4}{6}$$

$$= \frac{5}{6}$$

b)

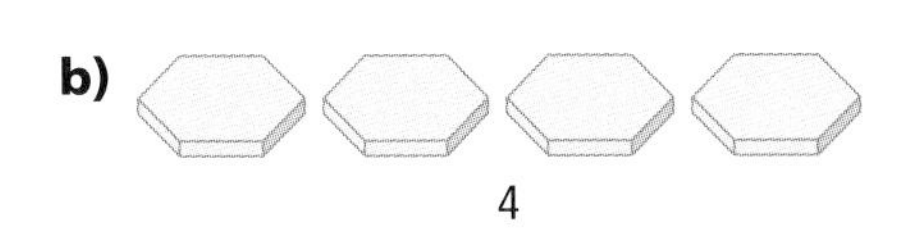

I need to remove

$\frac{2}{3}$

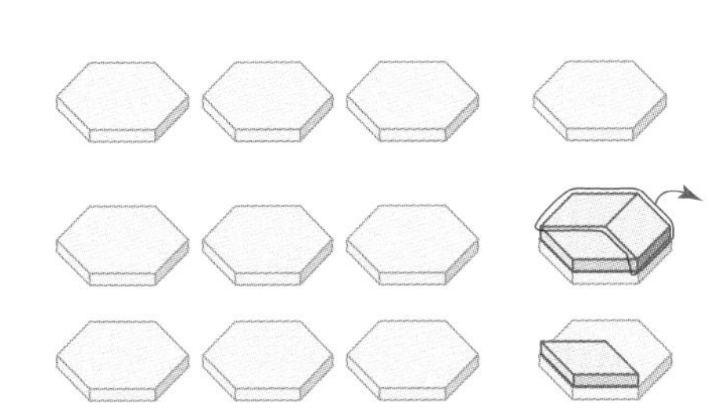

$$4 = 3 + 1$$

$$\text{So, } 4 - \frac{2}{3} = 3 + \frac{3}{3} - \frac{2}{3}$$

$$= 3\frac{1}{3}$$

Example 5: How Much Is Left Over?

After the class lunch, there were $2\frac{5}{8}$ cheese pizzas and $\frac{3}{4}$ of a vegetarian pizza left. How much pizza was left altogether?

Solution

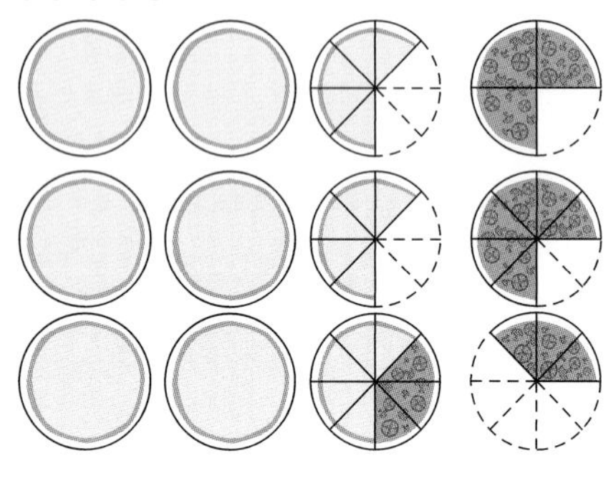

$$2\frac{5}{8} + \frac{3}{4}$$
$$= 2\frac{5}{8} + \frac{6}{8}$$
$$= 2\frac{11}{8}$$
$$= 3\frac{3}{8}$$

$\frac{11}{8} = \frac{8}{8} + \frac{3}{8} = 1\frac{3}{8}$

Strategies
Draw a diagram.

There are $3\frac{3}{8}$ pizzas left altogether.

Example 6: Add and Subtract Fractions Mentally

Calculate.

a) $\frac{4}{5} + \frac{1}{10}$ **b)** $4 - \frac{1}{3}$ **c)** $3\frac{7}{8} + 2\frac{1}{4}$

Solution

a) Since one denominator is a multiple of the other denominator, this can be done mentally.

Express $\frac{4}{5}$ as a fraction in tenths: $\frac{4}{5} = \frac{8}{10}$ (× 2 numerator, × 2 denominator)

Strategies
Apply a previously learned procedure.

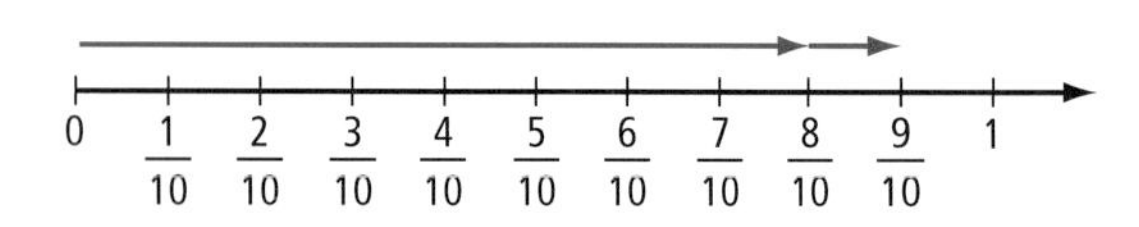

So, $\frac{4}{5} + \frac{1}{10} = \frac{8}{10} + \frac{1}{10}$.

Add the numerators: $\frac{8}{10} + \frac{1}{10} = \frac{9}{10}$

b) $4 - \frac{1}{3}$

You can mentally subtract a simple fraction from a whole number.

$4 - \frac{1}{3} = 3 + 1 - \frac{1}{3}$

Subtracting $\frac{1}{3}$ from one whole gives $\frac{2}{3}$.

$1 - \frac{1}{3} = \frac{3}{3} - \frac{1}{3} = \frac{2}{3}$

So, $4 - \frac{1}{3} = 3\frac{2}{3}$.

Check: The LCM of 1 and 3 is 3. Use a $\frac{1}{3}$ number line to subtract.

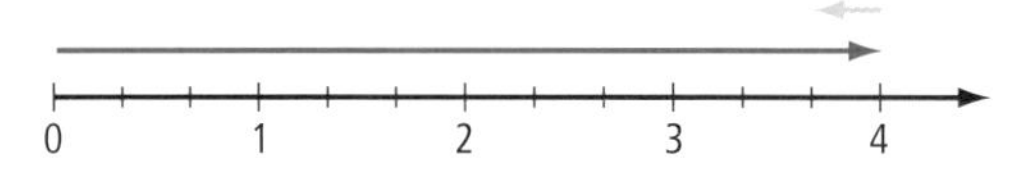

c) **Make "1" Mental Math**

Use Fraction Factory drawings to help you, if necessary.

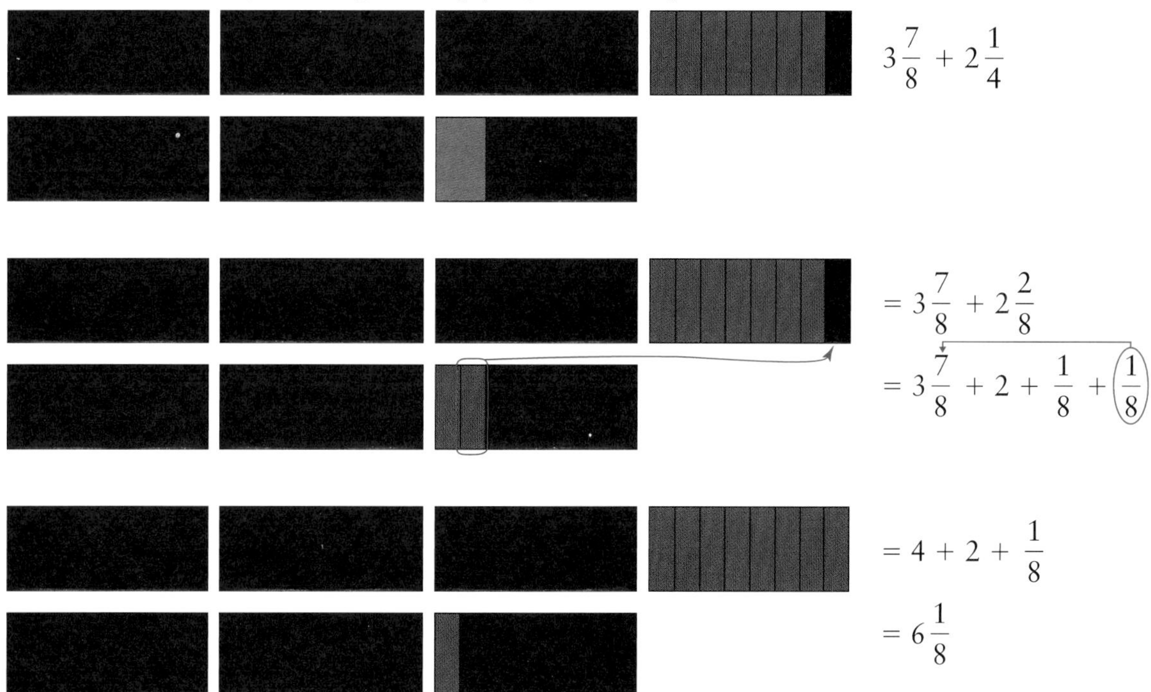

$3\frac{7}{8} + 2\frac{1}{4}$
$= 3\frac{7}{8} + 2\frac{2}{8}$
$= 3\frac{7}{8} + 2 + \frac{1}{8} + \frac{1}{8}$
$= 4 + 2 + \frac{1}{8}$
$= 6\frac{1}{8}$

Communicate the Key Ideas

1. Explain the error in Armand's thinking. Use diagrams to support your reasoning.

2. Draw a diagram for each stage in this addition and explain each stage.

Stage 1: $2\frac{2}{3} + 1\frac{1}{2}$

Stage 2: $2\frac{4}{6} + 1\frac{3}{6}$

Stage 3: $3\frac{7}{6}$

Stage 4: $4\frac{1}{6}$

3. Explain how to use regrouping to find the difference.

$5\frac{3}{10} - 3\frac{4}{5}$

4. Your friend missed class yesterday. Explain how to find the **lowest common denominator (LCD)** of three fractions, and why you might need to find this number. In your explanation, use an example with three fractions whose denominators are not multiples of one another.

lowest common denominator (LCD)
- the lowest common multiple of the denominators of two or more fractions

Check Your Understanding

1. Use benchmarks to estimate each sum.

a) $\frac{2}{9} + \frac{4}{7} + \frac{1}{10}$

b) $4\frac{1}{7} + 3\frac{1}{2} + 6\frac{1}{10}$

2. Estimate each sum.

a) $\frac{1}{8} + \frac{1}{9} + \frac{1}{10}$

b) $\frac{6}{19} + \frac{8}{25} + \frac{11}{30}$

3. Use mental math to add or subtract.

a) $\frac{1}{5} + \frac{4}{5}$

b) $\frac{1}{2} - \frac{1}{4}$

c) $3 + \frac{1}{3}$

d) $5 - \frac{1}{4}$

4. Write an addition sentence to represent combining the fractional sections shaded in each figure or the pattern and fraction blocks. Do not use improper fractions.

a) 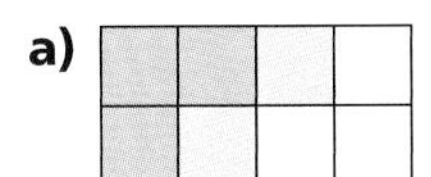

b) 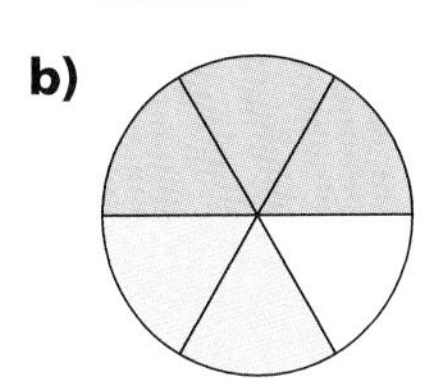

c) Let a yellow hexagon represent 1.

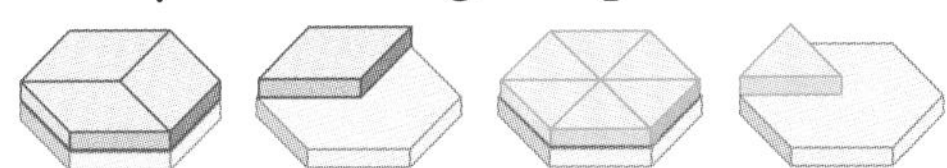

d) Let a double pink hexagon represent 1.

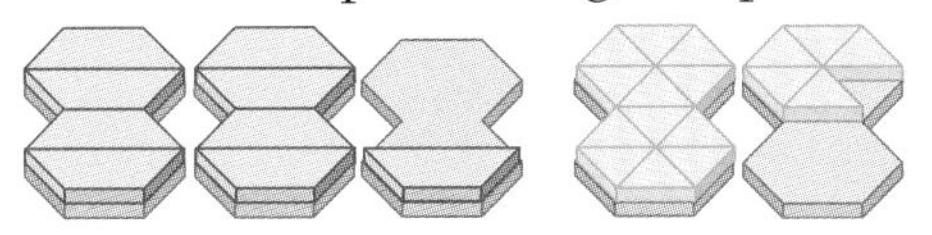

5. Write a subtraction sentence to represent the fraction of each figure that remains. Do not use improper fractions.

a) 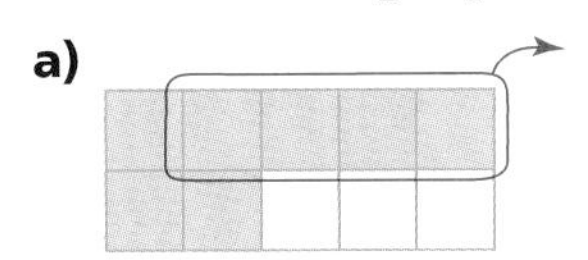

b) 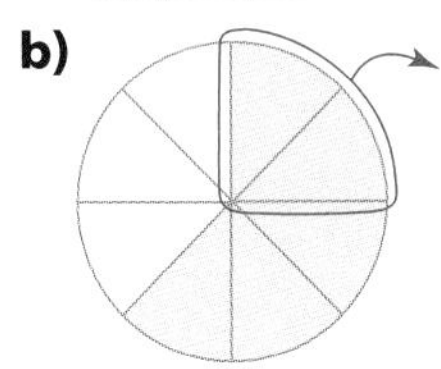

c) Let a yellow hexagon represent 1.

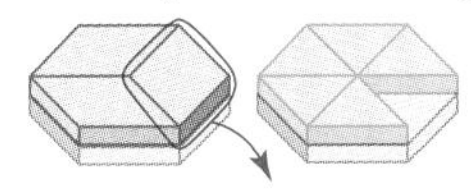

d) Let a double pink hexagon represent 1.

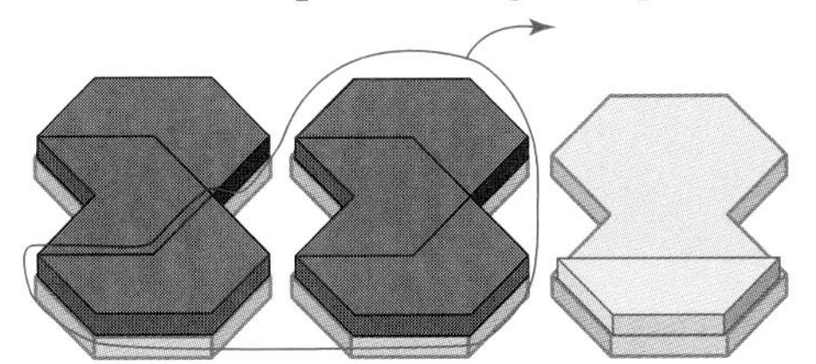

6. Let a double pink hexagon represent one whole. Use fraction and pattern blocks to determine each sum. Record your work in your notebook. Check using a number line.

a) $\frac{1}{2} + \frac{1}{4}$

b) $\frac{5}{12} + \frac{1}{2}$

c) $\frac{1}{2} + \frac{1}{6}$

d) $1\frac{3}{4} + \frac{1}{12}$

e) $2\frac{2}{3} + 3\frac{1}{2}$

f) $4\frac{1}{2} + 1\frac{1}{6}$

7. Let a black Fraction Factory piece represent one whole. Use Fraction Factory pieces to determine each difference. Record your work in your notebook. Check using a number line.

a) $\frac{5}{6} - \frac{2}{3}$

b) $\frac{9}{10} - \frac{3}{5}$

c) $\frac{6}{8} - \frac{1}{4}$

d) $2\frac{9}{10} - 1\frac{2}{5}$

e) $3\frac{3}{5} - 1\frac{1}{2}$

f) $3\frac{1}{2} - 2\frac{1}{6}$

8. Write the addition sentence each diagram represents. Write your answer in simplest form.

a) Let a double pink hexagon represent 1.

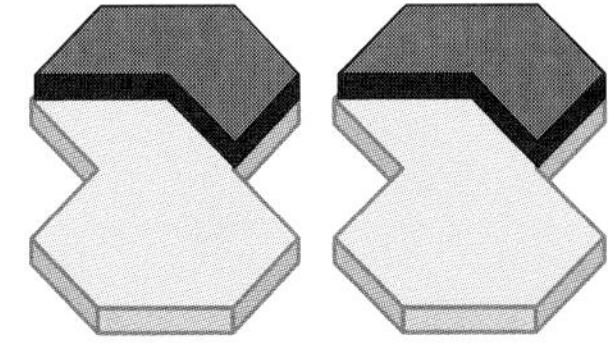

b) 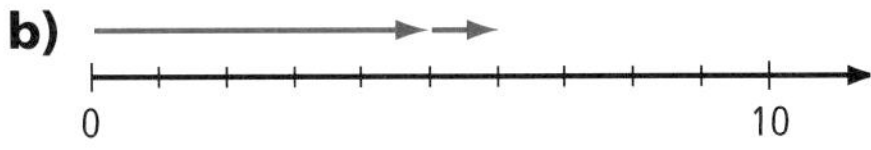

c) 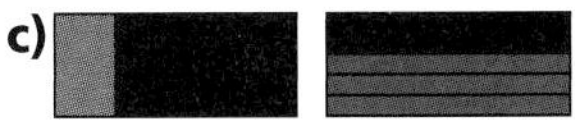

d) 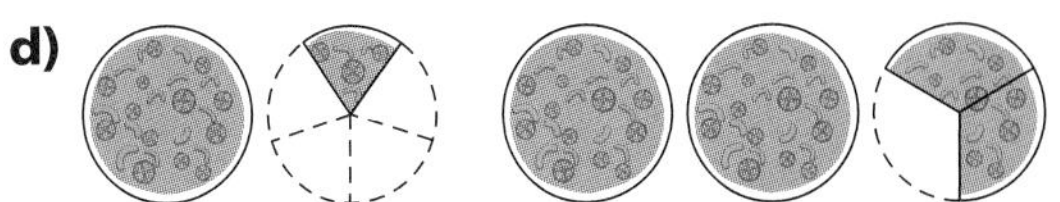

e) 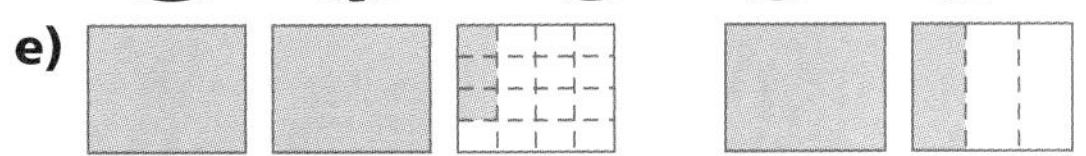

f) 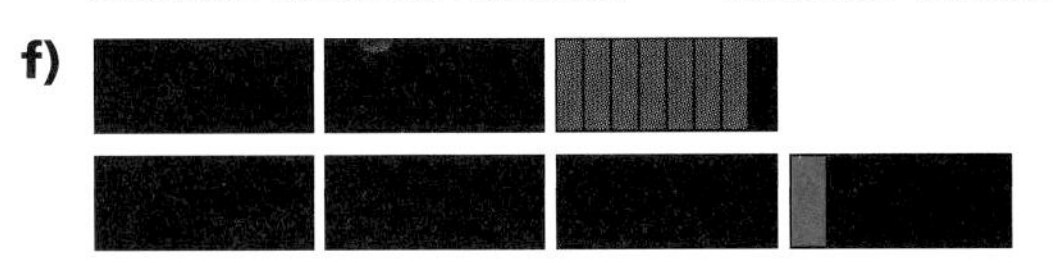

9. Subtract. Use Fraction Factory pieces to help you.

a) $\frac{1}{2} - \frac{3}{8}$
b) $\frac{3}{5} - \frac{4}{10}$
c) $\frac{7}{10} - \frac{1}{2}$
d) $4\frac{4}{6} - 2\frac{1}{3}$
e) $7\frac{2}{3} - 5\frac{1}{6}$
f) $6\frac{2}{3} - 4\frac{2}{6}$

10. Calculate. Use manipulatives to help you.

a) $\frac{1}{2} + \frac{1}{3}$
b) $\frac{7}{8} - \frac{1}{2}$
c) $\frac{1}{12} + \frac{1}{2}$
d) $3\frac{3}{4} - 1\frac{2}{3}$
e) $2\frac{3}{8} + \frac{1}{4}$
f) $\frac{7}{8} - \frac{1}{4}$

11. Calculate. Use pictures to help you.

a) $\frac{2}{3} - \frac{1}{3}$
b) $\frac{3}{4} + \frac{1}{5}$
c) $\frac{5}{6} - \frac{1}{5}$
d) $\frac{2}{3} + \frac{1}{10}$
e) $\frac{2}{4} - \frac{1}{3}$
f) $\frac{9}{10} - \frac{3}{8}$

12. Leah picked two types of apples. She picked $1\frac{3}{5}$ bushels of Golden Delicious apples and $2\frac{1}{2}$ bushels of Honeycrisp apples. Estimate how many bushels of apples Leah picked. Explain your estimation strategy.

13. Write two fractions, with the same denominator, whose sum is $\frac{5}{6}$. Use tiles, pattern and fraction blocks, or drawings to model your answer.

14. Write one mixed number and one fraction, with the same denominator, whose difference is $\frac{2}{3}$. Use tiles, pattern and fraction blocks, or drawings to model your answer.

15. Write two fractions, with different denominators, whose sum is $1\frac{1}{4}$.

16. Write two fractions, with different denominators, whose difference is $\frac{1}{3}$.

17. Calculate. Use manipulatives to help you, and record your models.

a) $8\frac{1}{2} - 3\frac{5}{6}$

b) $2\frac{7}{8} + 3\frac{1}{2}$

c) $4\frac{3}{4} + 2\frac{1}{2}$

d) $3\frac{1}{10} - 1\frac{2}{5}$

18. Theo's cat ate $2\frac{1}{3}$ cans of cat food in one week and $3\frac{1}{2}$ cans the next week. How many cans did he eat in the two weeks?

19. Joan picked $2\frac{2}{3}$ boxes of bakeapple berries. She used $1\frac{1}{2}$ boxes to make jam. How many boxes of berries are left over?

20. Explain why the following solution is not reasonable.

$$\frac{5}{8} - \frac{1}{2} = \frac{5 - 1}{8 - 2}$$
$$= \frac{4}{6} \text{ or } \frac{2}{3}$$

21. Explain why the following solution is not reasonable.

$$\frac{5}{8} + \frac{1}{2} = \frac{5}{8} + \frac{1 \times 4}{2 \times 4}$$
$$= \frac{5}{8} + \frac{4}{8}$$
$$= \frac{9}{16}$$

22. For each question, write a number sentence, and then find the result.

a) Carol used $1\frac{3}{8}$ cups of flour to make a cake. Jacques used $\frac{7}{8}$ cup of flour to make cookies. How much flour did they use altogether?

b) In part a), how much more flour did Carol use than Jacques?

c) William used $\frac{3}{4}$ cup of water, $\frac{1}{2}$ cup of milk, and $1\frac{1}{3}$ cups of flour to make pancakes. How many cups of ingredients did he use in total?

23. Find the missing fraction.

$$\frac{1}{3} + \frac{2}{5} = \frac{5}{15} + ■$$

24. a) Which pairs of fractions have a sum of $\frac{3}{4}$?

$\frac{1}{2}, \frac{3}{8}, \frac{1}{4}, \frac{1}{8}, \frac{5}{8}$

Show how you know.

b) Name two other fractions that have a sum of $\frac{3}{4}$.

25. Which sum is greater: $\frac{1}{2} + \frac{1}{3}$ or $\frac{1}{2} + \frac{2}{5}$? Explain, without using common denominators.

26. Which difference is greater: $\frac{3}{4} - \frac{1}{5}$ or $\frac{3}{4} - \frac{3}{10}$? Explain, without using common denominators.

27. Justina and Pierre each solved the same problem.

Justina's Solution

$$\frac{1}{6} + \frac{3}{4} = \frac{2}{12} + \frac{9}{12}$$
$$= \frac{11}{12}$$

Pierre's Solution

$$\frac{1}{6} + \frac{3}{4} = \frac{4}{24} + \frac{18}{24}$$
$$= \frac{22}{24}$$

Who is incorrect, if anyone? Explain the difference between their solutions.

Chapter Problem

28. For the Truro Tulip Festival, Julie decided to plant $\frac{1}{6}$ of her garden with red tulips and $\frac{2}{5}$ with yellow tulips.

a) What fraction of the garden area did she plant with red and yellow tulips?

b) She wants to complete the garden with purple, pink, and white tulips. What fraction of the garden area does she still have to plant?

29. The school recycling bin was $\frac{1}{8}$ full at the beginning of the week. The bin was $\frac{5}{12}$ full by mid-week. By the end of the week, the bin was $\frac{5}{6}$ full.

a) How much was added to the bin in the first half of the week?
b) How much was added to the bin in the last half of the week?
c) What fraction of the bin was empty by mid-week?
d) What fraction of the bin was empty at the end of the week?

30. A swimming pool has eight lanes.

During practice, swimmers use three lanes for front crawl, and half of the lanes are used for the butterfly stroke.

a) Write an expression for the total fraction of lanes used during practice.
b) Find the number of lanes used during practice.
c) If $\frac{5}{8}$ of the pool is used during a race, what fraction of the lanes are not used?

31. The Eustace family have a strawberry pie and an apple pie that are the same size. On Monday, they ate $\frac{3}{4}$ of the strawberry pie and $\frac{3}{8}$ of the apple pie. On Tuesday, they ate another $\frac{1}{6}$ of the strawberry pie and another $\frac{1}{2}$ of the apple pie.

a) Which pie did they eat more of on Monday? How much more? Draw a diagram to illustrate your answer.
b) How much of the strawberry pie was eaten in total?
c) How much of the apple pie was eaten in total?
d) If you combine the remainders of both pies, estimate how much pie was left in total.

32. For each question, show your work using pictures and symbols.

a) For the grade 8 awards dinner, $8\frac{1}{2}$ gallons of punch were provided, of which $3\frac{5}{6}$ gallons was used. How much punch was left over?
b) Ginny took 6 pounds of butter out of the fridge to soften. She used $4\frac{1}{4}$ pounds in her baking. How much butter did she not use?
c) Carlos had $7\frac{1}{2}$ quarts of motor oil in his garage. After filling his car engine, he had $3\frac{7}{8}$ quarts left. How much oil did he use?

33. Calculate.

a) $\frac{1}{6} + \frac{1}{4} + \frac{1}{2}$

b) $\frac{8}{9} - \frac{2}{3} - \frac{1}{6}$

Extend

34. Geraldine ran for $\frac{1}{2}$ h on Monday, $\frac{5}{6}$ h on Wednesday, $\frac{2}{3}$ h on Friday, and $\frac{3}{4}$ h on Saturday.

a) Estimate and then calculate how many hours she ran altogether.

b) For how many more minutes did she run on Wednesday than on Saturday?

c) Geraldine followed this schedule every week during February. Explain how you would find the total number of hours she ran that month.

35. Gilles and Marthe were comparing their work on addition of fractions. Gilles said he always found a common denominator by multiplying the two denominators together. Marthe said she found a common denominator by trying consecutive multiples of the larger denominator until she found the first one that was divisible by both denominators. Try both methods using the following questions. Decide which method is more efficient.

a) $8\frac{19}{24} + 9\frac{31}{36}$

b) $5\frac{12}{17} - 3\frac{5}{23}$

36. Kent collects coins from different countries. Of his coins, $\frac{1}{2}$ came from Germany, $\frac{1}{4}$ came from Korea, $\frac{1}{8}$ came from Russia, $\frac{1}{16}$ came from Spain, and the remaining three coins came from Japan. How many coins does he have in all, and how many are from each country?

Puzzler

If a vowel is worth $\frac{1}{4}$ and a consonant is worth $\frac{1}{5}$ then the word MATH is worth

$\frac{1}{5} + \frac{1}{4} + \frac{1}{5} + \frac{1}{5} = \frac{17}{20}$

a) Find a word worth the value of 2.

b) Find a word worth $\frac{3}{5}$. What assumption are you making?

2.2 Multiply Fractions

Focus on...
- multiplying fractions using manipulatives, pictures, and symbols
- multiplying fractions using mental math

During a track and field competition, many events take place on the field inside the track. Suppose $\frac{2}{3}$ of the field is used for field events and $\frac{1}{4}$ of the field event area is used for the long jump. What fraction of the field is used for the long jump?

Discover the Math

Materials
- sheet of paper
- pencil crayons
- ruler

Part A: Area Model

1. Fold a rectangular piece of paper one way into thirds to represent the field. Open it up and shade the part of the rectangle used for field events.

2. Fold the same paper into fourths the other way. Open it up and use a different colour to shade the part of the rectangle used for the long jump. Make sure you are shading over part of the area used for field events.

3. **a)** How many sections has the paper been folded into altogether?
   b) How many shaded sections overlap?
   c) How does this number represent the fraction of the field used for the long jump?
   d) What fraction of the field is used for the long jump?
   e) Write a number sentence to represent what was done in parts a) to d).

4. Recall that the answer to a multiplication statement is called the product. Use paper folding or drawings to find each product. Write a complete number sentence for each product.

   a) $\frac{1}{3} \times \frac{3}{5}$ **b)** $\frac{2}{3} \times \frac{1}{2}$

   c) $\frac{2}{5} \times \frac{1}{2}$ **d)** $\frac{1}{2} \times \frac{1}{4}$

5. Use Fraction Factory pieces to find each product. Write a complete number sentence for each product. Part a) has been completed for you.

   a) $\frac{3}{4} \times \frac{1}{2}$

   Step 1: Lay out $\frac{1}{2}$.

   Step 2: Trade for pieces that will allow you to make fourths, in this case $\frac{4}{8}$.

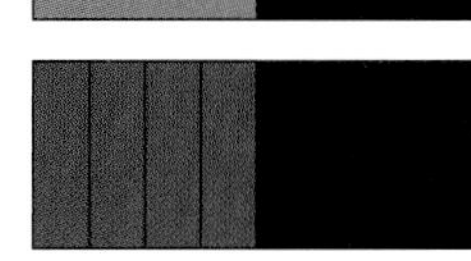

   Step 3: Then keep three-fourths of the set.

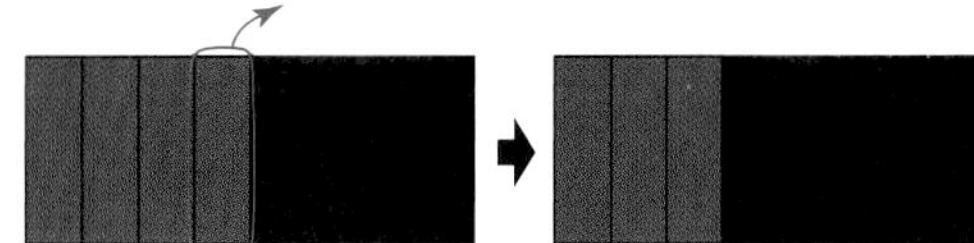

   The product is $\frac{3}{4} \times \frac{1}{2} = \frac{3}{8}$.

   b) $\frac{1}{2} \times \frac{5}{6}$

   c) $\frac{2}{3} \times \frac{1}{3}$

6. **Reflect** Look closely at the number sentences you wrote in questions 4 and 5. Write a rule for multiplying a proper fraction by another fraction.

Part B: Multiply a Fraction by a Whole Number

Materials
- square dot paper

1. Maria ran for $\frac{3}{4}$ h each day for 7 days. What is her total running time?

   Let a circular clock face represent 1 h. Draw a shape to represent $\frac{3}{4}$ h.

   Draw your shape 7 times. How many fourths parts did you draw in total? Rearrange the parts to make as many wholes as possible. Write a number sentence to express your total as an improper fraction and as a mixed number.

   Strategies
   How else could you model $\frac{3}{4}$ h?

   a) Complete the number sentence $7 \times \frac{3}{4} = ■$.

   b) What is Maria's total running time?

   c) Does your rule from question 6 in Part A work for multiplying a fraction by a whole number? Explain.

   d) Model and write a complete number sentence for each product.

   i) $5 \times \frac{2}{3}$ **ii)** $4 \times \frac{2}{3}$ **iii)** $8 \times \frac{5}{6}$

Multiply a Mixed or Whole Number by a Fraction

2. Annie sells oatcakes at the Saturday morning market at the Antigonish Farmer's Market. Near the end of the morning, she had $7\frac{1}{2}$ packages of oatcakes left. Peggy bought $\frac{2}{3}$ of them. How many packages of oatcakes did Peggy buy? Use a small rectangle to represent a package of oatcakes. Draw $7\frac{1}{2}$ packages. Draw 3 baskets below the packages. Divide the packages equally among the 3 baskets. You may need to subdivide 1 or 2 packages. Circle 2 baskets to represent $\frac{2}{3}$. Find the total number of packages in the 2 baskets.

   a) Complete the number sentence $\frac{2}{3} \times 7\frac{1}{2} = ■$.

   b) How many packages did Peggy buy?

   c) Does your rule from question 6 in Part A work for multiplying a mixed number by a fraction? Explain.

   d) Model and write a complete number sentence for each product.

   i) $\frac{1}{4} \times 12$ **ii)** $\frac{2}{3} \times 9$ **iii)** $\frac{3}{4} \times 5\frac{1}{3}$

Multiply a Mixed Number by a Mixed Number

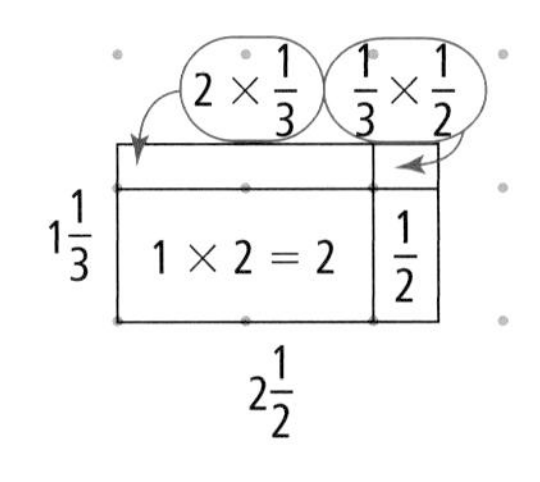

3. The rectangle measures $1\frac{1}{3}$ units by $2\frac{1}{2}$ units. What is the area of the rectangle in square units? On dot paper (2 cm or larger), carefully draw a model of the rectangle like the one shown. Find the areas of the four smaller rectangles, and then combine them to find the total area.

a) Complete the number sentence $1\frac{1}{3} \times 2\frac{1}{2} = ■$.

b) What is the area of the rectangle in square units?

c) Does your rule from question 6 in Part A work for multiplying a mixed number by a mixed number? Explain.

d) Model and write a complete number sentence for each product.

i) $1\frac{2}{3} \times 3$ **ii)** $2\frac{3}{4} \times 1\frac{1}{2}$ **iii)** $2\frac{2}{3} \times 1\frac{1}{3}$

Part C: Multiply Fractions and Mixed Numbers

1. Does the rule you established work for multiplying fractions by whole numbers, and multiplying mixed numbers by mixed numbers?

The table lists all the completed number sentences you found in Part B. Copy and complete the table by applying the rule. To apply the rule, you will first need to express the whole numbers and mixed numbers as improper fractions. Compare your answers with those you found in Part B.

Number Sentence	Apply the Rule
$7 \times \frac{3}{4}$	$\frac{7}{1} \times \frac{3}{4} = \frac{7 \times 3}{1 \times 4} = \frac{21}{4} = 5$ and $\frac{1}{4} = 5\frac{1}{4}$
$5 \times \frac{2}{3}$	
$6 \times \frac{2}{3}$	
$8 \times \frac{5}{6}$	
$\frac{2}{3} \times 7\frac{1}{2}$	
$\frac{1}{4} \times 12$	
$\frac{2}{3} \times 9$	
$\frac{3}{4} \times 5\frac{1}{3}$	
$1\frac{2}{3} \times 3\frac{1}{2}$	
$2\frac{3}{4} \times 1\frac{1}{2}$	
$2\frac{2}{3} \times 1\frac{1}{3}$	

2. Reflect In your own words, write a rule for multiplying fractions and mixed numbers. Include an example to help you explain.

Example 1: Multiply Fractions Mentally

Evaluate mentally.

a) $\frac{2}{5} \times \frac{2}{3}$ **b)** $4 \times 1\frac{1}{4}$

Solution

a) $\frac{2}{5} \times \frac{2}{3} = \frac{4}{15}$

I can multiply the numerators and denominators mentally.

b)

$$4 \times 1\frac{1}{4} = (4 \times 1) + \left(4 \times \frac{1}{4}\right)$$
$$= 4 + 1$$
$$= 5$$

I can multiply these mentally, by multiplying each part of the mixed number by the whole number.

Example 2: Half of Three-Quarters

You had three-quarters of a sheet of construction paper. You gave your friend half of that piece. How much of the original sheet of paper did you give to your friend?

Solution

I gave my friend $\frac{1}{2}$ of $\frac{3}{4}$ of the sheet of construction paper.

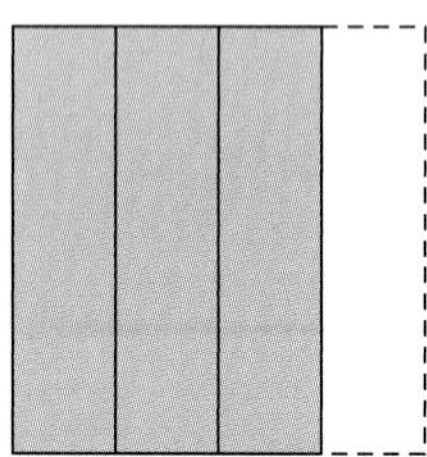

I can use a grid to represent the original sheet of paper. I have three-quarters of the original sheet, shown in blue.

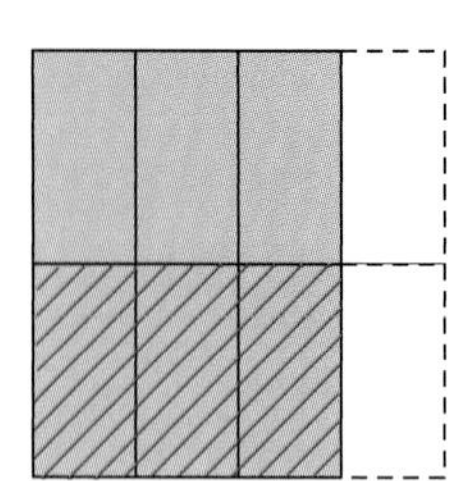

To represent half of the blue portion, I shaded half of the blue portion red. I gave my friend the portion shaded in red. This represents three $\frac{1}{8}$ parts of the original sheet.

So, $\frac{1}{2} \times \frac{3}{4} = \frac{3}{8}$.

Communicating Mathematically

In math, "of" means "multiplied by."
For example, $\frac{1}{2}$ of $\frac{3}{4}$ means $\frac{1}{2} \times \frac{3}{4}$.

Strategies

Can you think of another way to solve this problem?

Example 3: Multiply a Fraction by a Fraction

Mr. MacNamara has $\frac{2}{3}$ acre of land. He wants to plant potatoes on $\frac{3}{4}$ of his land. How much land will be planted with potatoes?

Solution

Evaluate.

$\frac{3}{4} \times \frac{2}{3}$

Method 1: Use a diagram.

Draw a diagram to show $\frac{2}{3}$.

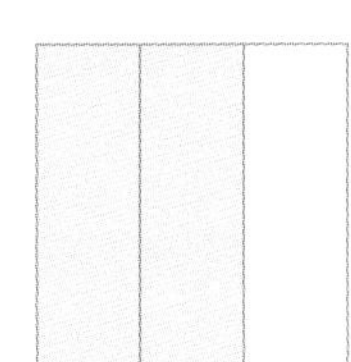

Divide the whole diagram into quarters.

Colour $\frac{3}{4}$ of the shaded region.

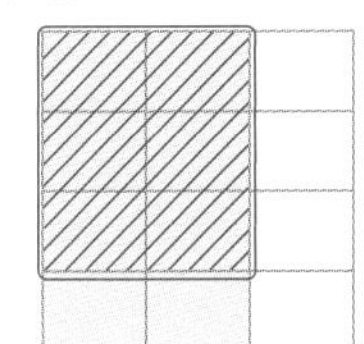

I have highlighted $\frac{6}{12}$ of the diagram.

The double-coloured area shows the product $\frac{3}{4} \times \frac{2}{3} = \frac{6}{12}$.

Mr. MacNamara will have $\frac{6}{12}$ acre of farmland planted with potatoes.

Method 2: Multiply the numerators and denominators.

$$\frac{3}{4} \times \frac{2}{3} = \frac{3 \times 2}{4 \times 3}$$

$$= \frac{6}{12}$$

That's the number of parts circled in the diagram.

That's the number of parts in the whole.

The diagram shows this statement is true. Six shaded rectangles out of 12 rectangles are circled.

Mr. MacNamara will have $\frac{6}{12}$ acre of land planted with potatoes.

Method 3: Simplify before multiplying.

You could also say that $\frac{3}{4} \times \frac{2}{3} = \frac{\overset{1}{\cancel{3}}}{\underset{2}{\cancel{4}}} \times \frac{\overset{1}{\cancel{2}}}{\underset{1}{\cancel{3}}} = \frac{1 \times 1}{2 \times 1} = \frac{1}{2}$.

Example 4: Multiply a Fraction by a Whole Number and Vice Versa

a) $3 \times \frac{3}{4}$ **b)** $\frac{2}{3} \times 4$

Solution

a) $3 \times \frac{3}{4}$ **Use a number line. $3 \times \frac{3}{4}$ can be presented by 3 jumps of $\frac{3}{4}$.**

$= \frac{3}{1} \times \frac{3}{4}$

$= \frac{3 \times 3}{1 \times 4}$

$= \frac{9}{4}$ or $2\frac{1}{4}$

After 3 jumps of $\frac{3}{4}$, you end at $2\frac{1}{4}$.

b) $\frac{2}{3} \times 4$ **Use fraction blocks. Let a double pink hexagon represent 1.**

$= \frac{2}{3} \times \frac{4}{1}$

$= \frac{2 \times 4}{3 \times 1}$

$= \frac{8}{3}$ or $2\frac{2}{3}$

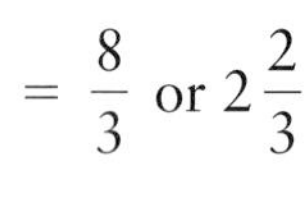

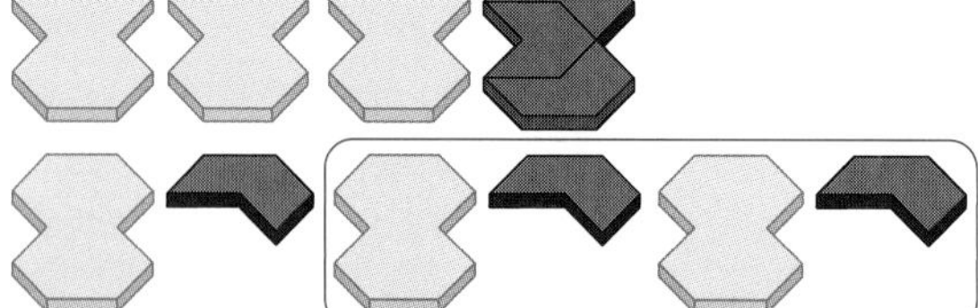

Example 5: Multiply a Mixed Number by a Whole Number

Miguel worked $3\frac{1}{2}$ h each day for 5 days during the March Break. How many hours did he work altogether?

Solution

Method 1: Use a number line.

Represent $3\frac{1}{2}$ on a number line. Then, show 5 jumps of $3\frac{1}{2}$.

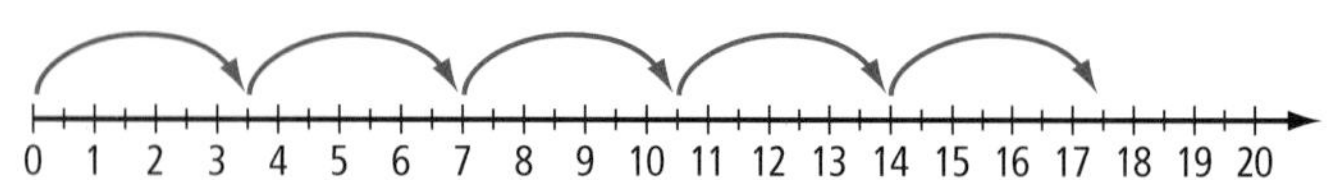

$5 \times 3\frac{1}{2} = 17\frac{1}{2}$

Miguel worked $17\frac{1}{2}$ h altogether.

Method 2: Change the mixed number to an improper fraction.

$5 \times 3\frac{1}{2} = 5 \times \frac{7}{2}$

Express 5 as a fraction. Then, multiply the numerators and the denominators.

$$\frac{5}{1} \times \frac{7}{2} = \frac{5 \times 7}{1 \times 2}$$

$$= \frac{35}{2} \text{ or } 17\frac{1}{2}$$

Miguel worked $17\frac{1}{2}$ h altogether.

Strategies
What previously learned procedures are you applying here?

Method 3: Use the distributive property.

$$5 \times 3\frac{1}{2} = (5 \times 3) + \left(5 \times \frac{1}{2}\right)$$

$$= 15 + 2\frac{1}{2}$$

$$= 17\frac{1}{2}$$

Miguel worked $17\frac{1}{2}$ h altogether.

Example 6: Multiply a Mixed Number by a Mixed Number

Evaluate.

$2\frac{2}{3} \times 2\frac{1}{4}$

Solution

$$2\frac{2}{3} \times 2\frac{1}{4} = \frac{8}{3} \times \frac{9}{4}$$ **Write the mixed numbers as improper fractions then multiply.**

$$= \frac{8 \times 9}{3 \times 4}$$

$$= \frac{72}{12}$$

$$= 6$$

Communicate the Key Ideas

1. Mary evaluated $\frac{2}{3} + \frac{3}{4}$ as shown.

$$\frac{2}{\cancel{3}_1} + \frac{\cancel{3}^1}{4} = \frac{2}{4}$$

$$= \frac{1}{2}$$

Explain her error.

2. **a)** Explain how an area model can help you calculate the product of two fractions. Give an example.
   b) Explain how you can find the product of the two fractions using numbers and symbols.

3. Does $\frac{1}{4}$ of $\frac{1}{2}$ have the same meaning as $\frac{1}{2}$ of $\frac{1}{4}$? Illustrate and explain your answer.

Check Your Understanding

1. Use mental math to multiply.

a) $5 \times \frac{4}{5}$ **b)** $\frac{1}{2} \times \frac{1}{4}$ **c)** $\frac{2}{3} \times \frac{9}{13}$ **d)** $5 \times 2\frac{1}{3}$

2. Write the multiplication sentence that each diagram represents.

a)

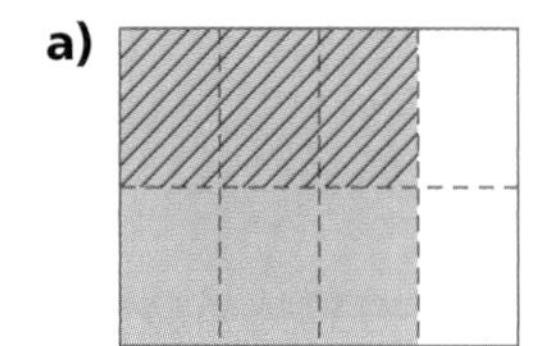

b)

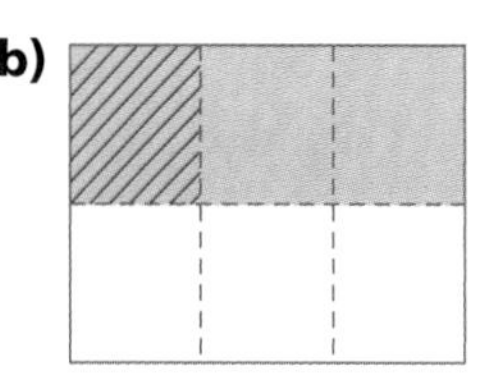

c)

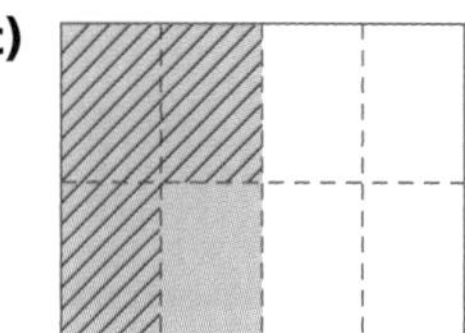

d)

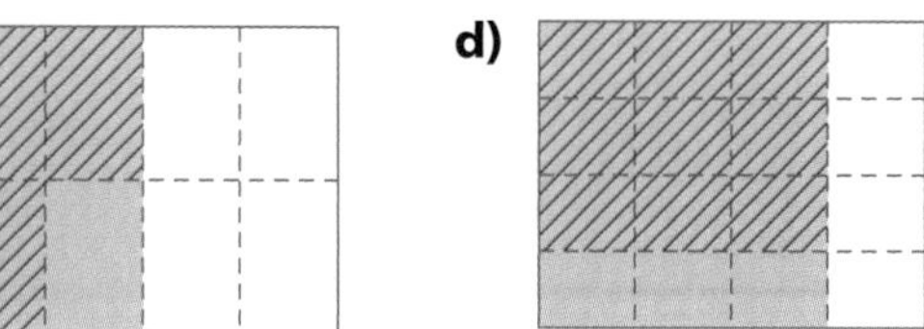

e)

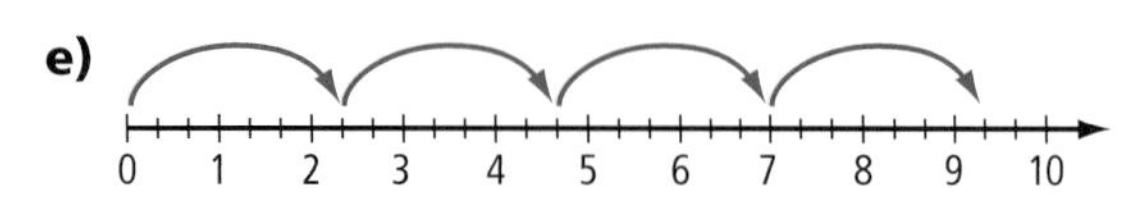

f)

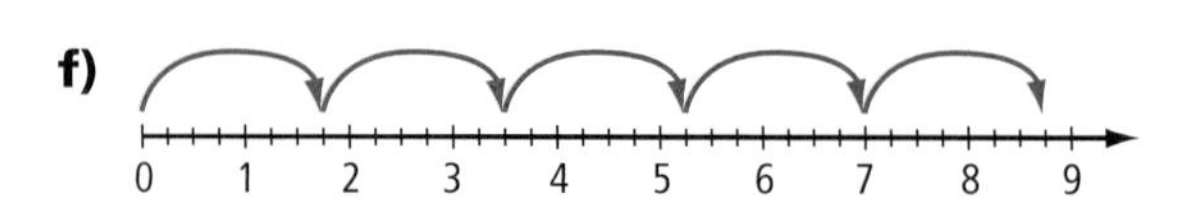

g)

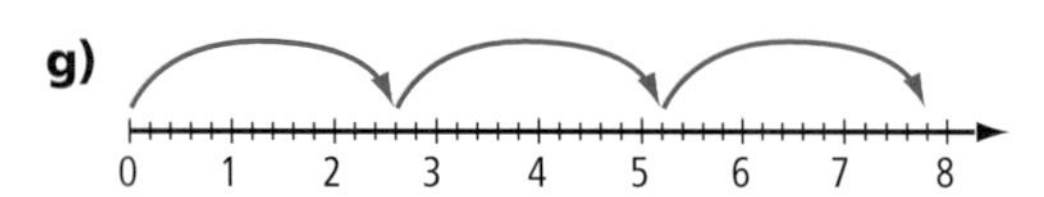

3. Draw a picture to represent each multiplication statement. Find the product.

a) $\frac{1}{2} \times 2$ b) $8 \times \frac{1}{4}$

4. Evaluate. Use models or drawings to help you explain your answer.

a) $\frac{3}{4} \times \frac{1}{5}$ b) $\frac{3}{5} \times \frac{5}{6}$

c) $\frac{2}{5} \times \frac{3}{5}$ d) $\frac{9}{2} \times \frac{3}{8}$

5. Evaluate. Use a model to help you explain your answer.

a) $6 \times \frac{2}{6}$ b) $2 \times \frac{5}{6}$

c) $4 \times 1\frac{1}{2}$ d) $3 \times \frac{2}{6}$

e) $\frac{3}{4} \times 8$ f) $\frac{2}{3} \times 5$

g) $3\frac{1}{3} \times 4$ h) $\frac{5}{6} \times 12$

6. Evaluate. Use the area model to help you explain your answer.

a) $2 \times \frac{5}{6}$ b) $2\frac{1}{2} \times \frac{3}{4}$

c) $1\frac{2}{3} \times 6$ d) $1\frac{2}{3} \times 2\frac{1}{4}$

7. It rained and snowed yesterday at the Halifax International Airport for $4\frac{1}{2}$ h. If it snowed for $\frac{2}{3}$ of the time, how many hours did it rain?

8. Crystal swam $5\frac{1}{3}$ lengths of the pool and Brenda swam $1\frac{1}{2}$ times as far as Crystal. How many lengths did Brenda swim?

9. Jayden cycled 1 lap around the cycling track in $2\frac{1}{4}$ min. Estimate how long it will take her to cycle 10 laps.

10. Giulia completed $\frac{1}{5}$ of her music practice and Claudia completed $\frac{2}{3}$ of her music practice. A complete music practice for Giulia is $2\frac{1}{2}$ h and $1\frac{3}{4}$ h for Claudia. Who has practised more? How much more?

11. a) In a grade 8 class, $\frac{2}{3}$ of the students have brown eyes. Of the students with brown eyes, $\frac{1}{2}$ have brown hair. What fraction of the students have brown eyes and brown hair? Draw a picture to show how you found your answer.

b) In the same grade 8 class, $\frac{2}{5}$ of the students wear glasses. Of the students who wear glasses, $\frac{1}{4}$ are boys. What fraction of the students are boys who wear glasses? Draw a picture to show how you found your answer.

12. Write two different multiplication statements that have a product of $\frac{4}{9}$.

13. Siddhartha and Chandra order a small pizza. Siddhartha eats $\frac{1}{2}$ of it and Chandra eats $\frac{1}{3}$ of it.

a) What fraction of the pizza did they eat altogether?
b) What fraction of the pizza was left over?

14. Elena and her friends ordered a 24-slice party pizza. They ate $\frac{2}{3}$ of it.

a) How many slices did they eat altogether?
b) Elena's brother Marcus ate $\frac{1}{2}$ of the leftover pizza. What fraction of the 24 slices did Marcus eat?
c) Draw a picture to show that your answers are correct.

15. Alene had $\frac{3}{5}$ of an extra large submarine sandwich. She gave Richard $\frac{3}{4}$ of her portion.

a) Did Richard receive more or less than one whole sandwich? Explain how you know.
b) Did Richard receive more or less than $\frac{3}{10}$ of one whole sandwich? Explain how you know.

16. Every year, the school participates in a read-a-thon.

This year, about $\frac{4}{7}$ of the class collected pledge money for the event. Close to $\frac{3}{8}$ of these students raised over \$50 each. There are 28 students in the class.

a) What fraction of the class raised over \$50 each?
b) What fraction of the class raised less than \$50?
c) Estimate how many students raised over \$50.
d) How would you find the minimum amount of money raised?

17. CJ calculated that the product of $\frac{2}{3} \times 1\frac{1}{2}$ is 1. Is CJ correct? Explain, using pictures and words.

18. **a)** Place the numbers 1, 2, 3, and 4 in the boxes to get the greatest possible answer. Explain your strategy.

$$\frac{■}{■} \times \frac{■}{■}$$

b) Place the numbers 6, 7, 8, and 9 in the boxes in part a) to get the least possible answer. Explain your strategy.

19. To "halve and double" is a Mental Math Strategy that works well with whole numbers. For example, 15×24 has the same product as 30×12 and 60×6. Hannah wondered if the same strategy would work for fractions. Use these multiplication statements to investigate the strategy.

a) $\frac{8}{9} \times 2\frac{1}{2}$

b) $\frac{24}{25} \times 1\frac{1}{4}$

20. Use pictures and words to explain how to multiply a fraction by a whole number.

Chapter Problem

21. A neighbourhood group is planting a community garden. Half of the garden will have fruit and the other half will have vegetables. In the vegetable section, $\frac{1}{3}$ will be carrots, $\frac{1}{4}$ will be beans, $\frac{1}{5}$ will be potatoes, and the rest will be squash.

a) What fraction of the entire garden will be carrots?

b) What fraction of the entire garden will be potatoes?

c) What fraction of the entire garden will be squash?

22. a) Find the product of 4×3 and of $\frac{2}{3} \times \frac{1}{2}$.

b) Is the product of two whole numbers that are greater than 1, always greater than, or always less than, each of the factors? Use pictures, symbols, or words to justify your answer.

c) Is the product of two proper fractions always greater than, or always less than, each of the factors? Use pictures, symbols, or words to justify your answer.

23. Find each product.

a) $\frac{4}{5} \times \frac{2}{3} \times \frac{1}{2}$ **b)** $\frac{3}{4} \times \frac{1}{2} \times \frac{1}{3}$

24. a) Use diagrams to explain why

$2 \times \frac{1}{2} = 1$,

$3 \times \frac{1}{3} = 1$,

$4 \times \frac{1}{4} = 1$,

$5 \times \frac{1}{5} = 1$, and so on.

b) What fraction multiplied by $\frac{4}{3}$ has a product equal to 1? This fraction is called the *reciprocal* of $\frac{4}{3}$.

Communicating Mathematically

An inverted fraction is called a **reciprocal** of the original fraction. Reciprocals are two numbers that have a product of 1. For example, $\frac{2}{3}$ and $\frac{3}{2}$ are reciprocals.

$\frac{2}{3} \times \frac{3}{2} = \frac{2 \times 3}{3 \times 2} = \frac{6}{6} = 1$

Extend

25. Evaluate.

$\frac{2}{3} \times \frac{3}{4} \times \frac{4}{5} \times \frac{5}{6} \times \frac{6}{7} \times \frac{7}{8} \times \frac{8}{9} \times \frac{9}{10}$

26. Model $3\frac{1}{2} \times 2\frac{1}{2}$ using

a) drawings of pies

b) a number line

2.3 Divide Fractions

Focus on…
- dividing fractions using manipulatives, pictures, and symbols
- dividing fractions using mental math

To plant the school garden, each grade gets 6 full flats of flowers. The grade 8 classes are split into groups and each group gets $\frac{1}{4}$ of a flat of flowers to plant. Into how many groups are the grade 8 classes divided?

Discover the Math

Materials
- at least 6 sheets of paper
- scissors

Part A: How can you divide fractions?

1. Use 6 sheets of paper to represent the 6 flats of flowers. Cut each piece of paper into quarters.

a) Write a division statement to represent this situation.
b) Complete this sentence to show the meaning of your division statement. "How many ■ are in ■?"
c) How many pieces of paper do you have now?
d) Write a detailed division sentence to reflect the process.

2. Use a drawing to show each division statement. Copy and complete the table.

Division Statement	Drawing	Detailed Division Sentence
$3 \div \frac{1}{2}$		$3 \div \frac{1}{2} = \frac{6}{2} \div \frac{1}{2} = \frac{6}{1} = 6$
$3 \div \frac{1}{4}$		$3 \div \frac{1}{4} = \frac{12}{4} \div \frac{1}{4} = \frac{12}{1} = 12$
$3 \div \frac{1}{3}$		
$3 \div \frac{1}{6}$		
$3 \div \frac{2}{3}$		$3 \div \frac{2}{3} = \frac{9}{3} \div \frac{2}{3} = \frac{9}{2} = 4\frac{1}{2}$
$4 \div \frac{3}{4}$		
$3 \div \frac{2}{5}$		
$2 \div \frac{3}{5}$		

3. Recall that the answer to a division statement is called the quotient. Predict the quotient for each division statement.

a) $3 \div \frac{1}{5}$ **b)** $3 \div \frac{3}{4}$

4. Reflect

a) Refer to question 2. Describe how your drawing and detailed division statement model the division statement in column 1.

b) Explain how to rewrite each whole number as an improper fraction.

c) Use your answers to parts a) and b) to create a procedure to help you divide a whole number by a fraction.

5. Reflect In this question, why is the remaining piece (the white piece) called $\frac{1}{2}$, even though it is actually $\frac{1}{3}$ of a piece of paper?

$3 \div \frac{2}{3}$		$3 \div \frac{2}{3} = \frac{9}{3} \div \frac{2}{3} = \frac{9}{2} = 4\frac{1}{2}$

Example 1: Use Mental Math to Divide Fractions

Use mental math to divide.

a) $\frac{3}{4} \div 3$ **b)** $\frac{2}{3} \div \frac{1}{3}$ **c)** $3 \div \frac{1}{4}$

Solution

a)

I need to divide $\frac{3}{4}$ into three equal parts. There are three $\frac{1}{4}$ parts in $\frac{3}{4}$.

So, $\frac{3}{4} \div 3 = \frac{1}{4}$.

b)

How many $\frac{1}{3}$ parts are in $\frac{2}{3}$? There are two $\frac{1}{3}$ parts in $\frac{2}{3}$.

So, $\frac{2}{3} \div \frac{1}{3} = 2$.

c)

How many $\frac{1}{4}$ parts are in 3? There are four $\frac{1}{4}$ parts in one whole; in three wholes, there are 3×4 or twelve $\frac{1}{4}$ parts.

So, $3 \div \frac{1}{4} = 12$.

Example 2: Divide a Fraction by a Fraction Using Common Denominators

Evaluate.

$\frac{3}{4} \div \frac{1}{3}$

Solution

Rewrite both fractions as twelfths.

$\frac{3}{4} \div \frac{1}{3} = \frac{9}{12} \div \frac{4}{12}$

To find how many sets of $\frac{4}{12}$ can be made from $\frac{9}{12}$, divide the numerator of the first fraction by the numerator of the second fraction.

$\frac{3}{4}$ 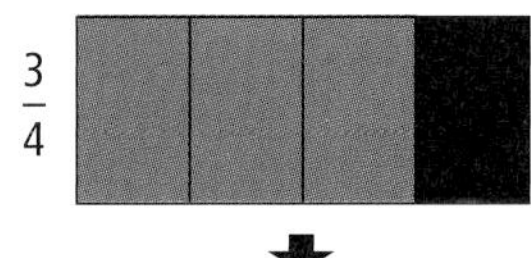

$\frac{9}{12}$ 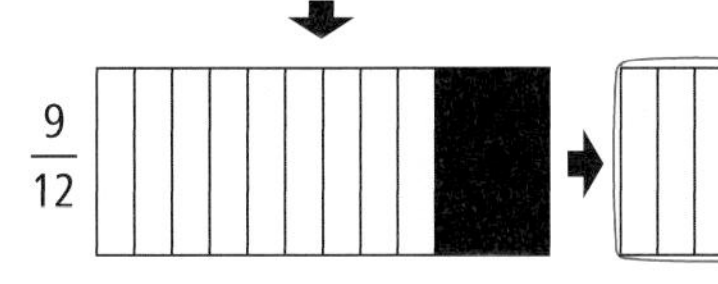

The LCM of denominators 4 and 3 is 12. Use $\frac{1}{12}$ pieces to cover the purple pieces.

$$\frac{3}{4} \div \frac{1}{3} = \frac{9}{12} \div \frac{4}{12}$$
$$= \frac{9}{4}$$
$$= 2\frac{1}{4}$$

$\frac{9}{4} = 9 \div 4$

Example 3: Divide a Fraction by a Whole Number

Evaluate.

$\frac{1}{4} \div 2$

Solution

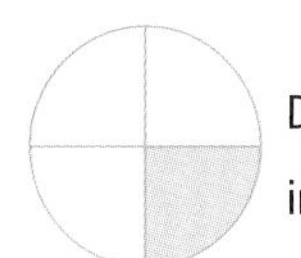

Divide $\frac{1}{4}$ equally into 2 parts.

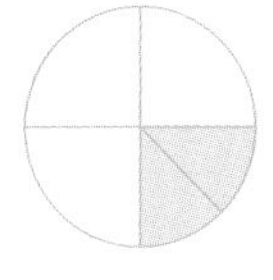

Each part represents $\frac{1}{8}$ of the whole.

Strategies
Draw a diagram.

$\frac{1}{4} \div 2 = \frac{1}{8}$

Example 4: Divide a Whole by a Fraction

You have 3 apples and want to cut each apple in half. How many pieces will you have?

Solution

Method 1: Draw a diagram.

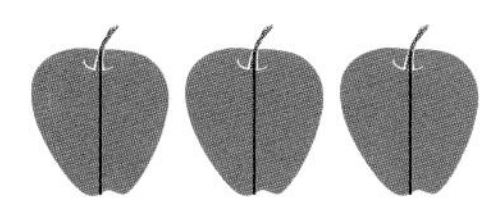

So, $3 \div \frac{1}{2} = 6$.

I cut each apple in half. There are 6 halves in total.

I will have 6 pieces.

Method 2: Use a number line.

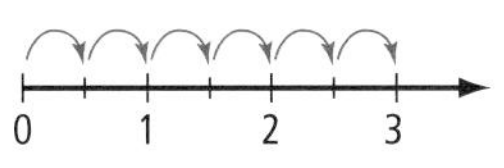

So, $3 \div \frac{1}{2} = 6$.

How many jumps of $\frac{1}{2}$ are there from 0 to 3?

I will have 6 pieces.

Example 5: Divide a Mixed Number by a Fraction

a) You have $2\frac{1}{4}$ bags of trail mix. If you eat $\frac{1}{4}$ of a bag of trail mix each day, how many days will your supply last?

b) Evaluate.

$2\frac{1}{4} \div \frac{5}{8}$

Solution

a) $2\frac{1}{4} \div \frac{1}{4}$

- Draw $2\frac{1}{4}$ rectangles.
- Divide them into fourths.
- Count the number of fourths.

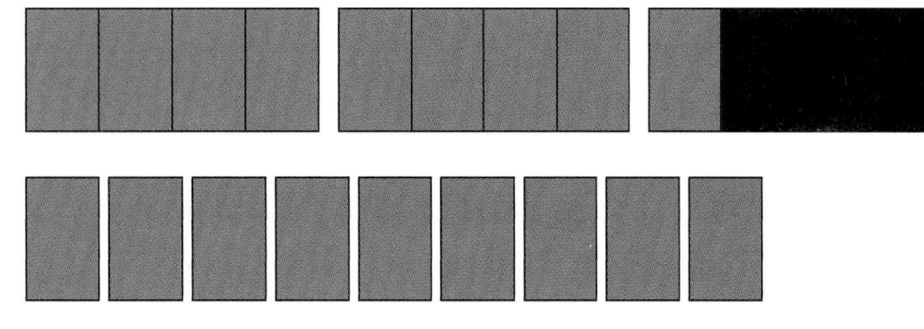

I divided each bag into fourths. There are 9 fourths in total.

$$\begin{aligned} \text{So, } 2\frac{1}{4} \div \frac{1}{4} &= \frac{9}{4} \div \frac{1}{4} \\ &= \frac{9}{1} \\ &= 9 \end{aligned}$$

My supply will last 9 days.

b) $2\frac{1}{4} \div \frac{5}{8}$

- Rewrite both fractions using a common denominator, and then change them to improper fractions.
- Divide the numerator of the first fraction by the numerator of the second fraction. Notice that the remainder is $\frac{3}{5}$, not $\frac{3}{8}$. This is because you are making sets of 5 eighths and the last 3 parts make $\frac{3}{5}$ of a set of 5 eighths.

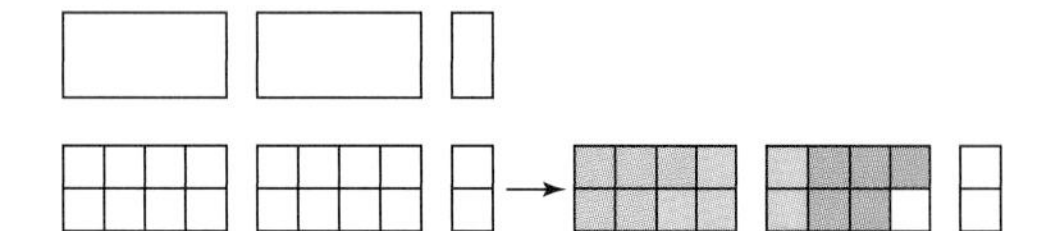

$$2\frac{1}{4} \div \frac{5}{8} = 2\frac{2}{8} \div \frac{5}{8}$$
$$= \frac{18}{8} \div \frac{5}{8}$$
$$= \frac{18}{5}$$
$$= 3\frac{3}{5}$$

Discover the Math

Part B: Develop the Math

Kenzie had modelled many division problems using drawings and common denominators. His grandmother showed him a shortcut in which you skip the common denominator step, but his teacher would not allow him to use it until he could explain it. He worked the following problems, and then noticed a pattern.

i) $\frac{7}{9} \div \frac{1}{2} = \frac{14}{18} \div \frac{9}{18} = \frac{14}{9} = 1\frac{5}{9}$

ii) $1\frac{1}{3} \div \frac{3}{4} = \frac{4}{3} \div \frac{3}{4} = \frac{16}{12} \div \frac{9}{12} = \frac{16}{9} = 1\frac{1}{4}$

iii) $3\frac{1}{2} \div 1\frac{1}{4} = \frac{7}{2} \div \frac{5}{4} = \frac{14}{4} \div \frac{5}{4} = \frac{28}{10} = 2\frac{8}{10}$ or $2\frac{4}{5}$

iv) $2 \div \frac{3}{4} = \frac{2}{1} \div \frac{3}{4} = ■$

v) $7 \div 1\frac{1}{2} = ■$

a) Explain how Kenzie evaluated each division statement.

b) Describe the pattern or shortcut that you think he noticed.

c) Kenzie's grandmother had mentioned that the shortcut was related to reciprocals. Can you explain the link to reciprocals? Discuss with a partner.

d) Solve Kenzie's problems using the shortcut.

Communicate the Key Ideas

1. Use a model to show $5 \div \frac{2}{3}$.

2. $4 \div \frac{3}{5} = 6\frac{2}{3}$

Explain why the fraction part of $6\frac{2}{3}$ is in thirds and not in fifths.

3. Alene solved the following division problem using the method shown. If she had estimated her answer before solving the problem, how could she tell that her solution was incorrect? Explain, and correct the error in her thinking.

$$\frac{9}{10} \div \frac{2}{5} = \frac{10}{9} \times \frac{2}{5}$$
$$= \frac{20}{45}$$

Check Your Understanding

1. Divide, using mental math.

a) $2 \div \frac{1}{4}$ **b)** $\frac{3}{8} \div \frac{1}{8}$ **c)** $\frac{4}{5} \div 4$

2. Match each picture to a division statement, if possible, and then find the quotient.

a) 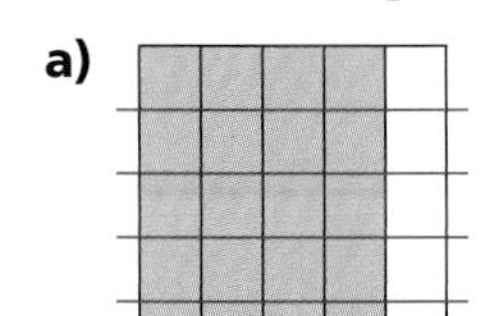

b) 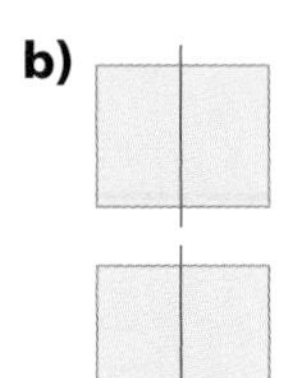

c) 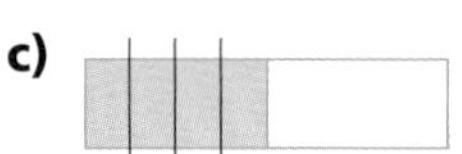

d) 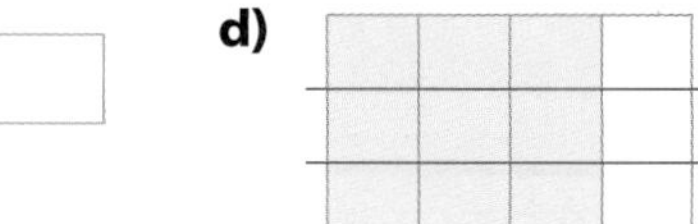

e) 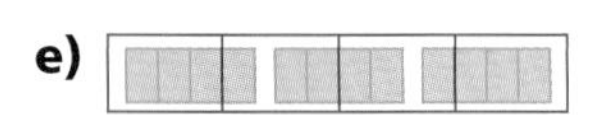

f) 

A $2 \div \frac{1}{2}$ **B** $3 \div \frac{3}{4}$ **C** $\frac{1}{2} \div 4$ **D** $\frac{4}{5} \div 5$ **E** $2 \div \frac{2}{3}$

3. Which diagram in question 2 did not match a division statement? Write a division statement for it and find the quotient.

4. Write a division sentence for each diagram.

a) 0 1 2 3

b) 0 1 2 3 4 5

c) 0 1 2 3

d) 0 1 2 3

e) 0 1 2 3 4 5

5. Use diagrams to help you find each quotient.

a) $6 \div \frac{1}{3}$ **b)** $5 \div \frac{1}{2}$

c) $6 \div \frac{3}{4}$ **d)** $3 \div \frac{2}{3}$

6. Divide.

a) $\frac{1}{2} \div 3$ **b)** $\frac{3}{5} \div 4$

c) $\frac{4}{3} \div 7$ **d)** $\frac{3}{4} \div 8$

7. Divide.

a) $\frac{3}{4} \div \frac{1}{2}$ **b)** $\frac{3}{4} \div \frac{1}{8}$

c) $\frac{4}{7} \div \frac{5}{7}$ **d)** $\frac{2}{3} \div \frac{1}{2}$

8. Divide.

a) $\frac{5}{6} \div \frac{2}{3}$ **b)** $\frac{3}{8} \div \frac{1}{5}$

c) $\frac{3}{5} \div \frac{9}{10}$ **d)** $\frac{3}{4} \div \frac{2}{3}$

9. Divide.

a) $1\frac{5}{6} \div \frac{2}{3}$ **b)** $2\frac{3}{4} \div \frac{2}{5}$

c) $4\frac{2}{3} \div \frac{5}{12}$ **d)** $3\frac{1}{3} \div \frac{1}{2}$

10. Explain to a classmate the importance of using models and pictures to show if a solution to a division statement involving fractions is correct. Provide an example to help you explain.

11. Muriel earned \$300 from mowing lawns over the summer. She wants to spend her earnings on fall clothes. Her favourite store is having a sale.

Clothes and More!

- shoes reg. price \$59.99, $\frac{1}{3}$ off
- jeans reg. price \$69.99, buy one pair, get second pair $\frac{1}{2}$ off
- sweaters reg. price \$35.98, $\frac{1}{4}$ off
- skirts reg. price \$24.99, $\frac{1}{5}$ off
- shirts reg. price \$27.99, buy second one for $\frac{1}{4}$ off, buy a third for $\frac{1}{2}$ off, buy a fourth for $\frac{3}{4}$ off

She wants to buy a pair of shoes, two pairs of jeans, one sweater, one skirt, and four shirts.

a) Estimate how much money she has left after each purchase.

b) Calculate the amount of money she will have left after buying all nine items.

c) Does she have enough to pay for the taxes on her purchases? (HST is approximately 15%.) Explain why or why not.

d) Explain your estimation strategy.

12. Betty swims $5\frac{1}{2}$ laps each day. This morning, she swam $\frac{1}{4}$ of her laps.

a) How many laps has she swum?
b) How many more laps does she have to swim today?
c) How many laps does Betty have to swim in seven days?

13. Fourteen friends share $8\frac{3}{4}$ bags of trail mix equally. What fraction of a bag does each friend get?

14. It takes one cashier $3\frac{1}{2}$ min to check out a customer's groceries at the local grocery store.

a) How many customers can one cashier serve in 1 h?
b) If 4 cashiers are working, how many customers will be served in 1 h?
c) Is the total number of customers accurate? What could change that number?

15. Joel can run one lap around the track in $4\frac{1}{2}$ min.

a) How many laps can he run in 30 min?
b) How many laps can he run in 30 min if he decreases his speed so that he runs one lap in $5\frac{3}{4}$ min?

16. Carlos runs $7\frac{3}{5}$ laps around the school track every morning. This morning, he ran $\frac{2}{3}$ of his laps before it started to rain heavily.

a) How many laps did he run?
b) How many more laps should he run to complete his morning routine?
c) How many laps will Carlos run in six days?

17. Allie assembles bicycles for the Brand Bicycle Company. She takes $\frac{3}{4}$ h to assemble one bicycle.

a) How many bicycles can Allie assemble in a $7\frac{1}{2}$ h work day?
b) If Allie works $37\frac{1}{2}$ h a week, how many bicycles can she assemble?
c) A new model of bicycle takes $\frac{2}{3}$ h to assemble. How many of these new bicycles can Allie assemble in a day?
d) The company would like Allie to assemble 60 of the new bicycles this week. How many hours of overtime will Allie need to work?

18. Dimitry works $12\frac{3}{4}$ h per week at the local library. During the first week of March, he was able to work only $\frac{5}{8}$ of his time due to illness.

a) How many hours did he work?

b) How many more hours should he work to complete his full week's time?

c) How many hours per day will he have to work in the second week of March to make up for time lost during the previous week?

19. Use two methods to find each quotient.

a) $\frac{3}{8} \div \frac{7}{8}$ **b)** $\frac{1}{4} \div \frac{5}{8}$

c) $\frac{4}{5} \div \frac{1}{2}$ **d)** $\frac{2}{5} \div \frac{3}{4}$

Extend

20. You can calculate the approximate diameter of any circle using the following formula.

$$\text{Diameter} = \frac{\text{circumference}}{\frac{22}{7}}$$

Calculate the approximate diameter of each object.

a) a garbage can having a circumference of 330 mm

b) a plate having a circumference of 44 cm

Did You Know?

The word *fraction* comes from the Latin "fractio" which means to break.

Puzzler

This is a tangram.

a) What part of the whole square does each shape represent?

b) If the large triangle represents 1 or the whole, what part of the large right triangle does each shape represent?

Did You Know?

The ancient Chinese (around 100 C.E.) had an interesting way of simplifying proper fractions.

If the numerator and denominator could be halved, then both were halved repeatedly.

Example: $\frac{20}{28} = \frac{10}{14} = \frac{5}{7}$ (÷ 2, ÷ 2 applied to numerator and denominator)

If the numerator and denominator could not be halved then a series of differences were calculated until a common divisor for both the numerator and denominator was found.

Example: $\frac{35}{91}$

91 – 35 = 56
56 – 35 = 21
35 – 21 = 14
21 – 14 = 7
14 – 7 = 7 **Common divisor.**
7 – 7 = 0

So, $\frac{35 \div 7}{91 \div 7} = \frac{5}{13}$.

2.4 Fractions and the Order of Operations

Focus on...

- applying the order of operations to fractions and mixed numbers, using pencil and paper, and a calculator

You often have to correctly answer a skill-testing question to win a contest. Which student will win the contest? How do you choose the order to apply operations when evaluating an expression?

Example 1: Evaluate Expressions With Fractions

Use the **order of operations** to evaluate $\frac{5}{6} - \frac{1}{2} \times \frac{1}{3}$.

order of operation

- correct sequence of calculations for evaluating an expression
  - **B** Brackets
  - **E** Exponents
  - **D**, **M** Division and Multiplication, in order from *left* to *right*
  - **A**, **S** Addition and Subtraction, in order from *left* to *right*

Solution

$\frac{5}{6} - \frac{1}{2} \times \frac{1}{3}$ **Multiply.**

$= \frac{5}{6} - \frac{1 \times 1}{2 \times 3}$

$= \frac{5}{6} - \frac{1}{6}$ **Subtract.**

$= \frac{4}{6}$

$= \frac{2}{3}$

Strategies

Apply procedures for order of operations you learned before.

Example 2: Evaluate Expressions With Fractions and Brackets

Use the order of operations to evaluate $\left(\frac{5}{6} - \frac{1}{2}\right) \times \frac{1}{3}$.

Solution

$\left(\frac{5}{6} - \frac{1}{2}\right) \times \frac{1}{3}$ **Brackets.**

The LCD for $\frac{5}{6}$ and $\frac{1}{2}$ is 6, so $\frac{1}{2}$ can be expressed as $\frac{3}{6}$.

$= \left(\frac{5}{6} - \frac{3}{6}\right) \times \frac{1}{3}$

$= \frac{2}{6} \times \frac{1}{3}$ **Multiply.**

Multiply the numerators. Multiply the denominators.

$= \frac{2}{18}$

$\frac{2}{18}$ can be written in simplest form as $\frac{1}{9}$.

$= \frac{1}{9}$

Did You Know?

It is thought that the order of operations dates back to the 1600s. Mathematicians of the time agreed on the order so they would all get the same answer for the same question.

Communicate the Key Ideas

1. Andre and Linh each found a different answer to the skill-testing question $\frac{1}{3} - \frac{1}{4} \times \frac{1}{2}$.

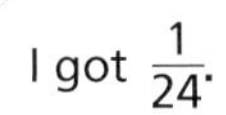

I got $\frac{5}{24}$.

Who is correct? What mistake might the other person have made?

2. Eli, Raquel, and George solve $\frac{2}{3} - \frac{1}{3} + \frac{1}{2}$.
Eli adds $\frac{1}{3}$ and $\frac{1}{2}$ first and gets a final answer of $-\frac{1}{6}$. Raquel subtracts first, and gets a final answer of $\frac{5}{6}$. George adds $-\frac{1}{3}$ and $\frac{1}{2}$ first. What should George's final answer be? What conclusion can you make from this?

3. Explain why the order of operations is necessary.

Check Your Understanding

1. Copy each expression. Circle the step you would do first to evaluate the expression.

a) $\frac{1}{2} - \frac{1}{4} \div \frac{1}{2}$

b) $\frac{3}{4} \times 2 \div \frac{1}{2}$

c) $\frac{8}{9} - \frac{1}{2} + \frac{1}{3}$

d) $\left(\frac{3}{8} + \frac{1}{4}\right) \div \frac{2}{3}$

2. Evaluate. Show your steps.

a) $\frac{2}{3} + \frac{1}{4} - \frac{1}{6}$

b) $\frac{4}{5} - \frac{1}{2} + \frac{7}{10}$

c) $1\frac{2}{3} \times \frac{3}{8} - \frac{1}{12}$

d) $1\frac{9}{12} - \frac{3}{4} \times \frac{2}{5}$

3. Evaluate.

a) $\frac{1}{2} \times \frac{1}{5} + \frac{1}{3} \div \frac{3}{4}$

b) $\frac{5}{6} - \left(\frac{1}{2}\right)^2$

4. Evaluate.

a) $\left(\frac{4}{5} + \frac{2}{3}\right) \times \frac{5}{2}$

b) $\frac{2}{5} \times \left(\frac{7}{8} + \frac{3}{4}\right)$

c) $\left(1\frac{3}{4} - \frac{1}{2}\right) \div \frac{1}{8}$

d) $\frac{7}{8} \div \left(\frac{1}{2} + \frac{1}{4}\right)$

5. Evaluate.

a) $\frac{2}{3} \times \frac{1}{3}\left(\frac{1}{5} + \frac{1}{4}\right) \div \frac{7}{6}$

b) $\frac{5}{9} + \left(\frac{5}{6} - \frac{1}{3}\right) \div \frac{1}{4}$

c) $5 - \left(\frac{2}{9} + \frac{1}{6}\right) \div \frac{7}{6}$

d) $3 + \left(\frac{8}{9} - \frac{2}{3}\right) \div \frac{2}{3}$

6. Evaluate.

a) $\frac{4}{3} + \frac{7}{8} \times \frac{3}{7}$

b) $\left(\frac{3}{5} - \frac{2}{7}\right) \times \frac{3}{4}$

c) $\frac{3}{10} + \frac{2}{5} \div \frac{1}{3}$

d) $\left(\frac{5}{6} + \frac{1}{4}\right) \times \left(\frac{2}{3} + \frac{1}{2}\right)$

7. Insert brackets to make each number sentence true. Justify your answer.

a) $\frac{1}{4} + \frac{1}{3} - \frac{2}{5} \div \frac{1}{11} = 2\frac{1}{60}$

b) $\frac{3}{4} \times \frac{2}{3} - \frac{1}{2} + 4 = 4\frac{1}{8}$

8. Insert <, >, or = to make each number sentence true.

a) $1\frac{1}{2} - \left(\frac{1}{3} \times 3\right)$ ■ $2\frac{3}{4} \div \frac{1}{4}$

b) $\left(2\frac{3}{4} + 1\frac{1}{2}\right) \div \frac{1}{4}$ ■ $\left(3\frac{3}{4} - 2\frac{1}{2}\right) \div \frac{1}{8}$

c) $\left(\frac{5}{8} + \frac{1}{4}\right) \div \frac{1}{2}$ ■ $1\frac{1}{2} \div \frac{9}{10} - \frac{7}{8}$

9. Nathalie's brother made this bowl of popcorn for Nathalie and her three friends, who are watching a movie.

Her brother takes $\frac{1}{4}$ of the popcorn for himself. Nathalie must share the rest of the popcorn equally with her friends. Write and evaluate two different number sentences for this problem. How much of the bowl of popcorn does each person get? Show your work.

10. Copy each equation. Use brackets where necessary to make the equation true. Show how you found your answer.

a) $12 \times \frac{1}{3} + \frac{1}{2} = 10$

b) $\frac{2}{3} - \frac{2}{7} \div \frac{6}{7} = \frac{1}{3}$

c) $\frac{3}{4} \div \frac{5}{6} - \frac{3}{4} \times \frac{2}{3} = 6$

d) $\frac{5}{6} \times \frac{3}{5} + \frac{3}{4} - \frac{1}{3} = \frac{11}{12}$

Chapter Problem

11. Jodie's backyard has a vegetable garden in one corner and a flower garden in the other corner. Each garden takes up $\frac{1}{6}$ of her backyard. She grows tomatoes in $\frac{1}{4}$ of the vegetable garden and roses in $\frac{1}{2}$ of the flower garden.

a) Draw a diagram to represent Jodie's backyard.

b) Write an expression to calculate the fraction of her backyard that Jodie uses to grow tomatoes and roses.

c) Evaluate your expression from part b).

d) How would your expression from part b) change if you wanted to calculate how much more of her backyard Jodie uses for roses than for tomatoes?

12. Is Ravi correct? Explain.

Pierre, $\frac{1}{3} \times \frac{1}{4} \div \frac{1}{5}$ has the same answer as $\frac{1}{3} \div \frac{1}{4} \times \frac{1}{5}$.

13. Carol takes $\frac{1}{3}$ h to prepare an apple pie, $\frac{3}{4}$ h to prepare a lemon meringue pie, and $\frac{3}{5}$ h to prepare a strawberry rhubarb pie. She wants to make 10 of each type of pie for a dinner party that she is catering.

a) Write an expression to determine how many hours it will take Carol to prepare all the pies.

b) Evaluate your expression from part a). Show your work.

c) Once the pies are baked, she cuts each pie into sixths. She saves $\frac{1}{2}$ of a pie of each kind for her assistants and serves the rest. If each dinner guest will receive one piece of pie, how many guests are expected to attend? Write an expression and show how you solved this.

14. a) Make up a skill-testing question that involves fractions and uses up to three operations. Share your question with your class. Keep a copy of your answer.

b) Choose three questions from the class set. Revise each question by adding or removing brackets.

Did You Know?

There is a reason why contest winners have to answer a skill-testing question ... it is a Canadian law! According to Canada's *Criminal Code*, giving away prizes based only on chance is illegal. A contest must involve a skill, such as answering a question. Then, the contest is not considered a game of chance.

15. You get a job installing tires on cars in an assembly line. You get paid \$5.00 for each set of tires you install. It takes you $\frac{1}{4}$ h to put on a set of tires. If you worked for $8\frac{1}{2}$ h, how much would you expect to make?

16. You decide to make school banners for your friends. You find $5\frac{1}{2}$ yd of cloth in the closet. Each banner requires $\frac{3}{4}$ yd of cloth. How many banners can you make? How much material will you have left over, if any?

Extend

17. Evaluate the following expressions, if

$$a = \frac{2}{3}, b = \frac{1}{2}, \text{ and } c = \frac{3}{4}.$$

a) $a \times (b + c)$

b) $a \times b \times c$

c) $b \div c \times a$

18. The letters and numbers typed using a computer take up different amounts of space. The following are the guidelines:

- I, l, 1, and punctuation use $\frac{1}{2}$ space
- m, w, ×, =, +, −, and ÷ use $1\frac{1}{2}$ spaces
- all other numbers and lowercase letters use 1 space
- M and W use $2\frac{1}{2}$ spaces
- all other uppercase letters use 2 spaces
- the space between two words uses 1 space

a) How many spaces would your full name occupy?

b) How many spaces would the sentence, "How are you?" occupy?

c) Write a sentence that would use between 20 and 25 spaces.

19. The formula to convert a temperature in degrees Fahrenheit, *F*, to degrees Celsius, C, is:

$$C = \frac{5}{9}(F - 32)$$

Convert these temperatures to degrees Celsius.

a) Orlando, Florida: 92°F

b) Detroit, Michigan: 75°F

c) Anchorage, Alaska: 25°F

Puzzler

A man drove from Whoville to Beavertown. On day one he travelled $\frac{1}{3}$ of the distance. On day two he travelled $\frac{1}{2}$ of the remaining distance. On day three he travelled $\frac{2}{3}$ of the remaining distance. On day four, after covering $\frac{3}{4}$ of the remaining distance, he was still 6 km away from Beavertown. How many kilometres had he covered so far?

Puzzler

What fraction of the rectangle is shaded yellow?

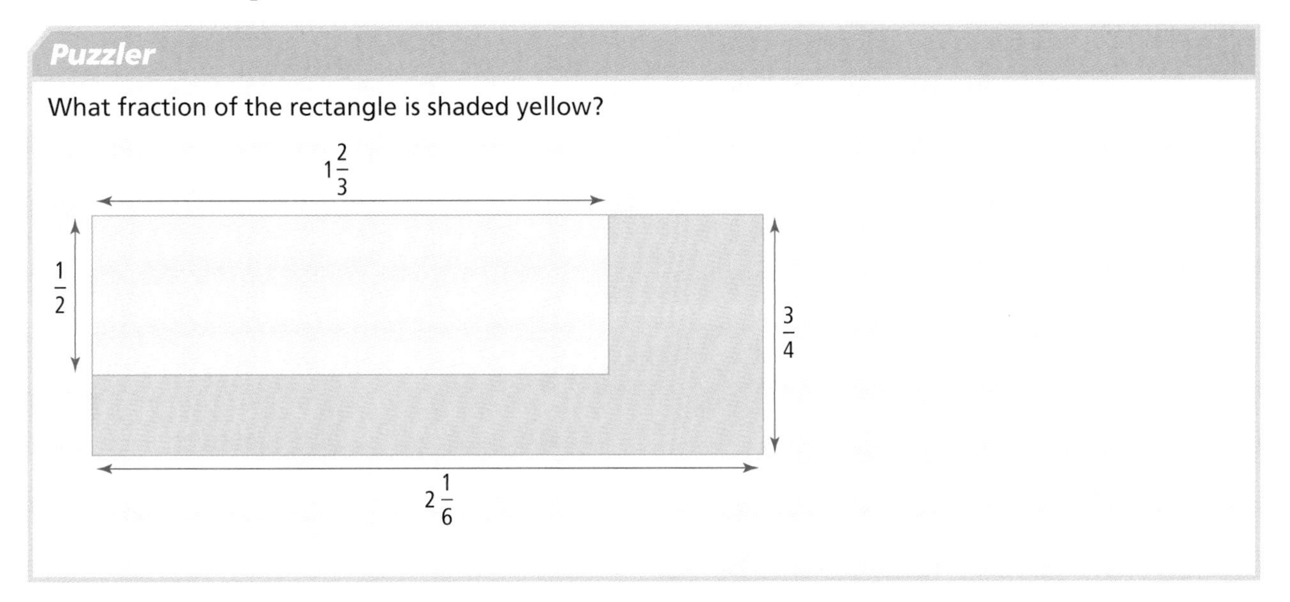

CHAPTER 2

Review

1. Evaluate. Use mental math when appropriate.

   a) $\frac{1}{5} + \frac{2}{5}$ b) $2 - \frac{1}{3}$

   c) $3\frac{9}{10} + 2\frac{3}{10}$ d) $\frac{4}{5} - \frac{1}{10}$

2. For question 1, did you use mental math to add or subtract the fractions? Explain how you made your decision. For which addition statement did you use the Make "1" strategy? Show your work using this strategy.

3. Evaluate. Use a model to help you.

   a) $\frac{2}{3} + \frac{1}{4}$ b) $\frac{7}{8} - \frac{1}{2}$

   c) $2\frac{5}{6} - \frac{3}{4}$ d) $\frac{1}{3} + 3\frac{5}{6}$

4. Ronnie had to model $\frac{3}{5} + \frac{1}{2}$. He recorded his models and presented this solution.

$\frac{3}{5} + \frac{1}{2}$

$= \frac{4}{7}$

Is Ronnie correct? Justify your answer.

5. Write the multiplication expression for each picture, and then find the product.

   a) 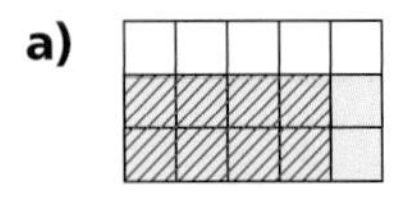

   b)  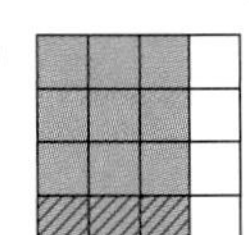

   c) 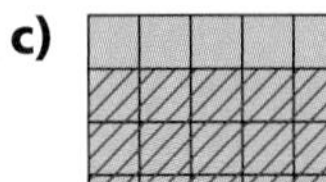

6. Evaluate. Use mental math.

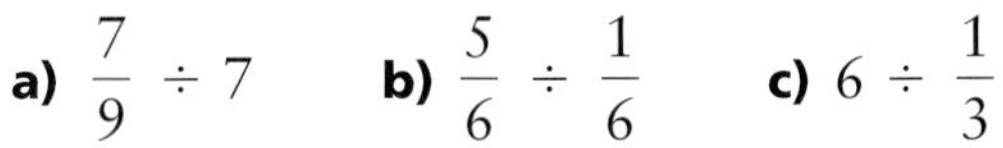

   a) $\frac{7}{9} \div 7$ b) $\frac{5}{6} \div \frac{1}{6}$ c) $6 \div \frac{1}{3}$

7. Evaluate.

   a) $\frac{3}{5} \times \frac{1}{2}$ b) $\frac{2}{3} \times \frac{7}{9}$

   c) $\frac{4}{6} \times \frac{1}{7}$ d) $1\frac{1}{4} \times \frac{2}{3}$

8. Write a division sentence for each diagram.

   a) 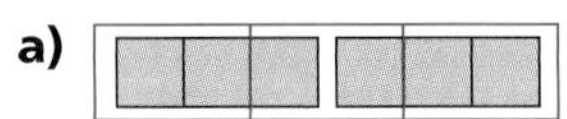

   b) 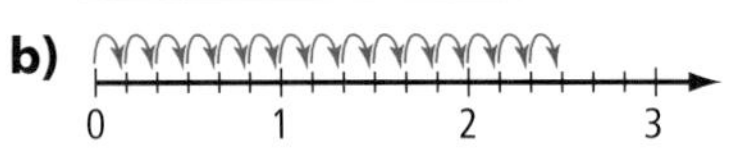

   c) 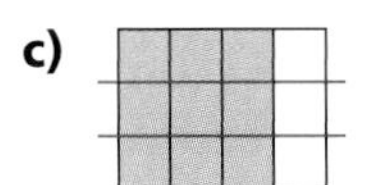

   d) 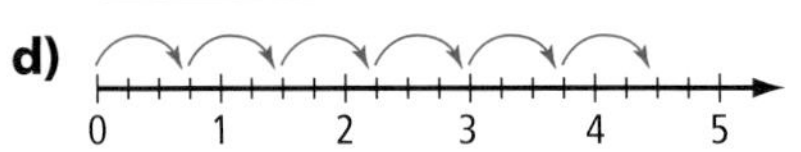

9. Evaluate. Show your work and express your answer in simplest form.

   a) $\frac{3}{4} + \frac{5}{8}$ b) $2\frac{7}{9} + \frac{1}{3}$

   c) $\frac{7}{8} - \frac{2}{3}$ d) $3\frac{1}{2} - 2\frac{5}{6}$

   e) $\frac{4}{9} \times 2\frac{2}{3}$ f) $5 \times \frac{3}{8}$

   g) $\frac{3}{4} \div \frac{1}{5}$ h) $2\frac{3}{4} \div 1\frac{1}{3}$

10. Write a mathematical sentence involving two fractions, with different denominators, whose

    a) sum is $\frac{3}{4}$ b) quotient is $\frac{5}{4}$

    c) difference is $\frac{1}{3}$ d) product is $\frac{1}{2}$

11. Evaluate.

a) $\frac{1}{3} \times \left(\frac{2}{5} + \frac{3}{4}\right)$

b) $\left(\frac{5}{7} - \frac{3}{5}\right) \div \frac{1}{2} \times \frac{7}{9}$

c) $\left(\frac{5}{8} + \frac{1}{4}\right) \times \left(\frac{2}{3} + \frac{6}{7}\right)$

12. Copy each equation. Fill in the ■ to make it true.

a) $■ \times \frac{3}{4} = \frac{1}{4}$

b) $\frac{3}{16} \div ■ = \frac{3}{4}$

13. Write a story problem for each multiplication statement.

a) $\frac{1}{2} \times 8$ **b)** $8 \times \frac{1}{2}$

14. Queen Elizabeth Junior High held an election for student council. A total of 360 students voted. The table shows the results.

Candidate	Fraction of Total Votes
Libby	$\frac{1}{4}$
Noah	$\frac{2}{5}$
Tansey	$\frac{7}{20}$

How many votes did each candidate receive?

15. Copy each equation. Add brackets to make it true.

a) $\frac{5}{6} \times \frac{3}{5} + \frac{1}{2} - \frac{11}{12} = 0$

b) $\frac{1}{4} + \frac{2}{5} - \frac{3}{10} \times 2 = \frac{7}{10}$

16. Aden bought a new stereo system and will pay for it in installments.

It cost \$399 including taxes. He made a down payment of $\frac{1}{3}$ of the total cost and must pay the rest in 4 equal payments. How much is each payment?

17. Precious Stone Mining stock went up in value by $\frac{3}{8}$ units every day until it reached a total increase of $4\frac{1}{2}$ units. Over how many days did the increase occur?

18. Clearances Galore wants to clear its holiday inventory. Every item in the store will be reduced by $\frac{1}{3}$ each week for one month. For example, a \$100 item will cost approximately \$70 after one week, approximately \$50 the next, and so on. If a personal stereo is priced at \$149, estimate the sale price after one month.

CHAPTER 2 Practice Test

Selected Response

Select the best answer.

1. Which subtraction sentence does the model represent?

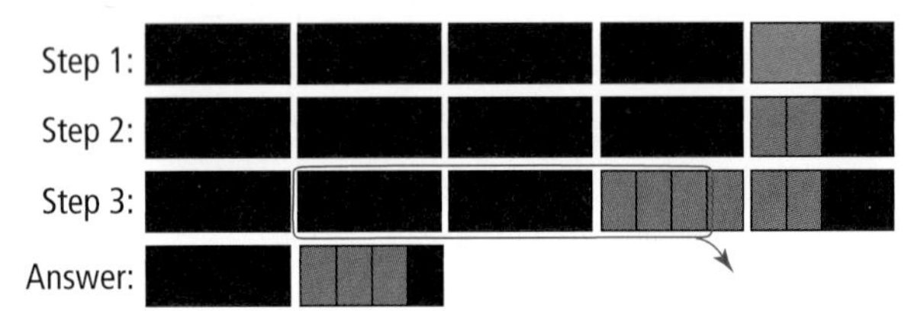

A $5 - 2\frac{1}{2}$

B $4\frac{1}{2} - 2\frac{1}{2}$

C $4\frac{1}{2} - 2\frac{3}{4}$

D $5 - 2\frac{3}{4}$

2. Which number line represents $3\frac{1}{2} \div \frac{3}{4}$?

A

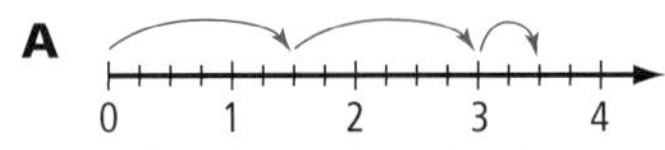

B

C

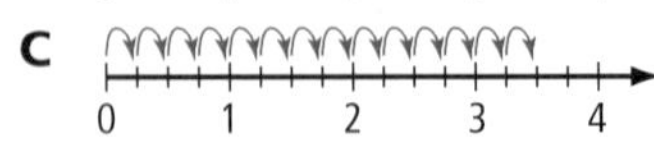

D

3. Which grid represents $\frac{2}{3} \times \frac{3}{6}$?

A

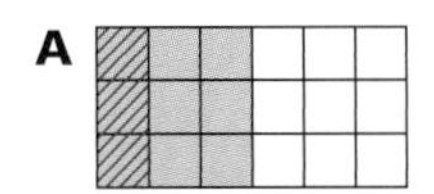

B

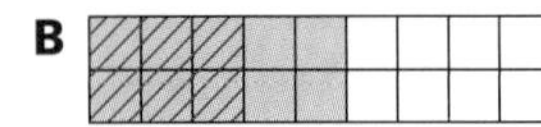

C

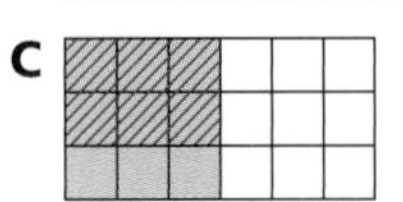

D

4. Which fraction would make $\frac{3}{5} \times \left(\frac{■}{■} + \frac{2}{3}\right) = \frac{7}{10}$ true?

A $\frac{1}{2}$ **B** $\frac{1}{3}$ **C** $\frac{3}{4}$ **D** $\frac{6}{7}$

Short Response

Show all of your steps.

5. Evaluate. Use mental math.

a) $2\frac{1}{4} + \frac{3}{4}$

b) $1\frac{1}{3} - \frac{2}{3}$

c) $5\frac{8}{9} + 2\frac{4}{9}$

d) $2\frac{2}{6} + \frac{1}{2}$

6. Jordan spent $\frac{3}{4}$ h planting flowers in his garden. It took him $\frac{1}{8}$ h to plant each flower. How many flowers did he plant? Use pictures in your answer.

7. Catherine used $1\frac{7}{10}$ tubes of red paint and $2\frac{3}{8}$ tubes of green paint for her art project.

a) How much more green paint than red paint did she use?

b) How much paint did she use in total?

8. Andy takes $4\frac{2}{3}$ min to type one page using a computer.

a) Estimate how many minutes Andy will take to type 10 pages.

b) How many pages could Andy type in 18 min?

9. Create a story problem for each fraction sentence. Calculate the answer. Check, using a model or pictures. Show your work.

a) $\frac{5}{8} - \frac{1}{4}$ **b)** $3\frac{5}{6} + \frac{1}{4}$

c) $1\frac{3}{5} \times \frac{1}{2}$ **d)** $2\frac{3}{4} \div \frac{1}{8}$

10. Evaluate.

a) $\left(\frac{3}{4} + \frac{2}{3}\right) \div 1\frac{1}{4}$ **b)** $\left(\frac{1}{2}\right)^2 - \frac{1}{4} \times \frac{2}{5}$

Extended Response

Provide a complete solution.

11. Over the summer, Alec worked at a grocery store $9\frac{1}{5}$ h per day, 4 days a week. He also spent $3\frac{1}{2}$ h per day mowing lawns 2 days a week.

a) How many hours a week did Alec work?

b) What would be Alec's total weekly salary if he earns \$11/h at the grocery store and \$7.50/h mowing lawns? Alec must pay \$105/week in taxes on his earnings from the grocery store.

c) Alec needs to save $\frac{3}{4}$ of his salary after taxes for university. How much will he save over the 8 weeks of summer?

d) If Alec gives half of the remaining $\frac{1}{4}$ of his weekly salary to charity, how much will he have left over?

Chapter Problem Wrap-Up

1. The grade 8 classes are asked to suggest a design for the school flower garden. Your class is responsible for $\frac{1}{4}$ of the entire garden. This portion must contain four different colours of flowers.

a) Create a design for your class's part of the garden.

b) What fraction of the entire garden is each colour? Show how you know.

2. Create a design for your own garden. Use at least four different types of flowers or vegetables. Write as many fraction statements as you can about your garden. Use each of the four operations (+, −, ×, ÷) at least once in each statement. Use pictures, words, and symbols in your report.

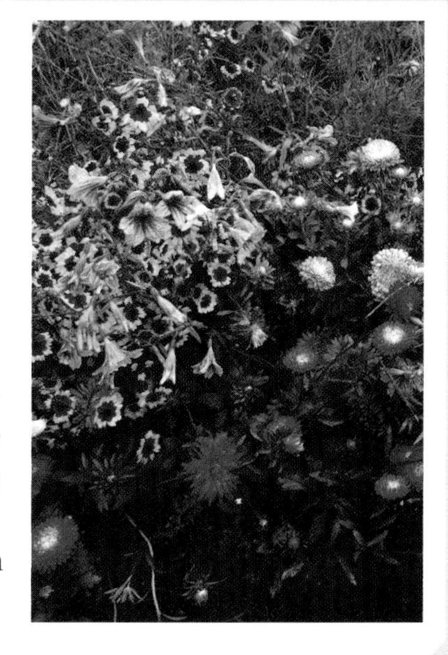

Making Connections

Game Boards

All over the world, since ancient times, games have been played on square boards. The simplest is Xs and Os, played on a 3 by 3 grid.

One of the more complex is Go (Chinese Wei-Qi) in which tiles are used to capture regions of the 18 by 18 grid.

Variations of games have developed over time in different parts of the world.

North American checkers, played on an 8 by 8 grid, is essentially the same as Polish draughts, which is played on a 10 by 10 grid.

North American chess uses an 8 by 8 grid. The Japanese have three variations: Shogi uses a 9 by 9 grid, Chu Shogi uses a 12 by 12 grid, and Tai Shogi is played on 25 by 25 grid.

Snakes and Ladders is played on a 10 by 10 grid, and Scrabble® on a 13 by 13 grid.

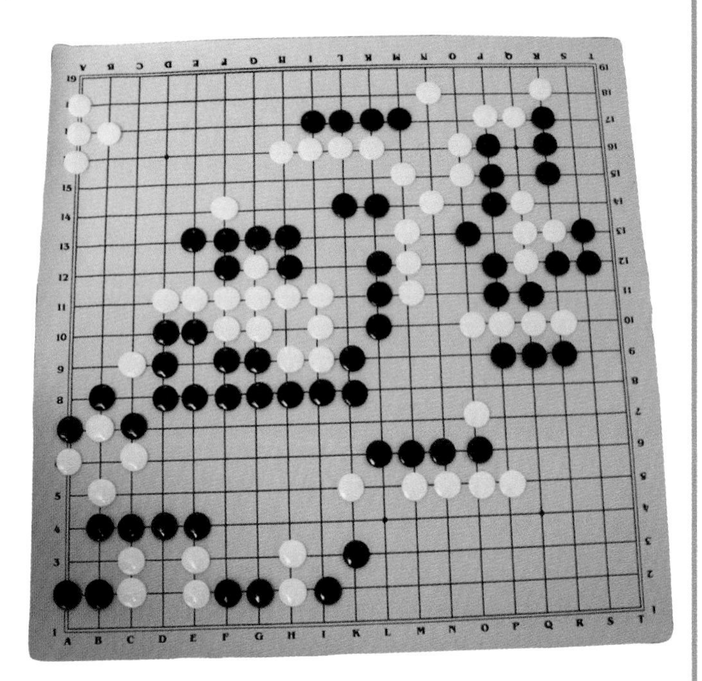

1. What is the diagonal distance across a Go board if each of its small squares has a side length of 2 cm?

2. Research to find other games that are played on square boards. Write a brief description of each game for display on a bulletin board.

Making Connections

What's math got to do with drain access covers?

Have you ever thought about why the metal covers on drains are round? Why are none of them square or rectangular?

Pythagoras can help provide the answer. If the cover were square or rectangular, then the diagonal length across the hole would be longer than the sides. A worker could accidentally drop the cover through the hole. A circular cover cannot fall through.

Write a paragraph, or prepare a demonstration with visual aids, to explain this to a younger student.

Design a Park

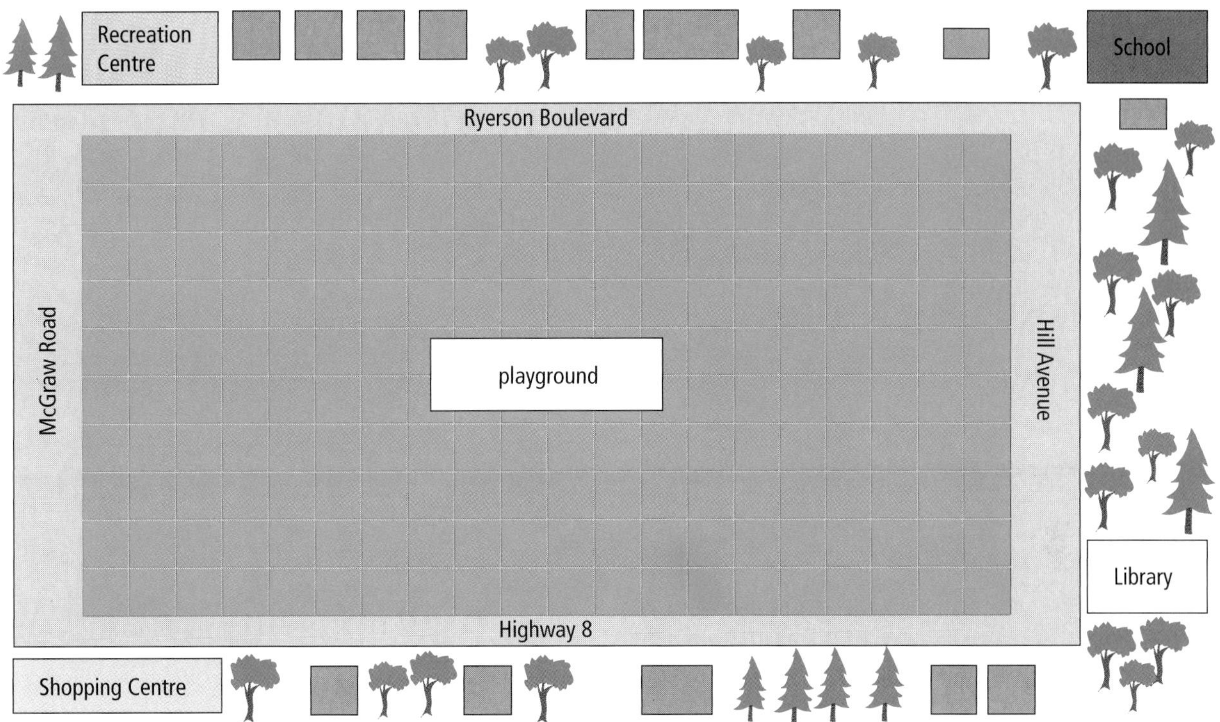

> **WANTED**
>
> Innovative design for playground with square wading pool and sports field.
>
> **Requirements:**
> - to be located in the green space
> - each feature should cover between one tenth and one fifth the area of the green space
> - features can cover no more than one third of the area of the green space
> - wading pool to cover no more than one quarter of the area of the playground
> - features must be 50 m from any road
> - each feature must be accessible by a diagonal path to at least two streets
> - there must be a path from the playground to the sports field

The map shows a large park in a community.

The town has advertised for a design for two features in this park: a sports field and a playground with a square wading pool. The newspaper advertisement shows the requirements.

Choose a location and size for the park features. Then develop a design for presentation to the town council. Include all measurements. Write a report that explains how your design meets all of the requirements.

- understand and apply the minimum sufficient conditions for producing unique triangles
- understand and apply the minimum sufficient conditions to prove triangles are congruent
- understand and apply the properties of transformations to make 2-D constructions
- understand and apply the properties of regular polygons

Key Words

unique
congruent
rotation
translation
reflection
orientation
transversal
regular polygon
exterior angle

CHAPTER 3

Geometry I

Designing and sewing quilts provide an artistic outlet for many people. A complex design can take days to complete but produces a quilt that is a work of art. Many quilt designs incorporate various types and sizes of polygons that are reflected and rotated within the quilt square to create symmetrical designs. Then, the squares can be replicated and attached side-by-side, forming translated images of one another. In this chapter, you will explore the properties of shapes and transformations. These properties are essential knowledge for anyone using geometric figures in design and manufacturing.

Internet Connect To learn more about the history of quilting in Nova Scotia, go to **www.mcgrawhill.ca/links/math8NS** and follow the links.

Chapter Problem

Examine the quilt square. What shapes do you see? How did the designer create a design that appears three-dimensional? At first glance the image looks symmetrical; however, it is not. What did the designer do so that the design is not symmetrical?

In this chapter you will learn how to create congruent triangles and regular polygons, and to transform shapes using translations, reflections, and rotations. Then you will use your knowledge to design your own quilt pattern.

Get Ready

Regular Polygons

A **regular polygon** has all sides congruent and all angles congruent. These are regular polygons.

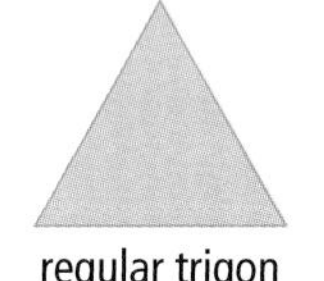
regular trigon (equilateral triangle)

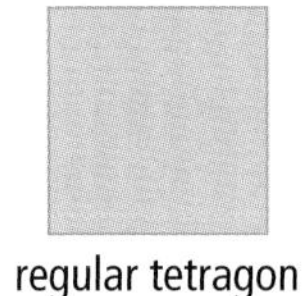
regular tetragon (square)

regular pentagon

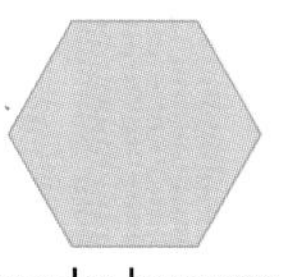
regular hexagon

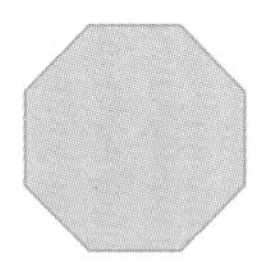
regular octagon

They can also be named using the number of sides and the suffix "-gon". For example, another name for a regular hexagon is regular 6-gon. Regular polygons have both **reflective** and **rotational symmetry**. For example, an equilateral triangle has three lines of reflective symmetry and rotational symmetry of order 3.

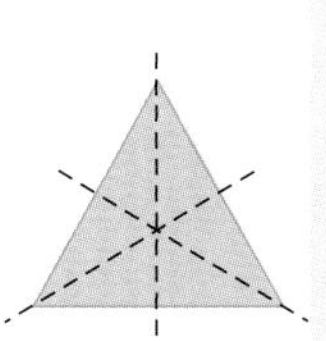

1. Copy and complete the table for regular polygons with 4, 5, 6, 8, and 12 sides.

Number of Sides in Regular Polygon	Drawing	Common Name(s)	*n*-gon Name
3		triangle	3-gon

2. **a)** Refer to the table in question 1. Draw the lines of reflective symmetry for each regular polygon.

b) Describe the reflective symmetry and the rotational symmetry of each polygon. What patterns do you see?

Congruent Polygons

Congruent figures have the same shape and size. All corresponding sides are equal in length and all corresponding angles are equal in measure.

These polygons are congruent.

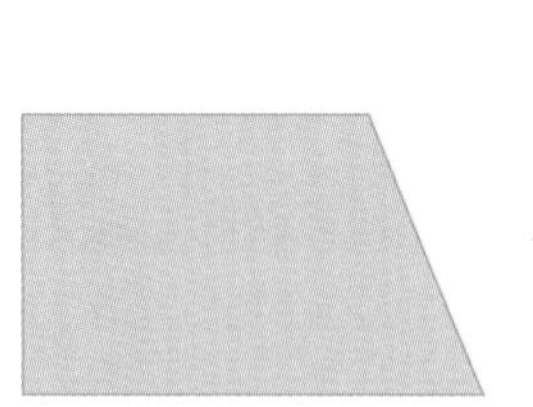

3. Which pairs of figures are congruent?

a)

b)
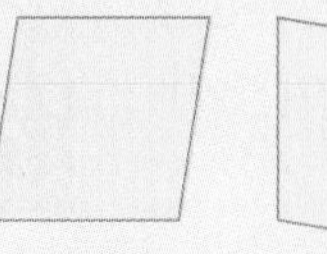

c)
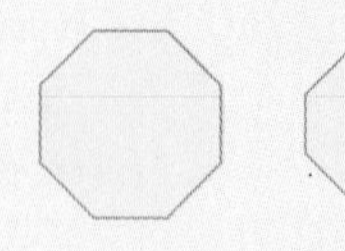

d)
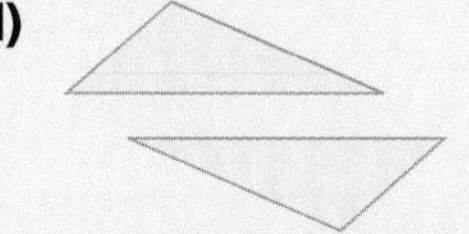

Transformations

Translations, reflections, and rotations of a polygon all result in an **image** polygon that is congruent to the original or **pre-image**. The vertices of the image are usually labelled with the same letters as the corresponding vertices of the pre-image. The addition of symbol ′ (pronounced prime) is used to indicate the relationships with the corresponding vertices.
△A′B′C′ is a translation of △ABC.
△A″B″C″ is a reflection in the x-axis of △ABC.
△A‴B‴C‴ is a 90° rotation about (0, 0) of △ABC.

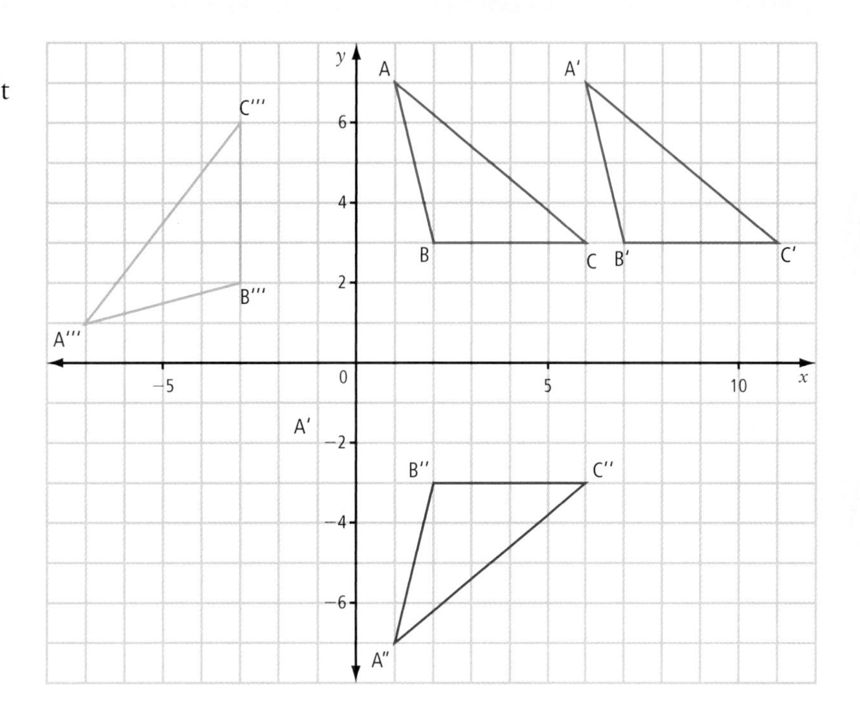

For questions 4 to 6, draw your answers on the same set of axes.

4. On grid paper, draw △RST with R at (−4, 1), S at (−1, 2), and T at (−3, 4). Draw the image of △RST translated 5 units right and 4 units up. Label the image using prime symbols. List the pairs of corresponding congruent sides and angles of the triangles.

5. Draw the image of △RST reflected in the x-axis. Label this image using double prime symbols (″). List the pairs of corresponding congruent sides and angles of △RST and △R″S″T″.

6. Draw △DEF with D at (4, −1), E at (1, −2), and F at (3, −4). △DEF is a transformation of △RST. Describe this transformation.

3.1 Unique Triangles

Focus on...
- making informal deductions about the minimum sufficient conditions to guarantee the uniqueness of a triangle

How much information do you need to give your classmates so that each person draws the same triangle?

Do they need to know the lengths of all three sides and the measures of all three angles? How much information is needed so that the triangle is **unique**? You will discover that triangles are special polygons and that there is a minimum amount of information needed to draw a unique triangle.

unique (triangle)
- a triangle that is one-of-a-kind
- any other triangle constructed using the same specific set of instructions will be **congruent** to the original triangle

congruent figures
- two figures are said to be congruent if they have the same shape and size
- all corresponding sides are equal in length and all corresponding angles are equal in measure

Discover the Math

What do you need to know to draw a unique triangle?

Work with a partner. Construct your triangles without your partner watching.

Materials
- Geostrips®
- Bullseye compass
- protractor
- pattern block triangle paper

Construct a Triangle Given the Measure of Three Sides

1. You each need an identical set of three Geostrips® of different lengths.

a) Use the strips to construct a triangle. Draw the triangle by tracing the inside of your model.

b) Compare your triangle to your partner's. Are they congruent? If they are, is it possible to use the strips to construct a triangle that is not congruent to your triangle? Explain.

c) Repeat parts a) and b) using a different set of three Geostrips®.

2. You each must construct a 7-cm line segment. Label one line segment AB and the other BA.
   a) At point A, use a Bullseye to draw an arc with radius 5 cm. At point B, draw an arc with radius 6 cm. Label the point of intersection of the arcs point C. Use straight line segments to join points A and C, and points B and C.
   b) Measure and record the side lengths and angle measures of your triangle.
   c) Compare your triangle to your partner's. Are they congruent? Are the angle measures the same, even though no angle measures were given?

3. **Reflect** If you know all three side lengths of a triangle, will all other triangles constructed with the same side lengths be congruent to the original (meaning the original triangle is unique)? Explain.

Construct a Triangle Given the Measure of Two Angles

4. a) Construct a triangle with two angles measuring 35° and 65°. What is the measure of the third angle? Explain how you know.
   b) Compare your triangle to your partner's. Are they congruent? If they are, is it possible to construct a triangle with the given angle measures that is not congruent to your triangle? Explain.

5. **Reflect**
   a) If you know the measure of two angles in a triangle, how can you determine the measure of the third angle?
   b) If you know the measure of all three angles in a triangle, will you have a unique triangle? Explain.

Construct a Triangle Given the Measure of Two Sides and an Angle

6. You each need an R1 and an R2 Geostrip®.
   a) Join the two strips to form a 60° angle. Use any other strip to complete the triangle and check that the angle you made is still 60°. Which strip did you use for the third side? Draw the triangle by tracing the inside of the model. Mark the 60° angle. This is called the included angle because it is between the two sides.

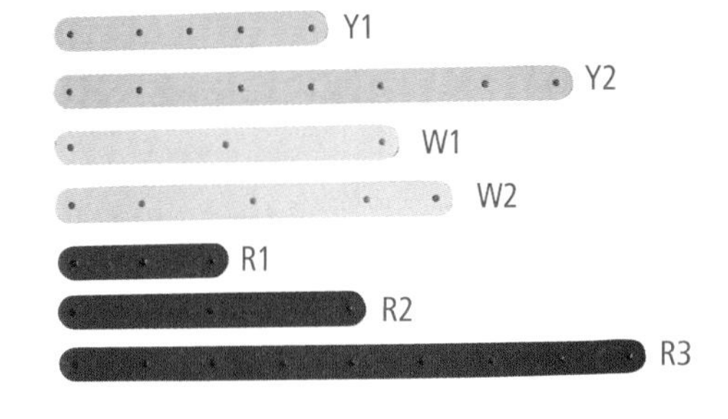

   b) Compare your triangle to your partner's. Did you both use the same Geostrip® to complete the triangle? If so, which one?
   c) Repeat parts a) and b) using a Y1 and a W2 Geostrip® and an angle of 130°.

7. a) Write a definition of *congruent* and *unique* in your own words.
b) A triangle has side lengths of 5 cm and 8 cm and the measure of the included angle is 55°. Suppose every person in your class drew a triangle with these measures. Discuss with your partner whether or not all the triangles would be congruent. Explain your reasoning. If you think the triangles would be congruent, would they be unique? Explain, using your definitions from part a).

8. Fred and Gurda both drew $\triangle DEF$ with $\angle D = 32°$, DE = 5 cm, and EF = 3 cm. Are the triangles congruent? How are the triangles alike? How are they different?

Fred's Triangle

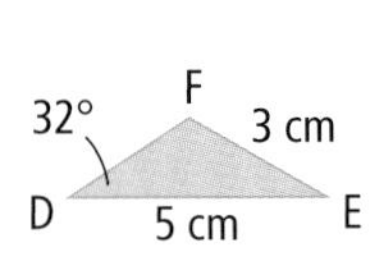

Gurda's Triangle

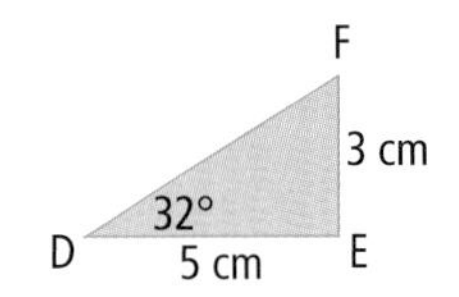

Communicating Mathematically

The angle between two known sides is called the **included** or **contained angle**.

9. Reflect

a) If you know two side lengths of a triangle and the measure of the angle between the two sides, do you have a unique triangle? Explain.
b) Is the same true if the known angle is not between the two sides? Explain.

Construct a Triangle Given The Measures of Two Angles and a Side

10. You each must construct an 8-cm line segment. Label one line segment PQ and the other QP.

a) At point P, construct a 40° angle and extend the arm of the angle. At point Q, construct a 60° angle and extend the arm of the angle. Label the point of intersection of the two arms point R. Record the measures of PQ, ∠P, and ∠Q on your triangle.
b) Compare your triangle to your partner's. Are they congruent? Are the measures of PR and QR the same for both triangles?

11. Reflect If you know the measure of two angles and the side length between the two angles, do you have a unique triangle? Explain.

12. Reflect There are six measures in a triangle: three sides and three angles. Knowing a combination of three measures can produce a unique triangle. Describe the combinations of three measures that will produce a unique triangle and the combinations that will not. Use a table to organize your work. Provide diagrams to help you explain.

Example 1: Is a Triangle Unique?

Is each triangle unique or not unique? If it is not unique, draw a different triangle that has the same measurements as those marked on the original.

a)

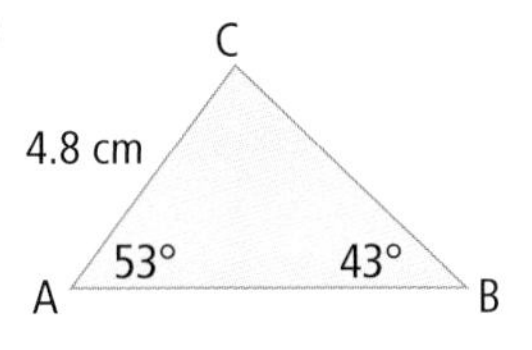

b)

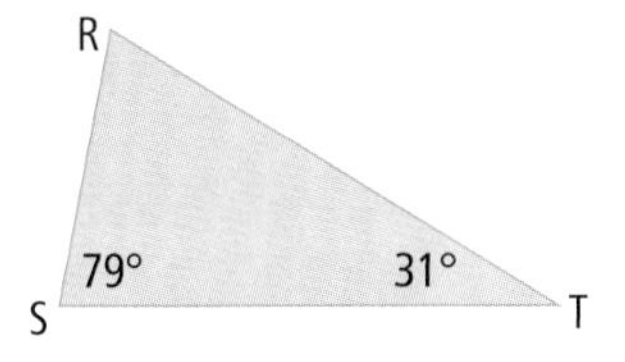

Solution

a) For △ABC, the length of one side and the measure of two angles are given. This combination of measurements produces one and only one triangle. Therefore, the triangle is unique.

b) For △RST, only the measures of two angles are given. The measure of the third angle is 70°. There are many **similar** triangles that you can draw that will have these same angle measurements but different side lengths. Here is an example.
So, this triangle is not unique.

J
79° 31°
K L

Strategies
Use what you learned previously.

The sum of the interior angles of a triangle is 180°. I can find the measure of the third angle if I know two angle measures.

similar figures
- figures that have the same shape but may have different sizes

Example 2: Describe Unique Triangles

What additional information will you need to produce a unique triangle?

a) △RST with RS = 5 cm and ∠R = 64°

b) △JKL with ∠K = 53° and ∠L = 86°

Solution

a) *Method 1: Use the measures of two sides and an included angle.*

If the length of RT were known, △RST would be unique because the measures of two sides and the included angle would be known.

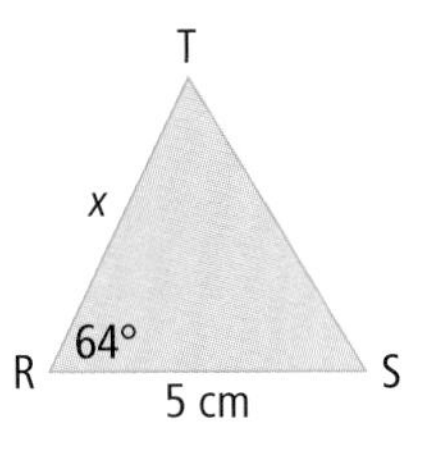

Method 2: Use the measures of two angles and an included side.

If the measures of ∠S were known, △RST would be unique because the measures of two angles and the included side would be known.

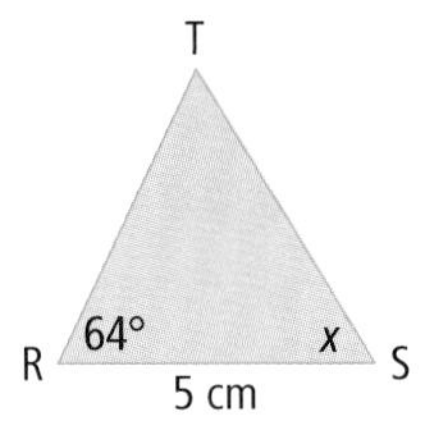

Method 3: Use the measure of one side and two angles.

If the measure of ∠T were known, △RST would be unique because the measures of one side and two angles would be known.

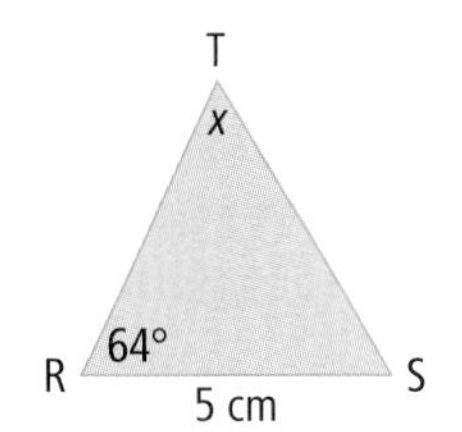

I could find the measure of ∠S by subtracting the angle measures from 180°.

Strategies
Draw a diagram.

b) If the length of any one of the three sides were known, $\triangle JKL$ would be unique because the measures of one side and two angles guarantee a unique triangle.

Communicate the Key Ideas

1. Who is correct? Explain where and why the other person went wrong in their thinking.

Justina, I drew a triangle that has side lengths 6 cm, 7 cm, and 8 cm. My triangle is unique.

Sorry, Ravi. I can draw the same triangle. So, your triangle is not unique.

2. Are these triangles unique? Explain.

a) $\triangle DEF$ with $\angle D = 42°$, $\angle E = 75°$, and $DE = 12$ cm

b) $\triangle LMN$ with $LM = 8$ cm, $MN = 9$ cm, and $\angle L = 37°$

c) $\triangle ABC$ with $\angle A = 36°$, $\angle B = 47°$, and $\angle C = 97°$

Check Your Understanding

1. Carla drew this triangle. She says that it is unique. Is she correct? Explain.

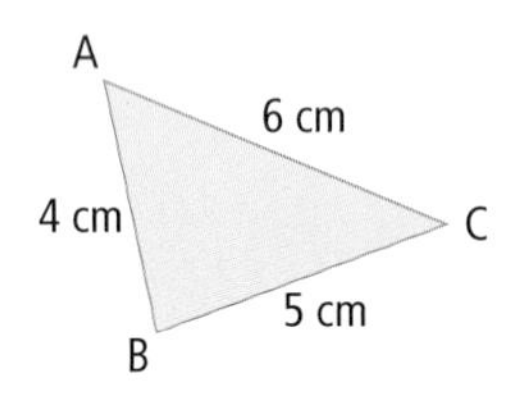

2. Draw $\triangle DEF$ with $DF = 8$ cm, $\angle D = 35°$, and $\angle F = 95°$. Is your triangle unique? Explain. Compare your triangle with several classmates to check.

3. When asked to construct $\triangle PQR$ with $PQ = 12$ cm, $QR = 15$ cm, and $\angle P = 27°$, Gina and Jackie drew congruent triangles. Are there any other possible triangles with these measurements that are not congruent to the ones that were drawn? Explain.

4. Show all the unique triangles you can make from $\triangle LMN$ by adding one more measurement. Explain why each triangle is unique.

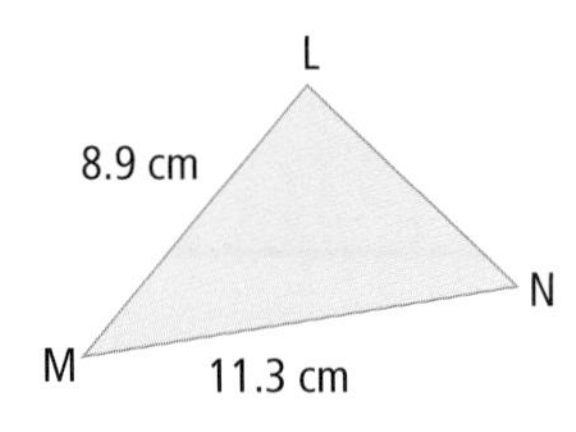

5. Explain how this could be true.

I have drawn two triangles. Both have a side length of 8.5 cm and two angles of 46° and 65°, but they look different.

6. Determine if each triangle is unique. Explain how you know.

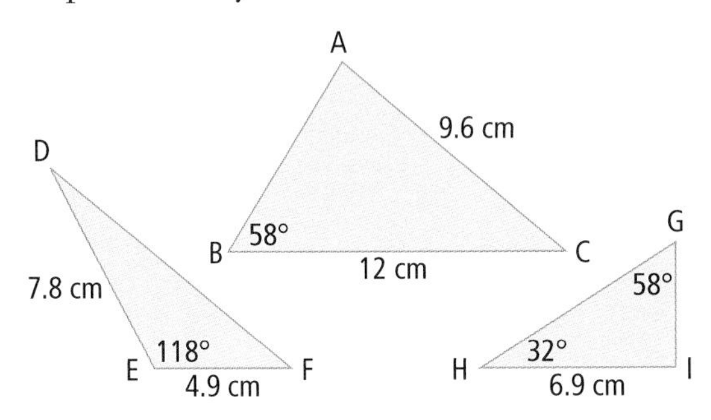

7. △STU has ∠S = 90°, ST = 5 cm, and TU = 13 cm. Sue Ellen says, "If ∠S were not 90°, I would not be sure that △STU was unique, but because ∠S = 90°, I am guaranteed △STU is unique." Explain what Sue Ellen means. Why does this 90° angle make the triangle a special case?

8. The Parks Department wants a triangular flowerbed planted. Two of the sides of the flowerbed should be 20 m and the third side should be 16 m. Could you use ropes and pegs to lay out a design that meets the specifications? Explain how you know.

9. If you know the lengths of the four sides of a quadrilateral, will you have a unique quadrilateral? Experiment by building quadrilaterals from four Geostrips® of different lengths. Explain your results. Use what you know about unique triangles.

10. Determine if each triangle is unique. Explain how you know. Include diagrams.

a) △ABC where AB = 4.2 cm, BC = 6.4 cm, and AC = 3.2 cm.
b) △DEF where ∠D = 60° and ∠E = 110°.
c) △JKL where ∠J = 90°, JK = 4 cm, and KL = 5 m.
d) △QRS where ∠R = 45°, ∠Q = 74°, and QS = 4.5 mm.
e) △XYZ where ∠X = 20°, XY = 6.0 m, and YZ = 3.1 m.

Chapter Problem

11. This Pineburr quilt square uses triangles to form a pattern.

How many different triangles can you see? How many of each type of triangle are there? Draw an example of each type of triangle.

Extend

12.

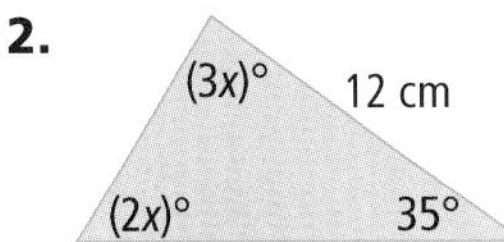

My triangle is unique.

No, it's not because x is an unknown number.

It doesn't make any difference because the value of x is actually known.

Who is correct, Andre or Alene? Explain your reasoning.

3.2 Prove Triangles are Congruent

Focus on...
- learning the minimum sufficient conditions to prove triangles congruent
- applying congruency of triangles

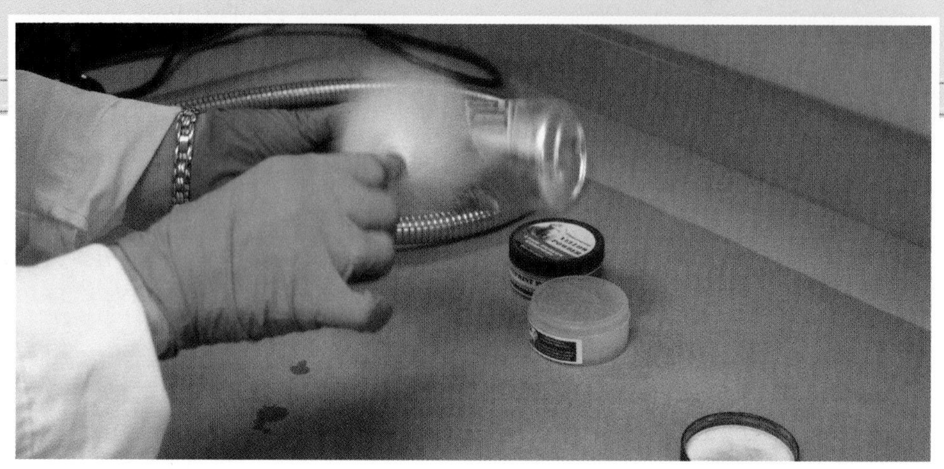

Forensic scientists, police detectives, and prosecutors all gather evidence, and combine this evidence with their prior experiences to make logical arguments to prove their cases.

As a mathematician, you need to learn to organize your thoughts in a logical order to convince others why you know something is true. How could you convince someone that a triangle with two 60° angles is equilateral?

In previous grades, I learned that the sum of the interior angles of any triangle is 180°. So, if I know that two angles of a triangle are 60°, I can logically reason that the third angle is also 60°. Then I know this triangle is equilateral because having three 60° angles is a property of equilateral triangles.

So, how can you use what you know about unique triangles to prove two triangles are congruent?

Did You Know?

Euclid was a Greek mathematician best known for his work on geometry published in *The Elements*. This work has influenced the development of Western mathematics for more than 2000 years.

Discover the Math

Materials
- rulers
- protractors
- Bullseye compass

Part A: Use Reasoning to Convince Others

When you know one relationship, you can often use what you know to find another relationship.
Work with a partner. For questions 1 and 2, take turns convincing each other. Draw diagrams to help you.

1. a) If you know that all three sides of $\triangle DEF$ are 7 cm long, then how could you use this information to convince your partner that $\angle D = 60°$?

b) If you know that in $\triangle ABC$, $\angle A = 67°$ and $\angle C = 23°$, then how could you use this information to convince your partner $AC^2 = AB^2 + BC^2$?

2. a) If you know that two straight lines intersect and one of the angles formed at the intersection is 60°, then how could you use this information to find the measures of the other three angles? Explain how you found your answers.
   b) If you know that a quadrilateral has four lines of reflective symmetry, then how could you use this information to determine what else you know about the quadrilateral? Explain.

3. If you know that △DEF and △ABC are congruent, then how could you use this information to find the measures of DE, EF, and DF? of ∠D, ∠E, and ∠F? Explain how you found your answers.

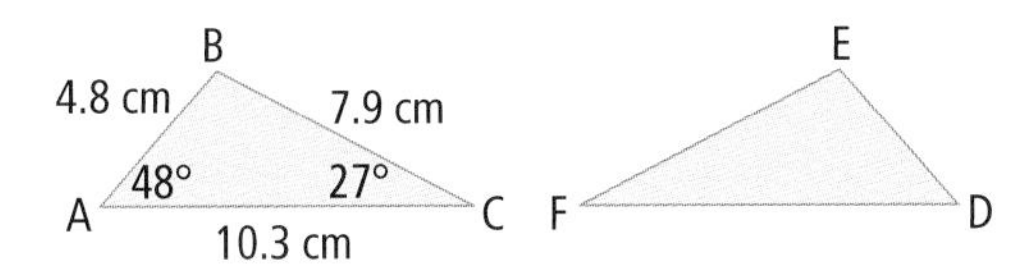

4. Answer each question as true or false. Give reasons for your answers to convince your partner. Draw diagrams to help you.
   a) If a triangle is isosceles, then one of its angles is 60°.
   b) If all the angles of a triangle have different measures, then all the sides have different measures.
   c) If two of the angles of a triangle are 31° and 54°, then the triangle is acute.
   d) If the sides of a triangle are 3 cm, 4 cm, and 6 cm, then the triangle is unique.

Communicating Mathematically

The "If-Then" thinking you are doing in questions 1 to 3 is called **deductive reasoning**. It is also the type of reasoning used to present evidence in criminal cases.

Part B: What are the minimum sufficient conditions to prove two triangles are congruent?

What makes a triangle unique will help you determine when triangles are congruent.

Work with a partner.

1. a) Each construct a triangle using an R1, a W1, and a Y1 Geostrip®. Is the triangle unique? How do you know? Draw the triangle by tracing the inside of the model.
   b) Compare your triangle to your partner's. Are they congruent? How do you know? If they are congruent, could you use the same three strips to construct a triangle that is not congruent to your partner's? Explain.
   c) If you know that △PQR has side lengths of 5 cm, 7 cm, and 10 cm, then is the triangle unique? Explain. If you know that △LMN has side lengths of 5 cm, 7 cm, and 10 cm, then is the triangle unique? Explain. What is the relationship between △PQR and △LMN? Why?

2. a) Join an R2 and a Y1 Geostrip® to form an angle of 45°. Add a third Geostrip® to complete the triangle. Is the triangle you made unique? How do you know?

b) Compare your triangle to your partner's. Are they congruent? How do you know? If they are congruent, could you use the three strips, keeping the 45° angle, to construct a triangle that is not congruent to your partner's? Explain.

c) If you know that in $\triangle DEF$, $DE = 4.5$ cm, $DF = 5.8$ cm, and $\angle D = 47°$, then is $\triangle DEF$ unique? Explain. If you know that in $\triangle JKL$, $JK = 4.5$ cm, $JL = 5.8$ cm, and $\angle J = 47°$, then is $\triangle JKL$ unique? Explain. What is the relationship between $\triangle DEF$ and $\triangle JKL$? Explain with reasons.

3. Both Pierre and Linh drew a triangle with angle measures of 45°, 75°, and 60°.

My triangle has the same angle measures as yours. It's the same size so if two triangles have all corresponding pairs of angles congruent, then the triangles are congruent.

It's a coincidence that I drew the same sized triangle as you. Having corresponding pairs of angles congruent does not necessarily guarantee that the two triangles will be congruent.

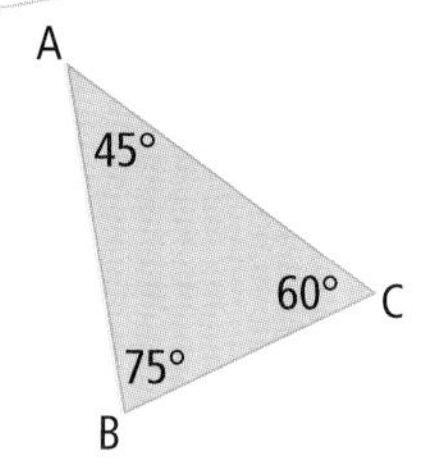

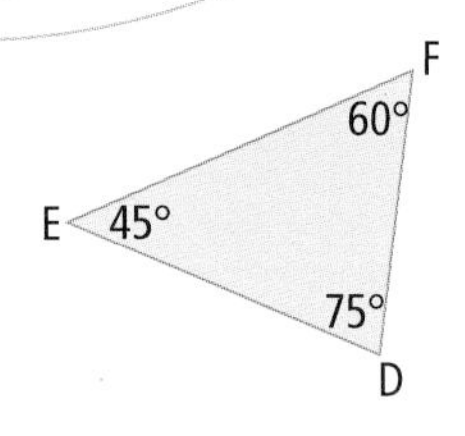

Who is correct? Explain.

4.

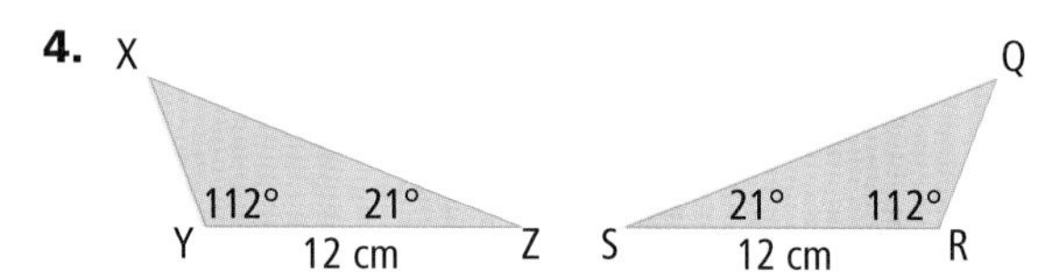

a) Are both triangles unique? Explain.

b) Are the triangles congruent? If they are, what reasons would you give for their congruency?

5. a) Join an R3 and a W1 Geostrip® to make an angle of 30°. Attach a B1 Geostrip® to the end of W1 and somewhere along R3 to create a triangle while keeping the 30° angle.

B1

B2

b) Compare your triangle to your partner's. Are they congruent? How do you know? If they are congruent, could B1 be attached somewhere else along R3 to create a triangle that is not congruent to your triangle? Explain.

c) △GHI has GH = 8.9 cm, HI = 7.5 cm, and ∠G = 38° and △LMN has LM = 8.9 cm, MN = 7.5 cm, and ∠L = 38°. Can you be certain the triangles are congruent without drawing them? Explain.

d) Describe under which conditions you cannot prove two triangles are congruent even though you know that two pairs of corresponding sides and one pair of corresponding angles are congruent.

6. a) Use a protractor to measure ∠C and ∠D in △CDE, and ∠P and ∠Q in △PQR.

b) Measure the length of CE and PR.

c) From your measurements, can you be certain the triangles are congruent? Explain.

d) Compare how the triangles relate with how the triangles in question 3 relate. How are they different? alike?

E R C P D Q

7. Reflect Describe the possible side and angle combinations of comparisons that you can use to convince someone that two triangles are congruent. Refer to questions 3 and 6 to help you.

Communicate the Key Ideas

1. Is Alene correct? Draw two congruent triangles, marking the appropriate measurements to illustrate her claim.

Two triangles are congruent if two pairs of corresponding sides and the corresponding included angles are congruent.

2. Is Armand correct? Draw two congruent triangles, marking the appropriate measurements, to illustrate his claim.

Two triangles are congruent if two pairs of corresponding angles and one pair of corresponding sides are congruent.

3. David draws two triangles that he claims are congruent. Explain how you can check his claim by making a minimum number of measurements.

Example 1: Use Logical Reasoning

Statement: If a triangle has angles of 72° and 36°, then it has one line of reflective symmetry.

Use logical reasoning to convince a classmate that the statement is true.

Solution

The sum of the interior angles of a triangle is 180°.
The third angle must be 180° − (72° + 36°) = 72°.
The unknown angle is 72°.

To show that the triangle has one line of symmetry, I need to prove the triangle is isosceles.

Strategies
Draw a diagram.
Use logical reasoning.

If exactly two angles in a triangle have the same measure, then the triangle is isosceles.

The isosceles triangle has one line of reflective symmetry. This is true of all isosceles triangles.

So, the statement is true.

Example 2: Prove Triangles are Congruent

In △LMN, LM = 7.5 cm and LN = 4.0 cm.
In △PMN, PM = 7.5 cm and PN = 4.0 cm.

Convince a classmate that △LMN ≅ △PMN.

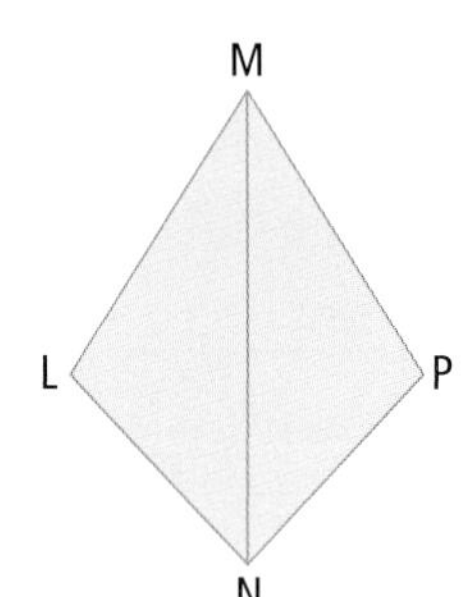

Communicating Mathematically

The symbol ≅ means "is congruent to".

Solution

Label the diagram with the known side lengths.

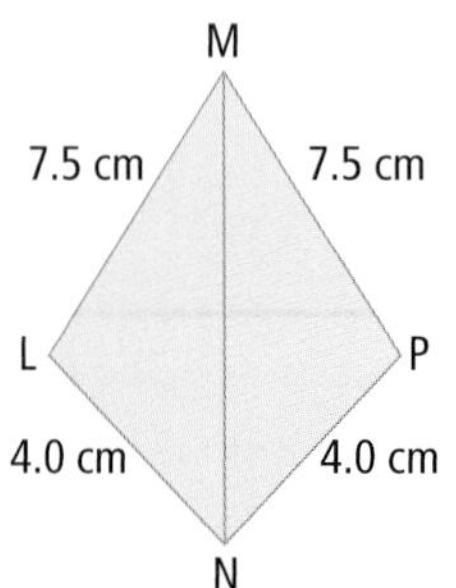

If I know that three sides of one triangle are congruent to the corresponding sides of a second triangle, then the triangles must be congruent.

I know that two pairs of corresponding sides are congruent, LM and PM, and LN and PN.

The third side of each triangle is MN. It is common to both △LMN and △PMN, so it is congruent to itself.

Therefore, △LMN ≅ △PMN because the three pairs of corresponding sides are congruent.

Check Your Understanding

1. In $\triangle PQR$, if $\angle P = 45°$ and $\angle R = 45°$, then convince a classmate that $\triangle PQR$ is an isosceles right triangle. Draw a diagram to help you. Record your reasoning.

2. If $\triangle WXY \cong \triangle PRT$, then what corresponding parts of the triangles are congruent? Draw a diagram to help you. Record your reasoning.

3. Which pairs of triangles are congruent? What reasons would you give in each case? For the pairs that are not congruent, explain why.

a)

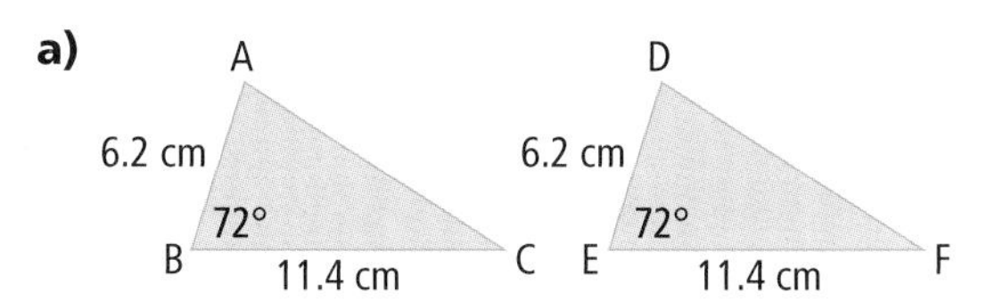

b)

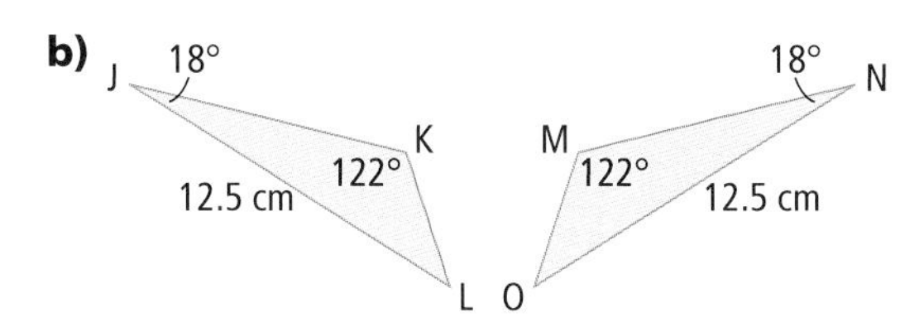

c)

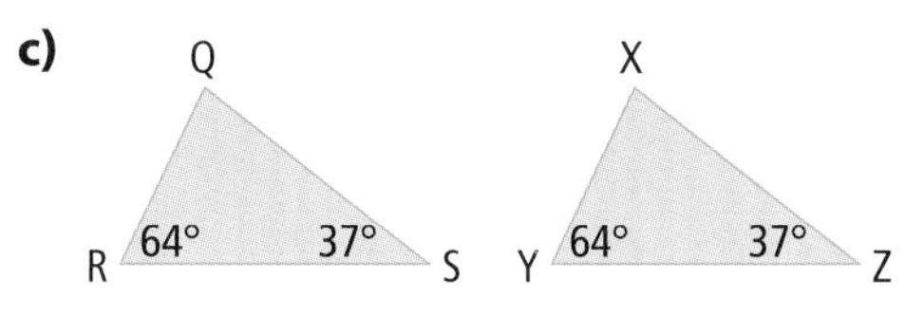

d)

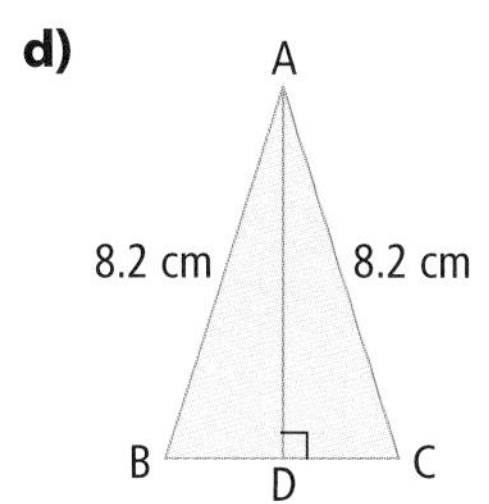

4.

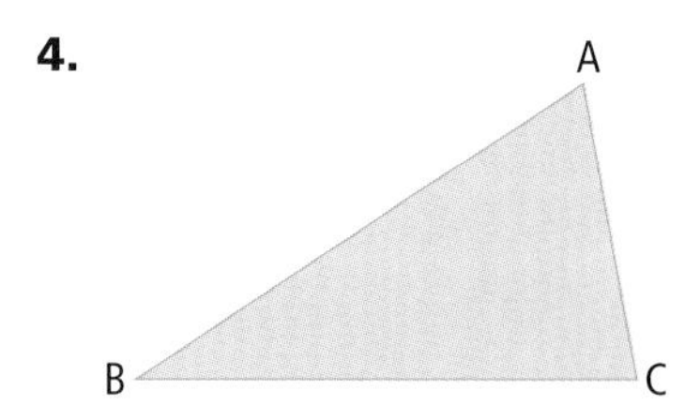

a) Take the fewest measurements of $\triangle ABC$ needed to draw a copy of it. Construct the copy and label it $\triangle XYZ$.

b) What measurements did you take and what instruments did you use?

c) Label your triangle with the measurements.

d) Why are the original triangle and the copy congruent, given the measurements you used?

e) What is the minimum number of measurements that you need to construct this triangle? Discuss your answer with a partner.

5. Both Mr. Smith and Mrs. Jones have triangles made from masking tape on their classroom walls. How could you use a metre stick to convince the teachers that the triangles are, or are not, congruent? Why could you not use just a protractor to convince the teachers?

6. Two triangles like the one in question 4 are painted on a mural outside your school. Your teacher asks you to verify that the triangles are congruent but your ladder can only reach the bottom of each triangle. How can you convince your teacher that the triangles are congruent?

7. If in quadrilateral DEFG, DE = 8 cm, DG = 8 cm, and FD bisects ∠EDG, then use logical reasoning to convince a classmate that △DEF ≅ △DGF. Record your reasoning.

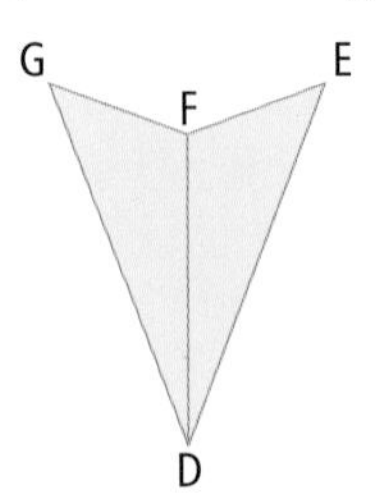

8. Refer to question 7. Suppose ∠G = 36°, ∠E = 36°, and DF bisects ∠EFG. Use logical reasoning to convince a classmate that DE = DG.

Strategies
Use previously learned knowledge.

9. Explain why knowing the measures of two pairs of corresponding sides and one pair of corresponding angles (not included between the two sides) is not proof that two triangles are congruent.

10. In the diagram, DE = 7 cm, AB = 7 cm, DE ∥ AB, ∠E = 39°, and ∠A = 21°. Use logical reasoning to convince a classmate that △DEC ≅ △ABC. Record your reasoning.

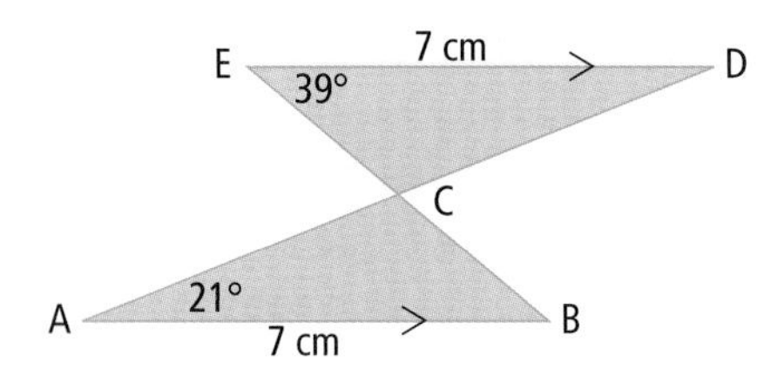

Communicating Mathematically

The symbol > on a diagram is used to show that two lines are parallel.

The symbol ∥ means "is parallel to".

11. Two neighbours built triangular decks on the backs of their houses. Explain how you could use a long piece of rope to check if the decks are congruent.

Chapter Problem

12.

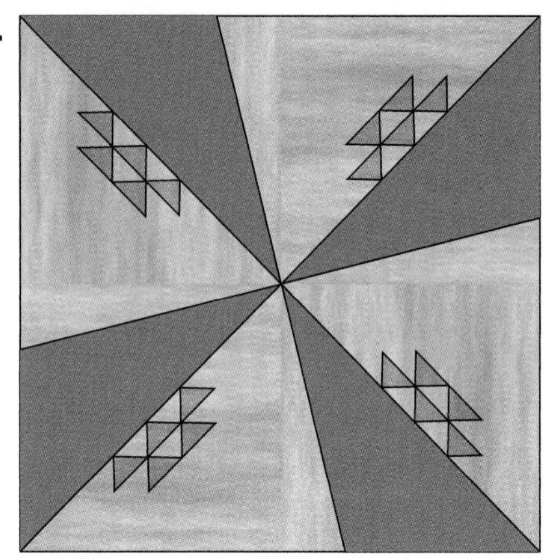

a) Are all the small blue triangles in this quilt square congruent? How can you be certain? Explain which parts you could compare.

b) Are the four large blue triangles congruent? How can you be certain? Explain which parts you could compare.

c) Is the quilt square symmetrical? Explain.

Extend

13. In △RST, RS = 15.6 cm, ∠R = $(4x - 20)°$, ∠T = $(3x + 20)°$, and ∠S = 40°. In △JKL, JK = 15.6 cm, ∠J = $(3x)°$, ∠L = $(4x)°$, and ∠K = 40°. Are the triangles congruent? Explain how you know.

Making Connections

Triangles are often used in architectural designs because they are very stable, rigid figures. They make buildings and structures difficult to collapse or blow over.

3.3 Properties of Transformations

Focus on...
- discovering the properties of transformations
- performing a variety of 2-D constructions using the properties of transformations

Carpentry, quilting, and other crafts require careful and precise work with transformations. Examine a piece of furniture, a quilt, a china pattern, a wallpaper design, or a stained glass window. You will see translations, reflections, and rotations at work.

What do craftspeople need to know about translations, reflections, and rotations in order to use them effectively and efficiently?

Discover the Math

Materials
- centimetre square dot paper
- Bullseye compass
- protractor
- ruler
- transparent mirror
- tracing paper

Part A: Explore and Apply Properties of Translations

What will always be true about 2-D shapes that are translated images of each other?

Work with a partner.

1. The blue and green pentagons are translated images of the red pentagon (pre-image).

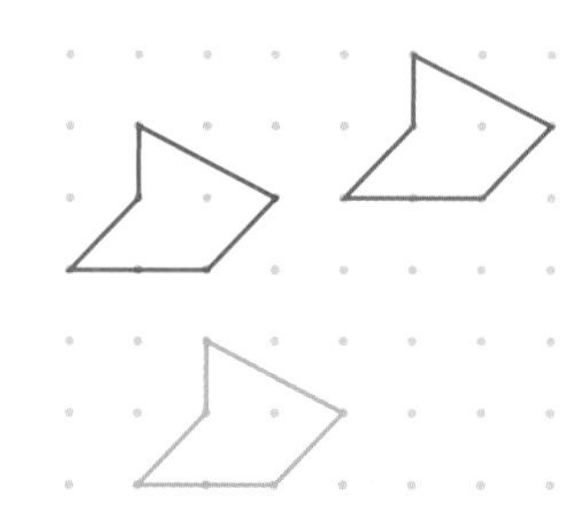

   a) What appears to be true between the corresponding angles in the red and blue pentagons? the red and green pentagons? How could you convince your partner this is true?

b) What appears to be true between the corresponding sides of the red and blue pentagons? the red and green pentagons? How could you convince your partner this is true?
c) Are the three pentagons congruent? Explain how you know.
d) On separate sheets of square dot paper, one partner copies the red and blue pentagons and the other partner copies the red and green pentagons. Use broken line segments to join the corresponding vertices of your two pentagons.
e) What appears to be the two relationships among the five broken line segments that join corresponding points? How could you convince your partner these relationships are true?
f) Leslie says that in a translation all points of the pre-image move the same distance. Do you agree with Leslie? Explain.

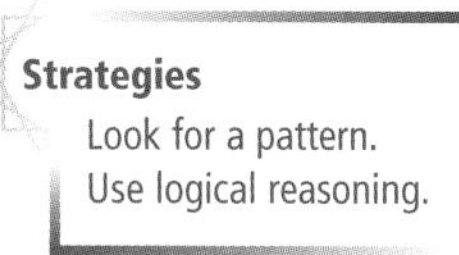
Strategies
Look for a pattern.
Use logical reasoning.

2. The blue pentagon is the image of the red pentagon translated 4 units right and 1 unit up. This translation can be represented by the mapping notation {4→, 1↑} or [4R, 1U].

The red pentagon could also be seen as the image of the blue pentagon translated 4 units left and 1 unit down. This translation can be represented by {4←, 1↓} or [4L, 1D].

a) Describe the translation of the red pentagon (pre-image) to obtain the green pentagon.
b) Describe the translation of the green pentagon (pre-image) to obtain the blue pentagon.

Communicating Mathematically

You can use arrows, letters, or mapping notation to indicate a translation.
For example:
2 units right is 2→ or 2R.
2 units left is 2← or 2L.
5 units up is 5↑ or 5U.
5 units down is 5↓ or 5D.
A translation 2 units right and 5 units up is {2→, 5↑} or [2R, 5U].

3. a) On a coordinate grid, draw a straight line that passes through the points A(−3, 4) and B(2, −3). Apply the translation {2←, 3↓} to the two points and draw a straight line through these two image points. Label the line A′B′. Why can you be certain that the two lines are parallel?
b) Apply the same translation to two different points on line AB. Did you produce the same parallel line A′B′? Explain why.
c) Draw a straight line that passes through points X(1, 2) and Y(7, 8). Locate and label the point T(4, 9). Choose any point on your line. Describe the translation that would map your selected point onto point T. Use words and mapping notation.
d) Choose another point on line XY. Apply the same translation to this point to find its image and label it V. Why can you be certain that the straight line through T and V will be parallel to your original line?

4. **Reflect** If △D′E′F′ is a translated image of scalene △DEF, then list everything you know about these two triangles.

Part B: Explore and Apply Properties of Reflections

What is always true about 2-D shapes that are reflected images of each other?

Work with a partner.

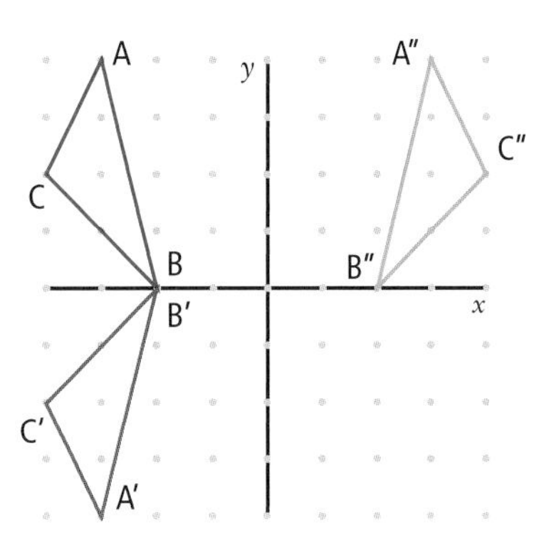

1. The red triangle (pre-image) is reflected in the x-axis to produce the blue triangle (image). The red triangle is reflected in the y-axis to produce the green triangle.
   a) Copy the diagram onto square dot paper.
   b) What seems to be true between corresponding angles of the blue and red triangles? the green and red triangles? How could you convince your partner that this is true?
   c) What seems to be true between the corresponding sides of the blue and red triangles? the green and red triangles? How could you convince your partner that this is true?
   d) Are the blue and green triangles congruent to the red triangle? Explain how you know.
   e) Did all the points on △ABC move when △ABC was reflected? Explain. For all of the points that moved, did they move the same distance under the reflection? Explain.
   f) Examine the distances from corresponding vertices to the mirror lines (x- and y-axes). What did you notice?
   g) Name the three triangles starting with A, A′, and A″, going clockwise. Compare the name of the red triangle to the names of the blue and green triangles. What do you notice?
   h) Use broken line segments to join several corresponding points between the red and green triangles. Explain why these line segments are not congruent. What seems to be true about the line segments joining corresponding points?
   i) What are the measures of the angles formed by the mirror line and all the line segments you drew in part h)? What does the mirror line do to each line segment you drew? Describe the relationship between the mirror line and the line segments joining corresponding points under a reflection.
   j) List all the properties of reflections learned in this question.

Mathematicians would say that the reflection in the y-axis caused a change in **orientation**. By comparing the red and green triangles, you see the top and bottom vertices remain in the same position while the vertex on the left in the red triangle is on the right in the green triangle. Compare the positions of the vertices in the red and blue triangles.

orientation
- the way the vertices of a polygon are ordered
- if you read a figure as A-B-C in a clockwise direction, and read its image as A-C-B in a clockwise direction, then the orientation of the image is the reverse of the pre-image

2. a) Plot points P(−2, −4), Q(0, 3), R(4, 1), and S(5, −3) on a coordinate grid. Join the points to form a quadrilateral.
   b) Draw a mirror line that passes through points (−2, 5) and (6, 5). What property of reflections could you use to plot the images of points P, Q, R, and S, without using a transparent mirror? Plot the image points and label them P′, Q′, R′, and S′. Join the points to create the image quadrilateral. Use a transparent mirror to check your work.
   c) Did the quadrilateral's orientation change as a result of the reflection? Explain. Since the mirror line was above the quadrilateral, and not to the right, what parts of the quadrilateral have changed their relative positions?
   d) List all the properties of reflections from question 1 that apply to quadrilaterals PQRS and P′Q′R′S′. Compare your answer with your partner and discuss the differences.

3. Fran says one triangle was rotated to produce the other. Jon says the triangle was reflected.
   a) By looking at the two triangles only, which property of reflections could Jon use to convince Fran that he is correct?
   b) Explain how you could use the properties of reflections to find the mirror line for the reflection shown.

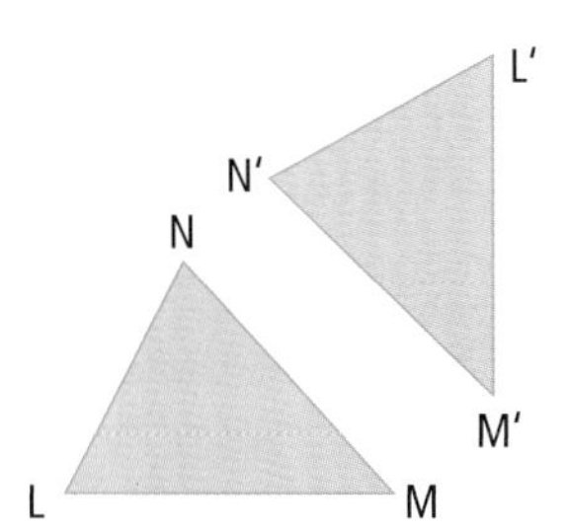

4. a) Draw a line segment AB. Place a transparent mirror between endpoints A and B and adjust it until you can see the image of point A fall on point B. Use a sharp pencil to draw a line segment along the bevelled edge of the transparent mirror. Label it CD. Explain why properties of reflections guarantee that line segment CD is the perpendicular bisector of AB.
   b) Draw a line segment DE. Place a transparent mirror between endpoints D and E and adjust it until you can see the image of point D fall on point E. Anywhere along the bevelled edge of the transparent mirror, except on DE, mark a point and label it F. Remove the mirror and join points for line segments DF and EF. Record and discuss why properties of reflections guarantee that △DEF is isosceles.

5. **Reflect** If △J′K′L′ is the reflected image of △JKL, then list everything you know about the two triangles.

Part C: Explore and Apply Properties of Rotations

What is always true about 2-D shapes that are rotated images of each other?

1. a) Draw scalene △LMN and point C as shown.

b) Use a dashed line segment to join points C and L. Use a Bullseye with C as a centre and draw an arc clockwise from L. Make a 60° angle at C between line CL and a line segment that intersects the arc at L′.

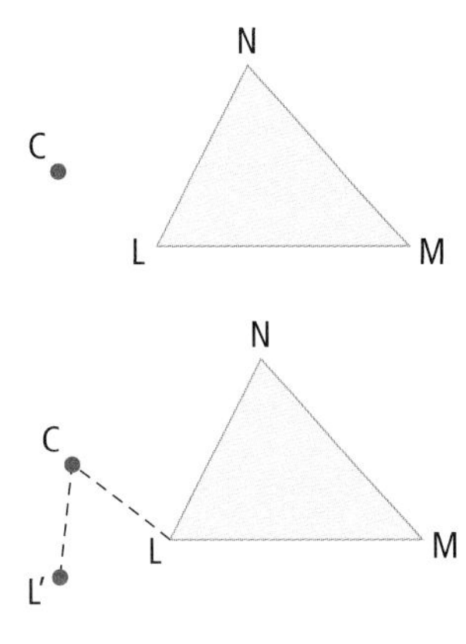

c) Use the same method to find the images of N and M for the same 60° rotation with centre C. Label these image points N′ and M′ and join the points to create △L′M′N′.

d) Do all corresponding sides and angles have the same measure?

e) Are the two triangles congruent? Explain how you know.

f) Do △LMN and △L′M′N′ have the same orientation? Explain.

g) Are CL and CL′ the same length? Compare the lengths of CM and CM′, then CN and CN′. What can you say about the line segments formed by joining the centre of rotation to corresponding points?

h) Do all points under the rotation move the same distance? Explain.

i) Why does ∠LCL′ = ∠NCN′ = ∠MCM′? Explain. What is the measure of these angles?

j) Join L and L′ and draw the perpendicular bisector of LL′. Extend the bisector. What point does it pass through? Repeat this process for MM′ and NN′. Do all three perpendicular bisectors pass through the same point?

k) List all the properties of rotations learned in this question.

2. a) △A′B′C′ is the image of △ABC under a rotation with centre P. Explain how you could find the angle and direction of rotation.

b) If the centre of rotation P had not been given, which property of a rotation could you use to find the centre of rotation? Explain.

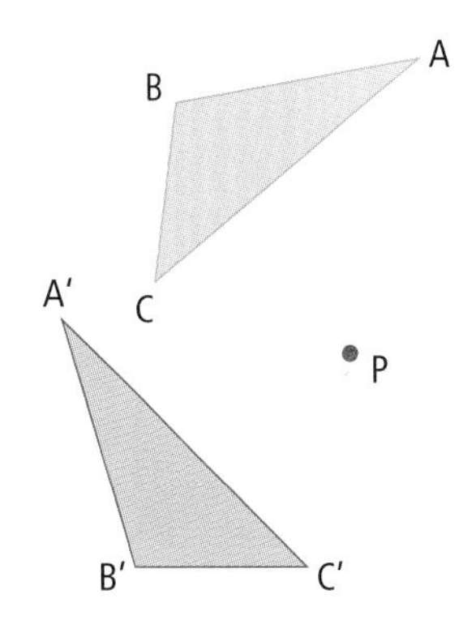

3. a) Copy this diagram onto square dot paper.
   b) Label the red quadrilateral as JKLM, and its blue image as J′K′L′M′.
   c) Are the quadrilaterals congruent? Explain how you know.
   d) Draw two line segments that join two pairs of corresponding points. Locate the point of intersection, P, of the perpendicular bisectors of the line segments. Check that the perpendicular bisectors of the other two line segments also intersect at P. What is this point of intersection? How do you know?
   e) Describe the rotation of JKLM to produce J′K′L′M′.
   f) Are the pairs of line segments that join corresponding points to P congruent? How do you know?
   g) Are all the angles formed by joining corresponding points to P congruent? How do you know?
   h) Suppose you trace JKLM and two points outside the figure, then place the tracing on top of the diagram. You place your pencil firmly on P then rotate the tracing. Would every point on the tracing be moved? Explain. Describe the movement of points under a rotation.
   i) Rotate JKLM 180° about P. What do you notice about the corresponding sides? What do you notice about the line segments between corresponding points?

4. **Reflect** △R′S′T′ is the image of △RST under a rotation of 60° counter-clockwise about centre point H. List everything you know about the two triangles and the point H.

5. **Reflect** Consider a translation, a reflection, and a rotation of a polygon. What properties do any two or three of these transformations have in common? Which properties are unique to each transformation?

Communicate the Key Ideas

1. Two shapes on a coordinate grid may be translations of each other. Explain what you should check to confirm this.

2. Two shapes are reflections of one another. Explain how you could use a ruler to find the mirror line for the reflection.

3. If you are given a polygon and a mirror line, explain how you could use a ruler to find the reflection image of the shape.

4. Two shapes may be rotational images of one another. Explain how you could use a ruler to confirm this.

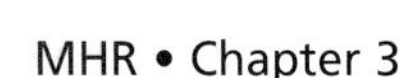

Example 1: A Translation or Not?

WXYZ looks like a translation of ABCD.

How could we know for certain?

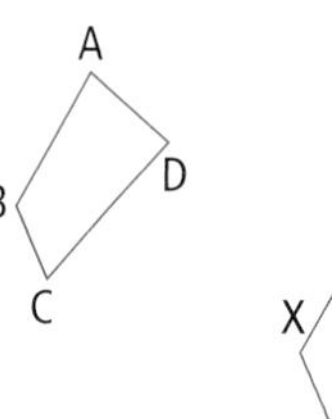

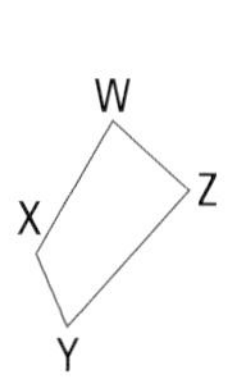

Solution

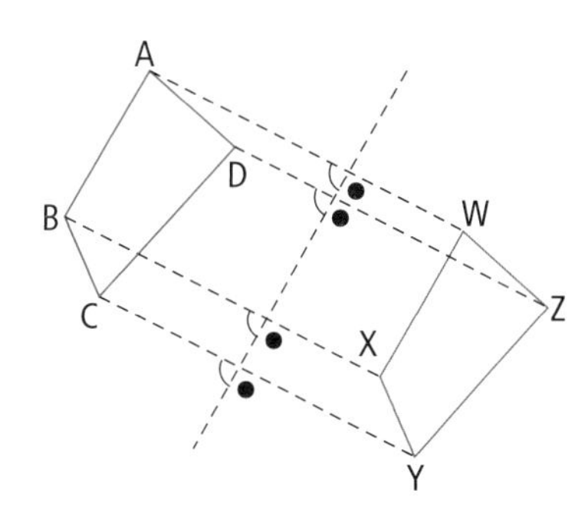

I know that translations have these properties.

- corresponding parts are congruent
- corresponding sides are parallel
- line segments joining corresponding vertices are congruent and parallel

The last point is true only for translations. Therefore, if I show it is true for this problem, then I know I have a translation.

I joined the corresponding vertices of the two shapes with broken line segments.

The line segments, AW, DZ, BX, and CY have the same length.

I drew a **transversal** through the four line segments and measured the **corresponding angles** along the transversal. Since these angles were congruent, AW ∥ DZ ∥ BX ∥ CY.

Since the line segments joining corresponding vertices are congruent and parallel, the property holds true for these quadrilaterals. I am certain that these quadrilaterals are translation images of one another.

transversal

- line that crosses two or more lines or line segments

corresponding angles

- angles either both above or below two lines on the side of a transversal. If the two lines are parallel, corresponding angles are congruent.

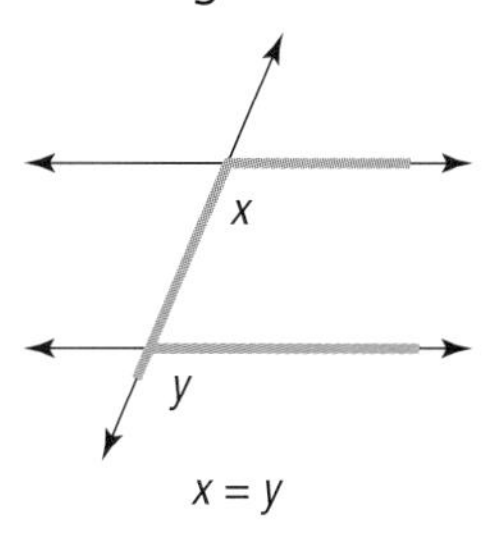

Example 2: Which Transformation?

$\triangle$ABC and $\triangle$A′B′C′ are images of one another. Which transformation was used?

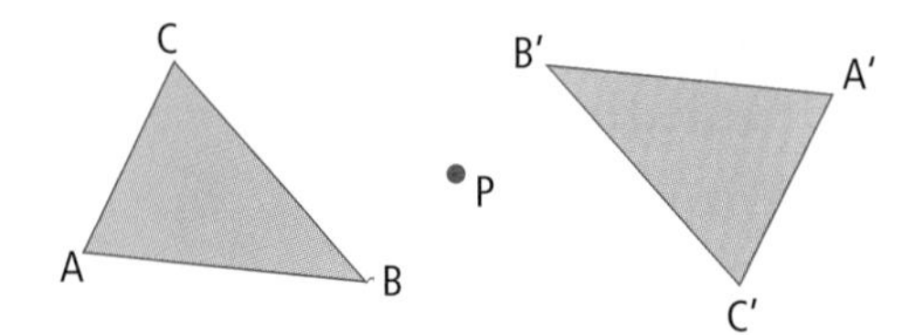

Solution

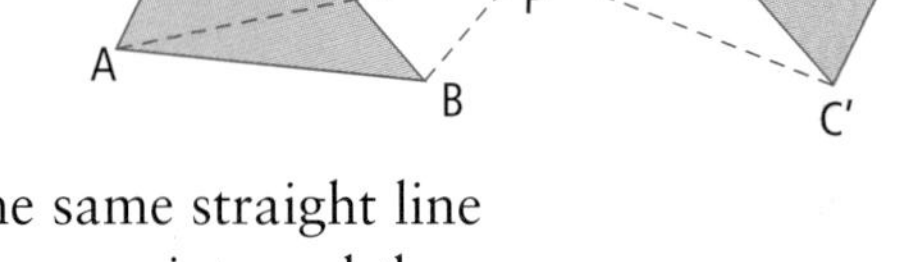

The orientation is the same, so this is not a reflection.

Check for a rotation with centre P.
Draw line segments from each vertex to point P.
A 180° rotation has these properties.

- corresponding parts are congruent
- corresponding sides are parallel or on the same straight line
- a pre-image point, the corresponding image point, and the centre of rotation form a straight line

The last point is true only for 180° rotations. Therefore, if you show it is true for this problem, then you know you have a 180° rotation.

By measuring, AP = A′P, BP = B′P, and CP = C′P.

From the diagram you can see that $\angle$APA′ = 180°, $\angle$BPB′ = 180°, and $\angle$CPC′ = 180°. Therefore, $\triangle$ABC was rotated 180° clockwise about centre P.

Check Your Understanding

1. A designer of wallpaper borders often starts with a simple shape and uses transformations of that shape to create the pattern.

a) Describe the transformations you see in the pattern.

b) Draw the smallest shape that the pattern could have been made from. Describe the transformation that could have been used on the shape to create the pattern.

2. The properties of transformations you discovered apply to any two 2-D shapes. Indicate which transformation(s) applies to each statement.

a) The shape and its image are congruent.

b) The corresponding sides of the shape and its image are parallel.

c) The corresponding angles of the shape and its image have the same measure.

d) The corresponding points of the shape and its image are equidistant from a fixed point.

e) The shape and its image have different orientations.

f) Line segments joining corresponding points of the shape and its image are parallel.

g) The corresponding points of the shape and its image are moved the same distance.

h) The perpendicular bisectors of the line segments joining corresponding points of the shape and its image intersect.

3. a) Draw a quadrilateral like the one shown.

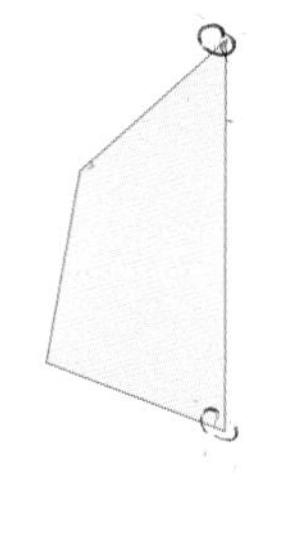

b) Use the right side of the quadrilateral as a mirror line and draw its reflected image. Do not use a transparent mirror. Explain the steps for drawing the image.

c) Which points did not change position under this reflection? Explain why.

d) The quadrilateral and its image form a polygon. What is the name of this polygon? Does it have symmetry? Explain how you know.

4. a) Draw a scalene △PQR with ∠P = 72°.

b) Trace your triangle onto tracing paper. Use the tracing to draw the image of △PQR under a clockwise rotation of 72° about centre P.

c) Did all the points of △PQR change position under this rotation? Explain.

d) Use the tracing to draw the images of △PQR under clockwise rotations of 144°, 216°, and 288° about centre P.

e) The triangle and its four images form a polygon. What is the name of this polygon? Does it have symmetry? Explain how you know.

5. a) Draw a scalene triangle. Use a transparent mirror to draw a reflected image of the triangle. Do not draw the mirror line.

b) Exchange your drawing with a partner. Use a ruler to find the mirror line on your partner's drawing. Check your results with a transparent mirror. Explain how you used the ruler to find the mirror line.

6. a) Trace this diagram.

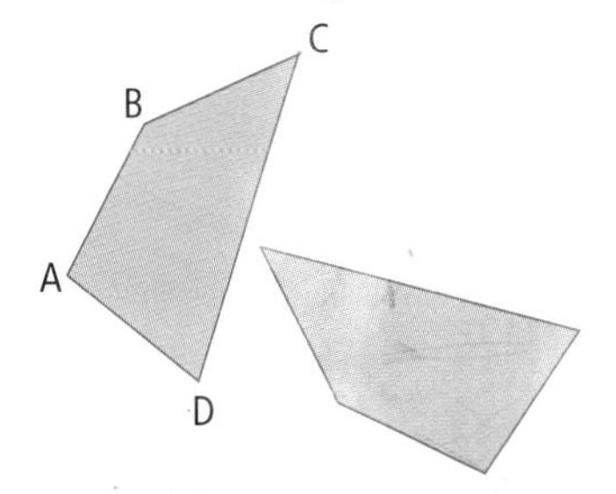

b) Is the blue quadrilateral *not* a translation or a reflection of ABCD? Explain how you know.

c) Label the corresponding vertices of the blue image as A′, B′, C′, and D′.

d) Find the centre of rotation and label it P. Describe the steps you followed.

e) What is the angle and direction of rotation about P that maps ABCD onto A′B′C′D′?

7. The blue square is an image of the black square.

a) Explain how the blue square could be
   i) a translation
   ii) a reflection
   iii) a rotation

b) Suppose the black square had vertices A, B, C, and D. Explain how you would know which transformation was used if the corresponding vertices A′, B′, C′, and D′ were labelled.

8. a) Draw a scalene triangle and a bisector of one of its angles as shown.

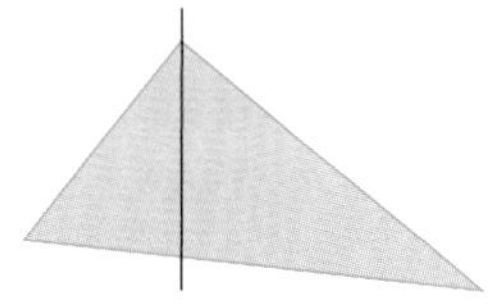

b) Use the angle bisector as a mirror line to find the image of your triangle.

c) Explain why you had to use the angle bisector as a *two-sided* mirror.

9. Are the following statements *always*, *sometimes*, or *never* true? Explain.
   - **a)** Under a reflection, all points on a shape change position.
   - **b)** Under a translation, the orientation of a shape changes.
   - **c)** Under a rotation, corresponding sides are equal and parallel.
   - **d)** Under a reflection, corresponding angles of a shape and its image have the same measures.
   - **e)** Under a rotation, all points on a shape move the same distance.

Chapter Problem

10. Describe the transformations in each quilt block that makes it symmetrical. Disregard the colours of the shapes.
   - **a)** Paper Airplane

   - **b)** V Block

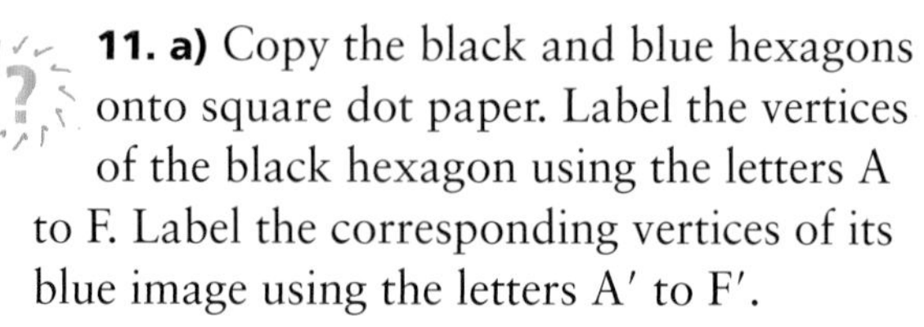

11. **a)** Copy the black and blue hexagons onto square dot paper. Label the vertices of the black hexagon using the letters A to F. Label the corresponding vertices of its blue image using the letters A′ to F′.

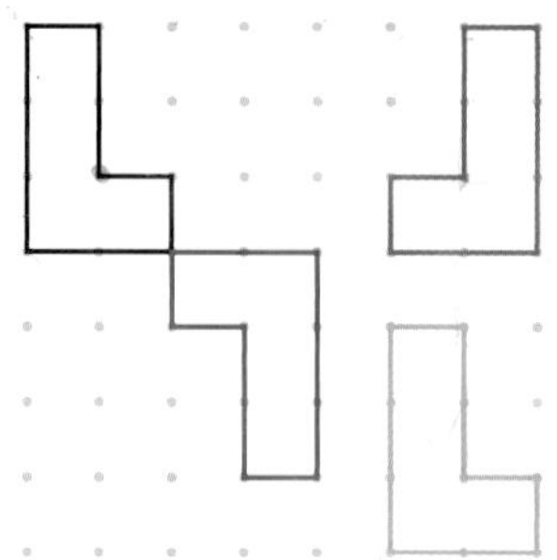

   - **b)** What transformation of the black hexagon produces the blue hexagon? Explain the properties of the transformation you chose to prove your choice is correct.
   - **c)** Repeat parts a) and b), using the
     - **i)** black and red hexagons
     - **ii)** black and green hexagons
   - **d)** What is the relationship between the blue and green hexagons? Explain how you know.

Extend

12. **a)** On a coordinate grid, draw a scalene triangle.
   - **b)** Perform each reflection. Examine the coordinates of the vertices of the pre-image and the image. Describe the pattern you see in the coordinates.
     - **i)** Reflect the pre-image in the x-axis.
     - **ii)** Reflect the pre-image in the y-axis.
     - **iii)** Reflect the image from part i) in the y-axis.
     - **iv)** Reflect the image from part ii) in the x-axis. What do you notice?
   - **c)** Compare your answers with several classmates. Did they get the same results?
   - **d)** What is the result of reflecting a shape in the x-axis and then in the y-axis?

3.4

Regular Polygons

Focus on...
- finding the properties of regular polygons
- applying the properties of regular polygons

Mathematicians through the ages have been fascinated by regular polygons. Regular polygons have all sides of equal length and all angles of equal measure, which means these shapes have many lines of symmetry. Artists use this symmetry in many types of designs, such as this decoration on a street sign from ancient Pompeii.

To learn more about Pompeii go to www.mcgrawhill.ca/links/math8NS and follow the links.

Triangle, square, and hexagon pattern blocks are examples of regular polygons. You could use these shapes (as well as trapezoid and rhombus blocks) to create interesting repeating designs or **tessellations** that could cover your desk with no gaps or overlaps.

What do you need to know about regular polygons to create a tessellation?

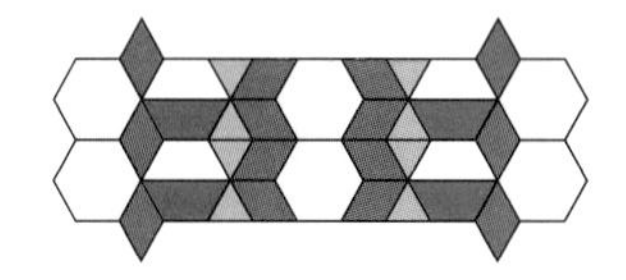

Communicating Mathematically

You can name a polygon using a number or a Greek or Latin prefix and the suffix "-gon" (except for the prefix quad). For example, a 3-gon is a three-sided figure. It is also called a trigon or triangle. A 4-gon is a four-sided figure. It is also called a tetragon or quadrilateral. An 11-gon is an eleven-sided figure. It is also called a hendecagon. These are the common names of the first 10 regular polygons.

3 sides: trigon or triangle
4 sides: tetragon or quadrilateral
5 sides: pentagon
6 sides: hexagon
7 sides: heptagon or septagon
8 sides octagon
9 sides: nonagon
10 sides: decagon
11 sides: hendecagon or undecagon
12 sides: dodecagon

Discover the Math

Materials
- Geostrips®
- BLM 3.4 Regular Polygons
- ruler
- protractor
- Bullseye compass
- tracing paper
- transparent mirror

Optional:
- The Geometry Template®
- BLM 3.4 DTM Polygon Table

What special properties do regular polygons have?

Work with a partner.

1. a) Join five Geostrips® of the same length end-to-end to form a pentagon. Take turns making four different pentagons by repositioning the Geostrips®. Trace the inside of your models to record the pentagons.

b) How are your pentagons alike? different?

c) Are your pentagons concave or convex? Explain.

d) Work together to make a regular pentagon. How do you know it is regular?

e) Is the regular pentagon concave or convex? Is it possible to create a concave regular pentagon? Explain.

2. Examine the regular polygons in the handout.

a) Describe the pattern Andre sees.

b) Use the pattern to predict which regular polygons would have pairs of parallel opposite sides. Write the number of pairs of sides.
- 14-gon
- 15-gon
- 20-gon
- 50-gon

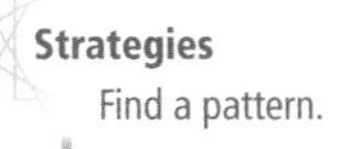
Strategies
Find a pattern.

Some of the regular polygons have pairs of opposite sides that are parallel and some of them don't even have opposite sides! But I see a pattern based on whether the number of sides of the regular polygon is even or odd.

3. a) Find the five lines of symmetry of a regular pentagon. Describe the relationship between the lines of symmetry and the sides of the regular pentagon.

b) The point of intersection of the lines of symmetry is the centre of the regular pentagon. Use a Bullseye set at the centre and extended to one of the vertices to draw a circle. What do you notice about the circle? Why do you think mathematicians call this a **circumscribed** circle?

c) Use a Bullseye set at the centre and extended to the midpoint of one of the sides of the pentagon to draw a circle. What do you notice about this circle? Why do you think mathematicians call this an **inscribed** circle?

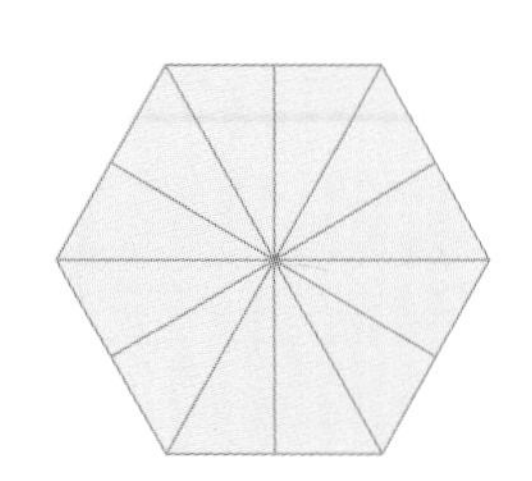

d) The six lines of symmetry of a regular hexagon are shown. How do their locations differ from those in the regular pentagon?

e) Copy the regular hexagon in your handout. Use a Bullseye to draw its circumscribed and inscribed circles.

f) Investigate Alene's claim by looking at the lines of symmetry of other regular polygons on your handout. Describe the pattern you find. Apply your pattern to describe the number and locations of the lines of symmetry of a regular 19-gon and regular 20-gon.

The location of the lines of symmetry depends on whether the number of sides is even or odd!

4. The opposite vertices of the regular octagon have been joined to create eight triangles in its interior.

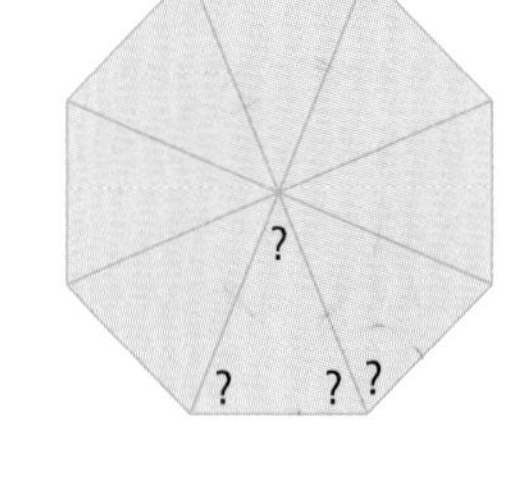

a) What type of triangles are they? How do you know? Are they all congruent? How do you know?

b) Each triangle has one angle at the centre of the regular octagon. These are called the central angles of the polygon. How could you find the measure of a central angle without measuring it with a protractor? Find the measure.

c) How could you find the measures of the base angles in a triangle without measuring them? Find the measures.

d) What is the measure of each interior angle of the regular octagon?

e) Copy and complete the table.

Regular Polygon	Number of Congruent Triangles in Interior	Measure of Each Central Angle	Measure of Each Base Angle of Triangle	Measure of Each Interior Angle of the Regular Polygon
square				
pentagon				
hexagon				
heptagon				
octagon	8	$360° \div 8 = 45°$	$(180° - 45°) \div 2 = 67.5°$	$67.5° + 67.5° = 135°$
nonagon				
decagon				
hendecagon				
dodecagon				
20-gon				
60-gon				
n-gon				

f) Explain how you can find the measure of the interior angles of any regular polygon. Use your explanation to find the measure of the interior angles of a regular 13-gon. Check to see that your explanation makes sense.

5. The **exterior angle** of a regular polygon is the angle formed by an extended side and the adjacent side.

a) What is the measure of each interior angle of a regular pentagon?

b) Explain how you can use the interior angle to find the measure of an exterior angle.

c) How do the central angle and the exterior angle of a regular pentagon compare? Is this relationship true in any regular polygon? Explain.

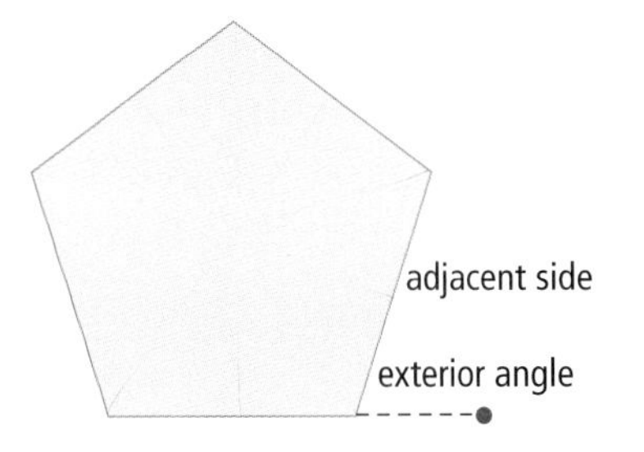

exterior angle
- angle formed by the extended side of a polygon and an adjacent side
- appears outside the polygon

d) Explain why it would be more useful to know the measure of the exterior angles of a regular polygon rather than the interior angles if you were cutting the shape from a piece of plywood.

6. **Reflect** You have a regular 15-gon and a regular 16-gon. List everything you know about these two polygons.

Communicate the Key Ideas

1. Explain reflective symmetry and rotational symmetry in your own words. Use a regular nonagon to help you.
2. Explain how the name of a regular polygon will tell you the locations of its lines of symmetry.
3. To find the interior angles of a regular polygon, you divide 360° by the number of sides. What is important about the number 360?
4. Why do some regular polygons have pairs of parallel sides and others do not?
5. Explain how to find the centre of any regular polygon.
6. In a regular polygon, what are the relationships among the number of sides of the polygon, the number of lines of symmetry, and the order of rotational symmetry?

Example 1: Determine the Angle Measures of a Regular Polygon

What is the measure of an interior angle and an exterior angle of a regular 100-gon?

Solution

Visualize a regular 100-gon.
By joining each vertex to the centre of the 100-gon, you create 100 congruent isosceles triangles.
Measure of a central angle = 360° ÷ 100
= 3.6°
Sum of the base angles in each triangle = 180° − 3.6°
= 176.4°

Since each interior triangle is isosceles, the base angles are congruent.
Measure of each base angle = 176.4° ÷ 2
= 88.2°

Two base angles of the isosceles triangles create one angle of the regular 100-gon. The measure of each angle of the regular 100-gon is 176.4°. Since each exterior angle is the supplement of an interior angle, its measure is 3.6°.

Example 2: Analyse a Regular Polygon

What would be useful to know about regular hexagons before you used them in a design?

Solution

1. Start by tracing two regular hexagons.
2. In one hexagon, find the midpoints of all six sides and join opposite midpoints with broken line segments. Join opposite vertices with broken line segments.
3. In the other hexagon, join each vertex to the centre, creating six triangles. Examine these regular hexagons.

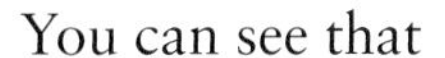

You can see that

- All six sides have the same measure.
- There are three pairs of parallel opposite sides.
- There are six lines of reflective symmetry: three lines through opposite vertices and three lines through the midpoints of opposite sides.
- Joining vertices to the centre creates six congruent, equilateral triangles.
- There is rotational symmetry of order 6.
- The central angles are 60° (360° ÷ 6).
- The interior angles are 120° (180° − 60°).
- The exterior angles are 60°.

When using regular hexagons in design work, each of these properties will be useful.

Did You Know?

Donald Coxeter was a world famous geometer from Toronto, Ontario. His work, extending over 70 years, included 12 books—at least four of them classics—and some 200 papers. His best-known book was *Introduction to Geometry* (1961).

Check Your Understanding

1. Copy and complete each statement to make it true.
   a) If a regular polygon has rotational symmetry of order 7, then ______.
   b) If a regular polygon has 11 lines of reflective symmetry, then ______.
   c) If a regular polygon has sides 5 cm in length and angles measuring 60°, then ______.
   d) If a regular polygon has angles measuring 162°, then ______.

2. a) Draw an isosceles triangle with two sides 6 cm in length and a 36° angle between them. Label the vertex with the given angle P.
   b) Use tracing paper to rotate the triangle 36° clockwise about P. Draw the image triangle.
   c) Rotate the image triangle another 36° clockwise about P and draw the second image triangle.
   d) Repeat part c) seven more times until you reach the original triangle (360° rotation).
   e) The triangle and its images form a polygon. What is the name of the polygon? Explain how you know.

3. a) Trace the regular hexagon, heptagon, octagon, and nonagon from your handout.

b) For each regular polygon, choose one vertex and divide the shape into a series of triangles by drawing diagonals from the vertex as shown. (The triangles should not overlap.)

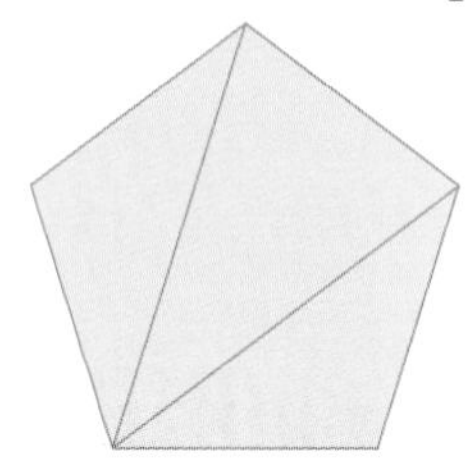

c) Copy and complete the table.

Regular Polygon	Number of Sides	Number of Interior Triangles	Total Number of Degrees	Measure of Each Interior Angle
pentagon	5	3	$3 \times 180° = 540°$	$540° \div 5 = 108°$
hexagon				
heptagon				
octagon				
nonagon				
20-gon				
n-gon				

d) Examine the table. What pattern did you see that helped you find the interior angle measure of the regular *n*-gon?

e) Draw a hexagon that is not regular. Divide it into triangles as in part b). Did you get the same number of triangles as you did for the regular hexagon? What is the total number of degrees in any hexagon?

f) Explain how you can find the total number of degrees in any *n*-gon.

4. a) Use a Bullseye and protractor to draw a regular nonagon around a centre point T. Make the distance from T to a vertex 5 cm.

b) Show the calculations and record the steps you followed to draw your nonagon.

5. List all of the properties of a regular decagon that would be useful to know before using the shape in a design.

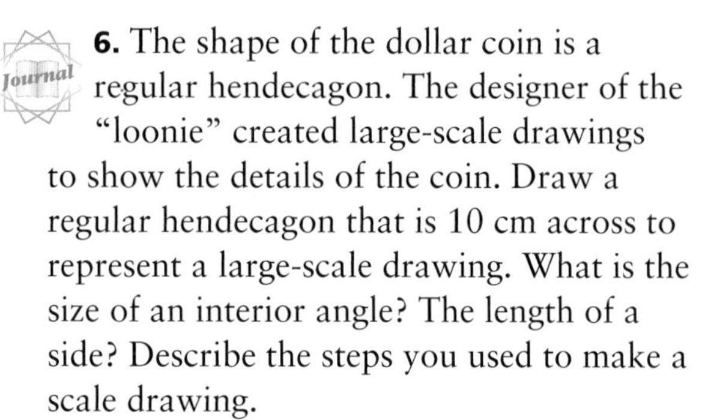

6. The shape of the dollar coin is a regular hendecagon. The designer of the "loonie" created large-scale drawings to show the details of the coin. Draw a regular hendecagon that is 10 cm across to represent a large-scale drawing. What is the size of an interior angle? The length of a side? Describe the steps you used to make a scale drawing.

7. If the measure of an interior angle of a regular polygon divides evenly into 360°, then the polygon will tessellate. Use this property to determine which regular polygons will tessellate.

- equilateral triangle
- square
- pentagon
- hexagon
- heptagon
- octagon
- nonagon
- decagon
- henadecagon
- dodecagon

8. The distance from the centre of a regular polygon to the midpoint of a side is called the **apothem**. TV is the apothem in the regular pentagon.

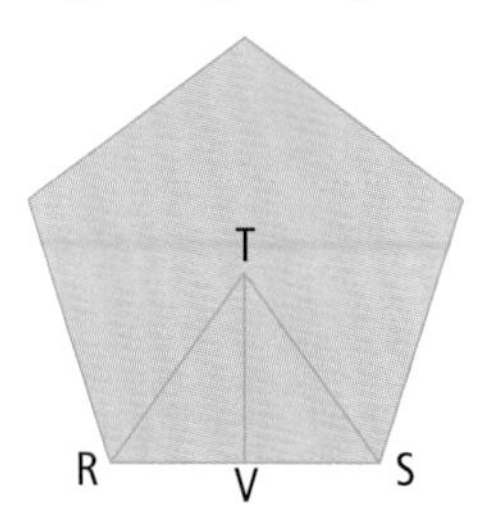

a) RT = 11 cm and TV = 9 cm. What is the length of a side of the regular pentagon?

b) Use the information in part a) to find the area of $\triangle$RST. Then, calculate the area of the regular pentagon. Explain your thinking.

c) What is the perimeter of the regular pentagon? Explain your method.

9. a) Draw an acute scalene triangle, then make a copy.

b) In the first triangle, construct the three perpendicular bisectors of its sides.

c) Draw a circle using the point of intersection of the perpendicular bisectors as a centre and the distance from this point to a vertex as a radius. What do you notice about the circle? The centre of the circle is called the **circumcentre** of the triangle. Why does this name make sense?

d) In the second triangle, construct the three angle bisectors.

e) Draw a circle using the point of intersection of the angle bisectors as a centre and the perpendicular distance from this point to a side as a radius. What do you notice about this circle? The centre of this circle is called the **incentre** of the triangle. Why does this name make sense?

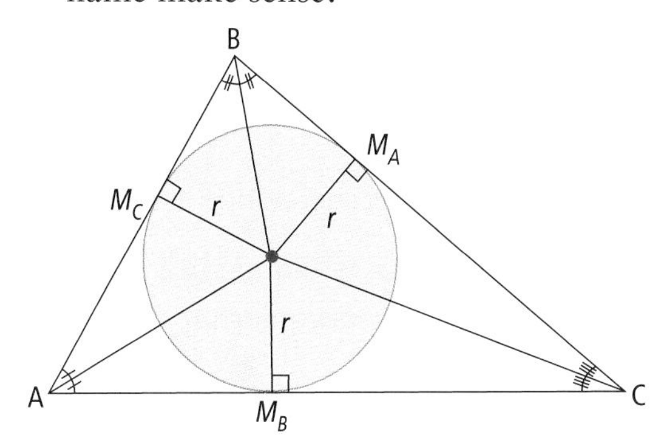

f) Explain why the centre of any regular polygon is both a circumcentre and an incentre.

10. List all the regular polygons (from equilateral triangle to dodecagon) that apply to each statement. Justify your answers.

a) I have seven congruent interior angles.

b) I have pairs of parallel opposite sides.

c) All my lines of symmetry are perpendicular bisectors of my sides.

d) I have three lines of symmetry that join opposite vertices.

e) I am the result of a number of rotations of an isosceles triangle about its vertex angle of 36°.

f) All my exterior angles are 30°.

g) I tessellate.

h) The number of my sides is a divisor of 360°.

Chapter Problem

11. a) List all the ways the design on this quilt block uses the number 7.

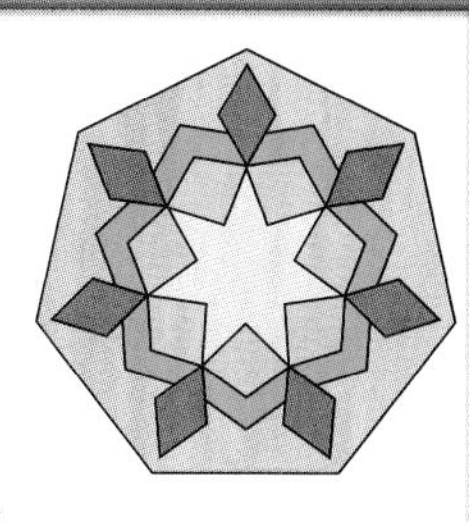

b) Design a quilt block similar to this one based on the number 5.

Extend

12. a) Draw a regular pentagon. Number the vertices from 1 to 5. Connect the vertices with line segments: the first vertex to the third vertex, the third to the fifth, the fifth to the second, the second to the fourth, and the fourth to the first.

b) This shape is called a star polygon. How many lines of reflective symmetry does the shape have? What is its degree of rotational symmetry?

c) Repeat parts a) and b) using a regular heptagon and a regular nonagon. How do the shapes you made compare? Can you predict the shape you would draw if you used a regular hendecagon?

d) Repeat parts a) and b) using a regular hexagon. How many line segments did you draw before returning to the first vertex?

e) Repeat parts a) and b) using the second vertex in the regular hexagon. What shape do you get?

Strategies
Try using a table to organize your work.

CHAPTER 3 Review

1. Which triangles are unique? Which are not? Justify your reasoning.

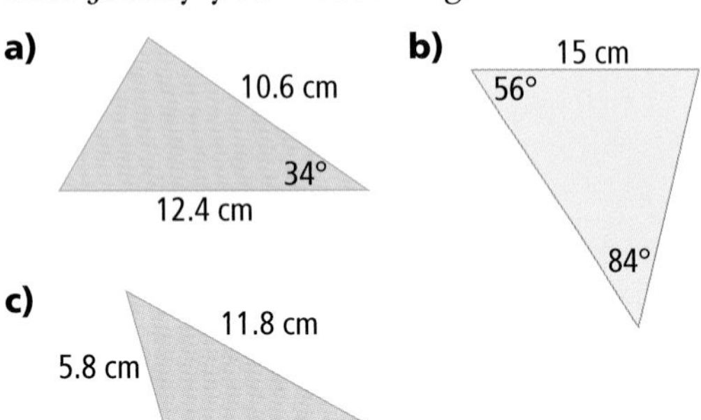

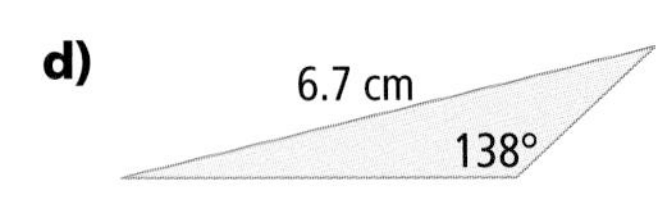

2. In $\triangle DEF$, $\angle D = 47°$, $DE = 15$ cm, and $\angle F = 75°$. In $\triangle LMN$, $\angle N = 75°$, $LM = 15$ cm, and $\angle L = 47°$. Use logical reasoning to convince a classmate that $DF = LN$.

3. **a)** You are given a triangle and its image under a reflection but the mirror line is not shown. Explain how you could use a Bullseye compass to locate the mirror line.
   b) You are given a triangle and its image under a rotation but the centre of rotation is not shown. Explain how you could use a Bullseye compass to locate the centre.

4. The front and back of an Australian 50¢ coin is shown.

   a) The coin is in the shape of what polygon?
   b) Calculate the measure of the interior angles using two different methods.
   c) Describe the relationships among the sides.
   d) What kinds of symmetry does the polygon have?

5. Which pairs of triangles are congruent from the given information? What reason would you give in each case? If they are not congruent, explain why.

a)

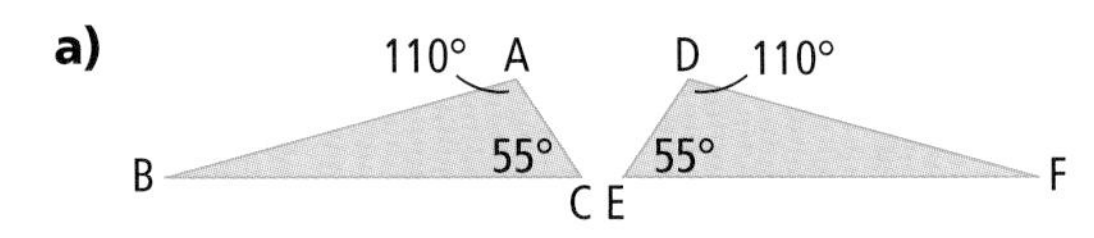

b)

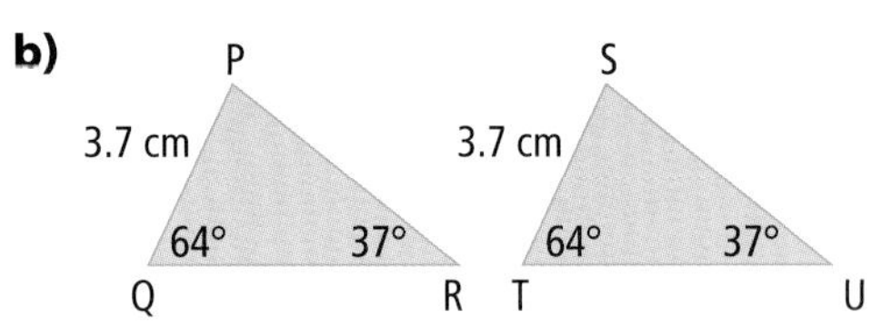

c)

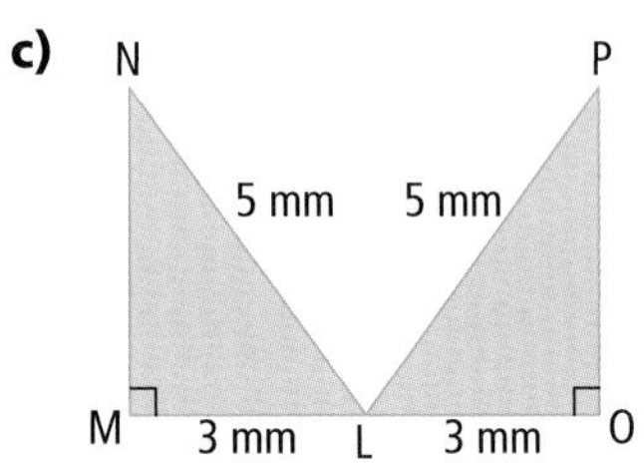

d)

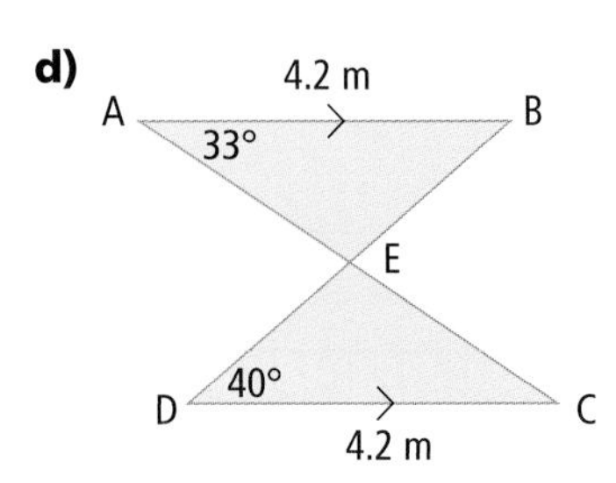

Use these instructions for question 6 and 7. Copy the triangle onto square dot paper. Label the vertex of the largest angle P, and, going clockwise, label the other two vertices Q and R.

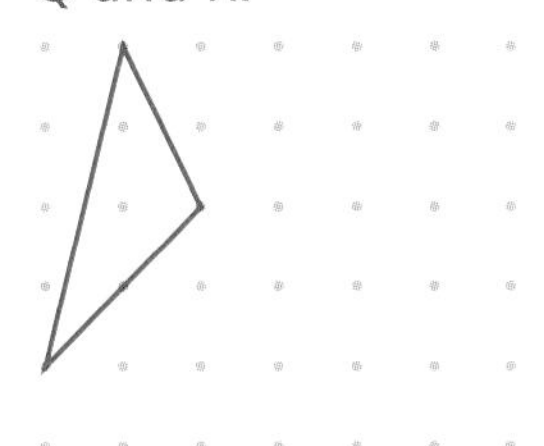

6. a) With centre Q, rotate △PQR 90° clockwise. Use prime symbols to label the corresponding vertices.

b) List all the relationships you know between the pre-image and the image.

7. a) Reflect △PQR in PQ and label the corresponding vertices using prime symbols.

b) List all the relationships you know between the pre-image and the image.

8. If you compare all the sides of two same-sized triangles, you may be able to prove the two triangles are congruent. But if you compare all the sides of two quadrilaterals, this will not prove they are congruent. Explain why.

9. Which transformation of the black quadrilateral produces the red quadrilateral? Explain how you know.

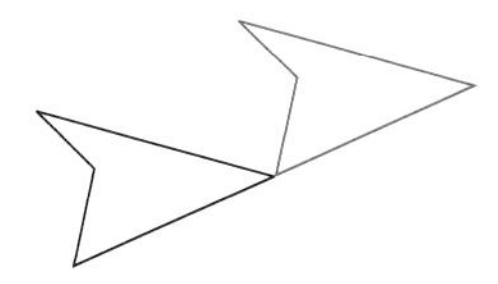

10. A quilt designer uses these instructions to draw a pattern for a quilt block with a nonagon.

- Start with a 12 cm by 12 cm square.
- Draw an inscribed circle in the square.
- Draw a line segment connecting the centre of the square to the point where the circle touches the top of the square.
- Use the top point as a starting point to draw the other vertices of the regular nonagon on the circle.
- Join each vertex to the centre of the square.

a) On square dot paper, follow the designer's instructions to draw the pattern.

b) Explain how you knew how to draw the inscribed circle.

c) Explain how you located the vertices of the regular nonagon on the circle.

d) The designer wants to use only two colours for his triangles. He wants his quilt block to have rotational symmetry. Is this possible? Explain.

11. a) Draw scalene △JKL.

b) Use tracing paper or a Bullseye compass to draw the image of △JKL rotated 180° about centre L. Label the corresponding points J′ and K′. What is the image of L? Why?

c) Is the centre and angle of rotation easier to see in this drawing than in the other rotation drawings for this section? Explain why. What is special about the direction of the rotation? Why?

d) Why do you know that JK = J′K′? What appears to also be true about JK and J′K′? Explain how you know for certain. Why is this only true for a 180° rotation?

CHAPTER 3

Practice Test

Selected Response

Select the best answer.

1. Which information about a triangle will *not* guarantee that the triangle is unique?

A the lengths of three sides
B the lengths of two sides and the measure of the included angle
C the lengths of two sides and the measure of an angle that is not included

2. The corresponding sides of a polygon and its image under a transformation are parallel. The transformation is

A a 90° rotation
B a translation
C a reflection
D none of the above

3. If the interior angles of a regular polygon are 60°, then the regular polygon is

A a nonagon
B a hexagon
C a pentagon
D an equilateral triangle

Short Response

Provide a complete solution.

4. Explain how to calculate the measure of an interior angle of a regular 30-gon.

5. You are given the measures of two angles of a triangle.

a) Explain why you do not know if the triangle is unique. Include a diagram to help you explain.
b) What additional information would you need to prove that the triangle is unique?

6. Nona says that the lines of symmetry of all regular polygons join opposite vertices. Is Nona correct? Explain and provide examples.

7. The Parks Department has planted a flowerbed in the shape of a scalene triangle on a hill near town. They want another flowerbed 1 m away. The second flowerbed is to be a reflection of the first. You are given the task of locating and marking the vertices of the second flowerbed. Explain how you and a partner could do this using ropes and stakes.

8. Identify each transformation. Explain how you know.

a) The pre-image and image are congruent. All points on the pre-image have moved, but by differing amounts. The pre-image and the image have different orientations.
b) The pre-image and image are congruent. All but one point on the pre-image has moved. The line segments joining corresponding points are not parallel.

Extended Response

Provide a complete solution.

9. a) Which pairs of triangles are congruent? Explain how you know.

i)

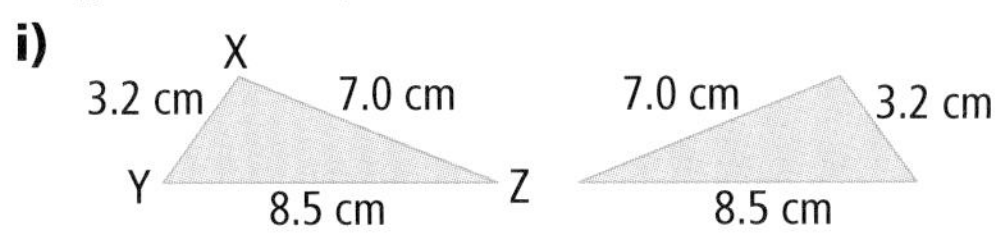

ii)

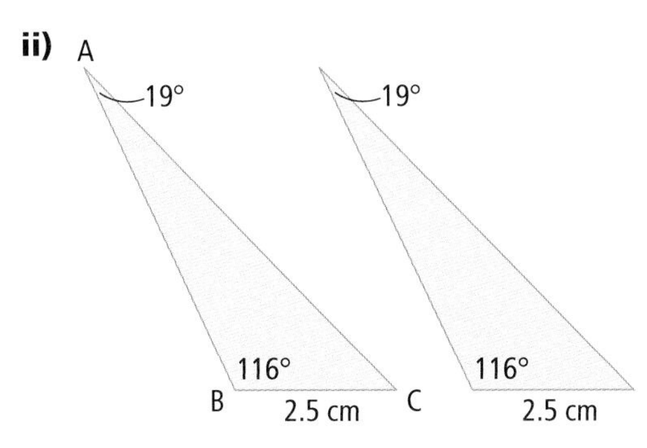

iii)

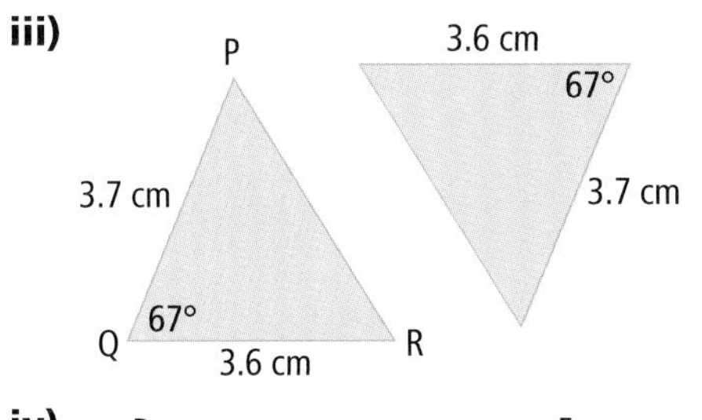

iv)

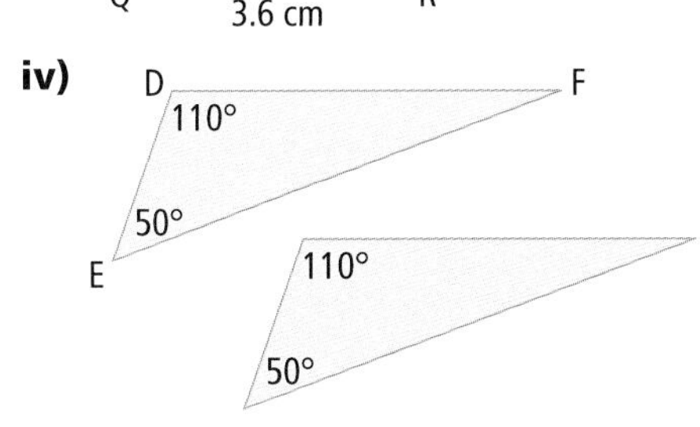

b) Copy each pair of triangles. Label the unlabelled triangle in each pair using prime symbols.

c) For each pair, state the congruencies between corresponding parts that are not labelled with measures.

d) For each pair of congruent triangles, describe the transformation used on the pre-image.

10. Reilly has drawn an octagon and a regular octagon. What will be different about the two polygons? Explain how you know.

11. An isosceles triangle has a vertex angle of 30°. Which regular polygon would you create if you rotated this triangle in a series of 30° turns about its vertex angle? Explain how you know.

12. Each interior angle of a regular polygon is greater than 140° but less than 150°. The polygon's lines of reflective symmetry are perpendicular-bisectors of its sides. Name the regular polygon. Justify your answer.

Chapter Problem Wrap-Up

Design a 12 cm by 12 cm quilt block. Use congruent triangles, regular polygons, transformations, and symmetry in your design. Colour your design to highlight the patterns you have made.

Chapters 1–3 Review

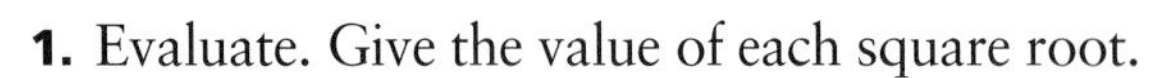

1. Evaluate. Give the value of each square root.

   a) $\sqrt{729}$ b) $\sqrt{144}$

   c) $\sqrt{0.36}$ d) $\sqrt{81 \times 121}$

2. A square has an area of 169 cm^2. What is the radius of the largest circle that can just fit around the square?

3. Find the value of each square root. Use prime factorization if the square root is exact. Use an estimate if the square root is not exact.

   a) $\sqrt{0.144}$ b) $\sqrt{1.44}$

   c) $\sqrt{144}$ d) $\sqrt{1440}$

4. The playing surface of a Scrabble® board has an area of 900 cm^2. What is the side length of each small square on the board? (Hint: The board is square and contains 225 congruent squares.)

5. List all of the properties of a regular dodecagon. Use a table to organize your information. Show your work.

6. Calculate.

   a) $\frac{1}{4} + \frac{3}{8}$ b) $5\frac{1}{2} - 1\frac{2}{3}$

   c) $4\frac{3}{5} - \frac{7}{10}$ d) $\frac{1}{6} + 3\frac{1}{5}$

7. Carol used $3\frac{1}{2}$ cups of flour to make a cake. Jacques used $2\frac{3}{5}$ cups of flour to make cookies. How much flour did they use altogether? How much more flour did Carol use than Jacques?

8. Multiply.

   a) $\frac{1}{3} \times \frac{9}{4}$ b) $6 \times \frac{1}{10}$

   c) $2\frac{1}{3} \times \frac{4}{7}$ d) $4 \times \frac{5}{8}$

9. Find the missing dimension using the Pythagorean relationship. Round your answer to one decimal place.

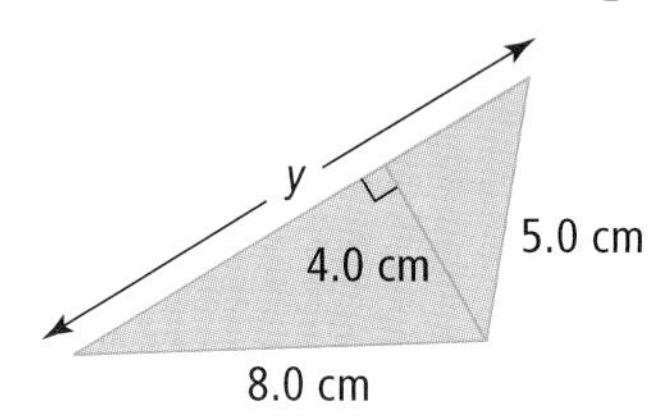

10. Divide.

    a) $5 \div \frac{1}{4}$ b) $2 \div \frac{1}{3}$

    c) $1\frac{1}{5} \div 2$ d) $\frac{3}{4} \div \frac{2}{3}$

11. Which transformation of the black quadrilateral produces the red quadrilateral? Explain how you could convince someone that your answer is correct.

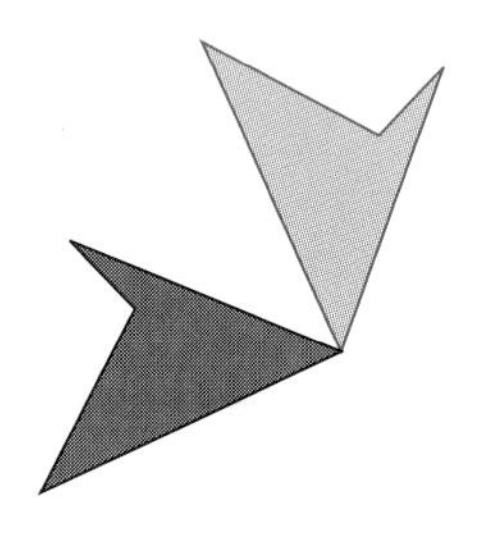

12. Betty runs $8\frac{3}{4}$ laps around the school track every morning. This morning, she ran $\frac{2}{5}$ of her laps before it started to rain heavily.

    a) How many laps did she run?

    b) How many more laps should she run to complete her morning routine?

13. In the diagram, AB ∥ EC. Use logical reasoning to convince a classmate that △ABC ≅ △ECD. Record your reasoning.

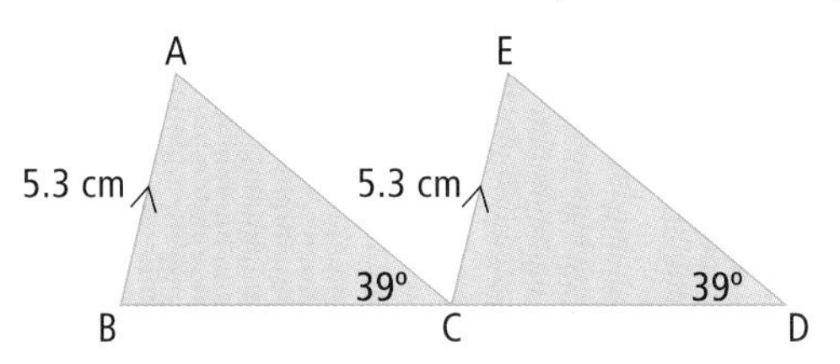

14. You get a job silk screening T-shirts. You get paid \$6.50 for each T-shirt that you silk screen. It takes you $\frac{3}{4}$ h to complete one T-shirt. If you worked for 7 h, how much would you expect to make?

15. Which pairs of triangles are congruent? What reasons would you give in each case? For the pairs that are not congruent, explain why.

a)

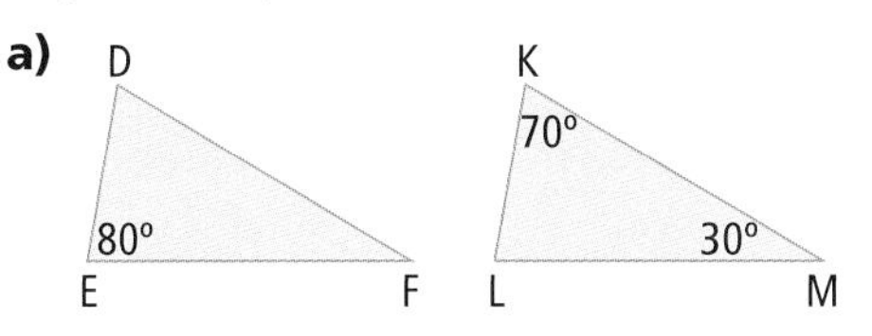

b)

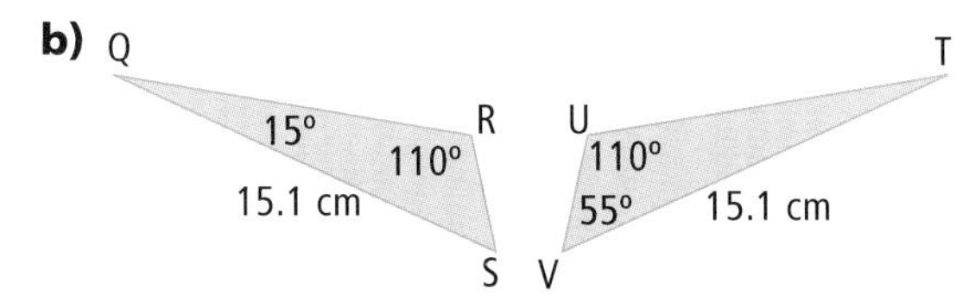

c)

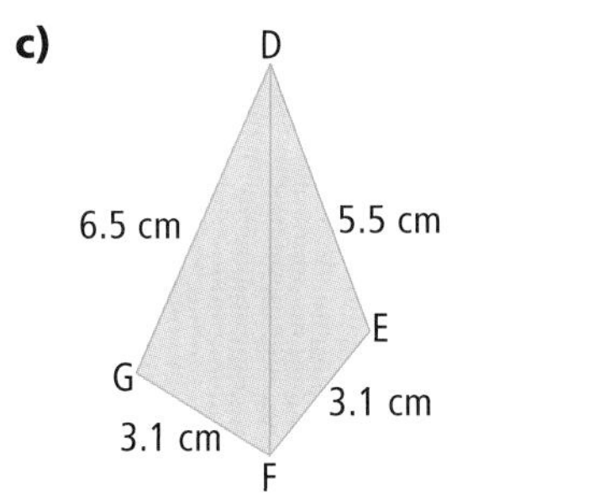

16. Which triangles are unique? Explain how you know. Include diagrams to help you explain.

a) △ABC where ∠A = 75° and ∠B = 45°.
b) △KLM where ∠K = 25°, KL = 7.0 m, and LM = 8.1 m.
c) △PQR where PQ = 8 cm, QR = 10 cm, and ∠P = 90°.
d) △STU where ∠S = 38°, ∠T = 64°, and SU = 8.9 mm.

17. Evaluate.

a) $\left(\frac{1}{2} + \frac{5}{8}\right) \div \frac{3}{4}$ **b)** $\frac{3}{4} + \left(\frac{1}{2}\right)^2$

c) $\left(\frac{1}{6} + 2\frac{1}{3}\right) \times \left(\frac{2}{3} + \frac{1}{7}\right)$

18. a) Copy the black and blue pentagons onto square dot paper. Label the vertices of the black hexagon using the letters A to E. Label the corresponding vertices of its blue image using the letters A′ to E′.

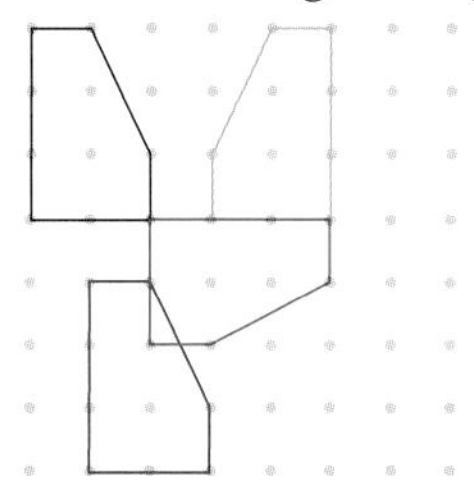

b) What transformation of the black pentagon produces the blue pentagon? Explain the properties of the transformation you chose to prove your choice is correct.

c) Repeat parts a) and b), using the
i) black and red pentagons
ii) black and green pentagons
iii) blue and black pentagons

- represent and apply fractional percents, and percents greater than 100, in fraction or decimal form, and vice versa
- solve proportion problems that involve equivalent ratios and rates
- solve problems involving proportions, using a variety of methods
- create and solve problems that involve finding *a*, *b*, or *c* in the relationship *a*% of *b* = *c*, using estimation and calculation
- apply percentage increase and decrease in problem situations
- solve measurement problems, using appropriate SI units

Key Words

unit rate
rate
ratio
proportion
fractional percent
percent increase
percent decrease
unit price

CHAPTER 4

Ratio and Proportion

At the Windsor-West Hants Pumpkin Festival in Windsor, Nova Scotia, pumpkin growers compete in the annual pumpkin weigh-off contest. In 1999, Windsor held its first Pumpkin Regatta: a race across Lake Pesaquid in giant hollowed-out pumpkins. What percent of a pumpkin has to be scooped out to make a boat? What can a participant's time crossing one-half of the lake's length tell you about the time they will take to complete the race? How does the number of paddle strokes affect a participant's speed? In this chapter, you will learn to calculate fractions, percents, ratios, and proportions so you can solve problems such as these.

Chapter Problem

Suppose you want to enter this year's pumpkin weigh-off contest. You find the following information about the world records for the heaviest pumpkin.

Year of World Record	Mass of Pumpkin (kg)
2006	682.7
2005	667.7
2004	657.3
2003	629.5

- Is there a pattern in the percent increase of mass each year?
- What is the mass of a pumpkin that is 108% of the mass of the 2006 world record pumpkin?
- An average-sized pumpkin used for making jack-o-lanterns is about 3.6 kg. What is the ratio of the mass of an average pumpkin to the mass of the 2006 world record pumpkin?

Get Ready

Percent

Percent means "out of 100".

1. One grid represents 100%. What percent do the shaded sections represent?

a) 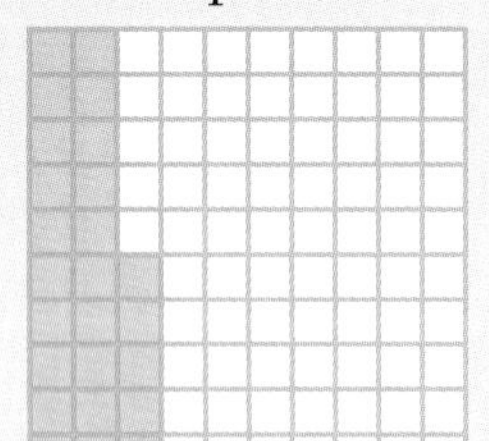

b) 

c)  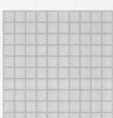 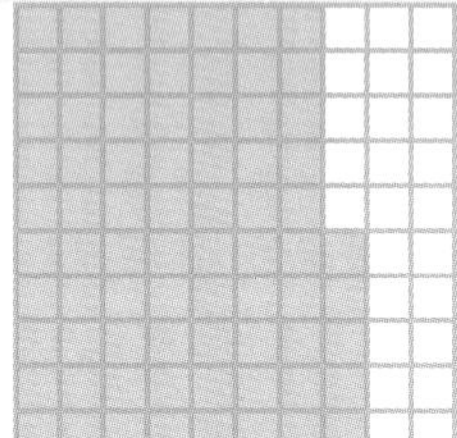

2. Suppose a pizza was cut into 15 slices, and 80% of the slices were eaten. How many slices were left over?

3. If 35% of the 200 people who attended a concert bought a souvenir program, how many people did not buy a program?

Unit Rates

A **rate** is a comparison of two quantities that have different units of measure. \$2.10 for 7 apples is a rate.

$$\frac{\$2.10 \div 7}{7 \text{ apples} \div 7} = \frac{\$0.30}{1 \text{ apple}} = \$0.30/\text{apple}$$

A **unit rate** is a rate in which the second term is 1. \$0.30/apple is a unit rate.

In the expression, \$0.30/apple, the "1" before "apple" is usually left out, but \$0.30/apple is understood to mean \$0.30 for 1 apple.

4. a) Cindy was paid \$45 for babysitting for 3 h. Calculate her hourly rate of pay.

b) Warren ran the 50-m dash in 8 s. What was his running speed per second?

c) Samia bought 12 hairbands for \$9.48. What is the price of 1 hairband?

d) Phillipe ate $4\frac{1}{2}$ hot dogs in 18 min. How many hot dogs did he eat per minute?

Express Fractions as Percents and Decimals

$\frac{2}{5} = 2 \div 5$ $\quad 0.4 = \frac{4}{10}$ or 40 hundredths $\quad \frac{7}{8} = 7 \div 8$ $\quad 0.875 = \frac{875}{1000}$ or 87.5 hundredths

$= 0.4$ $\quad = 40\%$ $\quad = 0.875$ $\quad = 87.5\%$

5. Copy and complete the table.

Fraction	Percent	Decimal
$\frac{9}{45}$		
$\frac{60}{100}$		
$\frac{5}{8}$		
	15%	
	29%	
	71.8%	
		0.556
		2.5
		0.048

Ratios

A **ratio** is a comparison of two numbers. For example, if a fruit basket contains 3 apples and 2 pears, then the ratio of apples to pears can be written as 3 to 2, 3:2, or $\frac{3}{2}$. In all these cases, you can say the ratio is "three to two." The *three* and *two* are called the *terms* of the ratio.

6. A parking lot contains 25 cars: 7 blue, 5 white, 5 silver, 4 green, 2 black, and 2 red.

a) What is the ratio of blue cars to white cars?

b) What is the ratio of red cars to silver cars?

c) What is the ratio of black cars to the total number of cars?

d) What is the ratio of green cars to the cars that are not green?

4.1 Fractions, Decimals, and Percents

Focus on…

- representing and applying percents in fraction or decimal form, and vice versa
- finding *a*, *b*, or *c* in the relationship $a\%$ of $b = c$
- applying percentage increase and decrease in problem situations

You see fractions, decimals, and percents in your everyday life in the form of price discounts, bank interest rates, sales commission, and so on. You might have heard some common phrases that include fractions or percents, such as a person who has 20/20 vision or someone who gives 110% in a game. What do these numbers mean, and are they mathematically sound?

Materials

- coloured pencils (red, blue, and four other colours)
- a sheet of hundred grids

Discover the Math

Part A: Represent Percents Less Than 1

Inflation rates are often expressed as a fractional percent that is less than 1.

1. a) On a hundred grid, shade one small square red. What fraction of the grid is red?

b) Express the fraction as a decimal, and then as a percent.

Communicating Mathematically

A **fractional percent** is a percent that contains a part of a percent in fraction form.
Example: $8\frac{2}{3}\%$ and $\frac{1}{4}\%$ are fractional percents.

2. On the same hundred grid, shade $\frac{1}{2}$ of a small square blue. What percent of the grid is blue? Explain what this means. Express the percent as a decimal.

3. On a new hundred grid, shade squares to represent these fractional percents. Use different colours.

a) $\frac{1}{3}\%$ **b)** $\frac{1}{4}\%$ **c)** $\frac{1}{5}\%$ **d)** $\frac{2}{5}\%$

4. a) On another hundred grid, shade squares to show 40%, 4%, and 0.4%, using different colours. Express each as a decimal and as a fraction in simplest form.
   b) How do 40%, 4%, and 0.4% differ in meaning?

Communicating Mathematically

The inflation rate is the rate of increase in the average price level (of goods and services).

It could also represent the rate of decrease in the purchasing power of money.

Part B: Represent Percents Greater Than 1 but Less Than 100

Bank rates for savings accounts, loans, and mortgages are often expressed as a fractional percent.

Suppose the bank pays $3\frac{1}{2}$% annual interest.

Materials
- hundred grids
- coloured pencils

1. a) On a hundred grid, shade three small squares. What percent of the grid does the shaded area represent? Write the percent as a decimal.
   b) On a second hundred grid, shade four small squares. What percent of the grid does the shaded area represent? Write the percent as a decimal.
   c) On a third hundred grid, shade $3\frac{1}{2}$% of the grid. Write the percent as a decimal.

2. Copy and complete the table.

Fractional Percent	Percent Expressed with a Decimal	Fraction With 100 in the Denominator	Decimal
$6\frac{3}{4}$%	6.75%	$\frac{6.75}{100}$	0.0675
$11\frac{1}{5}$%			
	19.25%		
			0.085

3. Suppose you deposit $100 into a bank account that pays $3\frac{1}{2}$% interest annually.
   a) How much is 3% of $100? How much is 4% of $100?
   b) Express each percent as a decimal and write a multiplication sentence for each part of part a).
   c) Use your answer in part a) to predict the amount of interest you would earn in one year.
   d) Explain how you can use the decimal equivalent of $3\frac{1}{2}$% to calculate $3\frac{1}{2}$% of $100.

Part C: Represent Percents Greater Than 100

Materials
- linking cubes

1. Use four linking cubes to build a 4 by 1 rectangular prism. The prism represents 100%.

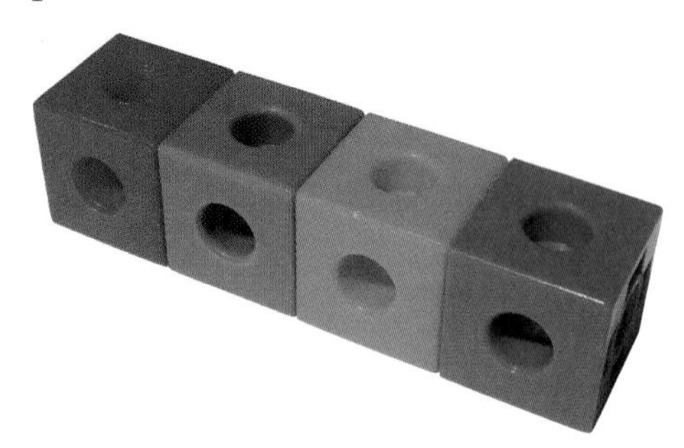

Strategies
What other concrete material could you use to model percent?

2. **a)** Build a second rectangular prism using only 50% of the linking cubes used in the rectangular prism from question 1.
   b) How can you tell you used the correct number of linking cubes?

3. Build a third rectangular prism that uses 50% more linking cubes than the original rectangular prism. This rectangular prism is 150% the size of the original prism, or 1.5 times larger.

4. **a)** How many linking cubes would you need to build a fourth rectangular prism that uses 100% more linking cubes than the original rectangular prism? Build it.
   b) What percent of the original is your new prism? What would be another way to say this?

5. Let represent 100%. Copy and complete the table.

Percent of Original	Drawing	Number of Cubes
50%		2
100%		4
125%		
		6
200%		
250%		
300%		

6. **Reflect** John's tomato plant grew 250% of its original size in 3 months. How can he represent the percent using a diagram and in words or numbers?

Example 1: Calculate and Convert Percents, Fractions, and Decimals

a) Express $5\frac{1}{4}\%$ as a decimal.

b) Express 0.6% as a fraction with a whole number numerator.

c) Find 110% of 88.

d) Calculate $2\frac{2}{5}\%$ of 350.

e) Express 0.0075 as a fractional percent.

Solution

a)
$$\begin{aligned} 5\frac{1}{4}\% &= 5.25\% \\ &= \frac{5.25}{100} \\ &= \frac{5.25}{100} \times \frac{100}{100} \\ &= \frac{525}{10\,000} \\ &= 0.0525 \end{aligned}$$

$\frac{1}{4}$ is the same as 0.25. Then $5\frac{1}{4}\% = 5.25\%$.

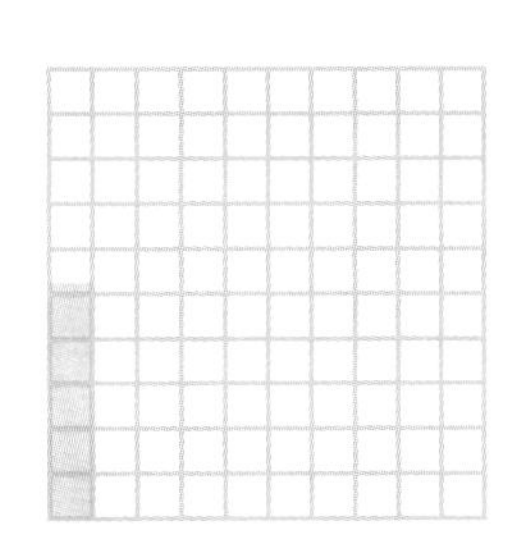

If a percent is written as a mixed number, convert the fraction part to a decimal first.

b)
$$\begin{aligned} 0.6\% &= \frac{0.6}{100} \\ &= \frac{0.6 \times 10}{100 \times 10} \\ &= \frac{6}{1000} \end{aligned}$$

Multiply the numerator and denominator by the same power of 10 to make the numerator a whole number.

c)
$$\begin{aligned} 110\% \text{ of } 88 &= \frac{110}{100} \times 88 \\ &= 1.1 \times 88 \\ &= 96.8 \end{aligned}$$

110% = 100% + 10%. 100% of 88 is 88 and 10% of 88 is 8.8. So, 110% of 88 is 88 + 8.8 = 96.8.

d)
$$\begin{aligned} 2\frac{2}{5}\% \text{ of } 350 &= 2.4\% \times 350 \\ &= \frac{2.4}{100} \times 350 \\ &= 0.024 \times 350 \\ &= 8.4 \end{aligned}$$

$\frac{2}{5} = 2 \div 5$
$= 0.4$

e)
$$\begin{aligned} 0.0075 &= \frac{75}{10\,000} \\ &= \frac{75 \div 100}{10\,000 \div 100} \\ &= \frac{0.75}{100} \\ &= 0.75\% \\ &= \frac{3}{4}\% \end{aligned}$$

Express as a fraction out of 100.

Example 2: Taxes on Purchases

In Nova Scotia, you must pay HST (Harmonized Sales Tax) on some purchases. The HST is 14% of the purchase price. A lamp costs \$25. Calculate the final price, including HST.

Solution

Method 1: Calculate the tax first, then add.

$$\begin{aligned}\text{Amount of tax} &= 14\% \text{ of original cost}\\ &= \frac{14}{100} \times 25\\ &= 0.14 \times 25\\ &= 3.5\end{aligned}$$

The tax is \$3.50.

> **Strategies**
> Break the problem into simpler parts.

$$\begin{aligned}\text{Final price} &= \text{Original cost} + \text{Amount of tax}\\ &= 25 + 3.5\\ &= 28.5\end{aligned}$$

The final price is \$28.50.

Method 2: Multiply the original cost by a percent greater than 100.

$$\begin{aligned}\text{Final price} &= 114\% \text{ of original cost}\\ &= 1.14 \times 25\\ &= 28.5\end{aligned}$$

The final price is \$28.50.

> The final price is the 100% of the original price plus 14% of the original price for the tax, or 114% in total.

Example 3: Build a Scale Model

Kavitha is building a model of the CN Tower, in Toronto. The tower is 553 m tall. How tall will her model be if the model is 0.1% of the actual size of the tower?

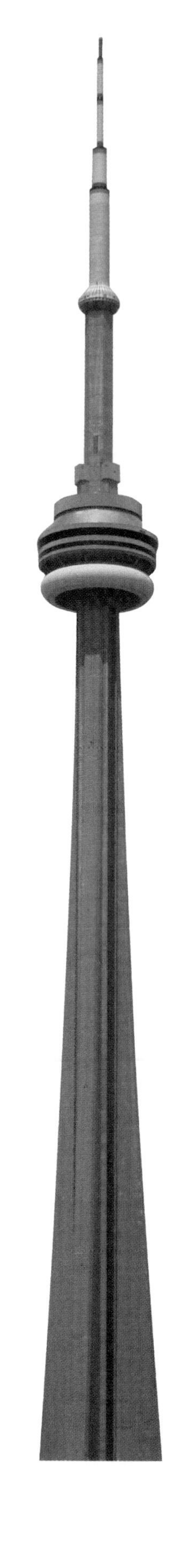

Solution

$$\begin{aligned}\text{Model size} &= 0.1\% \text{ of actual size}\\ &= \frac{0.1}{100} \times 553\\ &= 0.001 \times 553\\ &= 0.553\end{aligned}$$

The model will be 0.553 m, or 55.3 cm, tall.

> 1 m = 100 cm
> 0.553 m = 55.3 cm

Percent increase and decrease are ratios comparing the change in a quantity compared to the original quantity. A change greater than the original is a **percent increase**. A change less than the original is a **percent decrease**.

Percent Decrease

This year, there are 120 grade 8 students attending Truro Junior High. Last year, there were 135 grade 8 students at the school. This is a decrease in enrollment.

$$\begin{aligned}\frac{\text{Change in enrollment}}{\text{Last year's enrollment}} &= \frac{135 - 120}{135}\\ &= \frac{15}{135}\\ &= 0.111\\ &= 11.1\%\ \textbf{decrease}\end{aligned}$$

There was a 11.1% decrease in this year's enrollment compared to last year's enrollment.

Percent Increase

If next year's enrollment was 150 grade 8 students, this would be a percent increase.

$$\begin{aligned}\frac{\text{Change in enrollment}}{\text{This year's enrollment}} &= \frac{150 - 120}{120}\\ &= \frac{30}{120}\\ &= 0.25\\ &= 25\%\ \textbf{increase}\end{aligned}$$

Compared to this year's enrollment, there would be a 25% increase in next year's enrollment.

Example 4: Calculate Percent Decrease

A store sold a particular model of bicycle for \$360 during the spring. In the summer, the price was lowered to \$333.

a) What percent decrease does the new price represent?

b) Calculate the new price as a percent of the original price.

Solution

a) Amount of decrease = \$360 − \$333
= \$27

$$\frac{\text{Amount of decrease}}{\text{Original price}} = \frac{27}{360}$$
$$= 0.075$$ 27 ÷ 360
$$= 7.5\%$$

The new price is a 7.5% decrease from the original price.

b) *Method 1: Use percents and subtract.*

From part a), the new price is 7.5% less than the original price.
The original price is 100%.
$100 - 7.5 = 92.5$
So, the new price is 92.5% of the original price.

Method 2: Use a fraction.

Compare the new price to the original price.

$$\frac{\text{New price}}{\text{Original price}} = \frac{333}{360}$$
$$= 0.925$$
$$= 92.5\%$$

The new price is 92.5% of the original price.

Communicate the Key Ideas

1. Explain how you would convert a fractional percent, such as $\frac{1}{5}$%, to

a) a decimal

b) a fraction

2. Refer to the two methods in Example 2. Which method is more efficient in calculating the final price including the tax? Explain.

3. Michael has a savings account that earns $\frac{1}{2}$% interest monthly. How much interest will he earn if he deposits \$400 into the account for one month?

4. Let [two-square diagram] represent 100%.

a) What percent is represented by [three-square diagram]?

b) What *percent increase* is represented by [diagram]?

c) Draw a diagram to represent 275%.

5. a) Explain the difference between a percent increase and a percent decrease.

b) Is HST a percent increase or percent decrease? Explain.

Check Your Understanding

1. Express each percent as a decimal.

a) 177% **b)** $5\frac{3}{4}\%$

c) $12\frac{1}{2}\%$ **d)** 0.9%

2. a) Let [square] represent 100%. Draw diagrams to show 50%, 175%, and 300%.

b) Let [three-square diagram] represent 150%. Draw diagrams to show 75%, 100%, and 200%

c) Let [cube diagram] represent 400%. Draw diagrams to show 200%, 100%, and 600%. Use isometric dot paper.

3. Express each percent as a fraction with a whole number numerator.

a) 0.1% **b)** 0.8%
c) 2.3% **d)** 330%

4. Calculate. Check using hundred grids. Include your drawings.

a) 0.3% of 120

b) $\frac{1}{4}\%$ of 52

c) $8\frac{1}{2}\%$ of 400

d) 200% of 99

5. Express each decimal as a fractional percent.

a) 0.0025 **b)** 0.005
c) 0.008 **d)** 0.0675

6. What percent of each entire figure is shaded? Explain how you found your answer. Round your answer to two decimal places, where necessary.

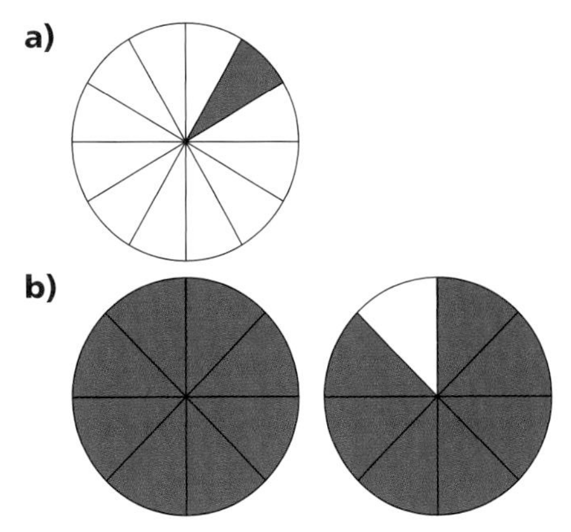

7. For each diagram

- What percent is shown?
- What value is represented?

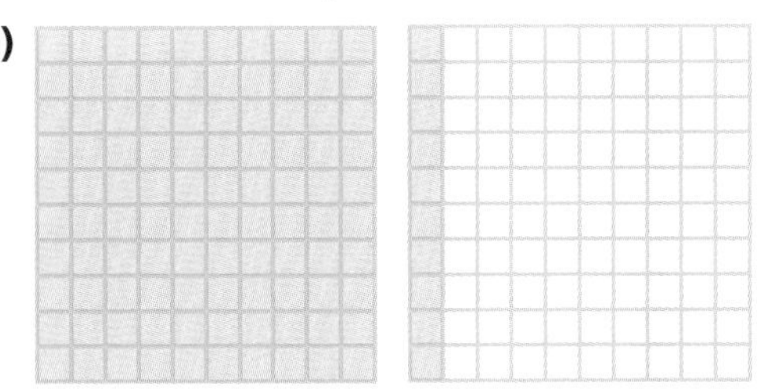

One grid represents 70.

b) 

The grid represents 44.

c) 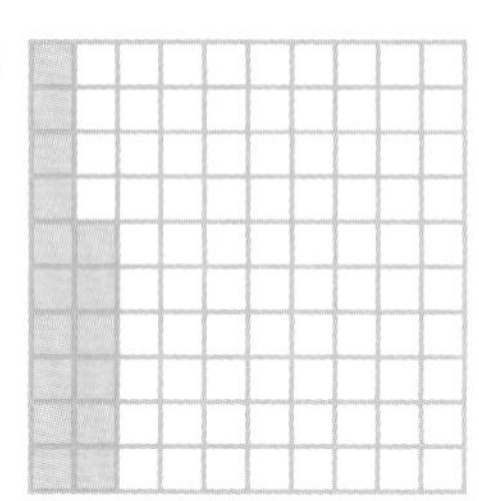

The grid represents 2.5.

8. Calculate the 14% HST on each item.

a) a CD that costs $17.50
b) a sweatshirt that costs $19.99
c) a haircut that costs $25
d) a sweater that costs $50 but is 30% off

9. Tyler opened a bank account that pays $\frac{3}{4}$% interest per month. How much interest will he earn if he deposits $500 into the account for one month?

10. Mario wants to draw a picture of his miniature airplane. He decides to make the drawing 325% the size of the airplane. How many times greater is the drawing compared to the airplane?

11. A movie ticket cost $9 last year. This year, it costs $10. What percent increase does this represent?

Chapter Problem

12. In preparation for the Pumpkin Regatta, Michelle selected a 300-kg pumpkin. She cut off the top of the pumpkin and removed the pulp and seeds, leaving walls about 5 cm thick. The pumpkin shell has a mass of 210 kg.

a) What fraction of the pumpkin's original mass was removed?
b) What percent of the pumpkin's original mass is the pumpkin shell?

13. The price of a sweater is $29.99.

a) Calculate the amount of HST on the sweater.
b) Calculate the total cost of the sweater, including tax.
c) What percent of the price is the total cost?

14. Alysha is building a model of the lighthouse at Peggy's Cove for her history project. The lighthouse is 15.2 m tall. Alysha wants to build her model so it is $2\frac{1}{2}$% the height of the original lighthouse. What is the height of the model, in centimetres?

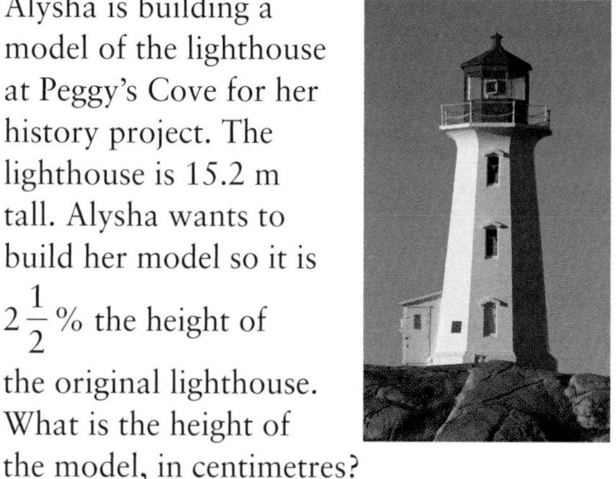

15. A pair of running shoes cost $72 last year. This year, they cost $60.

a) What percent decrease does this represent?
b) Calculate the new price as a percent of the original price.

16. Mona's brother has agreed to lend her some money so she can buy a dress for the prom. The dress costs $89.99 plus 14% HST.

a) Estimate the total cost of the dress.

b) Mona borrows $120 from her brother. She will pay her brother $2\frac{1}{2}$% interest on the loan. How much will Mona pay her brother?

17. Jonas wants to buy a watch for his mother's birthday present. A clearance sale this week reduced the price of a particular watch from $120 to $79.99.

a) What percent decrease does the sale represent?

b) Calculate the reduced price as a percent of the original price.

18. Baron, Maeve, and Winnie ran for student council president. Of the 950 students at the school, 672 voted.

a) Baron got approximately 12.8% of the votes. How many students voted for Baron? Explain.

b) Maeve got approximately 150% as many votes as Baron. How many students voted for Maeve? Explain.

c) There were no spoiled ballots (marked ballots that could not be counted). How many students voted for Winnie? Express the number of votes for Winnie as a percent of the number of votes for Baron.

Extend

19. At the music store, Stacey and Emtiaz are trying to determine if they have enough money to buy their favourite singer's latest CD. There is a sale of 20% off the original price of $24.99. The 14% HST is applied to the sale price.

a) Emtiaz say: "The final cost will be 94% of the original price, since a 20% sale minus 14% tax gives a total savings of 6%." Stacey disagrees. "The final cost will be less than 94% of the original price." Why might Stacey say this?

b) Calculate the cost of the CD, including the discount and HST.

c) What percent of the original price is the final price of the CD?

20. A pitcher contains 2 L of punch. 15% of the punch is grape juice. If 500 mL of orange juice is added to the pitcher, what percent of the punch is now grape juice?

21. Kevin's father said, "When I was young, you could buy a hot dog for 25¢ and a bottle of juice for 10¢."

a) What is the typical cost of these items today?

b) Calculate the percent increase in cost for each item.

22. a) A right triangle has side lengths of 3 cm, 4 cm, and 5 cm. Each of the side lengths is increased by 50%. Find the area of the original triangle and the area of the enlarged triangle. What is the percent increase in the area of the triangle?

b) Construct a right triangle with dimensions of your own choosing. Investigate what happens to the area of the triangle when you increase each side length by 50%. Compare your results to the results in part a).

c) Investigate what happens to the area when you increase the side length by 100%, then by 200%. Try some other percents. Look for a pattern.

4.2 Explore and Apply Proportion, Ratio, and Rate

Focus on...
- solving proportion problems that involve equivalent ratios and rates
- solving problems involving proportions, using a variety of methods

Suppose you want to make cookies for your school's bake sale. You need eight dozen cookies, but the recipe you have only makes two dozen cookies. How can you use proportions to determine how much of each ingredient you will need?

Discover the Math

Materials
- red and yellow counters

Part A: Equivalent Ratios
How can I use equivalent ratios to extend a pattern?

Sam wants to paint a small picket fence that surrounds the vegetable garden in his back yard. He has two different colours of paint: red and yellow. He wants to use a pattern of three red pickets then two yellow pickets per panel.

Let red and yellow counters represent the pickets in each panel.

1. Copy and complete the table.

Number of Panels	Number of Red Pickets	Number of Yellow Pickets	Total Number of Pickets
1	3	2	5
2			
3			
4			

2. Add a new column to your table.

   a) Write the ratio of red pickets to yellow pickets in one panel.

   b) Write the ratio of red pickets to yellow pickets in two panels.

3. What would be the ratio of red pickets to yellow pickets in five panels? six panels? Add this information to your table.

4. Write the ratios from question 2 a) and b) in fraction form. Are they equivalent? How do you know?

5. Compare the other ratios from the table to the first ratio. What can you conclude?

6. Suppose you have 24 red pickets.

   a) By what number are you multiplying the original number of red pickets?

   b) How many yellow pickets would you have?

7. Equal ratios form a **proportion**. If you know three numbers in the proportion, how can you find the fourth? Try the example. Explain how you found your answer.

   $\frac{3}{2} = \frac{27}{■}$

proportion

- a statement that says two ratios are equal. Can be written in fraction form, $\frac{1}{4} = \frac{4}{16}$, or in ratio form, 1:4 = 4:16.

8. Look at the set of counters you used to represent one panel.

   a) How many parts are in this ratio?

   b) What fraction of the total number of pickets is red? yellow?

9. **Reflect**

   a) How do you know when two ratios form a proportion? Give your own example.

   b) How are ratios similar to fractions? How are they different? Explain with your own examples.

Part B: The Meaning of Proportion

Materials
- centimetre grid paper

1. Copy the diagram onto centimetre grid paper.

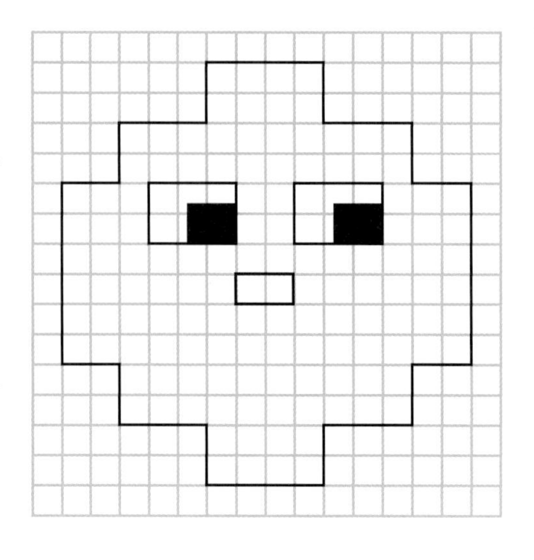

2. Predict how the diagram would change if each side were half as long. Check your prediction by drawing the new diagram. Was your prediction correct?

3. Predict how the diagram would change if each side were 1 cm shorter. Check your prediction by drawing the new diagram. Was your prediction correct?

4. **Reflect** Refer to your diagrams from questions 2 and 3. Which diagram is in proportion to the original diagram? Explain how you know.

Part C: Explore Rates
How can I compare quantities measured in different units?

Bus A, Bus B, and Bus C start from the bus station at the same time and travel in different directions.

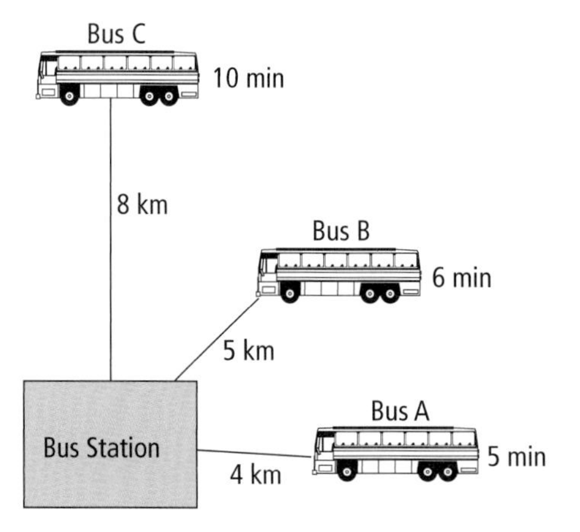

1. For each bus, write the distance travelled compared to the time taken as a **rate**.

rate
- a comparison of quantities measured in different units
- for example, 50 km/h (kilometres per hour) compares the distance travelled to the time taken

2. Which of the rates in question 1 are equivalent rates or are in proportion? Explain how you know.

3. Which bus is travelling at the greatest speed? Explain how you know.

4. If Bus A travels at the same speed throughout its route, how far will it travel in 20 min? in 30 min?

Example 1: Apply Ratios and Proportions to Patterning

Shandra is making a beaded necklace. The pattern will be 4 small blue beads and 3 large pink beads. Each section will be separated by a small silver bar. If there are a total of 140 beads, how many beads will be blue? How many beads will be pink?

Solution

Each section contains 7 beads: 4 blue and 3 pink.
The ratio of the number of blue beads to total number of beads in one section is 4:7.
If there are 140 beads in total, the proportion can be written as shown, where b represents the number of blue beads.

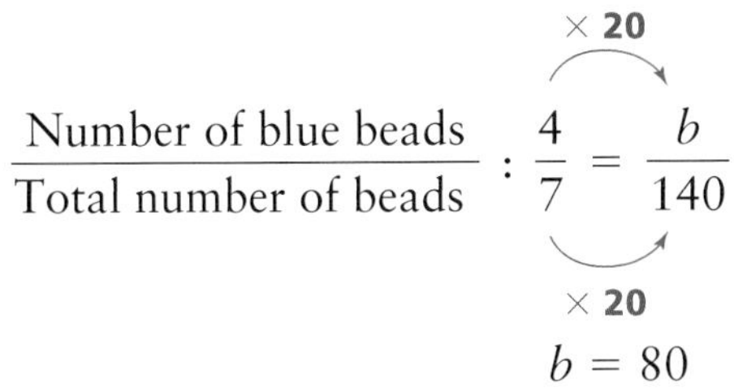

I have to multiply the denominator, 7, by 20 to get 140.

Then 4 × 20 = 80.

$b = 80$

There will be 80 blue beads.

The ratio of blue beads to pink beads is 4:3.

Since there is a total of 80 blue beads, the proportion can be written as shown, where p represents the number of pink beads.

$$\frac{\text{Number of blue beads}}{\text{Number of pink beads}}: \frac{4}{3} = \frac{80}{p}$$

$$p = 60$$

There will be 60 pink beads.

Alternatively, if you know the number of blue beads and the total number of beads, then you could find the number of pink beads by subtraction.

Number of pink beads
= Total number of beads − number of blue beads
= 140 − 80
= 60

Strategies

How could you solve this problem if you do not know about proportions? How is your method like the one shown here?

Example 2: Apply Ratios and Proportions to Recipes

Jasmine has a recipe for a fruit punch that requires 1 cup of grape juice to be mixed with 3 cups of cranberry juice.

a) How many cups of grape juice does Jasmine need for 12 cups of cranberry juice?

b) Suppose Jasmine wants to make 60 cups of punch for a party. How many cups of each type of juice will she need?

Solution

a) Compare two part-to-part ratios of the form "grape juice to cranberry juice."

$$\frac{1 \text{ cup}}{\text{grape juice}} : \frac{3 \text{ cups}}{\text{cranberry juice}} = \frac{x \text{ cups}}{\text{grape juice}} : \frac{12 \text{ cups}}{\text{cranberry juice}}$$

$$\frac{g}{c} : \frac{1}{3} = \frac{x}{12}$$

$$x = 4$$

In the denominator, 3 × 4 = 12; so in the numerator, 1 × 4 = 4.

For 12 cups of cranberry juice, Jasmine needs 4 cups of grape juice.

b) Compare two part-to-whole ratios of the form "grape juice to punch" and "cranberry juice to punch." The total amount of punch in the original recipe is 1 cup + 3 cups = 4 cups.

For the grape juice:
1 cup grape juice : 4 cups punch = x cups grape juice : 60 cups punch

$$\frac{g}{p} : \frac{1}{4} = \frac{x}{60}$$

$$x = 15$$

In the denominator: 4 × 15 = 60; so in the numerator 1 × 15 = 15.

Jasmine needs 15 cups of grape juice to make 60 cups of punch.
Cranberry juice = 60 cups of punch − 15 cups of grape juice
= 45 cups

Jasmine needs 45 cups of cranberry juice to make 60 cups of punch.

Example 3: Solve a Proportion Using the LCD

A pie recipe calls for 2 pears for every 3 apples used in the filling.

Andre and Justina are making the pie together. There are 5 pears. They decide to find the number of apples they should use, using different methods. Which method requires fewer calculations?

Solution

They both set up a proportion where x represents the number apples needed.

Andre's Solution

$$\frac{a}{p}: \frac{3}{2} = \frac{x}{5}$$

$$\frac{15}{10} = \frac{2x}{10}$$

$$15 = 2x$$

$$\frac{15}{2} = \frac{2x}{2}$$

Divide both sides by 2.

$$x = 7.5$$

So 7.5 apples are needed.

The LCD of $\frac{3}{2}$ and $\frac{x}{5}$ is 10. I can express each fraction with a common denominator of 10.

$$\frac{3}{2} = \frac{15}{10} \quad (\times 5) \qquad \frac{x}{5} = \frac{2x}{10} \quad (\times 2)$$

These are equivalent fractions.

Justina's Solution

$$\frac{a}{p}: \frac{3}{2} = \frac{x}{5}$$

2 × 2.5 = 5, so multiply 3 by 2.5.

$$\frac{3}{2} = \frac{x}{5} \quad (\times 2.5)$$

$$x = 7.5$$

So 7.5 apples are needed.

They both found the same answer. Andre found equivalent fractions using the LCD and then solved the proportion. Justina worked with the fractions in simplest form.

Justina's method was shorter.

Example 4: Apply Rates to Comparison Shop

Suppose store A is selling a 300-mL bottle of Shiny Hair shampoo for $1.99, and store B is selling a 500-mL bottle of the same brand of shampoo for $3.49.

What two units are being used? Which is the better deal?

Solution

The two units being used are millilitres (mL) and dollars ($).

Method 1: Use unit price.

$$\text{Unit price of shampoo at store A} = \frac{\$1.99}{300\text{ mL}}$$
$$= \$0.006\ 63/\text{mL}$$
$$= 0.663¢/\text{mL}$$

Use a calculator.

$$\text{Unit price of shampoo at store B} = \frac{\$3.49}{500\text{ mL}}$$
$$= \$0.006\ 98/\text{mL}$$
$$= 0.698¢/\text{mL}$$

Think: 3.50 ÷ 500
= 35 ÷ 5000
= 0.007

This answer is reasonable.

0.663¢/mL < 0.698¢/mL

Unit price at store A < Unit price at store B
Therefore, the shampoo sold at store A is a better deal.

> **Did You Know?**
>
> In grocery stores, unit prices are shown on the shelves so that customers can easily compare prices of products.

Method 2: Find the cost of equal amounts.

For 300 mL and 500 mL, a common amount could be 1500 mL.

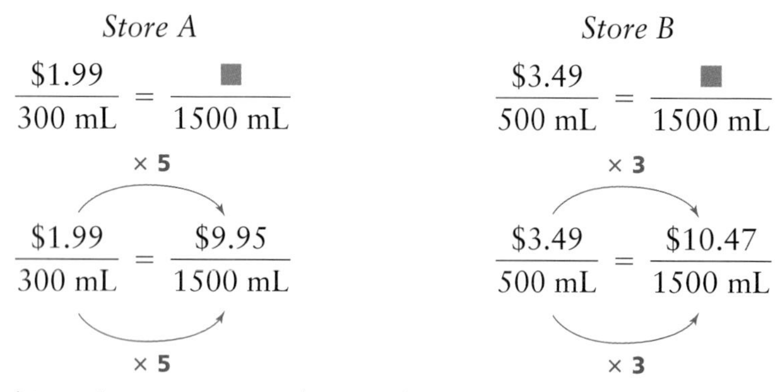

Store A

$$\frac{\$1.99}{300\text{ mL}} = \frac{■}{1500\text{ mL}}$$

× 5

$$\frac{\$1.99}{300\text{ mL}} = \frac{\$9.95}{1500\text{ mL}}$$

× 5

Store B

$$\frac{\$3.49}{500\text{ mL}} = \frac{■}{1500\text{ mL}}$$

× 3

$$\frac{\$3.49}{500\text{ mL}} = \frac{\$10.47}{1500\text{ mL}}$$

× 3

$9.95 for 1500 mL < $10.47 for 1500 mL

Price for 1500 mL at store A < Price for 1500 mL at store B
Therefore, the shampoo sold at store A is a better deal.

Communicate the Key Ideas

1. A proportion is a statement of equality that can be written as a ratio or as a fraction.
   a) What mathematical symbol must a proportion contain?
   b) How is a proportion different from a fraction? Explain using examples.

2. A proportion is written as $\frac{1}{2} = \frac{5}{10}$.
   a) Add 2 to the numerator and denominator of $\frac{1}{2}$. Does the equality still hold?
   b) Multiply the numerator and denominator of $\frac{1}{2}$ by 2. Does the equality still hold?
   c) Try adding, subtracting, multiplying, and dividing $\frac{5}{10}$ by different numbers. Does the equality still hold?
   d) What do your results tell you about the properties of proportions?

3. How is a rate similar to a ratio? How is it different? Explain using examples.

Check Your Understanding

1. Express each as a ratio.

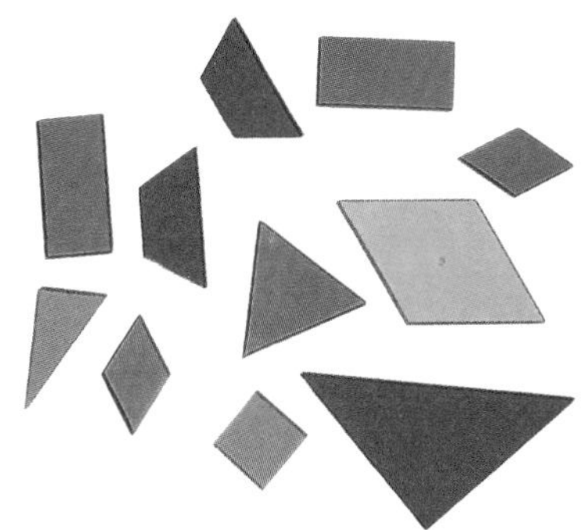

   a) green to orange
   b) red to blue
   c) blue to orange
   d) red to green
   e) orange to red
   f) blue to orange and red

2. In a bag of marbles, there are 8 blue marbles, 13 yellow marbles, and 17 red marbles. Express each as a fraction.
   a) number of blue marbles to total number of marbles
   b) number of yellow marbles to total number of marbles
   c) number of red marbles to total number of marbles
   d) number of marbles that are not red to total number of marbles

3. Express each ratio in simplest terms.

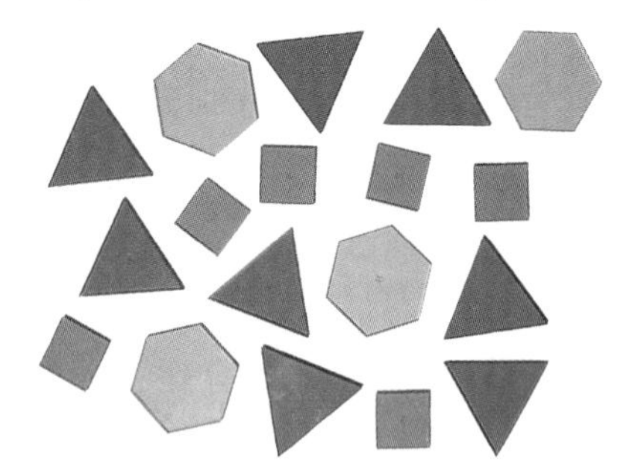

a) squares to triangles
b) triangles to hexagons
c) hexagons to squares

4. Find the missing number in each proportion.

a) ■:60 = 2:3 **b)** 5:10 = 1:■
c) 10:■ = 90:9 **d)** 0:55 = ■:44
e) 2:■ = 5:40 **f)** 1:5 = ■:7.5
g) ■:9 = 2:3 **h)** 2:■ = 5:4
i) 3:5 = ■:12.5 **j)** 4.5:3 = ■:13.5

5. Find the missing number in each proportion.

a) $\frac{1}{6} = \frac{■}{30}$ **b)** $\frac{2}{■} = \frac{80}{240}$
c) $\frac{5}{45} = \frac{1}{■}$ **d)** $\frac{■}{24} = \frac{6}{4}$
e) $\frac{48}{8} = \frac{6}{■}$ **f)** $\frac{56}{■} = \frac{112}{14}$

6. Store A sells a 250 mL bottle of hair conditioner for $3.49. Store B sells a 350 mL bottle of the same conditioner for $4.99. Which is the better deal?

7. Jorge ran 100 m in 20 s. For each part, list the assumptions that you are making.

a) How many metres did he run after 40 s?
b) How many metres did he run after 90 s?
c) How long did it take him to run 500 m?
d) How long did it take to him run 900 m?

8. Carl was paid $42.50 for a 5-h workday. What is his hourly wage?

9. Sara earns $48 working 5 h a day at the grocery store. One day, she worked an extra hour to earn more money. Write a proportion to help you determine the amount of money she earned that day.

10. A recipe for sponge cake calls for 6 eggs, 1 cup of sugar, and $1\frac{1}{4}$ cups of flour.

a) What is the ratio of cups of sugar to cups of flour?
b) If 9 eggs are used to make a larger cake, how much sugar and flour are needed?

11. Tim has his own recipe for barbeque sauce. In 250 mL of water, he mixes 90 mL of ketchup, 60 mL of sugar, and 15 mL of vinegar.

a) Write a ratio in simplest form to compare the amounts of sugar and vinegar used.
b) If Tim uses 150 mL of sugar, how much ketchup does he need?
c) If Tim uses 625 mL of water, how much of the other ingredients does he need?
d) How much of each ingredient would he need for 830 mL of ingredients in total?

12. Jess scored 6 goals in 10 hockey games.

a) Calculate his goals per game.
b) If he scores at the same rate for each game, how many games will he have to play to score his 20th goal?

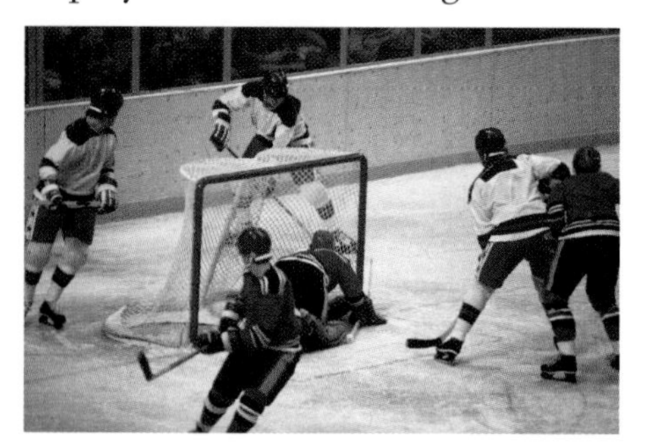

13. It took Carrie $\frac{3}{4}$ h to travel from home to work, a distance of 40 km.

a) At what speed was she driving, in kilometres per hour?

b) Express the speed in metres per second.

14. Apples are on sale at the corner store this week for 25¢ each. The supermarket sells the same type of apples at $2 for 7 apples. Which store has the better deal? Explain.

15. One brand of paper towels costs $1.79 for 90 sheets. Another brand costs $2.09 for 100 sheets. Find the unit price per sheet for each brand. Which brand is the better deal?

16. Compare each pair of products. Which is the better value in each case? Explain.

a) $1.50 for 600 mL of orange juice or $2.00 for 1 L of orange juice

b) $5.00 for 5 kg of potatoes or $7.50 for 15 kg of potatoes

17. Mac is going camping with his friends. He needs to buy some food for the trip. Determine which package is the best value, and how many of each item he can purchase given each budget.

a) $8.00 budget; $5.00 for 20 bottles of 600 mL water or $3.00 for 10 bottles of 600 mL water

b) $12.00 budget; $2.00 for a 200 g bag of apple chips or $1.50 for a 125 g bag of apple chips

18.

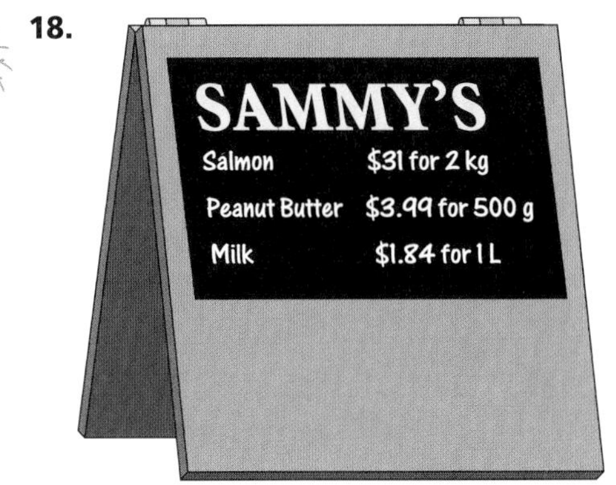

a) Jackie wants to know how much 5 kg of salmon would cost. She set up the following proportion $\frac{\$31}{2\text{ kg}} = \frac{\$\blacksquare}{5\text{ kg}}$ but does not remember what to do next. Explain the steps Jackie should follow to solve the proportion.

b) Show how Jackie could have found the cost of 5 kg, using unit price.

c) Another store is selling peanut butter for $2.50 for 300 g. Explain a method you would use to determine which is the better deal.

d) Scott wants to buy a 750 mL carton of milk. If a small carton is sold at the same rate as a 1-L carton, how much would 750 mL of milk cost? Explain how you found your answer.

Extend

19. A 5-kg bag of potatoes sells for $1.99. Most grocery stores will sell a 10- or 20-kg bag of potatoes for a smaller unit price than the 5-kg bag. Discuss the reasons for this pricing practice.

20. Sam says: "An increase from 5 to 20 is a 300% increase. So, a decrease from 20 to 5 must be a 300% decrease." What is wrong with his reasoning? Explain.

4.3

Solve Problems Involving Proportion, Ratio, and Rate

Focus on...
- solving problems involving proportions
- finding *a*, *b*, or *c* in the relationship $a\%$ of $b = c$
- applying percentage increase and decrease in problem situations

The scale on a map is an important tool that involves ratios and proportions. There are many other real-life applications that use proportions, ratios, and rates, such as comparing prices, altering recipes, and reducing or enlarging photographs.

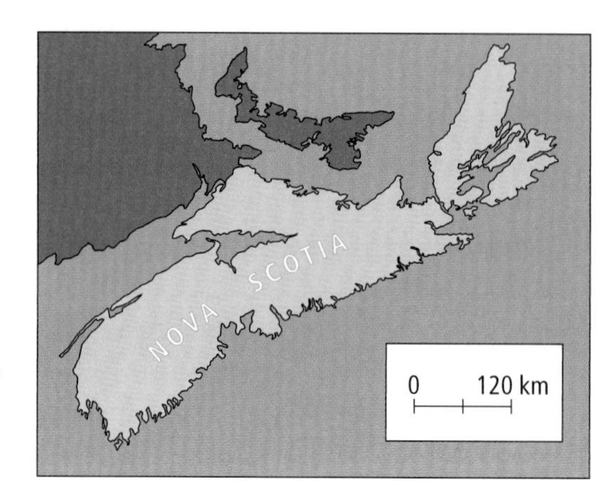

Discover the Math

Materials
- hundred grids

How can I use proportion to solve problems involving percents?

Patrick has a glass jar of marbles on his desk. Of those marbles, 58% are blue, his favourite colour. There are 174 blue marbles in the jar. What is the total number of marbles in the jar?

Strategies
Why is the hundred grid a good way to model the situation with the marbles?

1. a) Let a hundred grid represent all the marbles in the jar. Shade squares on the hundred grid to represent the percent of marbles that are blue.
b) What percent does one square represent?
c) How many squares on the grid are shaded? How many blue marbles does each shaded square represent?
d) How many marbles does each unshaded square represent?
e) How many marbles are in the jar?

2. Sixteen percent of the marbles in the jar are yellow. Show how to use your answers from question 1 to determine how many marbles are yellow.

3. Patrick decides to get a bigger jar and wants to increase the number of marbles to 225% of his current amount. Use hundred grids to model your work.
a) How many more grids would you need to model 225%?
b) How many marbles are represented by each grid?
c) How many marbles in total are needed to model 225%?

4. Most percent problems can be thought of as the relationship $a\%$ of $b = c$. In other words, the percent of the whole is the part.

$$a\% \quad \text{of} \quad b \quad = \quad c$$

In the marble problem, you know the part but not the whole, so you could use the relationship to solve for the whole.
Complete the steps for the 1% method.
58% of $b = 174$
1% of $b = 174 \div 58 = ■$
100% of $b = ■$

a) Solve the equation. Does it agree with your answer from question 1e)?
b) Solve question 2 using this method.

5. **a)** Write the relationship from question 4 as a proportion. Let w represent the whole.
b) Complete the proportion with the correct numbers.

$$\frac{\text{Part}}{\text{Whole}} : \frac{■}{100} = \frac{■}{w}$$
$$w = ■$$

c) Solve for the variable w to find the whole.
d) Does your answer agree with the grid and the equation methods?

Example 1: Find the Percent of a Given Number

A teacher returning a mathematics test tells her class that 80% of the students scored a grade of B or better on the test. If there are 25 students in the class, how many of them scored a grade of B or better on the test?

Solution

There are 25 students. 80% scored a B or better.

Method 1: Use a hundred grid.
Let a hundred grid represent the 25 students.

One square represents 0.25 students, since 100 squares represents 25 students.

Shade squares on the grid to represent 80%.

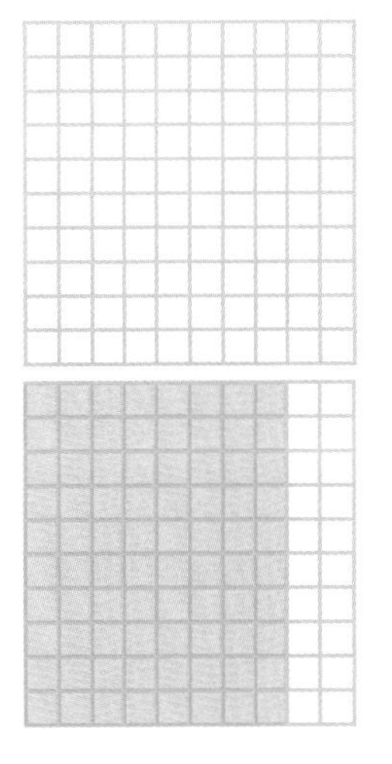

Check, using a hundred grid.
100% = 25 students
10% = 2.5 students
80% = 8 × 2.5 students
= 20 students

Determine the number that the shaded region represents. $80 \times 0.25 = 20$
Twenty students scored a B or better on the test.

Method 2: Use an equation.
Find 80% of 25 students. What percent of the whole is the part?

$a\%$ of $b = c$
80% of $25 = c$
$0.80 \times 25 = c$
$20 = c$

0.80 × 25 = 20

Twenty students scored a B or better on the test.

Method 3: Use a proportion.
Let x represent the number of students who scored a B or better.

$$\frac{\text{Part}}{\text{Whole}} : \frac{80}{100} = \frac{x}{25}$$

In the denominator, 100 ÷ 4 = 25, so 80 ÷ 4 = 20.

÷ 4

$$\frac{80}{100} = \frac{20}{25}$$

÷ 4

Twenty students scored a grade of B or better on the test.

Example 2: Find the Percent One Number Is of Another Number

Four years ago, Jimmy paid $15 000 for a car. Now he wants to trade in the car for a new one. The car dealer will pay Jimmy $6000 for his old car if he buys a new car from him. What percent of the original cost of the car is the trade-in offer?

Solution

Method 1: Use a hundred grid.
100 squares represents $15 000.
10 squares represents $1500.
40 squares represents $6000.
The trade-in offer is 40% of the original cost of the car.

Method 2: Use a part to a whole comparison.

$$\frac{\text{Part}}{\text{Whole}} : \frac{6000}{15\,000} = \frac{6}{15} = \frac{2}{5} = 0.4 = 40\%$$

$\frac{2}{5} = \frac{4}{10} = \frac{40}{100}$ or 40%

The trade-in offer is 40% of the original cost of the car.

Method 3: Use a proportion.
Let x represent the percent of the trade-in offer relative to the original cost of the car.

$$\frac{\text{Part}}{\text{Whole}}: \frac{6000}{15\,000} = \frac{x}{100}$$

In the denominator, 15 000 ÷ 150 = 100, so 6000 ÷ 150 = 40.

$$\frac{6000}{15\,000} \overset{\div 150}{=} \frac{40}{100}$$

÷ 150

$$x = 40$$

The trade-in offer is 40% of the original cost of the car.

Example 3: Find the Whole When a Specific Percent Is Given

Suzie bought a pair of roller skates marked 30% off. She paid \$140. What was the original price for the roller skates?

Solution

Method 1: Use parts.
Since 30% was taken off the original price, 70% of the original price remains.

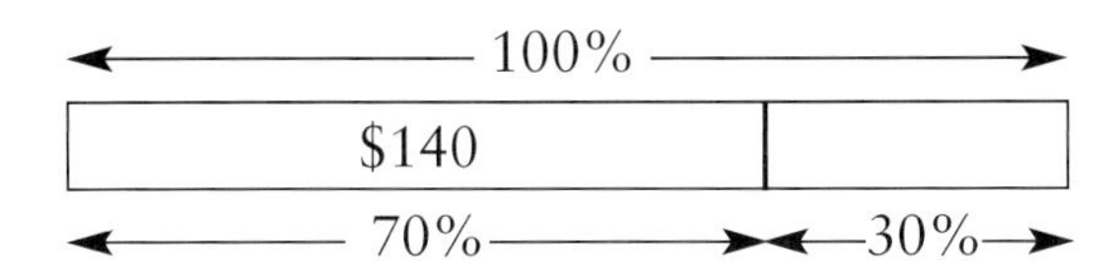

70% of the original price = \$140
10% of the original price = \$20
100% of the original price = \$200
The original price for the roller skates was \$200.

Method 2: Use an equation.
70% of what number is $140?
Let b represent the original price for the roller skates.

$$70\% \text{ of } b = 140$$
$$0.70 \times b = 140$$
$$b = \frac{140}{0.70}$$
$$b = 200$$

0.70 × 200 = 140, so b = 200.

The original price for the roller skates was $200.

Method 3: Use a proportion.
Let b represent the original price for the roller skates.

$$\frac{\text{Part}}{\text{Whole}}: \frac{70}{100} = \frac{140}{b}$$

× 2 (numerator), × 2 (denominator)

In the numerator, 70 × 2 = 140.
So, in the denominator, 100 × 2 = 200.

$$b = 200$$

The original price for roller skates was $200.

Example 4: Calculate Percent Increase and Percent Decrease

The price of goods increases as demand increases. One model of fan that originally cost $25 sold for $30 during a hot summer. In the autumn, the price of the fan dropped to $22.

a) What was the percent increase of the price in the summer?

b) What was the percent decrease from the original price to the price in autumn?

Solution

a) Amount of increase = $30 − $25
= $5

$$\frac{\text{Amount of increase}}{\text{Original price}} = \frac{5}{25} = \frac{20}{100} = 20\%$$

Multiply both the numerator and denominator by 4.

The price of the fan in the summer was a 20% increase from the original price.

b) Amount of decrease $= \$25 - \$22$
$= \$3$

$$\frac{\text{Amount of decrease}}{\text{Original price}} = \frac{3}{25}$$
$$= \frac{12}{100}$$
$$= 12\%$$

The price of the fan in the autumn was a 12% decrease from the original price.

Example 5: Scale Model

a) The minke whale can often be sighted along the coast of the Cabot Trail near Pleasant Bay, Nova Scotia. The minke whale is one of the smallest of the great whales. Females are usually longer than males. A distinctive feature of the minke whale is its narrow triangular pointed snout.

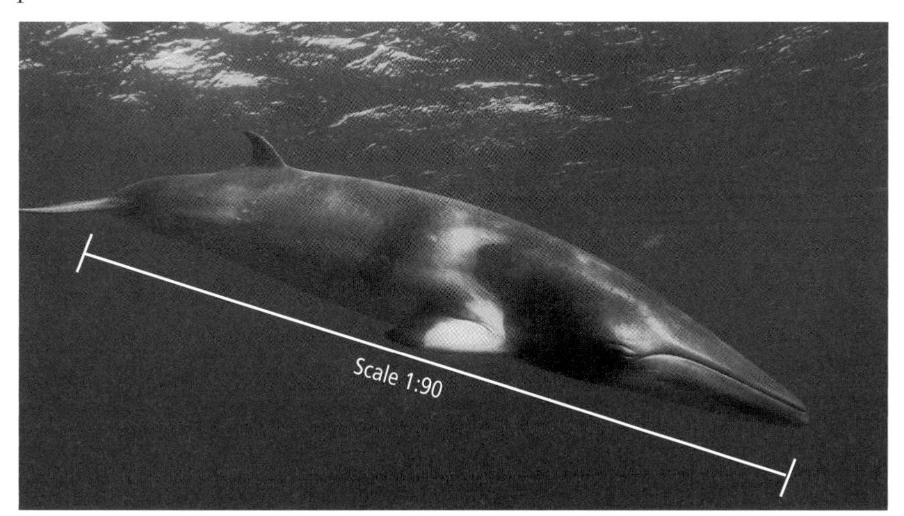

Internet Connect To learn more about minke whales, go to **www.mcgrawhill.ca/links/math8NS** and follow the links.

The photograph shows a **scale** of 1:90. This means 1 cm on the photograph represents 90 cm of the whale's actual length. What is the whale's actual length, in metres?

scale
- the size of a figure or drawing in relation to the original object

b) The *Stigmella ridiculosa* is the world's smallest butterfly, with an average wingspan of 2 mm. Suppose you want to make a scale drawing of the butterfly in the ratio 40:1, where 40 mm on the drawing represents 1 mm of actual length. What will be the wingspan of the butterfly on the drawing, in centimetres?

Solution

a) Measure the length of the whale in the photograph in centimetres.
The length of the whale is 10 cm.

Set up a proportion. Solve for the whale's actual length.

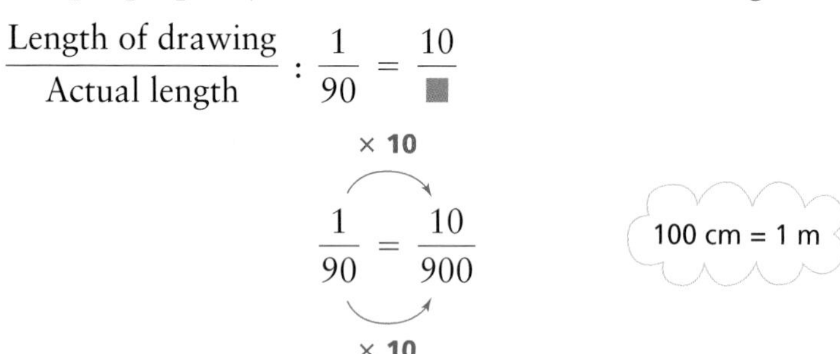

$$\frac{\text{Length of drawing}}{\text{Actual length}} : \frac{1}{90} = \frac{10}{■}$$

$$\frac{1}{90} \overset{\times 10}{=} \frac{10}{900}$$

The minke whale's actual length is 900 cm or 9 m.

b) Set up a proportion. Solve for the length of wingspan on the drawing.

$$\frac{\text{Wingspan in drawing}}{\text{Actual wingspan}} : \frac{40}{1} = \frac{■}{2}$$

$$\frac{40}{1} \overset{\times 2}{=} \frac{80}{2}$$

10 mm = 1 cm

The wingspan of the butterfly in the drawing will be 80 mm or 8 cm.

> **Strategies**
> Apply a previously learned procedure.

> ***Making Connections***
>
> Example 5a) is an example of a reduction.
>
> Example 5b) is an example of an enlargement.
>
> A reduction or enlargement of a figure is often called a dilatation.
>
> You will learn how to use ratios and proportions when studying dilatations.

Communicate the Key Ideas

1. Discounts are subtracted from a price, while taxes are added. In each situation, indicate whether the final price will be greater than, less than, or equal to the original price. Explain how you know.

a) 10% off a sweatshirt plus 14% tax
b) 25% off a bookcase plus 14% tax
c) 14% off a game plus 14% tax

2. Are store A and store B offering the same discount? Explain.

		Store A	Store B
a)	Jeans	30% off then another 20% off	50% off the original price
b)	Sweater	30% off then another 20% off	20% off then another 30% off

3. Consider the expression 25% of 72 = 18.

a) What number is the part? the whole?
b) Rewrite the expression as a proportion.
c) Rewrite a% of b = c as a proportion.

4. Jerry and Emma went to a winter clearance sale where there was no tax that day.
   a) Emma bought a coat for $91. It originally cost $140. What was the percent discount on the coat?
   b) Jerry bought a pair of gloves that were 40% off and paid $8. What was the original price of the gloves?
   c) Since no tax was charged and tax is usually 14%, how much did Jerry save?

Check Your Understanding

1. Draw a diagram to represent each statement.
   a) 50% of ■ = 12
   b) 20% of ■ = 8
   c) 60% of ■ = 24

2. **a)** The diagram represents 160% of a number. The result is 64. What is the number? Show your work.

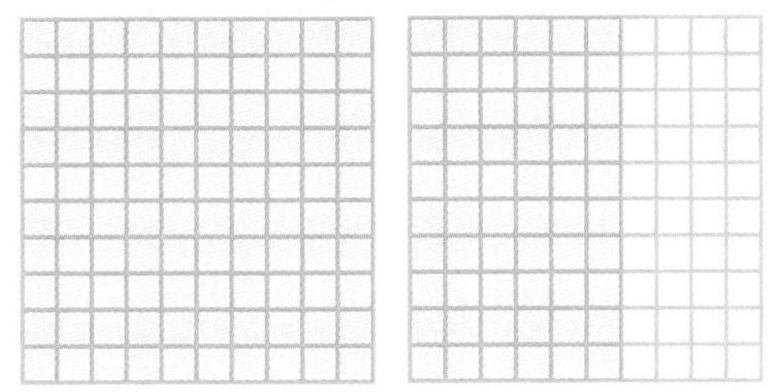

   b) One double pink hexagon represents a whole. What percent of the double pink hexagon is covered? Show your work.

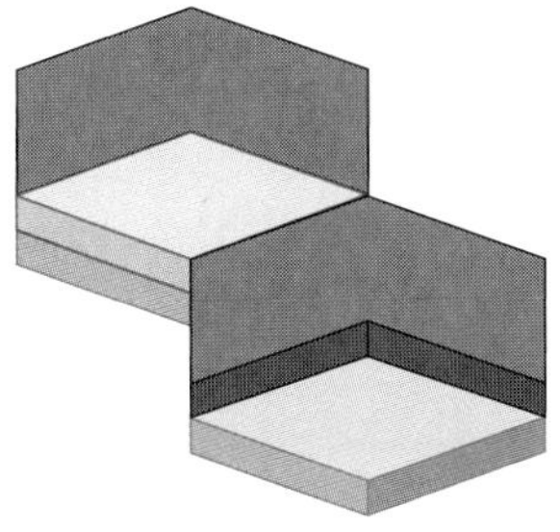

   c) The diagram represents a part of the whole. What part of the whole is represented by the yellow squares? Express your answer as a fraction in simplest form.

3. Calculate each discount.
   a) 50% off running shoes that cost $88
   b) 20% off a set of teacups that cost $24.99
   c) 15% off a T-shirt that costs $15
   d) 75% off gloves that cost $12.99

4. Calculate the final price of each item in question 3, including 14% HST.

5. A tennis racquet that sold for $45 last month is marked down to $36 this month. What percent decrease does this represent?

6. The scale on a map is 1:20 000. What is the actual distance between two cities that are 3 cm apart on the map? Give your answer in kilometres.

7. The blue whale is the longest animal on Earth.

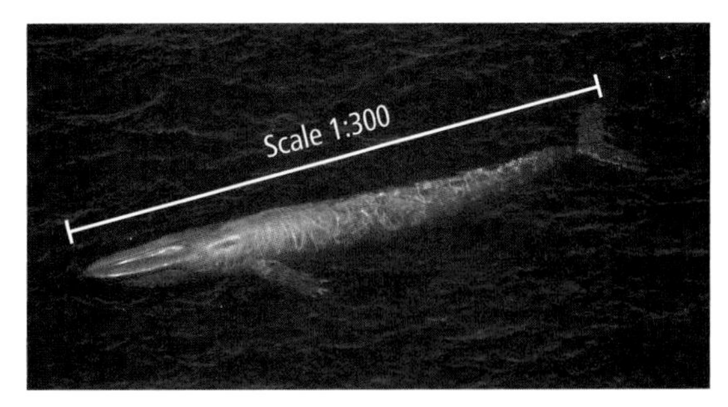

If the length of the whale in the photograph is 11 cm, what is the actual length of the whale to the nearest metre?

8. Aline, Yvette, and Linh each bought a pair of the same jeans marked 40% off. They each paid \$30. What was the original price for the jeans? Who is correct, if anyone? Correct the solutions where necessary.

Aline's Solution

$\frac{40}{30} = \frac{b}{60}$

$b = 80$

The original price was \$80.

Yvette's Solution

60%						40%			
\$3	\$3	\$3	\$3	\$3	\$3	\$3	\$3	\$3	\$3

40% off is \$12.
The original price was \$30 + \$12 or \$42.

Linh's Solution

60% of ■ = \$30
30% of ■ = \$15
10% of ■ = \$5
The original price was \$30 + \$15 + \$5 or \$50.

9. The pink-edged sulphur butterfly can be found in Nova Scotia in bogs and wooded areas where blueberries grow. The average wingspan of this butterfly is 3.3 cm.

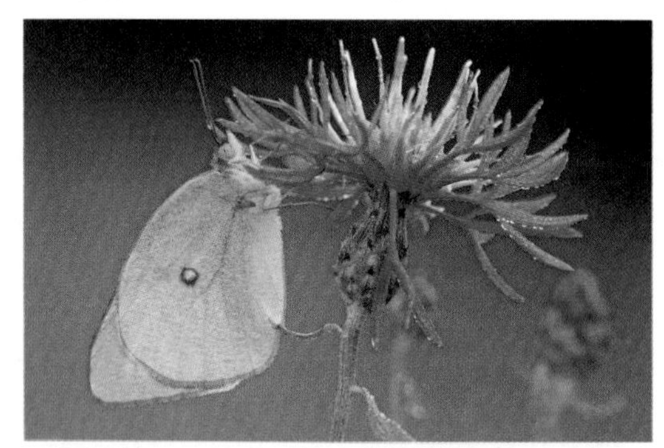

Suppose you drew the butterfly using a scale of 2.75:1. What would be the wingspan of the butterfly in the drawing?

10. The angles of a triangle are in the ratio 3:4:5. What is the size of each angle?

11. A recent census of a city indicates that the current population is 105% of the population recorded during the last census. What is the current population, if the city had a population of 40 000 in the last census?

12. A coat is selling for \$78. The sale sign says that it was reduced by 25%. What was the original price?

13. In the school election, Sarah ran for president and was elected with 348 votes. If she received 58% of the votes cast, calculate the total number of votes cast.

14. Each week, Rachel earns \$236 for 30 h of work at a hair salon. Her friend Jamal earns \$354 for 38 h of work at a pet store.
   a) Estimate each person's hourly rate of pay.
   b) Calculate each person's hourly rate of pay.
   c) How much more does Jamal make in an 8-h shift than Rachel? Express your answer as a percent.

15. Supergrow plant food contains nitrogen and potassium in the ratio of 3 to 4.
   a) How much nitrogen would be in a bag of fertilizer that contains 20 g of potassium?
   b) How many grams of nitrogen and potassium would be in a bag of fertilizer with a mass of 315 g?

16. Create a word problem for each of these statements.
   a) 0.5% of b = 20
   b) a% of 42 = 21
   c) 14% of 85 = c

Chapter Problem

17. The top ten results of the 2005 pumpkin weigh-off contest are shown.

Contestant	Town or City	Pumpkin Mass
Daryl Tingley	Fredericton, NB	567 kg
Lisa Wentzell	Bridgewater, NS	541 kg
Edmond Hempill, Joan Kent	Mt. Pleasant, NB	532 kg
Roger Wentzell	Bridgewater, NS	491 kg
Clifford Picketts	Kensington, PE	454 kg
Will Neily	Paradise, NS	425 kg
Bill Northup	Sussex, NB	411 kg
Howard Dill	Windsor, NS	410 kg
Lorie Shaw	Charlottetown, PE	407 kg
Dawn Northup	Sussex, NB	405 kg

a) What fraction of the pumpkins have a mass greater than 500 kg?

b) Compare the number of entrants from Nova Scotia to the number of entrants from other provinces, as a ratio.

c) What is the percent difference between the greatest and least mass?

18. A recipe for oil and vinegar dressing uses 180 mL of olive oil and 220 mL of vinegar.

a) Compare the volumes of the two ingredients, as a ratio, in simplest terms.

b) What percent of the total volume is olive oil?

c) For 270 mL of olive oil, what amount of vinegar is needed?

19. Patty's dad wants to buy her a computer for her birthday. At the local electronics store, computers are on sale for 20% off the original price. The 14% HST applies to computer purchases.

a) What is the sale price of a computer that regularly sells for $1300?

b) What is the final price of the computer, including tax?

c) Patty's brother found a webcam that had been marked down by 15%. The sale price was $47.00. What was the original price?

20. a) Measure the distance on the map between these places.

i) Yarmouth to Halifax

ii) Halifax to Sydney

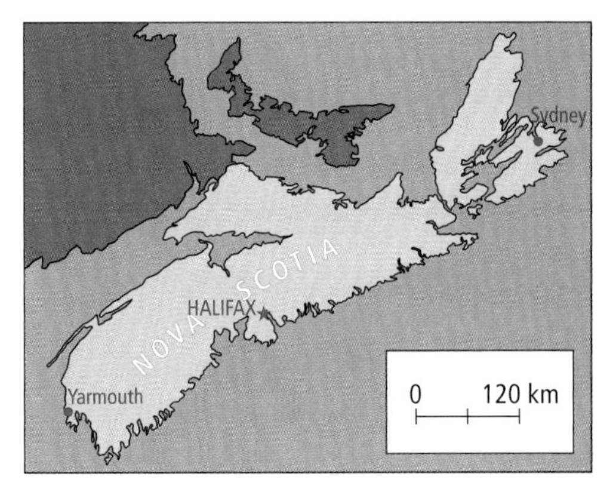

b) Determine the actual distances. Round your answers to the nearest hundred kilometres.

Extend

21. A car begins to depreciate in value after it is purchased. By the end of the first year, it is worth only 70% of the original price. After the first year, it loses about 20% of its value per year. A car is bought for $18 000.

a) What is the value of the car after the first year? after two years?

b) After three years, is the car still worth at least half of its original price? Explain.

c) What percent of the original price is the value of the car after three years?

CHAPTER 4 Review

1. Match each definition to one of the words below.

   fractional percent, percent increase, ratio, proportion, rate, unit price

   A a statement that two ratios are equal
   B a comparison of two quantities
   C a percent in which a fraction appears
   D a comparison of quantities measured in different units
   E the price of one unit, used in comparison shopping

2. Determine the number that is:

   a) 20% of 180 **b)** $\frac{1}{4}$% of 72
   c) 300% more than 15
   d) 175% less than 36

3. Find the missing number in each proportion.

   a) $\frac{1}{8} = \frac{\blacksquare}{56}$ **b)** $\frac{3}{\blacksquare} = \frac{21}{56}$
   c) $\frac{\blacksquare}{12} = \frac{3}{2}$ **d)** $\frac{32}{6} = \frac{4}{\blacksquare}$
   e) $\frac{5}{45} = \frac{2}{\blacksquare}$ **f)** $\frac{\blacksquare}{15} = \frac{3}{4}$

4. Elija is building a scale model of the Eiffel Tower. The tower is 300 m tall. He decides to build a model that is 0.05% the size of the actual tower. How tall will his model be, in centimetres?

5. Last year, Marie bought a radio for $36. This year, a new model costs $39. What percent increase does this represent? What percent of the original price is this year's price?

6. **a)** The diagram represents 115% of a number. The result is 92. What is the number? Show your work.

   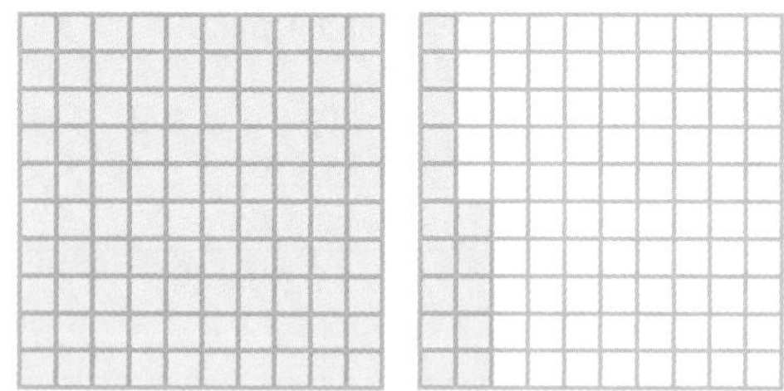

   b) One double pink hexagon represents a whole. What percent of the double pink hexagon is covered? Show your work.

   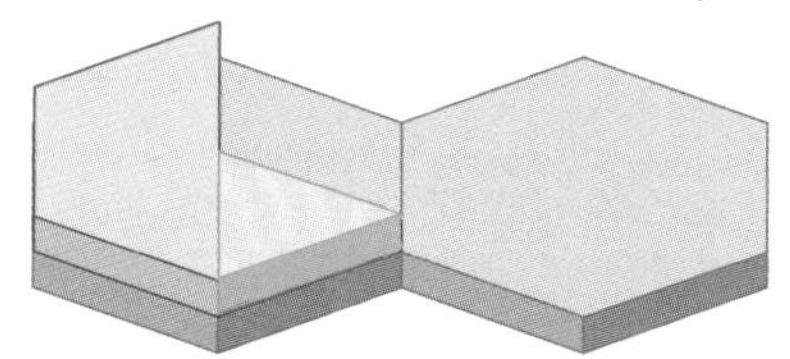

   c) The diagram represents a whole. What part of the whole is represented by the blue squares? Express your answer as a fraction in simplest form.

   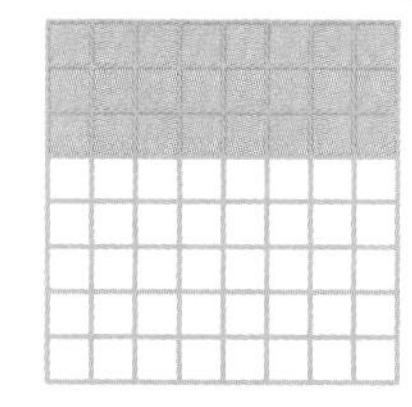

7. Maria opened a bank account that pays $\frac{2}{3}$% interest per month. How much interest would she earn on $400, deposited into the account for one month?

8. Jack wants to draw a picture of his miniature car. He decides to make the drawing 225% the size of the original. The length of the miniature car is 3.6 cm. What is the length of the car in the drawing?

9. In a drawer full of socks, there are 10 pairs of black socks, 5 pairs of grey socks, 2 pairs of blue socks, and 7 pairs of white socks. Express each of the comparisons as a ratio.

a) blue pairs to white pairs
b) black pairs to red pairs
c) white pairs to black pairs
d) grey pairs to total pairs of socks

10. Larry's recipe for pancakes requires 500 mL of flour, 375 mL of milk, 1 egg, 20 mL of sugar, and 10 mL of baking powder.

a) What is the ratio of flour to milk? Write the ratio in simplest terms.
b) If Larry uses 1 L of flour, how much of the other ingredients does he need?
c) Compare the quantities in part b) to the original recipe. What is the percent increase for each ingredient?

11. a) It took 25 min for Soon Ji to ride her skateboard home, a distance of 2 km. What is her average speed, per hour?
b) Garry earned $75 in 6 h. What is his hourly wage?
c) A 400-mL bottle of shampoo costs $3.49. What is the unit price of the shampoo?

12. A winter coat that originally cost $169 was marked down 40%. One week later, there was an additional 10% off the last sale price.

a) Calculate the discounted price of the coat, before HST.
b) What total percent decrease from the original price did the final sale price represent?

13. For each problem, set up a proportion and solve. Show your work.

a) The chain pickerel is found in the waters south of Colchester County, Nova Scotia. The length of the pickerel on a stamp will be 2.5 cm. The scale used is 1:20.4. How long is the actual fish?

b) A centipede is 12.7 mm long. A drawing of the centipede uses a scale of 18:1. What is the length of the centipede in the drawing?
c) Density is a rate that compares the mass of a substance to its volume. The density of a diamond is $\frac{3.5\text{ g}}{1\text{ cm}^3}$. The volume of a diamond is 7.6 cm^3. What is the mass of this diamond?
d) Carla and Dylan both have MP3 players. The ratio of Carla's number of songs to Dylan's is 2:5. After Dylan accidentally erased 36 songs from his MP3 player, Carla had twice as many songs as Dylan. How many songs did Dylan have originally?

14. To make an orange drink, Josh uses 4 cups of water and 5 scoops of powdered drink mix. Sandra uses 5 cups of water and 6 scoops of the powder mix. Will the drinks they make have the same strength? If not, which drink is weaker and by what percent? What assumptions did you make? Explain your reasoning.

CHAPTER 4

Practice Test

Selected Response

Select the best answer.

1. In a class of 27 students, the ratio of boys to girls is 5:4. How many girls are there?

A 8 **B** 12 **C** 15 **D** 10

2. What is the value of x in $\frac{2}{7} = \frac{x}{35}$?

A 10 **B** 30 **C** 12 **D** 15

3. What is 0.2% of 140?

A 2.8 **B** 28 **C** 0.28 **D** 280

Short Response

Provide a complete solution.

4. Find the missing number in each proportion.

a) $\frac{1}{4} = \frac{■}{28}$ **b)** $\frac{2}{■} = \frac{22}{77}$

c) $\frac{■}{16} = \frac{3}{2}$ **d)** $\frac{■}{15} = \frac{3}{5}$

e) $\frac{40}{6} = \frac{4}{■}$ **f)** $\frac{3}{7} = \frac{4.5}{■}$

5. a) The diagram represents 135% of a number. The result is 162. What is the number? Show your work.

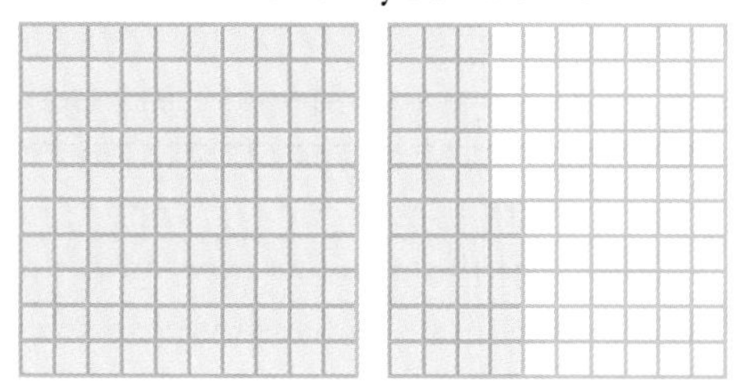

b) One double pink hexagon represents a whole. What percent of the double pink hexagon is covered? Show your work.

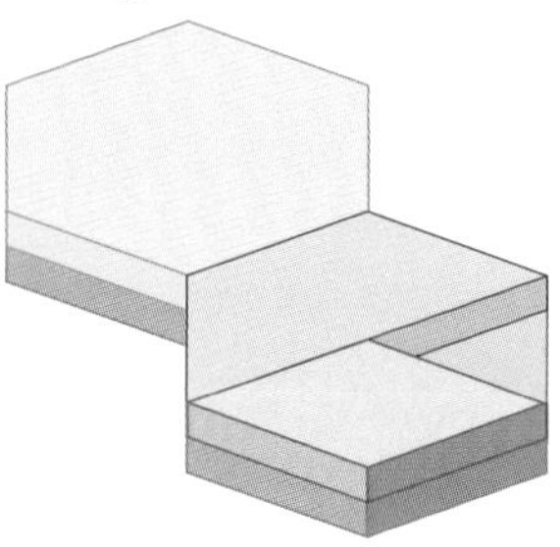

c) The diagram represents a whole. What part of the whole is represented by the blue squares? Express your answer as a fraction in simplest form.

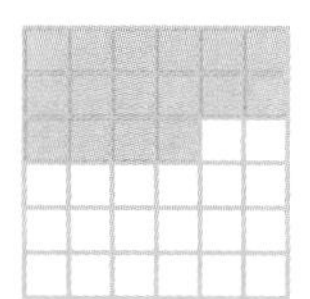

6. A large recipe for strawberry milkshakes calls for 300 g of strawberries, 500 mL of milk, and 1 L of vanilla ice cream.

a) What is the ratio of milk to ice cream?

b) If 450 g of strawberries are used, how much of the other ingredients are needed?

7. Calculate the percent increase or decrease in price for each item.

	Item	Price Five Years Ago ($)	Price Today ($)
a)	loaf of bread	0.97	1.29
b)	portable DVD player	349.00	120.00
c)	litre of milk	1.79	2.49
d)	MP3 player	500.00	99.99

8. a) Calculate the unit price for each item.

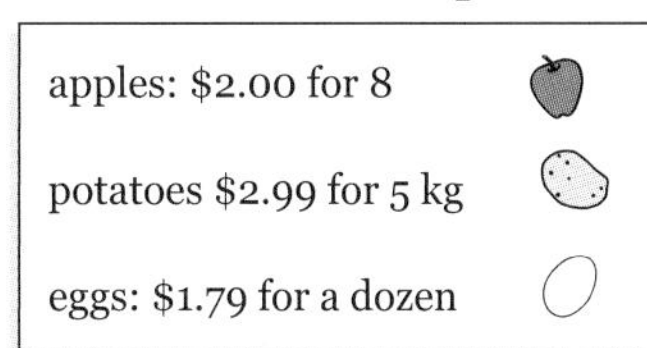

b) Determine the total cost for 10 apples, 2.5 kg of potatoes, and 2 dozen eggs.

9. Calculate the interest earned on each deposit for one month.

a) Nancy's bank pays $\frac{3}{4}$% interest per month. She deposited $600.

b) Jeremy's bank pays $\frac{2}{3}$% interest per month. He deposited $1200.

c) Brett's bank pays $\frac{5}{8}$% interest per month. He deposited $895.

10. Jamie collected 23 pledges for his charity fundraiser by noon. This represents 15% of the total number of pledges. Determine the total number of pledges.

11. A marine company wants to use the Atlantic cod as its logo. The logo will have a cod 4.2 cm long. The scale used is 1:48. What is the actual length of a cod?

Extended Response

Provide a complete solution.

12. At The Tech Store, a television set is marked at 25% off its original price of $600. At Gadgets Galore, the same television set had an additional 10% off its sale price. It was already priced at 20% off $650. Both televisions are subject to 14% HST.

a) Determine the discount and sale price of the television set at The Tech Store.

b) Calculate the total discount and sale price of the television set at Gadgets Galore.

c) Which store offers the better deal?

Chapter Problem Wrap-Up

In 2005, a record holding pumpkin pie had a mass of 916.25 kg after baking. It contained 409 kg of pumpkin. In 2007, the record was broken with a pie having a mass of 918.2 kg after baking.

a) What is the percent increase in mass of the pie in 2007 compared to the pie in 2005?

b) How many kilograms of pumpkin were used in the 2007 pie? What assumptions did you make?

- understand the variability of repeated samples
- develop and apply the concept of randomness
- conduct experiments and simulations to find probabilities of single and complementary events
- compare experimental and theoretical probabilities
- determine the mean, median, and mode
- determine the effect of variations in data on the mean, median, and mode
- construct appropriate data displays and make inferences
- evaluate data based on graphs and tables
- understand how data is used to establish broad probability patterns

Key Words

event
complementary events
random sample
theoretical probability
experimental probability
outlier
box-and-whisker plot
lower quartile
upper quartile
circle graph
line of best fit
interpolate
extrapolate

Data Management and Probability

Data management involves all areas of managing data as a useful resource, including the collection, analysis, interpretation, and presentation of data. Probability is a measure of how probable, or likely, it is that an event will happen. Examples of events that could be studied are the probability of a hurricane forming, or of a baby being born a girl or a boy, or of someone winning the lottery. Data management and probability can be used together to make predictions.

Chapter Problem

Your school is holding its annual fun fair to raise money for the year-end field trip. At one of the booths is a contest to guess the number of red jellybeans in a jar. It costs $1 to make a guess and the prize is a new bicycle donated by the local sports store.

The jar contains 800 jellybeans in four different colours: red, yellow, white, and green. There seem to be more red jellybeans than the other colours. When you guess, you are allowed to take a small sample out to look at, but must return all the jellybeans to the jar before the next person takes their turn.

How can you use data and probability to make a reasonable guess of the number of red jellybeans in the jar?

Get Ready

Tally Charts and Frequency Tables

Tally charts are used to record collected data.

Frequency tables are created when you count up the tallies.

Michelle organized her DVD collection according to genre.

Genre	Tally	Frequency	Percent
horror	卌 IIII		
romantic comedy	卌 卌 II		
action/adventure	卌 II		
documentary	IIII		

1. Copy and complete the frequency table.

2. How many DVDs does Michelle have in her collection?

Types of Graphs

A **line graph** shows **trends** in data, or a relationship between the two quantities or variables.

Age (weeks)	Hammy the Hamster's Mass (g)
0	0
1	5
2	8
3	15
4	22
5	29

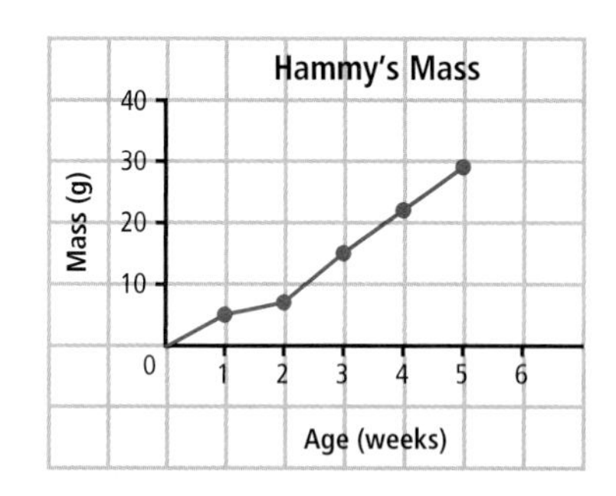

3. a) Describe the trend in the graph of Hammy's mass.

b) If this graph were extended for several more weeks, what do you think would happen to the line?

4. a) Display the data in the table using a line graph. Plot time on the horizontal axis and temperature on the vertical axis.

b) Use the graph to estimate the temperature of the water after 45 min had elapsed.

c) Extend the graph. Estimate the temperature after 180 min had passed. Explain your reasoning.

Time Elapsed (min)	Temperature of Water (°C)
0	20
30	14
60	8
90	2
120	−2
150	−2

In a **stem-and-leaf plot**, the stem represents the tens digit, and the leaf represents the ones digit.

The number of minutes a group of grade 8 students spent doing homework was recorded.
15, 20, 25, 25, 25, 30, 30, 35, 40, 45
The results are shown in a stem-and-leaf plot.

Stem (tens)	Leaf (ones)
1	5
2	0 5 5 5
3	0 0 5
4	0 5

5. The results for a math test, marked out of 50, are displayed in a stem-and-leaf plot.

Stem (tens)	Leaf (ones)
2	7 8
3	0 0 2 6 8
4	0 5 8
5	0

a) What is the lowest score?
b) What is the most common interval of scores? How do you know by looking at the stem-and-leaf plot?

6. The mathematics test scores (out of 100) achieved by 10 students are shown.
65 62 76 74 59 81 86 87 78 90
Display the data using a stem-and-leaf plot.

Measures of Central Tendency

A **central tendency** is a point or value in a range about which the rest of the data is considered to be balanced.

You know three measures of central tendency.

Mean $= \frac{\text{sum of values}}{\text{number of values}}$

Median is the middle value when a set of data is arranged in order from least to greatest.

Mode is the value that occurs most frequently in a set of data.

Dillon spent these numbers of minutes lifting weights over six days last week.
25 30 30 20 15 30

$$\begin{aligned}\text{Mean} &= \frac{\text{sum of values}}{\text{number of values}}\\ &= \frac{25 + 30 + 30 + 20 + 15 + 30}{6}\\ &= \frac{150}{6}\\ &= 25\end{aligned}$$

The mean exercise time is 25 min.

Get Ready

To find the median, arrange the numbers in order from least to greatest. Use a stem-and-leaf plot if there are many values in the set of data.

15 20 25 ↑ 30 30 30

median

The middle value is the mean of 25 and 30, or 27.5.
The median is 27.5 min.

The mode is the most common value; 30 min appears three times, so 30 is the mode.

The variability of a set of data is a measure of how closely the values tend to cluster around the measure of central tendency. **Range** is one measure of variability.

Range = greatest value − least value
= 30 − 15

The range of the data is 15. The difference between the range and the mean is approximately 25 − 15, or 10. The values here do not cluster closely around the measure of central tendency.

7. Refer to questions 5 and 6.
a) Calculate the mean, median, and mode of each set of marks.
b) Determine the range for each set of marks. Is there great variability in the data? Explain.

Use a Protractor

A protractor is a tool used to measure the size of an angle. Place the base of the protractor along one arm of the angle, with the vertex of the angle at the centre of the base. Read along the arc of degree numbers, beginning at zero, to mark or measure the position of the second arm of the angle.

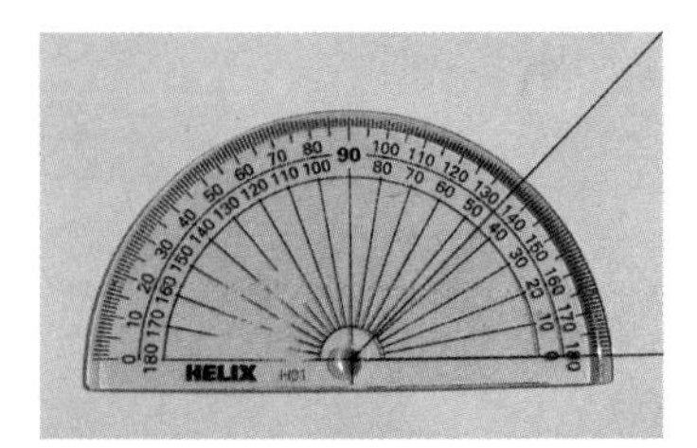

This angle measures 45°.

8. Measure each angle.

a)

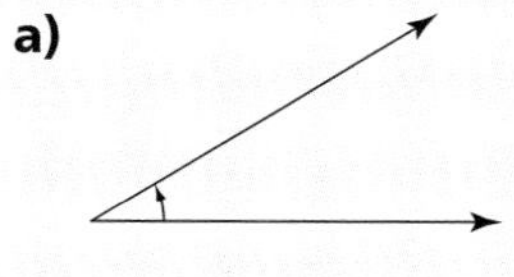

b)

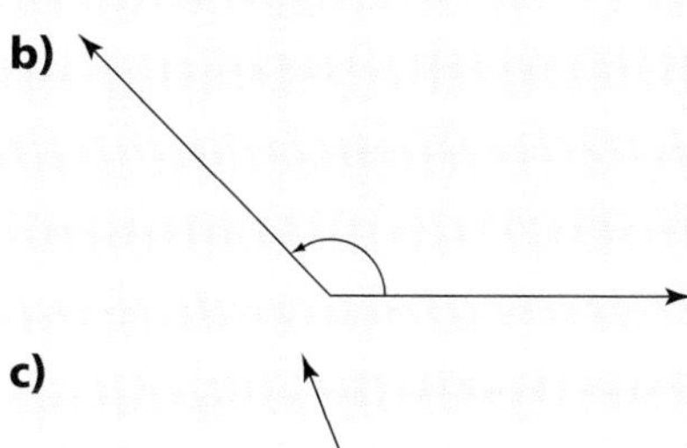

c)

9. Draw each angle using a ruler and a protractor.

a) 60°

b) 175°

c) 12°

Samples and Populations

When conducting a survey, you should consider the population that you want to survey.

If the population is too great to survey, consider using a representative sample to obtain your data.

10. Match each **sample** with the **population** it represents.

Sample

a) a group of children is asked to test new toys

b) three students from each class were randomly chosen and asked what music they would like to hear at the dance

c) a group of theatre patrons is asked what movie they like best

Population

A all people who watch movies

B all students in the school

C all children who play with toys

5.1 Collect, Organize, and Use Data

Focus on...
- understanding the variability of repeated samples
- applying the concept of randomness
- conducting experiments and simulations to find probabilities of single and complementary events

Have you ever been a spectator at a sports game? The excitement of the fans cheering in their seats can be contagious. Or perhaps you are a member of a sports team at school. Participation in sports can help you become physically fit, develop social skills and friendships, and learn the importance of being a team player.

A team name can often inspire players. Suppose a large school is choosing a name and a mascot for its sports team. How can the sports committee find out what name is the most popular among the students at school?

Discover the Math

Materials
- bag containing BLM 5.1 DTM Team Name Cards

How can I use sample data to make predictions about the population?

What factors affect the accuracy of these predictions?

Part A: Compare the Accuracy of Predictions Based on Large and Small Sample Sizes

Work in groups of three or four.

1. Suppose a population of 120 students are surveyed about what to name the school's sports team. They have a choice of three names: Bison, Eagles, or Camels. Get a bag from your teacher. The slips of paper in the bag represent the responses of the entire population. You will use sampling to predict the make-up of the population.

2. Choose a large sample from the population. Randomly select 40 slips of paper from the bag. Copy and complete the first three rows in the table. Put the slips of paper back into the bag.

	Camels	Eagles	Bison
Tally			
Frequency			
Fraction of Sample			
Predicted Number in Population			

3. Complete the last row in the table in question 2. Compare your results with the other groups in the class. Was there much variation among the results? Which group do you think has the most accurate prediction of the population's make-up? Why?

Strategies
Estimate the make-up of the population.

4. Choose a small sample from the population. Randomly select 10 slips of paper from the bag. Copy the table and complete the 1s columns for the first small sample. Put the slips of paper back into the bag. Repeat this sampling three more times and record the data.

	Camels				Eagles				Bison			
Sample	1	2	3	4	1	2	3	4	1	2	3	4
Tally												
Frequency												
Fraction of Sample												
Predicted Number in Population												
Mean of 4 Samples												

5. Complete the last two rows in the table in question 4. Compare the results of the first small sample to the mean of all four small samples and to your prediction from question 3 based on the large sample. Which sample do you think results in a more accurate prediction of the population's make-up? Share your results with the other groups in the class.

6. Use the results from all the samples to make one final prediction about the make-up of the population. Take all the slips out of the bag and count them. How close was your prediction? Did any group predict the population exactly?

7. **Reflect** What factors do you need to consider when selecting a sample that will provide a fair representation of the population?

Part B: Complementary Events

Materials
- bag of counters

1. Get a bag from your teacher.

2. Without looking, select one counter from the bag. Record its colour in a frequency table. Return the counter to the bag.

3. Repeat step 2 for a total of 20 selections.

4. Based on the 20 selections, determine the **experimental probability**, P, of each **event**.
   a) selecting a black counter: $P(\text{black}) = ■$
   b) selecting a red counter: $P(\text{red}) = ■$
   c) selecting a white counter: $P(\text{white}) = ■$

5. Based on your experimental data, what is the probability of not selecting a black counter?
   $P(\text{not black}) = ■$
   Selecting a black counter and *not* selecting a black counter are called **complementary events**. Either one or the other event will occur.

6. Based on your experimental data, find the sum of the probability of selecting a red counter and the probability of not selecting a red counter. $P(\text{red}) + P(\text{not red}) = ■$

7. Take all the counters out of the bag and count them. Use your results to calculate the **theoretical probability** of $P(\text{red})$ and $P(\text{not red})$. Explain how you found your answer.

8. **Reflect** Describe a situation in which the probability of an event and its complement would be the same.

9. Based on your experimental data, find the sum of the probability of selecting a red counter, the probability of selecting a white counter, and the probability of not selecting either a red or a white counter. $P(\text{red}) + P(\text{white}) + P(\text{not red or white}) = ■$

10. How else can you express the $P(\text{not red})$ in question 6 and $P(\text{not red or white})$ in question 9? Explain.

experimental probability
- a ratio that compares the number of favourable outcomes to the total number of trials in an experiment
- the experimental probability may be different from the theoretical probability

event
- a possible outcome of a probability experiment

complementary events
- two or more distinct events that represent the complete set of outcomes

theoretical probability
- a ratio that compares the number of favourable outcomes to the total number of possible outcomes
- to calculate theoretical probability, all outcomes must be equally likely

Example 1: Compare a Sample to a Population

Julie wants to find out the most popular after-school activity among grade 8 students.

One grade 8 class is surveyed. The sample data shows that playing sports is the most popular activity.

Activity After School	Tally	Frequency
play sports	𝍸 IIII	9
do homework	𝍸 I	6
watch TV	𝍸	5
surf the Internet	𝍸 II	7
other	III	3

The school's grade 8 population consists of 120 students. Use the sample data to predict how many grade 8 students in the school play sports after school.

Solution

Find the total number of students in the one class. This is the sample size.

$9 + 6 + 5 + 7 + 3 = 30$

There are 30 students in the sample.

Method 1: Use ratios and proportions.

9 out of 30 play sports after school.

This can be written as a ratio in fraction form: $\frac{9}{30}$

Write a proportion that compares the sample to the population.

The population size is 120.

$$\frac{9}{30} = \frac{\blacksquare}{120}$$

$30 \times 4 = 120$

$$\frac{9}{30} = \frac{36}{120}$$

× 4 (numerator), × 4 (denominator)

Multiply the numerator and denominator by 4.

Strategies
Set up an equation and solve it.

Based on the sample data, you can expect that 36 students of the school's grade 8 population play sports after school.

Method 2: Use percents.

9 out of 30 play sports after school.

This fraction can be written as a percent.

$$\frac{9}{30} = \frac{3}{10} = 0.30 = 30\%$$

Find 30% of the population of 120 students.

$$\begin{aligned} 30\% \text{ of } 120 &= 0.3 \times 120 \\ &= 36 \end{aligned}$$

I can do this mentally: 10% of the population is 12, so 30% will be 36.

Strategies
Can you do this calculation without using a percent?

Based on the sample data, you can expect that 36 students of the school's grade 8 population play sports after school.

Example 2: Randomness and Sample Size

Members of the student council are surveying the school population to determine what type of food they prefer to eat for lunch.

a) Some students eat their lunch in the cafeteria while others go home for lunch. Helen picks names from a hat to survey 20 students from the line in the cafeteria. Is this a **random sample** ?

b) After school, Denise surveys every fourth student as they leave the school's main doors. Is this a random sample?

c) Harvey surveys five of his friends after school. Will this sample be a representative sample that reflects the overall opinion of the entire student population? Explain your answer.

random sample

- a sample in which everyone in a population has an equal chance of being selected

Solution

a) Helen's sample is random, but only includes students who buy their food from the cafeteria. Those students who went home for lunch did not have an equal chance of being chosen, so her sample is not representative of the whole population.

b) Denise's sample is random and is representative of the school's population. Each student has an equal chance of being surveyed, since Denise, when choosing this sampling method, had no way of knowing the order in which the students would exit the school.

c) Harvey's sample size is too small. Using the results of a small sample to make predictions about a large population may lead to inaccurate conclusions. This sample is also biased. Those surveyed are Harvey's friends. Not everyone in the population has an equal chance of being chosen.

A sample is random if each person in the population has an equal chance of being selected. You can collect a random sample by

- selecting every *n*th person in a population
- selecting at random a limited number of people from the population
- selecting at random a number of people from each group (that is proportional to the size of the group) within a population
- selecting at random people from a certain group within a population (but you must limit your conclusions to that specific group)

Random samples are not always representative of a population.

Example 3: Probability of Complementary Events

Marcus and Mai-Ling are playing a board game that involves rolling a number cube. Marcus needs to roll a 4 to win. On his next turn, he will either roll a 4, or not roll a 4.

a) What is the probability that Marcus will roll a 4 on his next turn?

b) What is the probability that he will not roll a 4?

c) What is the sum of the complementary events?
$P(\text{rolling a }4) + P(\text{not rolling a }4) = ■$

Solution

a) There are 6 outcomes when a number cube is rolled. The probability of rolling a 4 is 1 out of 6, or $P(\text{rolling a }4) = \frac{1}{6}$.

b) When a number cube is rolled, there are 5 outcomes in which the number is not a 4. The probability of not rolling a 4 is 5 out of 6, or $P(\text{not rolling a }4) = \frac{5}{6}$.

Strategies
Can you think of another way to find P(not rolling a 4)?

c) $P(\text{rolling a }4) + P(\text{not rolling a }4) = \frac{1}{6} + \frac{5}{6} = 1$

The sum of the probabilities of complementary events is 1 because they represent the complete set of outcomes.

Communicate the Key Ideas

1. a) Explain why a sample might be used instead of a population when conducting a survey.

b) Why is a large random sample preferred over a small random sample?

2. Kareem tossed a coin 10 times. The results were:

H, T, T, H, T, T, T, H, T, T.

Since heads only showed up 3 out of 10 times, he thought the results must be wrong. If Kareem tossed the coin another 10 times, do you believe the results would be the same as the first sample? Explain why or why not.

3. Cindy conducted a survey at one of the local grocery stores. She asked 20 shoppers which fruit they liked best. Is this sample a good representation of the people living in the neighbourhood?

4. After grading the math test, Mr. James was concerned because the class average was only 62%. He asked three students who had scored over 80% on the test whether or not they felt the test was too difficult. Do you think this sample will accurately reflect the overall opinion of the class? Explain.

5. Give five examples of complementary events.

Check Your Understanding

1. Identify the sample and the population in each situation.
   a) A group of teens is asked what brand of running shoes they think most teenagers like to wear.
   b) A group of mothers at the mall is asked what brand of diapers they would recommend.
   c) Members of the student council are asked for suggestions on how the school should spend the budget for school improvements.
   d) Five doctors attending a convention are randomly chosen and asked for their expert opinion on a medical problem.
   e) Two auto mechanics are asked to provide an estimate for the cost of fixing the same car.

Use this information for questions 2 and 3.

A group of grade 8 students were surveyed about their favourite type of movie. The results are shown.

Type of Movie	Tally	Frequency
comedy	IIII	4
action/adventure	~~IIII~~ III	8
fantasy/science fiction	~~IIII~~ I	6
horror	~~IIII~~	5
animation/cartoon	II	2

2. Suppose there is a population of 100 grade 8 students at the school. Use ratios to predict how many students prefer to watch
   a) fantasy/science-fiction movies
   b) action/adventure movies

3. Suppose there is a population of 120 grade 8 students at the school. Use percents to predict the approximate number of students who prefer to watch
   a) horror movies
   b) comedies

4. A poll of 500 eligible voters across the province was conducted to determine which political party is favoured to win in the upcoming provincial election.

   The voters were selected at random.

Political Party	Number of Votes in the Poll
A	180
B	200
C	120

If the results of the poll are accurate, how many votes will each political party receive if 634 000 Nova Scotians vote on election day?

5. The names of all students at your school are placed into a box. Five names are chosen from the box without looking. Is this a random sample? Explain. Do you think it is representative of the school population?

6. Members belonging to the MADD (Mothers Against Drunk Driving) organization are surveyed to get their opinion on requiring new drivers to get their high school diploma before receiving their licence. Is this a random sample? Explain.

7. Fans at a hockey game are asked who they think will win the Stanley Cup. The results are used to predict the overall views of everyone who likes hockey. Is this a random sample? Explain.

8. Census data from 2000 people selected from the 10 provinces and 3 territories of Canada is used to determine the average household income in Canada. Is this a random sample? Explain.

9. Emily rolled a number cube 10 times and recorded her results: 3, 1, 3, 4, 6, 2, 1, 2, 3, 5. She concluded that she was more likely to roll a 3 than any other number on the number cube.
   a) Is that the result you would expect? Explain your answer.
   b) Roll a number cube 3 times. How do your results compare to Emily's?
   c) Roll a number cube 20 times. How do your results compare to Emily's?

10. A spinner has four congruent sections coloured red, yellow, blue, and green.

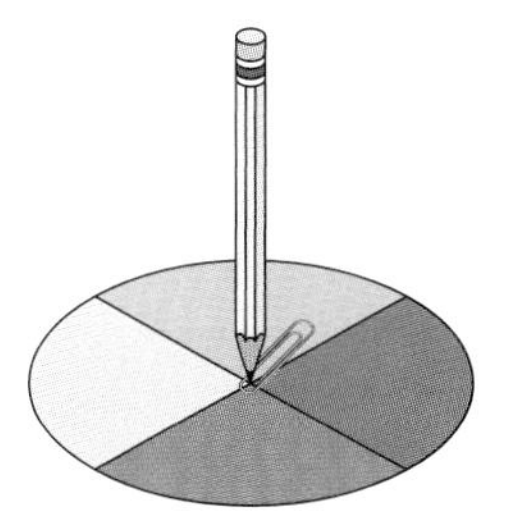

Doug spins it 25 times and the results are recorded in a tally chart.

Colour	Tally
red	||||
yellow	~~||||~~ |
blue	~~||||~~ |||
green	~~||||~~ ||

   a) What is the probability of spinning green based on the results in the chart? What is the probability of not spinning green?

   b) Based on his data, Doug concluded that it is easier to spin blue than any other colour. Examine the data and the spinner. How did Doug reach his conclusion? Write a note to Doug to help him understand the error in his thinking.

11. A survey at school randomly selected every fifth student, alphabetically, and asked: "What is your favourite dessert?" The results are shown.

Dessert	Number of Students
ice cream	205
pie/cake	110
fresh fruit	150
yoghurt	35

a) What is the probability that a student's favourite dessert is

i) fresh fruit?

ii) ice cream?

iii) not ice cream?

b) Give all the probabilities as percents.

12.

a) Toss a coin 3 times, 10 times, and 30 times. Record your results for each sample.

b) Based on the data for each sample, what is the probability that the next toss will be

i) tails?

ii) heads?

iii) not heads or tails?

c) Compare the experimental probability for heads for each sample. Are they the same? Does the sample size affect the experimental probability?

13. If the probability of precipitation is 30% tomorrow, what is the probability that there will be no precipitation tomorrow?

Chapter Problem

14. Gina, Thomas, and Dong-Soo are debating how large a sample they should take from the jar to help them guess the number of red jellybeans. Gina takes a sample of 10 jellybeans, Thomas takes 16 jellybeans, and Dong-Soo takes 25 jellybeans. Whose sample would help them make a more accurate estimate? Explain your reasoning.

15. Jack records all of the names of his grade 8 classmates on pieces of paper. He places all of the names in a hat. He selects 15 names, one at a time, without looking, and records the names. He then surveys his sample of students to see what they plan to do during their summer vacation. His results are shown in the table.

Summer Activity Plans	Tally
get a job	||||
relax (read, watch movies, go to the beach, etc.)	卌 |
travel	||
study for grade 9	|||

a) Was the sample a random sample? Explain.

b) If there are 30 students in Jack's grade 8 class, predict how many students want to get a job over the summer.

c) What is the probability of choosing a student from the sample who wants to travel over the summer?

d) What is the probability of choosing a student from the sample who wants to travel or just relax over the summer?

e) Suppose there are 105 grade 8 students in the whole school. Based on the data, how many students would you expect to be studying for grade 9 over the summer?

16. Conduct an experiment to determine the sum that appears most often when two number cubes are rolled. What is the probability of that event? Compare your answer with a classmate's. What would you expect to be the sum that appears most often? Why?

17. When an advertisement on television claims that 4 out of 5 dentists recommend a particular brand of toothpaste, it suggests that 80% of dentists recommend the same toothpaste.

a) Is this sample data reflective of the opinion of all dentists? Explain.

b) What suggestions can you make to validate or disprove this claim?

Extend

18. Work with a partner.

a) Hide your right hand behind your back. On the count of three, you and your partner show your right hands with 1, 2, 3, or 4 fingers extended. Record the total number of fingers extended.

b) Repeat part a) 10 times. Record the total number of fingers each time.

c) Calculate the experimental probability of each sum.

d) Repeat the experiment 20 more times. Calculate the probabilities for this experiment. Did any of the probabilities change? Why or why not?

e) If you repeated the experiment 100 more times, do you think your probabilities would change? Explain.

19. As part of Jennie's science project, she asked 100 students at her school to rank their concern about global warming on a scale of 1 to 5 (1 is not concerned, 5 is very concerned).

a) Describe two different methods to obtain a random sample from this population.

b) Is it possible to have a sample of 10 students whose mean response is 2.2 and another sample of 10 students whose mean response is 4.1 from the same population? Explain.

Puzzler

Joey buys a cup of hot chocolate that costs $1.07. He pays with a two-dollar coin. If the cashier gives him 8 coins for his change, what could these coins be?

5.2 Theoretical and Experimental Probabilities

Focus on...

- determining theoretical probabilities of single and complementary events
- comparing experimental and theoretical probabilities

A manufacturer of cosmetic products is promoting the sale of a new chocolate-scented hand cream, *ChocoNut*.

On the inside of 10 000 jar labels are printed the messages "Sorry, try again," "You win a free jar of ChocoNut," or "Grand prize: You win a backpack!".

There are 2250 jars of hand cream and 250 backpacks to be won. The manufacturer claims that the chance of winning a prize is 1 in 4. Does this mean you will definitely win something if you buy four jars?

Discover the Math

Materials

- bag containing BLM 5.2 DTM Contest Cards

How can I find the probability of an event occurring?

1. Get a bag from your teacher. The slips of paper in the bag represent the theoretical probabilities of winning a prize in the ChocoNut contest.

Strategies
Act it out.

2. What is the probability that you will win a prize? Without looking, draw one slip of paper from the bag. Record your result in a tally chart like the one shown. Put the slip of paper back in the bag. Repeat for a total of 10 selections, or trials.

Sorry, try again	You win a free jar of ChocoNut	Grand prize: You win a backpack!

3. Add the tallies for each message. Based on these results, determine the experimental probability of selecting each message.

4. Take all the slips of paper out of the bag and count them. Use your results to calculate the theoretical probability of selecting
- "Sorry, try again"
- "You win a free jar of ChocoNut"
- "Grand prize: You win a backpack!"

Explain your thinking.

5. Compare the theoretical probabilities with the experimental probabilities. Are these values the same?

6. Repeat the experiment for 20 selections and 50 selections. Compare your results with your classmates'. How do the theoretical and experimental probabilities compare when the sample size, or number of trials, is small? How do they compare when the sample size increases?

Example 1: Determine Theoretical Probabilities by Listing Outcomes

a) If two coins are tossed, what is the theoretical probability that both coins are tails? What is the theoretical probability that both coins are not tails?

b) If one coin is tossed three times, what is the theoretical probability that there are exactly two heads? What is the probability of not tossing exactly two heads?

Solution

a) Use an organized table to show all the equally likely outcomes when two coins are tossed.

		Coin 2	
		H	**T**
Coin 1	**H**	H, H	H, T
	T	T, H	T, T

Strategies
Account for all possibilities.

$$P(\text{event}) = \frac{\text{number of favourable outcomes}}{\text{total number of possible outcomes}}$$

There are four outcomes.

Of these, there is one favourable outcome in which both coins are tails.

The theoretical probability of tossing two tails is $P(\text{both tails}) = \frac{1}{4}$.

The theoretical probability that the coins are not both tails can be found in two ways.

Method 1: Count favourable outcomes.

$$P(\text{event}) = \frac{\text{number of favourable outcomes}}{\text{total number of possible outcomes}}$$

Look at the table on the previous page. There are three outcomes in which both coins are not tails.

So the probability of both not tails is $P(\text{both not tails}) = \frac{3}{4}$.

Method 2: Use complementary events.

$$\begin{aligned} P(\text{both tails}) + P(\text{both not tails}) &= 1 \\ P(\text{both not tails}) &= 1 - P(\text{both tails}) \\ &= 1 - \frac{1}{4} \\ &= \frac{3}{4} \end{aligned}$$

b) Use a tree diagram to show all the outcomes.
The tree diagram shows eight outcomes.
Of these, there are three outcomes in which there are exactly two heads.
The probability of tossing exactly two heads is
$P(\text{exactly two heads}) = \frac{3}{8}$.
The probability of not tossing exactly two heads is
$P(\text{not exactly two heads}) = \frac{5}{8}$.

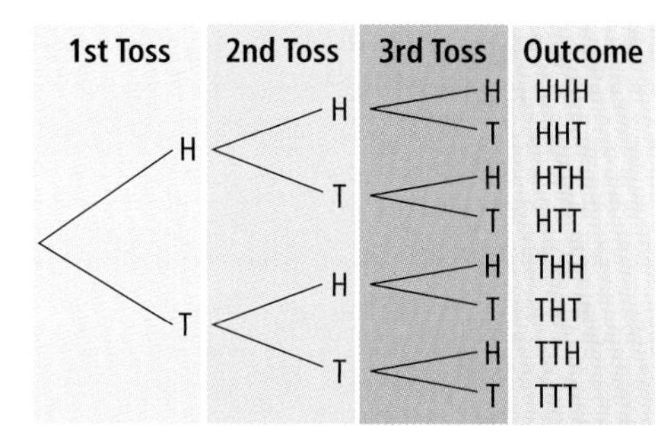

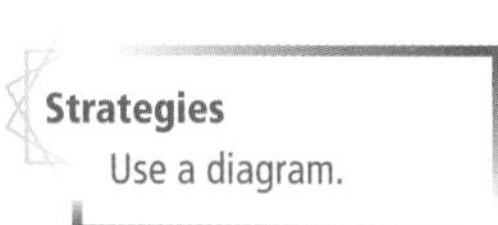

Example 2: Compare Experimental and Theoretical Probabilities

A number cube is rolled 20 times. The results are recorded in a tally chart.

a) Determine the theoretical probability of rolling each number.

b) Use the tally chart to calculate the experimental probability of
- rolling a 2
- rolling a 3

Number Rolled	Tally
1	II
2	IIII
3	III
4	𝍸
5	II
6	IIII

c) Compare the experimental and theoretical probabilities in parts a) and b). Are the values the same? Did you expect these results?

d) Janice and Roger want to play a board game. Each chooses a number and rolls a number cube. If a player's number turns up, that person makes the first play. Janice chooses the number 4 based on the results in the tally chart. Roger chooses 3. Whose number should turn up most often, if any, and why?

Solution

a) There are six numbers on a number cube: 1, 2, 3, 4, 5, 6

Each number has a 1 in 6 chance of being rolled.

$$P(\text{any number}) = \frac{\text{number of favourable outcomes}}{\text{total number of possible outcomes}}$$
$$= \frac{1}{6}$$

Strategies

Apply previously learned procedures for fractions and probability.

The theoretical probability of rolling each number is $\frac{1}{6}$.

b) A 2 was rolled 4 times out of 20.

The probability of rolling a 2 is $\frac{4}{20}$.

A 3 was rolled 3 times out of 20.

The probability of rolling a 3 is $\frac{3}{20}$.

c) The theoretical probability of rolling a 2 is $\frac{1}{6} = \frac{10}{60}$.

The experimental probability of rolling a 2 is $\frac{4}{20} = \frac{12}{60}$.

The theoretical probability of rolling a 3 is $\frac{1}{6} = \frac{10}{60}$.

The experimental probability of rolling a 3 is $\frac{3}{20} = \frac{9}{60}$.

The theoretical and experimental values are similar but they are not the same. I expected results like these. I thought the values might be close, but not identical.

d) The theoretical probability of rolling any number is $\frac{1}{6}$, so on average, both numbers 3 and 4 have an equal chance of being rolled. Janice's decision to choose number 4 was based on experimental probability. If the experiment were repeated for 20 more trials, 4 might not be the most frequently occurring number. Most times, the theoretical probability is a more accurate predictor of outcomes. Experimental probability is normally used when the theoretical probability is difficult or impossible to calculate.

Communicate the Key Ideas

1. Use an example to explain how you would find the experimental probability and the theoretical probability of an event.

2. Explain why the experimental probability of an event may be different from the theoretical probability.

3. Studies have shown that the probability of developing skin cancer in one's lifetime is 1 in 5.

a) This is experimental probability. How do you know?
b) What is the probability of not developing skin cancer?
c) Why is it not possible to find the theoretical probability of developing skin cancer?

Check Your Understanding

1. What is the theoretical probability of each event?

a) tossing tails on one coin
b) rolling a 5 on one number cube
c) landing on green on a spinner that has five equal sections: two coloured green and three coloured white

2. Create an organized table with six rows and six columns to show all the possible pairs of outcomes when two number cubes are rolled. What is the theoretical probability of

a) rolling a pair of 1s?
b) not rolling a pair of 1s?
c) rolling at least one 3?

3. The probabilities of being born a boy or a girl are equally likely.

a) Use an organized table or a tree diagram to find all the possible outcomes for the genders of two children in a family.
b) What is the theoretical probability of both children being boys?

Journal

4. Refer to question 3. Explain to a classmate how you could use a coin to simulate the genders of two children in a family. State which gender heads represents and which gender tails represents. Then toss the coin two times for 10 trials. Make a summary of your results. Explain what your results mean.

5. Two spinners both have three equal sections labelled 1, 2, and 3. Use an organized table or a tree diagram to list all the possible outcomes when both spinners are spun.

6. Suppose you roll two number cubes.

a) What is the theoretical probability of rolling a sum greater than 9? *Hint: Use the organized table in question 2 to help you.*

b) What is the theoretical probability of rolling a sum of 6?

c) What is the theoretical probability of not rolling a sum of 6?

7. Ryan is trying to decide which jeans and T-shirt to wear together. He has narrowed down his choices to a pair of black or blue jeans matched with a white, yellow, or grey T-shirt. Use a tree diagram to show all the possible combinations he can make.

8. A spinner has six equal sections: two coloured yellow, three coloured blue, and one coloured green.

a) What is the theoretical probability of landing on yellow?

b) What is the theoretical probability of not landing on yellow? Show two different ways to find this answer.

c) Create the spinner and spin it 20 times, recording the result each time.

d) Find the experimental probability of not landing on yellow using your results from the simulation. Compare this with your answer in part b). Are the probabilities the same or different?

e) What could you do to get experimental results that are closer to the theoretical probabilities in parts a) and b)?

9. A bag contains 7 blue marbles, 3 red marbles, and 5 black marbles.

a) What is the theoretical probability of randomly selecting a blue marble?

b) What is the theoretical probability of not selecting a blue marble? Show two different ways to find this answer.

10. A standard deck of playing cards contains 52 cards, divided into 4 suits: hearts, diamonds, clubs, and spades. Hearts and diamonds are red, while clubs and spades are black.

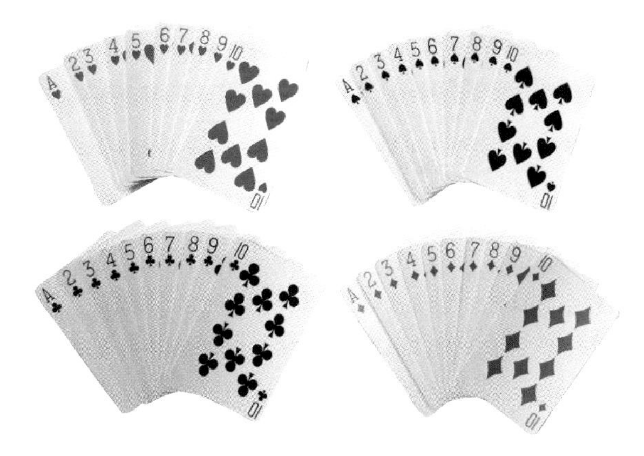

a) What is the theoretical probability of drawing a black card? a red card?

b) What is the theoretical probability of drawing a heart? a spade? a diamond? a club?

c) What is the sum of the theoretical probabilities of the complementary events in part b)?

d) Use complementary events to find the theoretical probability of not drawing a diamond.

11. Identify each as experimental or theoretical probability.

a) The number of different coloured cars passing an intersection is tallied to find the probabilities of people owning cars with specific colours.
b) If a coin is tossed, heads will occur 50% of the time.
c) According to the number of favourable outcomes, the number 2 on a number cube should occur once every six rolls.
d) Marcie rolled a 5 only once in her 12 turns.

12. A cereal company offers three prizes in boxes of its So Good cereal. In each production run, 75% of the boxes contain stickers, 20% contain a yo-yo, and 5% contain a music CD.

a) What is the theoretical probability that the prize is not a music CD?
b) Alan wants a music CD. To increase his chances of winning one, he buys 20 boxes of cereal. How many of each prize would you expect Alan to win?

13. At Pizza Palaisa, there is a special on the two-topping pizza. You can choose from pepperoni, sausage, mushrooms, and green peppers.

a) List all different combinations of two toppings. Be sure to record each combination only once.
b) According to the list of combinations, what is the probability of choosing a pizza with pepperoni on it?

14. A spinner is divided into coloured sections: orange, green, and purple. The probability of landing on orange is $\frac{1}{3}$, and the probability of landing on green is $\frac{1}{2}$.

a) What is the probability of not landing on the purple?
b) Use complements to find the probability of landing on purple.
c) How many equal sections could the spinner be divided into?

Chapter Problem

15. You and three friends take turns taking random samples of 10 jellybeans from the jar of 800 jellybeans. All the jellybeans are returned to the jar before the next sample is taken out. The table shows the number of different coloured jellybeans in each sample.

Sample Number	Red	Yellow	White	Green
1	6	1	3	0
2	4	3	1	2
3	5	2	2	1
4	4	4	1	1

a) For each sample, what is the experimental probability of picking a red jellybean?
b) For each sample, what is the experimental probability of not picking a red jellybean?
c) Use the sample data to predict the number of red jellybeans in the jar. Explain how you made your prediction.
d) Are you confident that you have predicted the correct number of red jellybeans? If not, what would make you more confident?

16. The label on a can of mixed nuts says it contains 50% peanuts, 25% cashews, 20% hazelnuts, and 5% Brazil nuts.

    a) You take a handful of 20 nuts from the can. How many of each type of nut would you expect to have in your sample?
    b) Is it possible for the can to have more than 50% peanuts? Explain why or why not.

17. Kim is teaching her dog, Rosco, to sit on command. Each time Rosco sits correctly, Kim rewards him with a dog treat. The box of treats contains three flavours. So far, Rosco has received 10 treats. The numbers he has eaten of each flavour are shown.

Flavour	Tally
chicken	|||
vegetable	𝍸
beef	||

    a) According to the tally, what is the probability that the next treat is chicken-flavoured?
    b) According to the tally, what is the probability that the next treat is not vegetable-flavoured?
    c) The label on the box claims that 25% of the treats are beef-flavoured. Is this claim supported by the results in the table? Explain your answer using the concepts of theoretical and experimental probability.

Extend

18. The student council is selling raffle tickets at the school dance. The prize is an MP3 player. One name will be drawn from all the tickets sold. While most students bought 1 or 2 tickets, Aisha decided to increase her chances of winning by buying 5 tickets, while her friend Dawn bought 12 tickets. A total of 150 tickets were sold.

    a) What is the theoretical probability that Aisha will win the prize?
    b) What is the theoretical probability that Dawn will win the prize?
    c) The theoretical probability of winning the prize with one ticket is less than 1%. Ming bought only one ticket and won the raffle. Aisha and Dawn complained that this was not possible and the draw must have been rigged. Is this true? Explain using your knowledge of experimental and theoretical probability.

19. What is the probability of tossing 4 heads if you toss a coin 4 times? Draw a tree diagram or an organized table to show the possible outcomes.

5.3 Explore the Effects on Mean, Median, and Mode

Focus on...
- determining the effect of variations in data on the mean, median, and mode

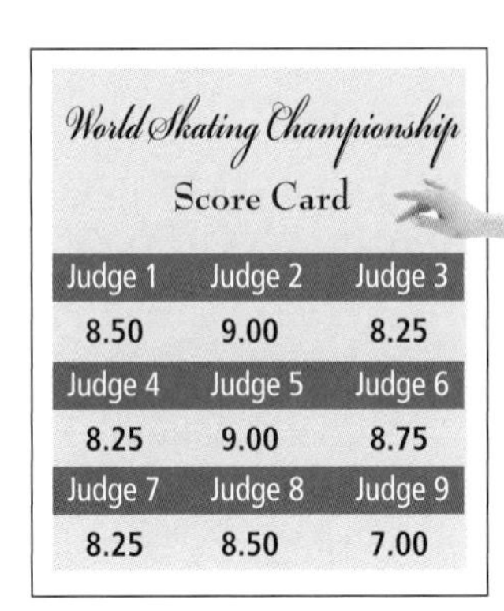
World Skating Championship

Score Card

Judge 1	Judge 2	Judge 3
8.50	9.00	8.25
Judge 4	**Judge 5**	**Judge 6**
8.25	9.00	8.75
Judge 7	**Judge 8**	**Judge 9**
8.25	8.50	7.00

Canadian senior pairs skaters Mylene Brodeur and John Mattatall. John was born in Tatamagouche, NS.

Measures of central tendency such as mean, median, and mode are values that represent averages. "Average" is a word that means a general or expected behaviour.

In professional figure skating, a skater's final score is determined by finding the mean of the judges' scores after the highest score and the lowest score have been eliminated. Can you suggest why this method is used?

Discover the Math

What happens to the mean, median, and mode for a set of data when the data changes?

1. Refer to the skating score card. List the scores in order from least to greatest. Eliminate the highest and lowest score. Find the mean, median, and mode of the seven remaining scores.

2. Which measure of central tendency (mean, median, or mode) best describes the figure skater's performance? Explain your choice.

3. Before the scores were finalized, the judges realized the performer had exceeded the time allowed, so 0.5 points were deducted from each score. List the revised scores and calculate the mean, median, and mode for this new set of data.

4. Compare the measures of central tendency for the original scores with those of the revised scores. What effect did subtracting the 0.5 points have on the mean, median, and mode of the set of data?

5. **Reflect** How do you think the mean, median, and mode are affected when the same number is added to, or subtracted from, each value in the set of data? Test your prediction by making up a set of data and finding the three measures of central tendency before and after the same number is added to each value in the set of data. Repeat the test by subtracting the same number from each value in the original set of data. Describe what happens to the measures of central tendency each time.

6. In the set of nine scores, there is one score that is not consistent with the others. This is called an **outlier** .

Sometimes the outlier might be the result of an error in measurement or a one-time occurrence caused by special circumstances. For this reason, all factors must be considered before deciding whether or not the outlier should be removed from the set of data.

a) Calculate the measures of central tendency for all nine scores.

b) Predict how removing the outlier from the set of data affects each of the measures of central tendency. Test your prediction by calculating the mean, median, and mode without the outlier.

c) Describe how the mean, median, and mode are affected when an outlier is excluded from the set of data. Which of the three measures of central tendency seem to be most affected by the outlier?

outlier

- an unusual value in a set of data that does not fit the pattern of the other data.

7. **Reflect** Why do you think the highest and lowest scores are removed from an ice skater's score card before the final score is calculated?

8. The judges are asked to enter their scores as a number out of 20. They decide to multiply their original scores by 2. What are the mean, median, and mode of the scores that will be submitted?

9. **Reflect** How do you think the mean, median, and mode are affected when each value in the set of data is multiplied or divided by the same number? Make up a set of data to test your prediction.

Example 1: Effect of an Outlier on Mean, Median, and Mode

Taiko wrote five math tests. His scores (out of 100) are shown.

80 82 62 79 82

a) Find the mean, median, and mode.

b) Which measure of central tendency best describes Taiko's typical performance on a test?

c) What are the mean, median, and mode if his least mark is removed from the set of data? Does this more accurately reflect Taiko's typical performance?

Solution

a) From least to greatest, the scores are 62, 79, 80, 82, 82.

$$\begin{aligned}\text{Mean} &= \frac{\text{sum of values}}{\text{number of values}}\\ &= \frac{62 + 79 + 80 + 82 + 82}{5}\\ &= \frac{385}{5}\\ &= 77\end{aligned}$$

Median = 80 Mode = 82

b) In this set of data,

- the mean is not a good measure of central tendency since $\frac{4}{5}$ or 80% of the data is greater than the mean
- the median is a good choice since the other values tend to be centred around the median
- the mode is not reliable because it is the greatest number in the set

c) The least mark is unusually low and could be an outlier. Removing it leaves four test scores: 79, 80, 82, 82

In this new set of data,

$$\begin{aligned}\text{Mean} &= \frac{79 + 80 + 82 + 82}{4}\\ &= \frac{323}{4}\\ &= 80.75\end{aligned}$$

Median = 81 Mode = 82

The mean and median are both within 2 points of Taiko's marks (without the outlier), so they are an accurate reflection of Taiko's typical performance.

Example 2: Constant Changes to Mean, Median, and Mode

The daily numbers of oranges Maggie purchased during one week for her daycare group are shown.

2, 12, 5, 7, 15, 12, 3

a) Find the mean, median, and mode.

b) Add 10 to each data value and find the mean, median, and mode.

c) Subtract 2 from each data value and find the mean, median, and mode.

d) Multiply each data value by 3 and find the mean, median, and mode.

e) Compare the mean, median, and mode for parts b) to d) with those from part a).

Solution

a)
$$\begin{aligned}\text{Mean} &= \frac{\text{sum of values}}{\text{number of values}}\\ &= \frac{2 + 3 + 5 + 7 + 12 + 12 + 15}{7}\\ &= \frac{56}{7}\\ &= 8\end{aligned}$$

Median = 7 Mode = 12

b) The new set of data is 12, 13, 15, 17, 22, 22, 25.

$$\begin{aligned}\text{Mean} &= \frac{12 + 13 + 15 + 17 + 22 + 22 + 25}{7}\\ &= \frac{126}{7}\\ &= 18\end{aligned}$$

Median = 17 Mode = 22

c) The new set of data is 0, 1, 3, 5, 10, 10, 13.

$$\begin{aligned}\text{Mean} &= \frac{0 + 1 + 3 + 5 + 10 + 10 + 13}{7}\\ &= \frac{42}{7}\\ &= 6\end{aligned}$$

Median = 5 Mode = 10

d) The new set of data is 6, 9, 15, 21, 36, 36, 45.

$$\begin{aligned}\text{Mean} &= \frac{6 + 9 + 15 + 21 + 36 + 36 + 45}{7}\\ &= \frac{168}{7}\\ &= 24\end{aligned}$$

Median = 21 Mode = 36

e) When 10 was added to each value in the set of data, the new mean, median, and mode were each 10 more than the mean, median, and mode of the original set of data.

When 2 was subtracted from each value in the set of data, the new mean, median, and mode were each 2 less than the mean, median, and mode of the original set of data.

When each value in the set of data was multiplied by 3, the new mean, median, and mode were each 3 times the mean, median, and mode of the original set of data.

Communicate the Key Ideas

1. When the same number is added to, or subtracted from, each value in the set of data, describe how you can find the new mean, median, and mode without calculating.
2. When each value in the set of data is multiplied or divided by the same number, describe how you can find the new mean, median, and mode without calculating.
3. Which measure of central tendency (mean, median, or mode) is most influenced by an outlier? Explain, using an example.

Strategies

Can you explain why these methods work? Look at the sum of the values in the sets of data. Use logical reasoning.

Check Your Understanding

1. Wendy bowled five games at the bowling alley. Her scores were 119, 128, 121, 122, and 119.
   a) Find the mean, median, and mode.
   b) Which measure(s) of central tendency best describe Wendy's typical bowling score? Explain why.

2. The results of six rolls of a number cube are given.

   1, 2, 5, 3, 2, 4

   a) Find the mean, median, and mode.
   b) Calculate the new mean, median, and mode when each data value is multiplied by 2.

3. The numbers of tie-dyed T-shirts Jamal sold over five days are shown.

   32, 28, 24, 50, 11

   a) Find the mean, median, and mode
   b) Predict what the new mean, median, and mode will be if 4 is added to each data value.
   c) Check your predictions by calculating the mean, median, and mode using the new set of data.

Chapter Problem

4. You decide to take six more samples of 10 jellybeans from the jar of 800 jellybeans to see if you can get a more accurate result.

Sample Number	Red	Yellow	White	Green
1	6	1	3	0
2	4	3	1	2
3	5	2	2	1
4	4	4	1	1
5	3	3	3	1
6	6	4	0	0
7	4	3	2	1
8	2	2	4	2
9	5	1	2	2
10	6	1	1	2

a) Determine the mean, median, and mode for each colour of jellybean.

b) Which measure of central tendency is least useful in finding the average number of jellybeans of each colour? Explain your answer.

5. Molly wrote six math tests. Her scores out of 100 are given.

75 72 73 42 77 72

a) Calculate the mean, median, and mode.

b) One of the scores is unusual. Suppose her teacher allows Molly to remove the lowest mark. How will this affect the mean and median?

c) Which measure of central tendency more accurately describes Molly's typical performance? Why?

6. The Halifax Regional Police kept track of the daily numbers of reported road accidents over four weeks.

12	14	10	17	15	11	14
13	13	20	15	12	9	16
14	12	41	13	17	12	12
20	13	16	10	18	14	15

a) Calculate the mean, median, and mode.

b) One of the amounts is unusually high. Remove this outlier and calculate the mean, median, and mode of the new data set.

c) Multiply each value in the set of data from part b) by 1.5. Calculate the new mean, median, and mode.

7. Jacob wants to improve his math mark before report cards are sent home. So far, his performance scores out of 100 are 65, 62, 67, 62, and 63. His report card mark is the mean of all performance scores.

a) What is Jacob's current math mark?

b) There is one more test. If Jacob scores 90, what is the new mean?

c) Should a score of 90 be considered an outlier? Explain why or why not.

8. a) Conduct a survey to find the average length of time it takes students to complete their math homework.

b) Determine the mean, median, and mode of your survey results.

c) What happens to the mean if a student does not finish the homework?

9. How will the mean, median, and mode change in each situation?

a) An outlier that is the greatest value is removed from the set of data.

b) An outlier that is the least value is removed from the set of data.

10. The most popular item sold at Farrah's Boutique is jeans with zippered pockets. The store owner keeps track of the jeans sold by size.

Jeans Size	Number Sold
4	1
6	8
8	13
10	16
12	6
14	2
16	2

a) What is the mean, median, and mode jeans size based on the data?
b) Which measure of central tendency is most informative to the store owner? Explain.
c) What would the mean, median, and mode be if twice as many pairs of jeans for each size are sold?

11. The wages for high school students working part-time was recorded.

Hourly Wage ($)	Number of Students Earning the Wage
8.00	3
8.50	4
9.00	10
9.50	8
10.00	6
10.50	3
11.00	1
11.50	0
12.00	2

a) What is the most common hourly wage?
b) What is the mean hourly wage?
c) What amount must be added to each wage for the current mean wage to change to $12/h?

12. In each situation, decide whether or not the outlier should be removed. Justify your choice in each case. Then calculate the mean for the set of data.

a) The gas prices at five gas stations in town are $0.87/L, $0.86/L, $0.84/L, $0.91/L, and $0.84/L.
b) The daytime high temperatures for the first eight days of winter 2006 in Halifax, NS, were 6.5°C, −0.7°C, 9.0°C, 9.3°C, 5.5°C, 5.9°C, 1.0°C, and −4.3°C.
c) A basketball player's points per game for the last 10 games are 26, 24, 27, 25, 24, 17, 28, 25, 27, and 24.

13. a) Construct a set of data where the mean is 10, the median is 7, and the mode is 8.
b) Construct a second set of data with a different number of values that meets the same criteria.

Extend

14. Mr. Lee works for the Department of Fisheries and Oceans. He counted the number of tagged fish in several different samples of 50 fish. Here are the results: mean = 15, median = 14.5, mode = 14. What are the mean, median, and mode numbers of untagged fish? Supply data to support your answer.

15. Last week, Mr. Brighton measured the heights of his seven prized oak seedlings. He noted that the range of the heights was 6.2 cm and that his tallest seedling measured 10.8 cm. The mean height was 7.4 cm, the median height was 7.6 cm, and the mode was 8 cm. What could be the heights of all seven seedlings?

5.4 Construct and Interpret Box-and-Whisker Plots

Focus on...
- constructing and interpreting box-and-whisker plots

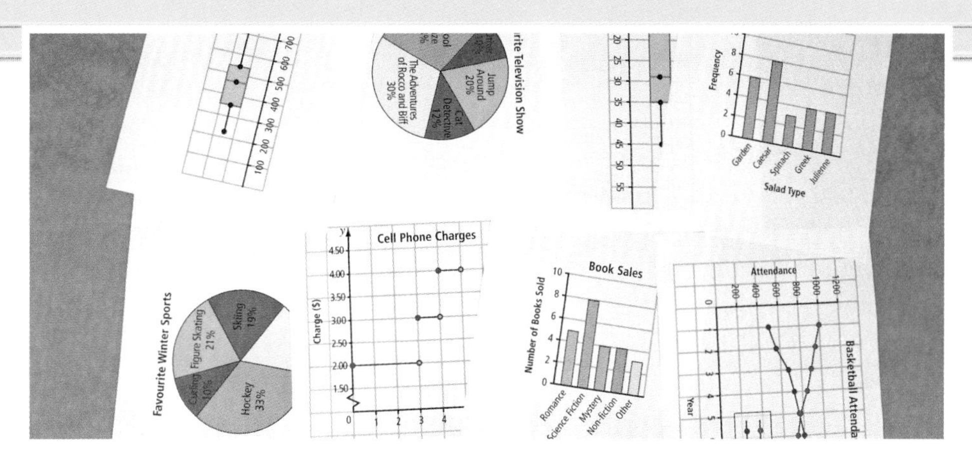

When you analyse a graph, you should ask yourself, "Do I have all the information?"

A circle graph shows data listed in categories that are parts of a whole.
A bar graph shows data listed in categories.
A line graph shows a trend in the data.

What other types of graphs are there and what information can they provide?

box-and-whisker plot
- a diagram that shows the median and range of a numeric set of data

lower quartile
- the median value of the first 50% of data points in an ordered set of data

upper quartile
- the median value of the second 50% of data points in an ordered set of data

A **box-and-whisker plot** shows how data is spread around the median of a set of data. It displays the data in four sections, each containing roughly 25% of the data points of the set. It is a convenient, visual way to show the median, the maximum and minimum values, and the distribution of a set of data. The box of the graph contains or represents at least 50% of the data and is bounded by a **lower quartile** and an **upper quartile** value.

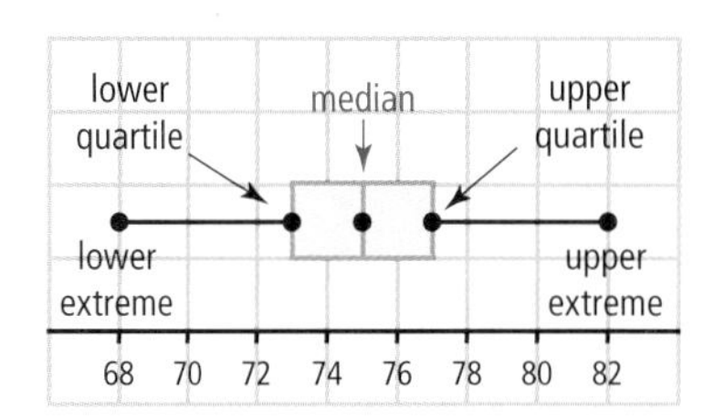

Discover the Math

How do I construct a box-and-whisker plot to show the distribution of students' heights in my class?

Materials
- centimetre grid paper
- ruler
- metre sticks

1. Work as a group to arrange the students in your class from shortest to tallest across the front of the classroom. Students of the same height will stand next to each other. The set of data consists of the heights of all the students in the class.

2. The median of the set of data is the first of five values needed to draw a box-and-whisker plot. Identify the median student in the ordered row of students. This student divides the class into two halves—the lower half and the upper half. Have this student hold up a metre stick vertically and keep it in the air. If there is an even number of students in your class, then the median will be the mean height of the two students standing in the middle of the row. Measure and record the height of the median student.

3. Identify and measure the height of the student in the middle of the lower half of students. This median of the lower half of the data set is called the lower quartile. It determines the left side of the box. Have the student representing this value hold up a metre stick and keep it up in the air.

4. Identify and measure the height of the student in the middle of the upper half of students. This median of the upper half of the data set is called the upper quartile. It determines the right side of the box. Have the student representing this value hold up a metre stick and keep it in the air.

5. The whiskers extend to the lower and upper extremes (i.e., the minimum and maximum values in the set of data). Measure and record the heights of the shortest and tallest student in the class.

6. Use grid paper to place the five values on a number line and label them. Draw a box-and-whisker plot with these values.

7. What percent of the set of data is contained in the box? *Hint: How many students stand between the median students of each half?* How many students have a height between the median height and the upper quartile? How many students have a height on the lower whisker?

8. Is the median located in the centre of the box or is it skewed to one side? What does this suggest about the values in the data set?

Journal

9. Draw a histogram showing the heights of the students in your class. Explain how a histogram and a box-and-whisker plot compare. Comment on the measures of central tendency and the range. Is there great variability in the data? Explain.

Example 1: Box-and-Whisker Plot

Scores for a test marked out of 25 are shown.

22, 14, 15, 20, 21, 21, 18, 19, 24, 8

a) Identify the median, the upper and lower quartiles, and the upper and lower extremes.

b) Draw a box-and-whisker plot with these values.

c) What can you say about the values in the set of data? Use the location of the median in the box-and-whisker plot, the quartiles, and the extremes.

d) Is there much variability in the data? Explain.

Strategies
Use a graph to analyse the data.

Solution

a) Arrange the scores from least to greatest.

8, 14, 15, 18, 19, 20, 21, 21, 22, 24

$$\text{Median} = \frac{19 + 20}{2} = 19.5$$

Lower quartile = 15
Upper quartile = 21

Lower extreme = 8
Upper extreme = 24

I need to find the median of the lower half of the data set: 8, 14, 15, 18, 19

I need to find the median of the upper half of the data set: 20, 21, 21, 22, 24

Strategies
Can you think of another way to find the mean of 19 and 20?

b)

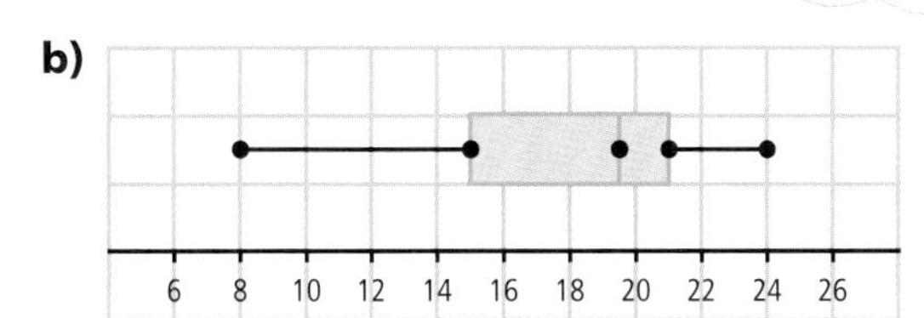

Strategies
What other kinds of graphs could you use to communicate the test results?

c) The lower whisker is longer than the upper whisker. This means that the lowest 25% of the students had grades that were more spread out than those in the highest 25%.

The lowest 25% of the scores were between 8 and 15. The highest 25% of the scores were between 21 and 24.

The median is slightly skewed toward the upper quartile. This suggests that most of the values in the data set are large with a few small ones. Also, half the class scored at least 20 out of 25.

d) Range = greatest value − least value
= 24 − 8
= 16

The range is 16, but the marks are skewed toward the higher values. One quarter of the data is closely clustered between 19.5 and 21.

The top half of the marks are close to each other, between 19.5 and 24, a 4.5 point spread. The lower half of the marks are more widely spread, between 8 and 19.5, an 11.5 point spread.
There is some variability in the data.

Stem-and-leaf plots are often used to collect and order data, and are very helpful in the construction of box-and-whisker plots.

Example 2: Draw a Box-and-Whisker Plot From a Stem-and-Leaf Plot

Use the data in the stem-and-leaf plot about typing speed (in words per minute) to draw a box-and-whisker plot.

Solution

There are 17 values in the data set. So the ninth value is the middle of the set of data.

Median = 36
Lower quartile = 23
Upper quartile = 46.5
Lower extreme = 10
Upper extreme = 54

There are eight values in the lower half of the data set. The median of the lower half is the average of the fourth and fifth values.

The median of the upper half of the data set is the average of the 13th and 14th values.

Stem (tens)	Leaf (ones)
1	0 2
2	1 1 5 7
3	2 3 6 6
4	1 3 5 8 9
5	2 4

5 10 15 20 25 30 35 40 45 50 55

Example 3: Draw a Box-and-Whisker Plot From a Table of Data

Use the data in the table to draw a box-and-whisker plot.

Number of Employees	Salary ($)
10	30 000
10	35 000
8	45 000
1	70 000

Solution

Imagine that the 29 employees are arranged in order according to salary.

Median:	35 000	**salary of employee 15**
Lower quartile:	30 000	**mean of the salaries of employees 7 and 8**
Upper quartile:	45 000	**mean of the salaries of employees 22 and 23**
Lower extreme:	30 000	
Upper extreme:	70 000	

To draw the box-and-whisker plot, choose a scale from 25 000 to 75 000.

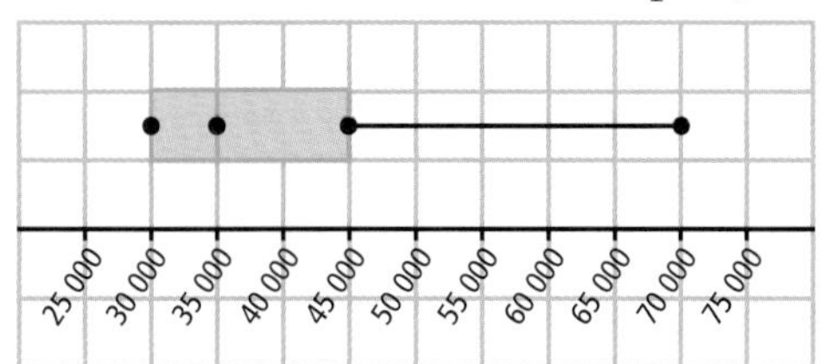

The plot has no lower whisker since the lowest 25% of the salaries are all the same.

Communicate the Key Ideas

1. What five values are needed to draw a box-and-whisker plot?

2. a) What percent of the set of data is contained in the box of a box-and-whisker plot?

b) What percent of the set of data lies in each whisker?

3. What information can you get from short whiskers? long whiskers? no whiskers?

4. Another graph that is used to illustrate the distribution of data is called a stem-and-leaf plot. How are box-and-whisker plots and stem-and-leaf plots similar? How are they different?

Check Your Understanding

1. Scores for a test marked out of 40 are shown.

31, 25, 28, 40, 32, 16, 35, 36, 27, 39, 23, 26, 18, 33, 37

a) Draw a box-and-whisker plot.

b) Complete the statements.

- The lowest 25% of the scores were from ■ to ■.
- The highest 25% of the scores were from ■ to ■.
- The range is ■.

2. The numbers of raffle tickets sold by the members of the student council are shown.

52, 30, 35, 17, 29, 41, 31, 32, 40, 15, 22, 25, 33, 40, 50, 36, 38, 27, 16, 20

a) Draw a box-and-whisker plot.

b) Complete the statements.

- The lowest 25% of the numbers were from ■ to ■.
- The highest 25% of the numbers were from ■ to ■.
- The range is ■.

3. a) Use the data represented in the stem-and-leaf plot to draw a box-and-whisker plot.

b) Based on the box-and-whisker plot, write three statements about the data.

Stem (tens)	Leaf (ones)
1	3
2	0 2 5
3	4 5 7 9
4	2 3 4 4 7 8
5	3 5 6
6	1 1

4. a) Use the data in the table to draw a box-and-whisker plot.

Number of Students	Number of Pages in Essay
4	5
1	6
5	8
7	9
1	10
1	15

b) Based on the box-and-whisker plot, write three statements about the length of the essays.

c) Which measure of central tendency would you use to indicate the typical number of pages in an essay? Explain.

d) What does the median say about the variability of the data?

5. Examine the box-and-whisker plot of percents of households who recycle in different Cape Breton communities.

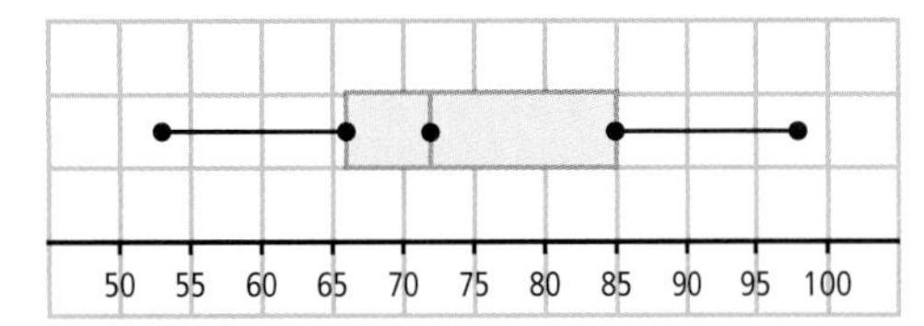

a) What was the median percent?
b) What was the greatest percent? What was the least percent?
c) Make a list of 15 numbers that this box-and-whisker plot could represent.

Chapter Problem

6. The correct number of red jellybeans in the jar is not revealed until the end of the fair. In the meantime, all guesses are listed on a chalkboard hung behind the jar. So far, 32 guesses are listed:

300, 400, 500, 240, 320, 320, 320, 320,
400, 400, 300, 350, 350, 350, 400, 400

400, 400, 450, 280, 320, 320, 360, 360,
400, 400, 400, 320, 384, 384, 416, 448

a) Draw a box-and-whisker plot to display the distribution of the guesses.
b) Analyse the graph to see what others thought about the correct number of red jellybeans. For example, what is the lowest number of red jellybeans believed to be in the jar? What other information does the graph tell you?
c) Is there great variability in the data? Explain.

7. The box-and-whisker plot represents the prices of 20 MP3 players being sold at The Funky Beat music store.

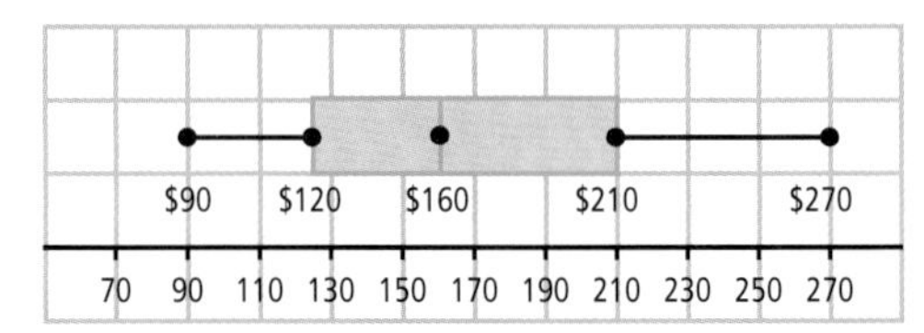

a) What is the median price for an MP3 player in the store?
b) How much is the most expensive MP3 player in the store?
c) About how many models of MP3 players cost $125 or more?

8. The two box-and-whisker plots show the numbers of points scored by Angela and Carla during the last 10 basketball games.

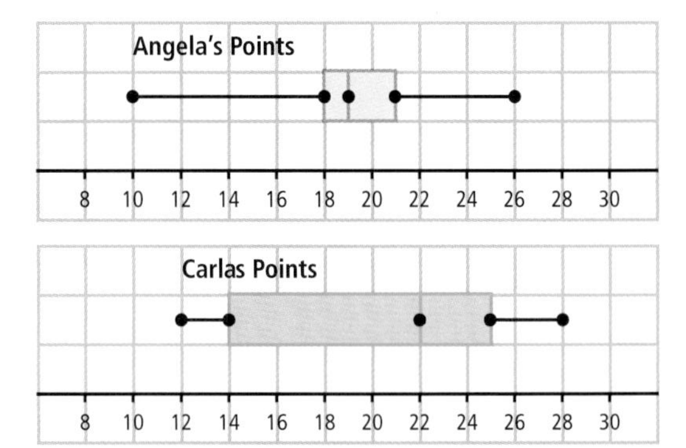

a) Which player has the higher median number of points scored in a game?
b) Which player has the higher maximum number of points scored in a game?
c) What is the minimum number of points scored for 75% of the games by Angela? by Carla?
d) Based on the data, which player would you choose to be on your team? Explain your choice.

9. The data shows the numbers of raisins found in the same-sized boxes of two brands of raisins.

Brand A: 18, 20, 21, 22, 22, 23, 23, 23, 24, 24, 25, 26, 26, 30

Brand B: 21, 23, 24, 24, 25, 25, 26, 26, 26, 26, 27, 28, 30, 32

a) Draw a box-and-whisker plot for each brand of raisins on the same grid.

b) Compare the two graphs. On average, which brand has more raisins in a box? Explain your answer.

c) What other information can you find from comparing the two box-and-whisker plots?

Extend

10. Rahul works at an electronics store. Part of his job includes keeping inventory of the numbers of stereo systems sold each month.

Jan.	Feb.	Mar.	Apr.	May	June
8	6	6	7	10	12
Jul.	**Aug.**	**Sept.**	**Oct.**	**Nov.**	**Dec.**
14	18	13	11	23	39

a) Identify the median, the upper and lower quartiles, and the upper and lower extremes.

b) Draw a box-and-whisker plot.

c) Draw a bar graph with the months displayed on the horizontal axis and the number of stereo systems sold on the vertical axis.

d) Compare the two graphs. What information is given by one type of graph but not by the other?

11. Minh works part-time at a convenience store. The number of hours per week that he worked for the past 12 weeks are shown.

18 15 12 13 16 17 8 3 10 6 12 11

a) Identify the median, the upper and lower quartiles, and the upper and lower extremes.

b) Draw a box-and-whisker plot for the data.

c) What can you conclude about the number of hours that Minh works per week?

d) Draw a scatterplot of the data.

e) Compare the two graphs. What information is given by one type of graph but not by the other?

Did You Know?

Sandra Lovelace Nichols is a Maliseet activist from New Brunswick who fought against a discriminatory clause in the Indian Act that removed a woman's Aboriginal status if she married a non-aboriginal man. (An aboriginal man could marry a non-aboriginal woman and still keep his status.) She took her case to the United Nations in 1977 but it was not until 1985, that Ottawa finally revoked the clause. She was awarded the Order of Canada in 1990 and the Governor General's Award in 1992. In September 2005, she was named to the Canadian Senate, as was lawyer Yoine Goldstein, raising the percent of women in the Senate to a record high of 36%.

5.5 Construct and Interpret Circle Graphs

Focus on...
- constructing and interpreting circle graphs

When you go grocery shopping, you may use advertisements to compare prices.

circle graph
- a graph in which a circle is used to represent a whole and is divided into sectors that show how data are divided into parts by percent. Also called a pie chart.

Circle graphs are good for comparing categories and seeing how each part compares to the whole.

A circle graph is also known as a pie chart. The circle is divided into sectors that show how the data are divided into parts by percent.

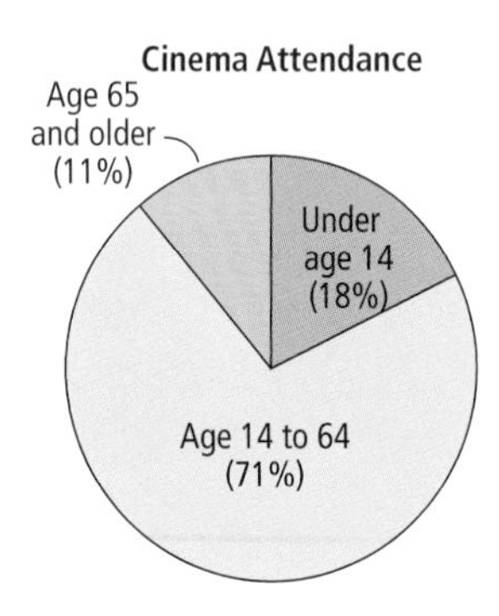

Discover the Math

Materials
- Bullseye
- coloured pencils
- pencil

Optional:
- BLM 5.5 DTM Percent Circle
- set of compasses
- protractor
- ruler

How can I draw a circle graph to display information?

Part A: Circle Graphs With Equal Sectors

A circle graph is similar to a spinner.

1. Use a Bullseye to draw a circle with four equal sectors. What fraction of the whole circle is each sector?

2. Colour two sectors red, one sector blue, and one sector green.

3. What information does the circle graph tell you about the percent of the circle that each colour represents? Explain how you know.

4. Examine the blue sector. Measure the sector angle.

5. There are 360° in a circle. What is the relationship between the size of the sector angle and the fraction of the whole circle that it represents?

6. Use the information in question 5 to draw a circle divided into five equal sectors. How large is each sector angle?

7. Colour two sectors yellow, one sector red, one sector blue, and one sector green.

8. What fraction of the whole circle is yellow? According to the method used earlier, what size should the sector angle be? Measure the size of the large yellow sector angle.

Journal

9. Suppose you change the circle into a spinner by adding a paper clip and using a pencil tip to anchor the clip. What information does your circle graph tell you about the percent chance of landing on each colour if you spun the spinner?

Part B: Circle Graphs With Unequal Sectors

Ryan surveyed 200 students about their most common use of the Internet.

1. Copy and complete the table.

Activity	Frequency	Percent (%)	Size of Sector Angle (°)
playing/downloading music	40	$\frac{40}{200} = 20$	
playing games	45		
chatting with friends	65		
getting help with homework	50		

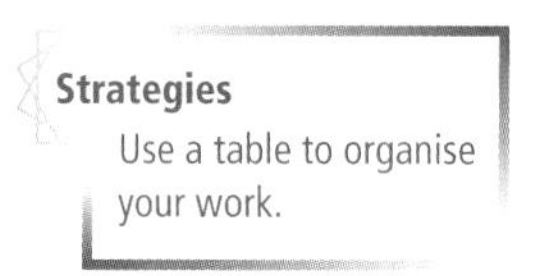
Strategies
Use a table to organise your work.

2. Use the size of the sector angle for each activity to construct a circle graph.

3. Label the sectors and give your graph a title.

4. **Reflect** Do you need to complete the table before you can construct the circle graph? Explain why or why not.

Example 1: Use a Circle Graph to Display Data

A grade 8 class of 32 students is surveyed about their favourite type of music. Draw a circle graph to display the data.

Type of Music	Number of Students
rock	12
hip hop	8
pop	8
country	4

Solution

- To draw a circle graph, the size of each sector angle must be determined. To do this,
  - express each category or event as a fraction of the whole
  - convert the fraction to a decimal (unless the fraction is easier)
  - since there are 360° in a circle, multiply the decimal or fraction value by 360 to find the size of the sector angle

Type of Music	Fraction of the Students Surveyed	Decimal	Sector Angle (°)
rock	$\frac{12}{32}$	$12 \div 32 = 0.375$	$0.375 \times 360° = 135°$
hip hop	$\frac{8}{32}$	$8 \div 32 = 0.25$	$0.25 \times 360° = 90°$
pop	$\frac{8}{32}$	$8 \div 32 = 0.25$	$0.25 \times 360° = 90°$
country	$\frac{4}{32}$	$4 \div 32 = 0.125$	$0.125 \times 360° = 45°$

I can do some of these calculations using the fraction instead of the decimal.
$\frac{8}{32} = \frac{1}{4}$, so $\frac{1}{4}$ of 360° = 90°

To check that the angle measures are correct, the sum of the angle measures should be 360°.

- Now, draw a circle and a line from the centre of the circle to a point on the circumference for its radius.
- Use a protractor to measure each sector angle. Colour the different sectors, and then label them with the categories they represent: rock, hip hop, pop, country. Include the percent for each category.
- Add an appropriate title.

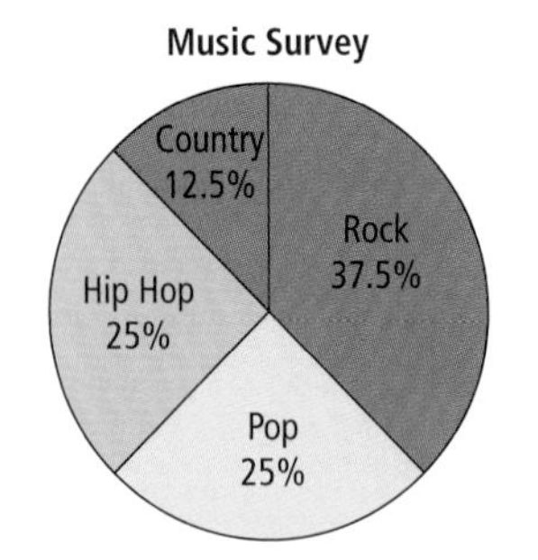

Strategies
What information is obvious in the circle graph but not obvious in the table?

Example 2: Interpret a Circle Graph to Obtain Information

This circle graph appeared in an advertisement in a magazine.

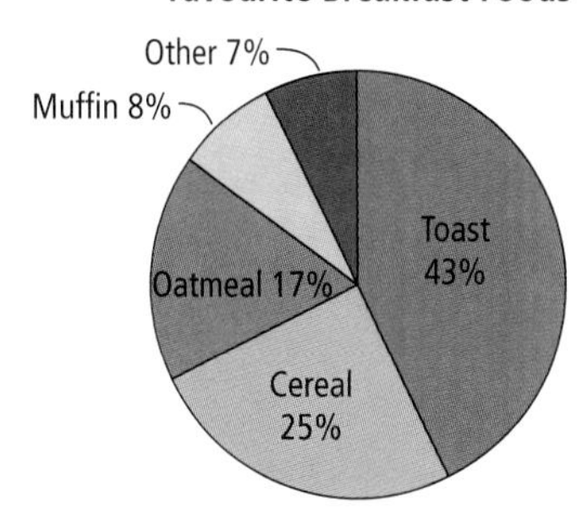

Use the information to predict how many people out of 200 would prefer each type of breakfast food.

Solution

Since the percent of each food preferred is shown by the circle graph, find the number of people who prefer

- toast: 43% of 200 = $0.43 \times 200 = 86$
- cereal: 25% of 200 = $0.25 \times 200 = 50$
- oatmeal: 17% of 200 = $0.17 \times 200 = 34$
- muffin: 8% of 200 = $0.08 \times 200 = 16$
- other: 7% of 200 = $0.07 \times 200 = 14$

I can calculate these mentally. 1% of 200 is 2. So 43% is 43 x 2 = 86.

To check that the numbers of people are correct, the sum of the numbers should be 200.

If 200 people were surveyed, you could expect 86 to prefer toast, 50 to prefer cereal, 34 to prefer oatmeal, 16 to prefer muffins, and 14 to prefer another breakfast food.

Communicate the Key Ideas

1. Explain when a circle graph would be used instead of a bar graph.

2. Describe how to find the sectors in a circle graph.

3. Explain how you could determine the percent of a category or event on a circle graph by measuring the size of the sector angle.

Check Your Understanding

1. Twelve teens are surveyed about how much money they spend on entertainment (arcade games, CDs/DVDs, movies, etc.) in one month.

Amount of Money Spent	Number of Teens	Fraction of the Group	Decimal	Sector Angle (°)	Percent (%)
less than $20	4	$\frac{4}{12} = \frac{1}{3}$			
approximately $20–$30	6				
more than $30	2				

a) Copy and complete the table.

b) Draw a circle graph to display the information. Include the percent of each sector. Give the graph a title. Describe what the circle graph shows about the data.

2. A group of teens is surveyed about the average amount of homework they have to finish each night.

Time Spent on Homework	Number of Teens	Fraction of the Group	Decimal	Sector Angle (°)	Percent (%)
0.5 h	6				
1.0 h	8				
1.5 h	7				
2.0 h	3				

a) Copy and complete the table.

b) Draw a circle graph to display the information. Include the percent of each sector. Give the graph a title. Describe what the circle graph shows about the data.

3. A group of teens is surveyed about their favourite type of TV show.

TV Show	Number of Teens
drama	40
comedy	17
reality	25
educational	15
other	3

a) Draw a bar graph and a circle graph to display the information. Give titles to the two graphs.

b) Use your graphs to determine the most popular type of TV show.

c) Use your graphs to determine the percent of teens who do not prefer reality TV.

d) Compare the two graphs. Explain which graph is more effective at displaying the data.

4. The circle graph shows sample survey data about the types of food students like to eat for lunch.

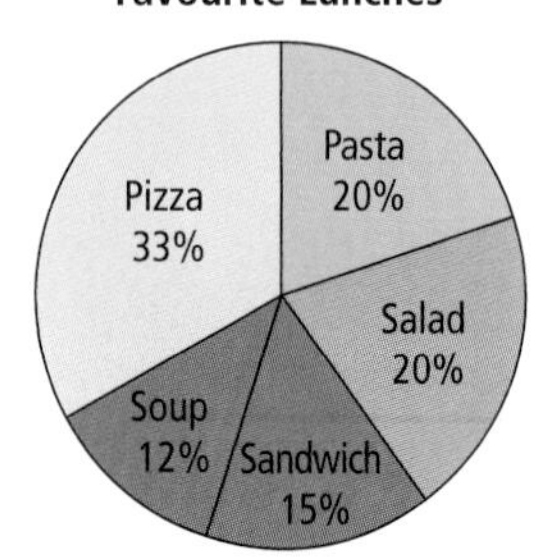

The school's population is 528 students. According to the data in the circle graph, approximately how many students in the school will prefer each type of food? Round your answers to the nearest whole number.

5. Consider the circle graph.

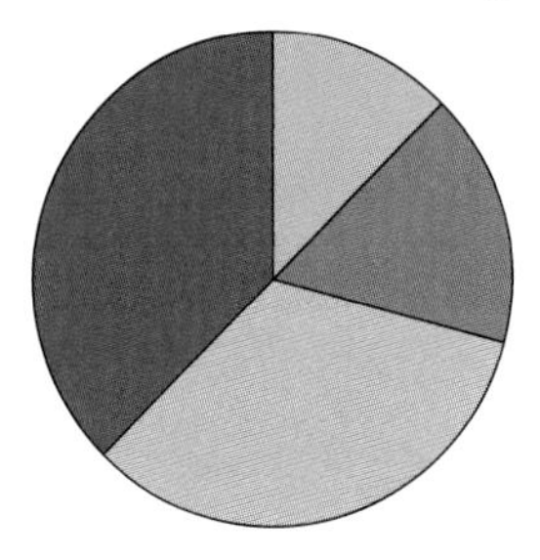

a) Determine the percent represented by each of the sectors.

b) What information could this circle graph be displaying? Copy the graph. Write a story problem to go with the graph. Be sure to label the graph. Give your problem to a classmate to solve.

6. Yesterday, Thomas went to the mall and spent all of his savings for the past four months. He bought a pair of sneakers for $90, a pair of jeans for $62, two CDs for $40, and lunch for $8.

a) Copy and complete the table.

Thomas's Purchases	Dollars Spent ($)	Fraction of Thomas's Savings	Decimal	Sector Angle (°)	Percent (%)
pair of sneakers	90	$\frac{90}{200}$	0.45	162	
pair of jeans					
CDs					
lunch					

b) Draw a circle graph to display the data. Include the percent of each sector. Give the graph a title.

7. Investigate how technology can be used to draw a circle graph. Write a list of step-by-step instructions.

8. Denise asked several families living in her apartment building what kind of pet they owned. Here are the results.

Pet	Number of Families
dog	12
cat	8
bird	2
fish	5
rabbit	1
other	2

a) What types of graphs could be used to display this data? Which type of graph would be best? Explain why.

b) Draw a circle graph to display the information. Give the graph a title.

c) There are 120 families living in Denise's apartment building. According to the data, how many families could be expected to own a dog? a fish?

Extend

9. a) Conduct a survey of students at your school to learn about a specific topic you find interesting (e.g., What are students' favourite pizza toppings? How should the school budget be used?, and so on).

b) Display the information on a circle graph.

c) Write a paragraph explaining the results of your survey and what information you can gather from the circle graph.

10. Find a bar graph in a newspaper, a magazine, or on the Internet. Use the information in the bar graph to create a circle graph that represents the same information. Which type of graph is better suited to the information displayed?

5.6 Construct and Interpret Scatterplots

Focus on...
- constructing and interpreting scatterplots and determine a line of best fit by inspection
- extrapolating and interpolating information from graphs

Statistics are collected every day to help us better understand and improve our way of life.

Graphing the data allows us to see trends that occur, such as changes in population, availability of resources, weather patterns, and so on.

Materials
- math textbooks
- stopwatches
- ruler
- string

Discover the Math

scatterplot
- a graph of ordered pairs of numeric data
- used to see relationships between two variables or quantities

How can I use a scatterplot to find trends in data?

Divide the class into two groups.

1. Conduct an experiment to investigate whether there is a relationship between the number of students in a group and the time it takes them to pass a textbook around in a circle.

2. Copy the table.

Group 1		Group 2	
Number of Students	Time (s)	Number of Students	Time (s)
4		4	
6		6	
8		8	
10		10	
12		12	
14		14	

3. Assign one student to be the timekeeper in each group. For the first round, have four students stand a circle. When the timekeeper says "Now," the first student passes the textbook to the left. Students should use both hands to pass or accept the textbook. When the textbook has been passed around the circle, the first student calls out "Now" and the timekeeper records the time. Repeat, adding two more students to the circle in each round, ending with the last round of 14 students.

4. Display the data for both groups using a scatterplot.
- Label the horizontal axis "Number of Students."
- Label the vertical axis "Time Taken to Pass the Textbook(s)."
- Add an appropriate title.

Strategies
A graph can help you see trends in the data.

5. Inspect the points on the graph. Do they seem to suggest a trend in the data? Describe this trend. Copy and complete this statement to help you: As the number of people added to the circle increases, the time it takes to complete one circuit ■.

6. Can you draw a line through the data points such that there is a balance of points above and below the line? Practise with a piece of string to find the best position before you draw the line onto the graph. This line is called the **line of best fit** and it represents the trend in the data.

7. Describe the line of best fit. Copy and complete this statement by circling the correct word:
The line of best fit [rises/falls] as you move from left to right.
How does the time change as the number of students increases by two?

8. Use the line of best fit to estimate how much time it would take to pass the textbook around if there were 9 students in the circle. This method of estimation is called **interpolation**.

9. Use the line of best fit to estimate how much time it would take to pass the textbook around if there were 20 students. This method of estimation is called **extrapolation**.

line of best fit
- the straight line that passes through, or as near as possible to, the points on a scatterplot
- the line shows the possible trend or relationship between the two quantities being measured

interpolate
- to estimate values lying between given data

extrapolate
- to estimate values lying outside the given data

Example 1: Draw a Scatterplot and Describe the Trend

The data in the table shows the distance a particular car can travel on a given number of litres of gasoline.

Volume of Gasoline Used (L)	Distance Travelled (km)
5	42
10	90
15	120
20	155
25	204
30	245
35	280

Draw a scatterplot of the data and describe the trend.

Solution

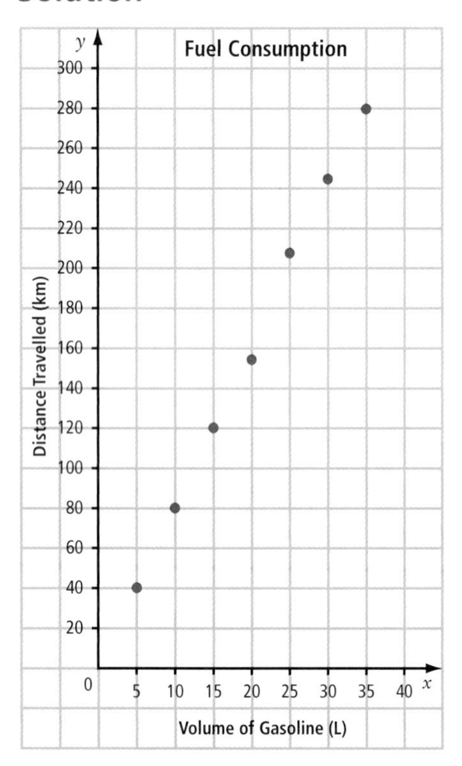

As the number of litres of gasoline consumed increases, the number of kilometres the car can travel increases, showing a strong positive relationship.

Example 2: Use a Line of Best Fit to Interpolate and Extrapolate Values

a) On the scatterplot in Example 1, draw a line of best fit so that the data points are balanced above and below it. Describe the line of best fit.

b) Use this line to estimate the distance travelled on 2 L and on 23 L of gasoline.

c) Determine the amount of fuel needed to drive 225 km. How much fuel is needed for a 300-km trip?

Strategies

How can you estimate these values without using a graph?

Solution

a) The line of best fit slopes upward and there is an increase of about 80 km for every 10 L of fuel consumed.

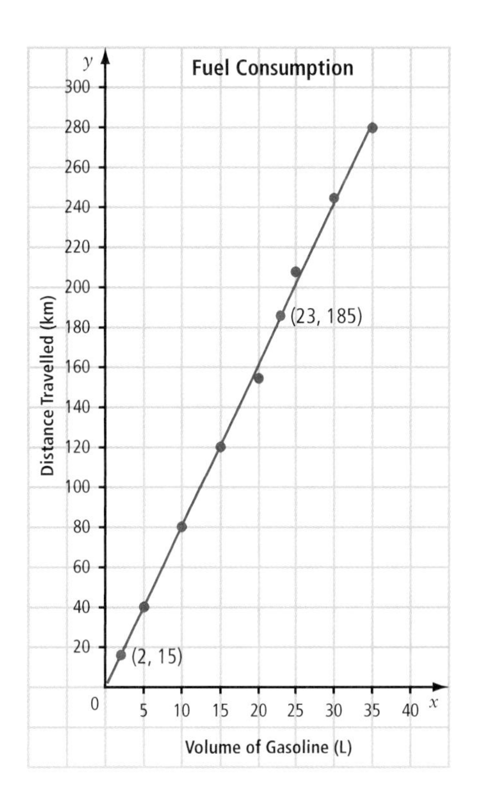

b) Using interpolation, the line of best fit indicates that the car will travel approximately 185 km on 23 L of gasoline.

Using extrapolation, the line of best fit is extended to indicate that the car will travel approximately 15 km on 2 L of gasoline.

c) Interpolation indicates that approximately 28 L of gasoline is needed to travel 225 km. Extrapolation indicates that approximately 40 L of gasoline is needed for a 300-km trip.

Communicate the Key Ideas

1. Why is the line through a scatterplot called the line of "best fit"?

2. Explain the difference between interpolation and extrapolation.

3. How is a scatterplot different from a circle graph and a box-and-whisker plot?

Check Your Understanding

1. Identify the scatterplots in which the lines of best fit have been drawn incorrectly.

a)

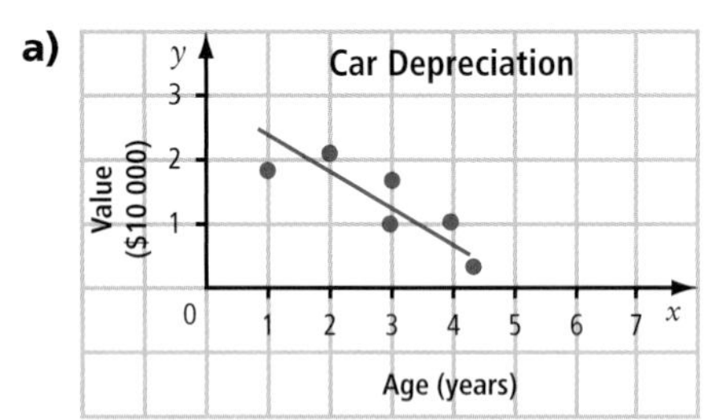

b)

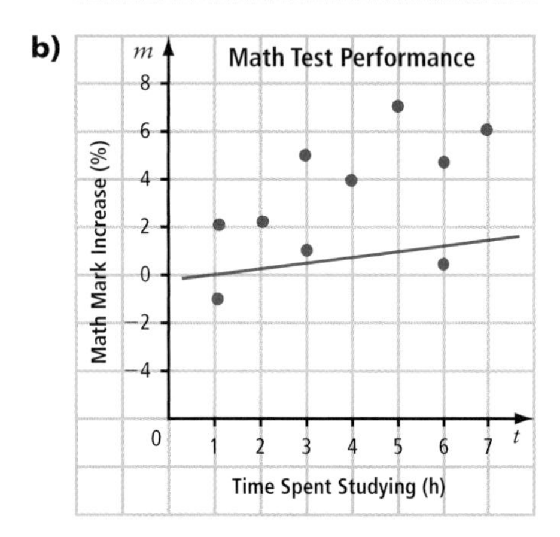

c)

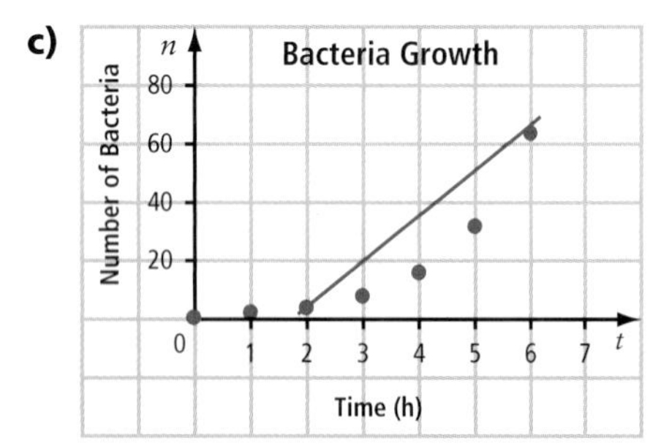

d)

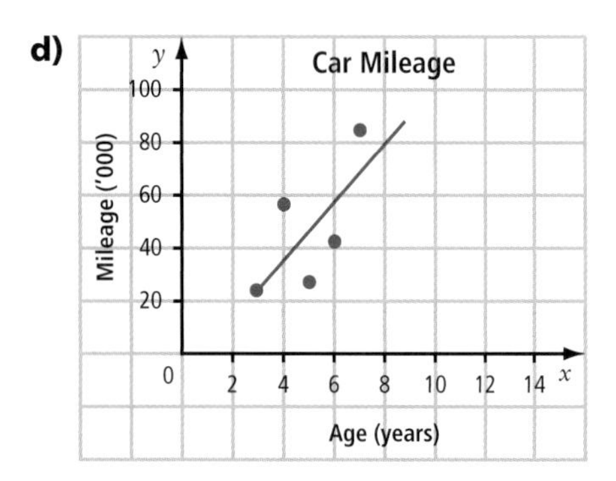

2. When there is no apparent trend or relationship between the two quantities being measured, a line of best fit cannot be drawn through the data points. For each scatterplot, identify whether or not a line of best fit can be drawn. Explain why.

a)

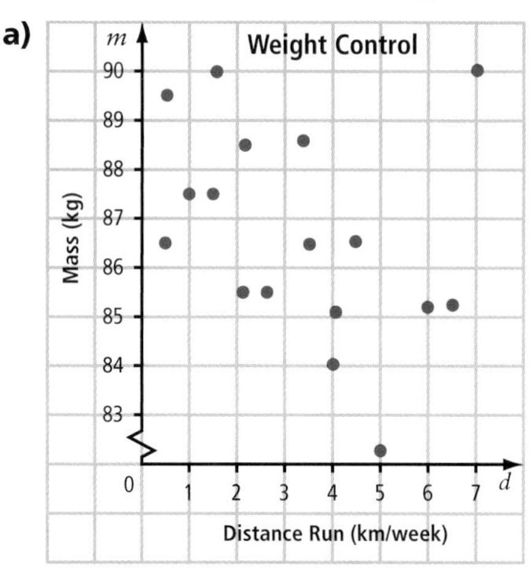

b)

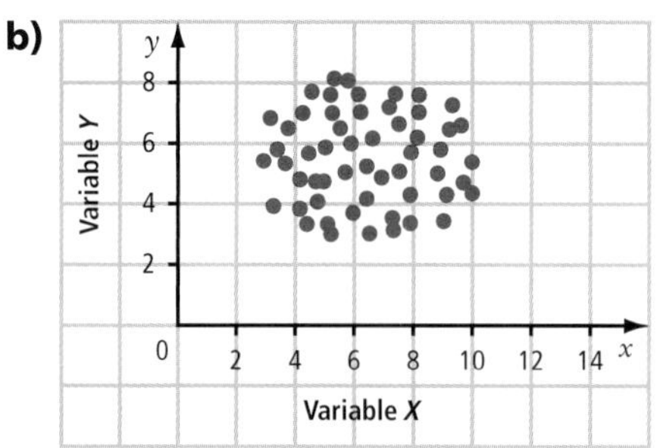

c)

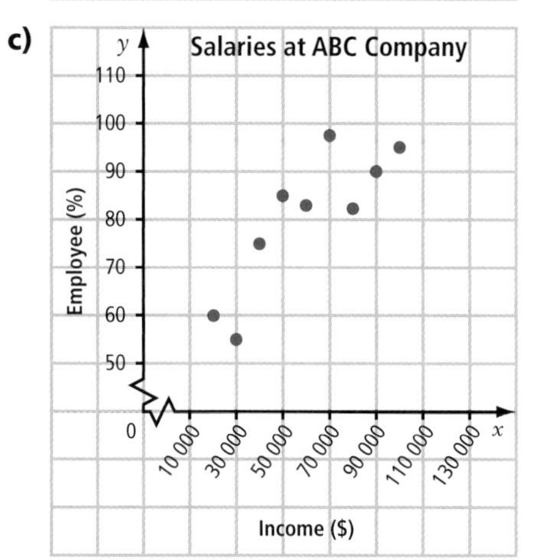

3. Place a string on each scatterplot as a line of best fit. Describe the trend that your line of best fit models.

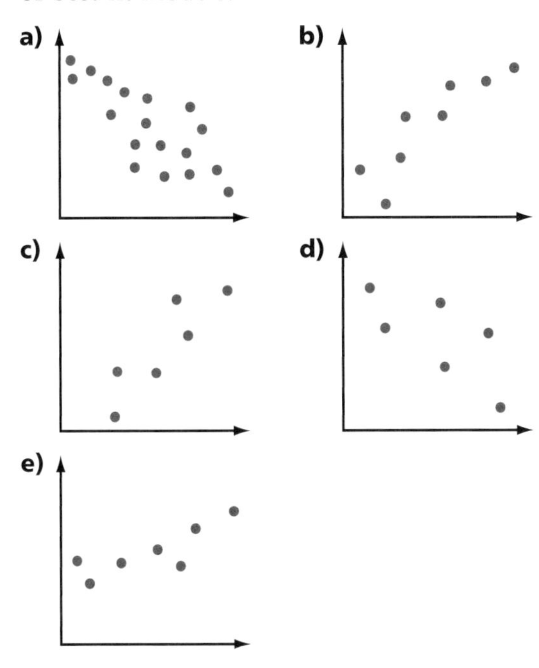

4. The data in the table compares the average heights of parents and their teenage children.

Height of Parent (cm)	Height of Teen (cm)
160	152
165	160
170	164
175	167
180	169
185	177
190	174

a) Draw a scatterplot of the data.

b) Is there a trend or relationship between the height of a parent and the height of the teen? If so, describe the relationship. If not, explain why not.

5. The scatterplot shows the relationship between height and mass of grade 8 students.

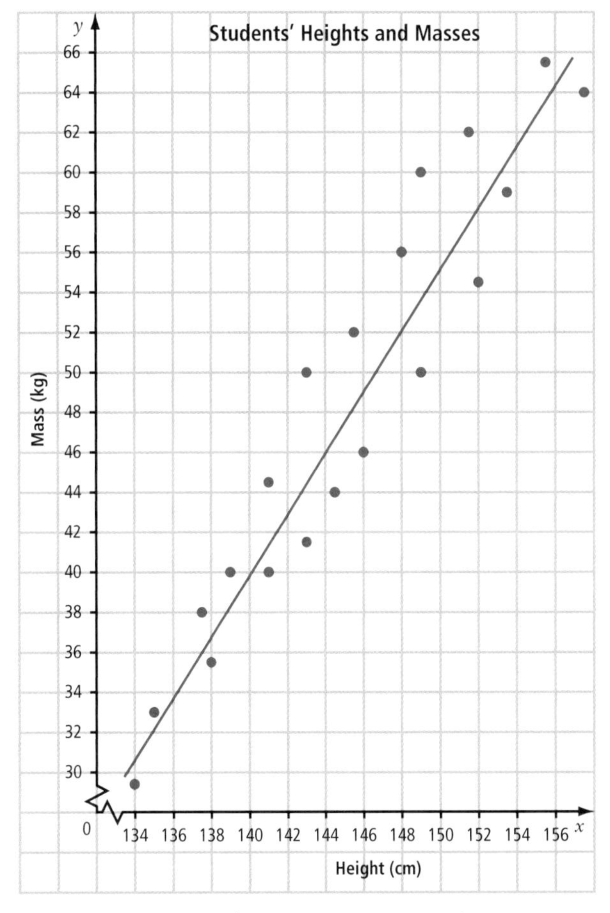

a) Use interpolation to estimate the mass of a student who is 150 cm tall.

b) Use extrapolation to estimate the mass of a student who is 130 cm tall.

c) Would you be surprised if a student had a height of 140 cm and a mass of 34 kg?

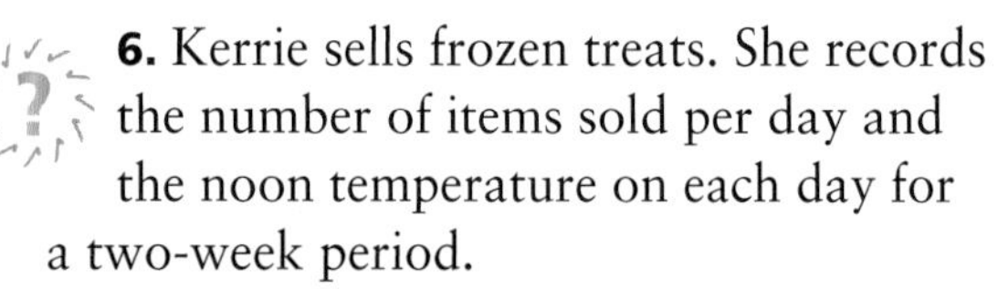

6. Kerrie sells frozen treats. She records the number of items sold per day and the noon temperature on each day for a two-week period.

Here are her results.

Day	Temperature (°C)	Number of Items Sold
1	15	4
2	15	6
3	17	8
4	18	9
5	17	12
6	19	10
7	19	13
8	21	14
9	20	15
10	23	22
11	23	25
12	25	28
13	24	26
14	25	21

a) Draw a scatterplot of the temperature and the number of items sold.

b) Draw a line of best fit.

c) Describe the relationship between the temperature and the number of items sold.

d) Use interpolation to estimate the number of items sold if the temperature was 16°C.

e) Use extrapolation to predict the number of items sold if the temperature was 27°C.

Chapter Problem

7. This data was collected from 12 people in the jellybean contest.

Sample Size	Number of Red Jellybeans
10	4
10	7
12	8
15	11
15	9
20	13
25	15
25	16
30	17
32	18
32	21
40	24

a) Graph the data as a scatterplot and draw a line of best fit.

b) Extrapolate the sample size to 800 to estimate the number of red jellybeans in the jar.

Extend

8.

Popcorn Sales

y, Number Sold (bags): 0, 20, 40, 60, 80, 100

x, Price ($): 1, 2, 3, 4, 5, 6, 7

a) Trace the scatterplot. Use a piece of string to estimate a curve of best fit. Draw the curve. Describe the trend in the data.

b) Estimate the number of bags sold at \$3.50.

c) Predict the number of bags sold at \$6.75 and at \$8.00. Explain how you found your answer.

9. Conduct a survey or an experiment to collect data to determine if there is a relationship between two variables. Choose a topic from the list or design your own experiment.

- shoe size and height
- study time and mark on a math test
- hand span and length of foot
- mass of paper airplane and distance of flight
- amount of water and plant growth

a) Record the data in a table.

b) Construct a scatterplot.

c) Discuss the relationship between the two quantities.

d) If applicable, draw a line of best fit on the scatterplot. Then, interpolate and extrapolate some data points and explain their meaning.

10. The table shows the number of bottles of water sold at a local convenience store over seven days in the month of May. The temperature for each day was recorded.

Temperature (°C)	Number of Bottles of Water Sold
9	3
11	10
14	12
20	20
25	30
27	40
26	35

a) Display the data using a scatterplot.

b) Draw a line of best fit. Extrapolate to estimate the number of bottles of water that would be sold if the temperature were 30°C.

c) How many bottles of water would you expect to be sold if the temperature were 4°C? Explain how you found your answer.

Puzzler

How many triangles are in this figure?

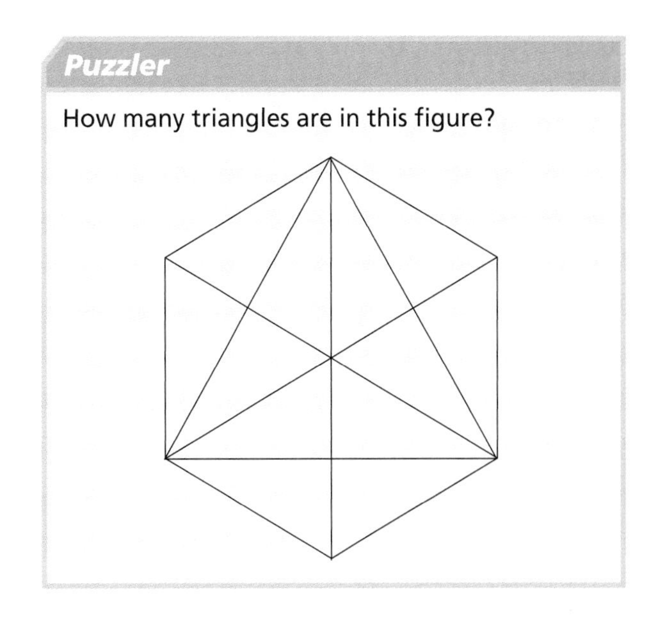

CHAPTER 5

Review

1. Identify the sample and the population in each situation.
   a) A grade 7 class was surveyed about their first week at school.
   b) Two teachers were asked whether they thought juice or coffee should be served at the staff meeting.
   c) Thirty eligible voters were randomly polled to see which party was most likely to win next week's election.
   d) A group of teens at the carnival was asked which ride they liked best.

2. a) Thirty elementary students are surveyed about their favourite toy. Draw a circle graph to display the data. Give the graph a title.

Type of Toy	Number of Students
stuffed animal	10
doll	5
action figure	8
musical instrument	4
other	3

   b) Which type of toy was chosen about 25% of the time? Explain how you know, using the circle graph.

3. Children at a grocery store were surveyed about their favourite kind of fruit.

Fruit	Frequency
apple	15
orange	9
grape	5
pear	8
other	3

   a) If the town's population of children is 600, how many children would prefer apples? pears?
   b) Is this an unbiased sample? Explain.

4. A standard deck of 52 playing cards is used.
   a) What is the theoretical probability of selecting an ace from the deck?
   b) What is the theoretical probability of not selecting an ace?
   c) Explain why selecting an ace and not selecting an ace are complementary events.

5. a) Use the data represented in the stem-and-leaf plot to draw a box-and-whisker plot.

Stem (tens)	Leaf (ones)
1	0 1 7 8
2	1 1 3 4 6 7 9
3	0 2 4 5 8
4	2 6 6 7
5	8 9

   b) Find the mean and label it in the box-and-whisker plot. What does the difference between the mean and median indicate?
   c) What is the range? Is there great variability in the data? Explain.

6. A number cube is rolled 25 times.

Result	Frequency
1	4
2	4
3	1
4	6
5	8
6	2

   a) What is the experimental probability that a 5 is rolled?
   b) If the number cube is rolled 120 times, how often would a 5 be rolled based on the experimental probability? What is the theoretical probability of rolling a 5?

7. A set of data is given.

7, 7, 8, 11, 14, 16, 17, 20

a) Find the mean, median, and mode.

b) If 5 is added to each data value, what effect does this have on the mean, median, and mode?

c) If 4 is subtracted from each data value, what effect does this have on the mean, median, and mode?

d) What effect does multiplying each data value by 2 have on the mean, median, and mode?

8. The table shows the population of Nova Scotia over a 40-year span.

Year	Population
1966	756 039
1971	788 965
1976	828 570
1981	847 442
1986	873 175
1991	899 942
1996	909 282
2001	908 007
2006	934 405

a) Round each population to the nearest ten thousand. Display the data using a scatterplot.

b) Draw a line of best fit. Extrapolate to estimate the population of Nova Scotia in 2016.

9. The circle graph shows sample survey data about the types of occupations that interest grade 12 students.

Grade 12 Interests

Teacher 2%
Writer 1%
Business Person 10%
Other 2%
Doctor 27%
Engineer 15%
Computer Programmer 22%
Lawyer 21%

a) Which occupation was chosen most often?

b) There are 120 graduates. How many want to become doctors? to go into business?

c) According to the graph, what is the probability that a grade 12 student will want to become a computer programmer?

10. Scores for a test marked out of 20 are shown.

17, 13, 15, 18, 13, 16, 14, 12, 20, 15, 14

a) Draw a box-and-whisker plot of the data.

b) The top 25% of students had scores between which two marks (as percents)?

11. Examine the box-and-whisker plot of test scores marked out of 100.

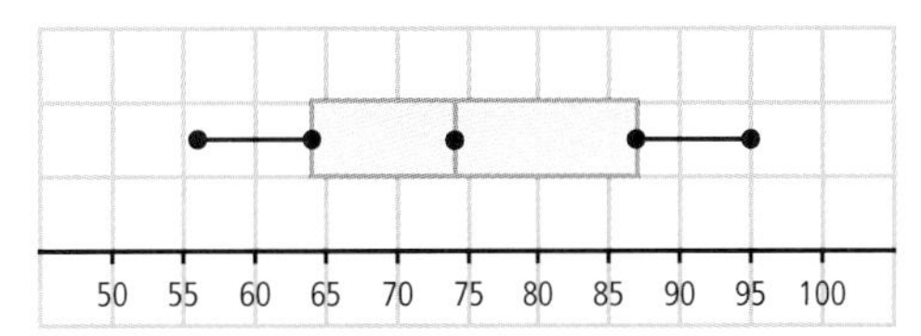

a) What is the median score?

b) What was the greatest mark on the test? What was the least mark?

c) What percent of the students received a mark above 64%?

d) What information is not shown that would be shown in a histogram?

12. Fifty students at Harry's school were asked to rank their concern about the famine situation in parts of Africa on a scale of 1 to 5 (1 is not concerned, 5 is very concerned).

a) Describe two different methods to obtain a random sample from this population.

b) Is it possible to have a sample of 10 students whose mean response is 3.1 and another sample of 10 students whose mean response is 4.0 from the same population? Explain.

CHAPTER 5 Practice Test

Selected Response

Choose the best answer.

1. The probability that an event will happen based on the number of favourable outcomes is known as

A experimental probability
B theoretical probability
C a single event
D complementary events

2. To extend the known data to make predictions beyond the given data is called

A sampling **B** fitting
C interpolating **D** extrapolating

3. A spinner has four equal sections: one blue, one yellow, and two green. The probability that the spinner does not land on yellow is

A $\frac{1}{2}$ **B** $\frac{1}{3}$ **C** $\frac{1}{4}$ **D** $\frac{3}{4}$

4. The mean of a set of data is 5. When 10 is added to each value in the data set, the new mean will be

A 10 **B** 15 **C** 16 **D** 17

Short Response

Provide a complete solution.

5. Joseph's scores for six math tests (marked out of 100) are: 74, 79, 77, 76, 41, 75.

a) Calculate the mean and median.
b) Identify the outlier in the set. What could have caused it? Justify whether this outlier should be removed.
c) Remove the outlier and calculate the new mean and median.

6. Heidi surveyed a sample of customers leaving an ice cream shop on their favourite flavour of ice cream. She displayed her results with a circle graph.

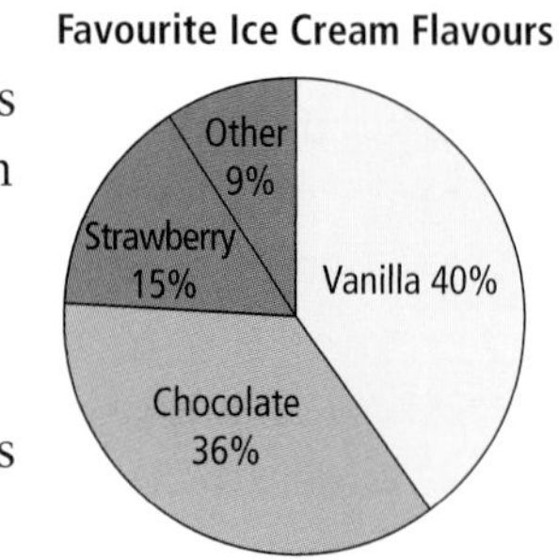

Use the graph to predict how many people out of 400 would prefer each flavour of ice cream.

7. Mary was asked what she did in a typical day.

a) Draw a circle graph to display the data.
b) What fractions of Mary's day do sleeping and school represent?

Activity	Number of Hours
sleeping	8.0
school	6.0
watching TV	2.0
playing on computer	1.5
homework	2.0
eating	2.5
other	2.0

8. The names of all the students at your school are placed in a box. Twenty-five names are chosen without looking. These names are placed in a bag. Of those 25 names, five are chosen without looking. Is this a random sample? Do you think it is representative of the school population?

9. Scores for a test marked out of 25 are shown.

16, 17, 20, 23, 18, 24, 15, 16, 12, 19, 21

a) Identify the median, the upper and lower quartiles, and the upper and lower extremes.

b) Draw a box-and-whisker plot.

c) What is the range of the top 75% of the marks?

10. a) Trace the scatterplot. Draw the line of best fit.

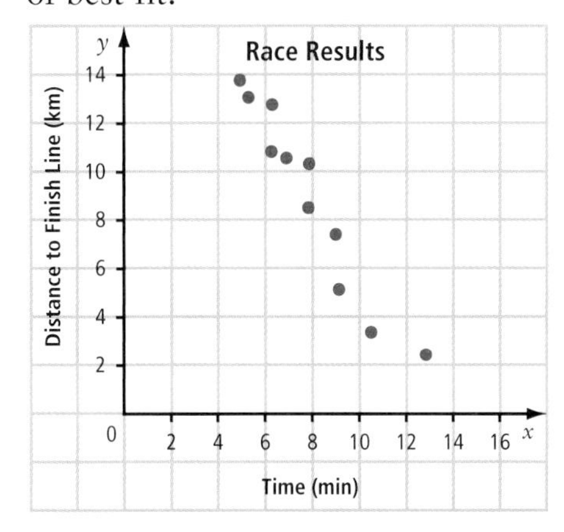

b) Use interpolation and extrapolation to determine the ordered pairs (10, ■) and (12, ■).

c) Write a sentence to explain the meaning of each ordered pair.

Extended Response

Provide a complete solution.

11. Two brands of light bulbs were tested. The lives, in hours, of eight bulbs of each brand are listed below. Draw box-and-whisker plots to compare brands. From this test, which brand of bulb is better? Justify your answer.

Brand A	950	967	835	1214	1130	891	1070	998
Brand B	1015	898	1147	935	946	893	1235	842

Chapter Problem Wrap-Up

Recall the different ways in which data and probability can be used to help you make a reasonable guess at the number of red jellybeans in the jar.

a) How does sampling determine the experimental probability of the number of red jellybeans in the jar?

b) What helpful information can be obtained from a box-and-whisker plot based on the different guesses made? Explain using the box-and-whisker plot you created for questions 6 on page 218.

c) Describe how a scatterplot that displays sample size versus number of red jellybeans can be used to extrapolate the number of red jellybeans in the jar. Explain using the scatterplot you created for question 7 on page 232.

d) How could you increase your chance of guessing the correct number of red jellybeans in the jar? Explain.

Making Connections

What does math have to do with vehicle crash tests?

Safety is an important factor when choosing a car to buy. Crash data collected by the Insurance Institute for Highway Safety show the overall improvement in crash ratings. About half of the vehicles tested in 1995 earned "marginal" or "poor" ratings. By 2002, most vehicles earned "good" ratings.

This type of graph is called a stacked bar graph. It is similar to a comparative bar graph. Instead of the bars being displayed side by side, the same bars are stacked on top of each other.

Why do you think a stacked bar graph was used to display the data?

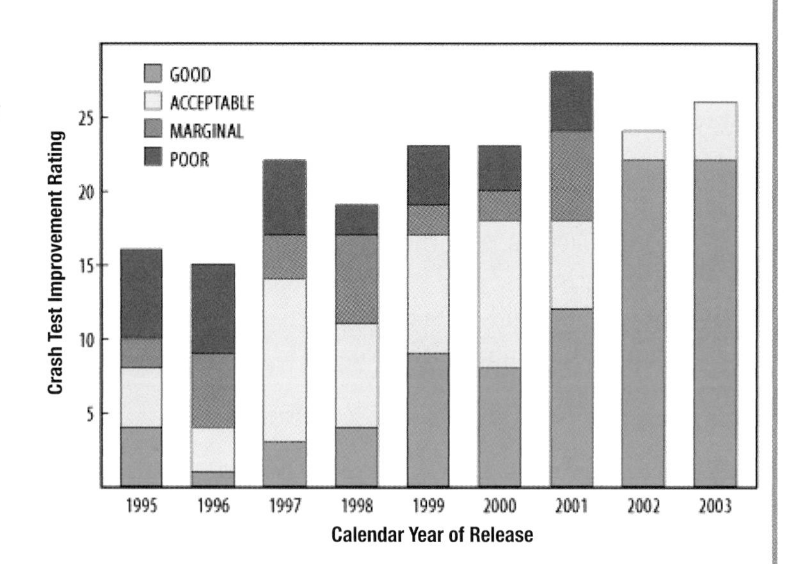

Making Connections

What does data analysis have to do with driving?

A Statistics Canada research paper called *Driving Characteristics of the Young and Aging Population* contains an analysis of the Canada Vehicle Survey data for the year 2000.

The report states that the youngest age group drives the greatest number of kilometres during the weekends (Fridays, Saturdays, and Sundays, 48%). Do you think the graph is misleading? Explain. How does the scale on the vertical axis emphasize the differences in daily driving distances?

Internet Connect To learn more about the driving habits of Canadians go to **www.mcgrawhill.ca/links/math8NS** and follow the links.

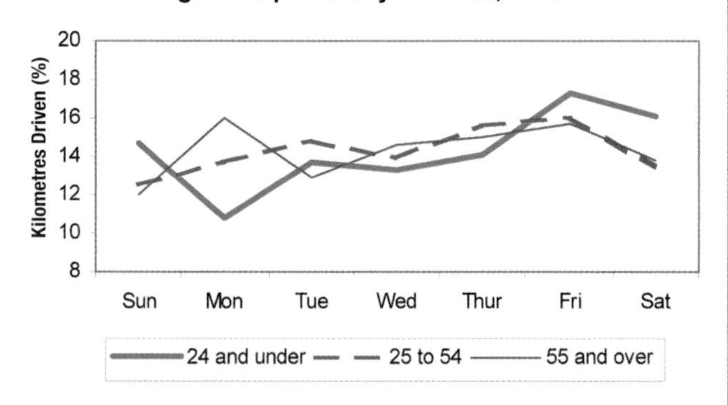

Source: Statistics Canada, Canadian Vehicle Survey, Transportation Division, 2000

TASK

Statistics in Everyday Life

You encounter statistics every day of your life:
- What is the probability of rain today?
- Every 10 s, someone is hurt in a car accident.
- 4 out of 5 doctors recommend ...

Learning more about statistics will help you better understand the world around you and allow you to make informed judgments. In this task, you will use the knowledge and skills you have learned throughout the chapter to complete a statistics project.

You will be involved in
- collecting data
- organizing data
- interpreting data
- communicating the results

The chance of winning this week's lottery is 1 in 10 million
POLLS SAY "3 OUT OF 5" VOTES WILL BE FOR RODNEY MACDONALD FOR PREMIER
Probability of rain today is 100%

1. Choose an area of interest to research. You might consider:
   - sports statistics
   - weather/environmental statistics
   - income and employment statistics
   - population and demography statistics
   - travel and tourism statistics

   Or choose another topic that interests you.

2. Make a list of questions you would like answered, such as:
   - How much rain does my community receive?
   - How does this compare with other communities?
   - What is the impact of the changing weather patterns?

Internet Connect **3.** Start collecting data. Design and conduct surveys or do research on the Internet. If using the Internet, go to **www.mcgrawhill.ca/links/math8NS** and follow the links for sites with Canadian data.

4. Organize the data you collected. Choose the most appropriate type of graph to display the data, such as a circle graph, a box-and-whisker plot, or a scatterplot.

5. Analyse the data to find trends or patterns. What do the trends mean?

6. Make a poster or a computer presentation of your findings. Write a summary explaining the purpose of your statistics research. Include all of your questions, charts, graphs, and analysis of the data in your report.

7. Share your presentation with the class.

- understand the meaning of negative exponents for base 10
- write numbers in scientific notation and in standard form
- compare and order rational numbers
- add, subtract, multiply, and divide positive and negative decimal numbers
- understand and apply the properties of operations on rational numbers
- solve and create problems using positive and negative decimal numbers

Key Words

power

exponent

negative exponent

rational numbers

CHAPTER 6

Rational Numbers

If you expand your skills in mathematics you open up many career opportunities. Areas of science like cell research, forensics, and nanotechnology require skills that involve more than working with whole numbers and integers.

In this chapter you will expand your knowledge with work in rational numbers.

Chapter Problem

The stock market consists of many companies that sell stocks or shares of their companies. If you buy a share from a particular company, you become part owner of that company.

Monday through Friday, stocks are bought and sold at a stock exchange. The price of a stock is affected by many different factors. The price changes constantly while the stock exchange is open.

Suppose you have $10 000 to invest in the stock market. Check the newspaper and select 5 to 10 different stocks from the Toronto Stock Exchange (TSX). This will be your stock portfolio. Obtain a Portfolio Performance Record sheet from your teacher. Record the name of each company and the present price of the company's stock. As you proceed through the chapter, you will be asked to record changes in the prices of your stocks and find the value of your portfolio.

Internet Connect To simulate a stock market situation of buying and selling, go to **www.mcgrawhill.ca/links/math8NS** and follow the links.

Get Ready

Place Value

Use a place value chart for 3824.791.

Thousands	Hundreds	Tens	Ones	Tenths	Hundredths	Thousandths
3	8	2	4	7	9	1
$3 \times 1\ 000$	8×100	2×10	4×1	7×0.1	9×0.01	1×0.001
3×10^3	8×10^2	2×10^1	4×10^0	$7 \times \frac{1}{10^1}$	$9 \times \frac{1}{10^2}$	$1 \times \frac{1}{10^3}$

$$\begin{aligned} 3824.791 &= 3 \times 1\ 000 + 8 \times 100 + 2 \times 10 + 4 \times 1 + 7 \times 0.1 + 9 \times 0.01 + 1 \times 0.001 \\ &= 3 \times 10^3 + 8 \times 10^2 + 2 \times 10^1 + 4 \times 10^0 + 7 \times \frac{1}{10^1} + 9 \times \frac{1}{10^2} + 1 \times \frac{1}{10^3} \end{aligned}$$

The number 3824.791 is in standard form.
The second last row in the chart shows the number written in expanded form.
The last row shows the number written in expanded form using powers of 10.

1. Write each number in expanded form using powers of 10.
a) 123 **b)** 47 098 **c)** 6 887 944
d) 27.49 **e)** 56 789.324 **f)** 0.0784

Exponents and Exponential Notation

Exponents are used to show the repeated multiplication of a number.

$25 = 5 \times 5 = 5^2$ ← exponent (5 is the base)

A number written using exponents is said to be in **exponential form**.
5^2 is the exponential form of 5×5.
A **power** is a number written in exponential form. The number 5^2 is a power of 5.

When you **evaluate** a power, you write it in **standard form**.
5^2 written in standard form is 25.

2. Write in exponential form.
a) $6 \times 6 \times 6 \times 6 \times 6$
b) $2 \times 2 \times 2$
c) $10 \times 10 \times 10 \times 10 \times 10 \times 10$

3. Evaluate.
a) 11^2 **b)** 3^4 **c)** 2^6

Scientific Notation

Scientific notation can be used to express large numbers in a shortened form. It is written as the product of two factors: a number greater than or equal to 1 but less than 10 and a power of 10.

Standard Notation	Expanded Form	Scientific Notation
24 000	$2.4 \times 10\ 000$	2.4×10^4 ← a power of 10; ↖ a number greater than or equal to 1 but less than 10
150 000 000	$1.5 \times 100\ 000\ 000$	1.5×10^8

4. Express in scientific notation.
a) 700 000
b) 14 400 000
c) 120 000 000 000
d) 1459
e) 623.4
f) 10.23

5. Express in standard notation.
a) 3×10^2
b) 8.1×10^8
c) 5.2×10^5
d) 5.314×10^2

Properties of Operations

The **commutative property** means you can change the order of the numbers when adding or multiplying. The numbers can move around, or "commute". The sum or product will remain the same.
Example: $6 + (-8) = (-8) + 6$ $\quad 5 \times 8 = 8 \times 5$

The **associative property** means you can regroup or "reassociate" the numbers when adding or multiplying. The sum or product will remain the same.
Example: $11 + (19 + 17) = (11 + 19) + 17$ $\quad (7 \times 5) \times 4 = 7 \times (5 \times 4)$

The **distributive property** means a product can be separated into parts. Multiplication "distributes" over addition and subtraction.
Example: $7 \times 92 = 7 \times 90 + 7 \times 2$ $\quad 6 \times 89 = 6 \times 90 - 6 \times 1$

6. Use the properties of operations to evaluate these statements mentally. Show your work.
a) $14 + 37 + 16$
b) $2 \times 17 \times 5$
c) $2.5 + (1.5 + 7.2)$
d) $(9 \times 8) \times \frac{1}{2}$
e) 7×73
f) $12 \times 23 - 2 \times 23$
g) $9.8 + (6.5 - 6.5)$
h) $(-12) \times (-3) \times 0 \times (-4)$

Get Ready

Compare and Order Fractions and Decimals

a) Fractions can be compared using benchmarks such as $\frac{1}{2}$ and 1.

Example: $\frac{3}{8} < \frac{5}{9}$ because $\frac{3}{8}$ is less than $\frac{1}{2}$ while $\frac{5}{9}$ is greater than $\frac{1}{2}$.

b) Fractions can be compared using common denominators.

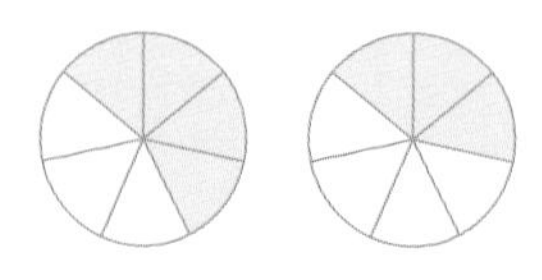

Example: $\frac{4}{7} > \frac{3}{7}$ because 4 equal parts will be more than 3 equal parts of the same whole.

c) Fractions can be compared using common numerators.

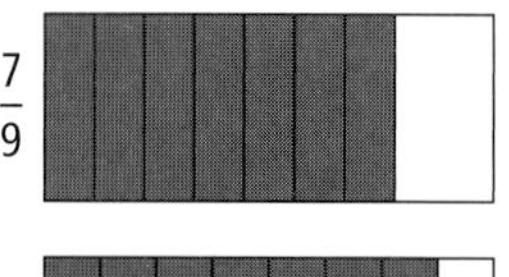

Example: $\frac{7}{9} < \frac{7}{8}$ because if you divided one whole into 9 equal parts and another equal whole into 8 equal parts, each of the 9 equal parts would be smaller than each of the 8 equal parts. So, 7 of the 9 equal parts would be smaller than 7 of the 8 equal parts.

d) Numbers can be compared by changing them to decimals and making place-value comparisons.

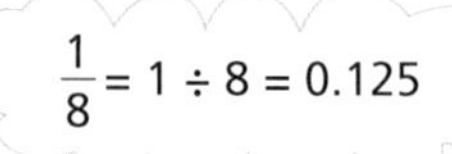

Example: $\frac{1}{8} < 0.13$ because $0.125 < 0.13$.

7. Which number in each pair is greater?

a) $\frac{5}{9}$ or $\frac{5}{10}$

b) $\frac{6}{11}$ or $\frac{4}{9}$

c) 0.3 or $\frac{1}{3}$

d) $\frac{5}{13}$ or $\frac{7}{13}$

8. Order the numbers from least to greatest.

$0.7, \frac{11}{10}, \frac{2}{3}, 0.69, \frac{3}{4}, 1\frac{1}{5}, \frac{2}{5}$

Work With Integers

Integers are the numbers in the set:

$\ldots, -5, -4, -3, -2, -1, 0, +1, +2, +3, +4, +5, \ldots$

All natural numbers, their opposites, and zero are integers.

You can use a number line to illustrate adding and subtracting integers.

To add $3 + (-7)$, start at 0.

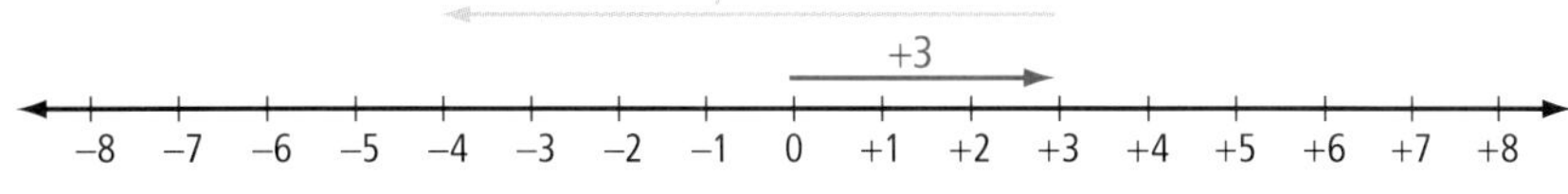

$3 + (-7) = -4$

To subtract $1 - (-3)$, you can apply the **zero principle**.

How can I take away -3?

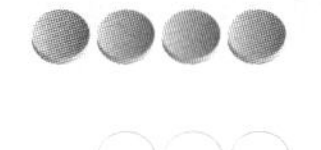

Add 3 pairs of opposites or zeros.

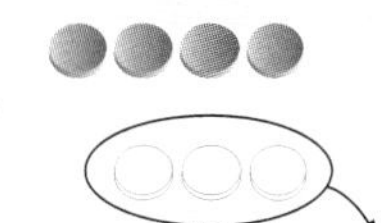

Take away -3.

The result is 4.

$1 - (-3) = 4$

When multiplying or dividing two integers:

- If the signs are the same, the answer is positive.
- If the signs are different, the answer is negative.

So:

$(-8) \times (-9) = +72$	$(-24) \div (-6) = +4$
$(+5) \times (+8) = +40$	$(+32) \div (+2) = +16$
$(-7) \times (+12) = -84$	$(-55) \div (+5) = -11$
$(+4) \times (-9) = -36$	$(+42) \div (-6) = -7$

9. Evaluate.

a) $(-3) + (-7)$
b) $4 + (-11)$
c) $(-5) - (-16)$
d) $(-7) - (+5)$
e) $12 \times (-6)$
f) $(-14) \times (-2)$
g) $(-36) \div (+4)$
h) $(-108) \div (-9)$

6.1 Negative Exponents

Focus on...
- the meaning of negative exponents

In mathematics, the idea of indicating repeated multiplication has been around for hundreds of years. Many different ways were used to write this, but it was not until the 17th century that a raised number became a common notation to indicate repeated multiplication. This raised number became known as an **exponent**.

The **negative exponent** has been in use for almost 400 years. Isaac Newton, in 1676, was one of the first people to use a negative exponent in his work. Today, negative exponents are used in almost every area of science and engineering.

exponent
- the use of a raised number to denote repeated multiplication of a base
- In 5^4, 4 is the exponent.

negative exponent
- an exponent whose value is a negative number

Discover the Math

How can you use decimal patterns to understand the meaning of a negative exponent?

1. a) Copy and complete the table.

Unit of Measure	Measure	Approximate Measurement in Standard Form (m)	Exponential Form
gigametre	diameter of the sun	1 000 000 000	$10^{\blacksquare}$
megametre	diameter of the asteroid *Ceres*	1 000 000	$10^{\blacksquare}$
kilometre	a 10-min walk	1 000	$10^{\blacksquare}$
hectometre	length of a soccer field	100	$10^{\blacksquare}$
decametre	length of a typical classroom	10	$10^{\blacksquare}$
metre	height of doorknob from floor	1	$10^{\blacksquare}$

b) Examine the third column of your table. Record the patterns that you see.

c) Examine the pattern in the exponents in the fourth column. Record the pattern that you see. Compare your observations with a partner's.

d) Assume the pattern continues. What you would expect to be the next three numbers in the third column? Discuss your prediction with a partner. Use the exponent feature on your calculator to check your prediction.

e) Extend the table in part a) by adding five more rows. Add this information to the table and complete the last column.

Strategies
Find a pattern and extend it.

Unit of Measure	Measure	Approximate Measurement in Standard Form (m)	Exponential Form
decimetre	diameter of softball	0.1	$10^{■}$
centimetre	diameter of marble	0.01	$10^{■}$
millimetre	width of grain of sugar	0.001	$10^{■}$
micrometre	length of bacteria	0.000 001	$10^{■}$
nanometre	width of an atom	0.000 000 001	$10^{■}$

2. a) Use base-10 blocks to represent decimal numbers. Copy and complete the table.

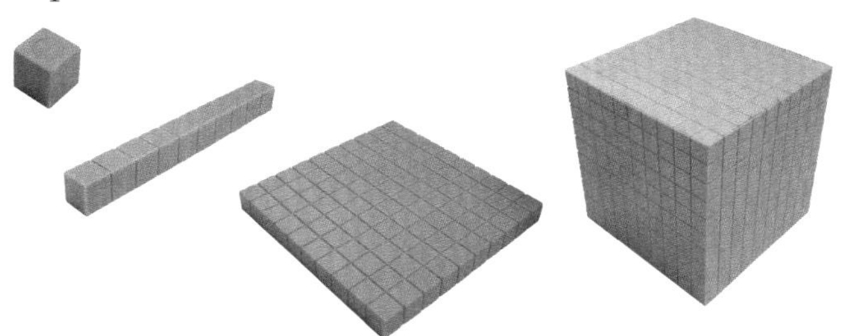

If a large cube represents 1, then a flat represents 0.1 or $\frac{1}{10}$, or 10^{-1}

Standard Form	Fraction Form	Expressed as a Fraction With an Exponent in the Denominator	Exponential Form
1	—	—	10^0
0.1	$\frac{1}{10}$	$\frac{1}{10^1}$	10^{-1}
0.01			
0.001			

Strategies
How many patterns can you find in the table?

b) What is 10^{-5} expressed as a fraction?

c) What connections do you see between corresponding entries in the last two columns?

d) Write 10^{-6} using a positive exponent.

e) What is the exponential form for 0.000 000 1?

3. *Think–Pair–Share!*

a) Consider the number line shown. Where would 10^{-1} and 10^{-2} be located? Copy and label the number line.

−1 0 +1

b) If you had to show an approximate location for 10^{-4}, would the scale be reasonable? Explain.

c) Describe where 10^{-4} should be located on a number line in relation to 10^{-1}.

Communicating Mathematically

The prefixes in the table in question 1 have specific meanings.
Giga means 1 000 000 000 or one billion.
Nano means 0.000 000 001 or one-billionth.

Prefixes Greater Than Giga
Tera means 1 000 000 000 000.
Peta means 1 000 000 000 000 000.
Exa means 1 000 000 000 000 000 000.

Prefixes Less Than Nano
Pico means 0.000 000 000 001.
Femto means 0.000 000 000 000 001.
Atto means 0.000 000 000 000 000 001.
Zepto means 0.000 000 000 000 000 000 001.
Yocto means 0.000 000 000 000 000 000 000 01.

4. Copy each statement. Insert the correct symbol >, <, or = to make each statement true. Write both terms in the same form, if necessary.

a) $\frac{1}{10^2} \blacksquare \left(\frac{1}{10}\right)^2$ b) $0.01 \blacksquare 10^{-3}$

c) $\frac{1}{10^5} \blacksquare \frac{1}{100\ 000}$ d) $10^{-3} \blacksquare 10^{-4}$

e) $10^{-5} \blacksquare 0.0001$ f) $0 \blacksquare 10^{-7}$

g) $10^{-3} \blacksquare \frac{1}{10^4}$ h) $\frac{1}{10^6} \blacksquare 10^{-6}$

5. **Reflect**

a) If a power of 10 has a negative exponent, why is the value of the power a positive number? Use examples or a table to help you.

b) If a power of 10 has a negative exponent, what does this mean in terms of its size? Give an example from question 1 or 2 to help you explain.

6. a) Express 10^{-8} in standard form.

b) What clues are there in the power to help you find the answer?

7. Examine the table. Describe to a classmate who was absent for this lesson the difference between a power that has a negative exponent and a power that has the opposite positive exponent.

$10^6 = 1\ 000\ 000$	$10^{-6} = \frac{1}{1\ 000\ 000}$
$10^3 = 1000$	$10^{-3} = \frac{1}{1000}$
$10^1 = 10$	$10^{-1} = \frac{1}{10}$

8. Copy and complete the statements.

a) What I understand about negative exponents is ▬▬.

b) What I might have difficulty with is ▬▬.

Example 1: Change Exponential Form to Standard Form

Express 10^{-7}, 10^{-8}, and 10^{-9} in standard form.

$$\left(\frac{1}{10}\right)^7 = \left(\frac{1}{10}\right)\left(\frac{1}{10}\right)\left(\frac{1}{10}\right)\left(\frac{1}{10}\right)\left(\frac{1}{10}\right)\left(\frac{1}{10}\right)\left(\frac{1}{10}\right) = \frac{1^7}{10^7} = \frac{1}{10^7}$$

Solution

Method 1: Express with a positive exponent.

$$10^{-7} = \left(\frac{1}{10}\right)^7 = \frac{1}{10^7} = \frac{1}{10\ 000\ 000} = 0.000\ 000\ 1$$

$$10^{-8} = \left(\frac{1}{10}\right)^8 = \frac{1}{10^8} = \frac{1}{100\ 000\ 000} = 0.000\ 000\ 01$$

$$10^{-9} = \left(\frac{1}{10}\right)^9 = \frac{1}{10^9} = \frac{1}{1\ 000\ 000\ 000} = 0.000\ 000\ 001$$

Check using a calculator.

10 (aᵇ) (±) **7** (=)   **10** (aᵇ) (±) **8** (=)   **10** (aᵇ) (±) **9** (=)

Method 2: Use repeated multiplication.

Since $\frac{1}{10}$ equals 0.1, you can use 0.1 as a factor as many times as the exponent shows. Then multiply using a calculator.

$$10^{-7} = 0.1 \times 0.1 \times 0.1 \times 0.1 \times 0.1 \times 0.1 \times 0.1 = 0.000\ 000\ 1$$

$$10^{-8} = 0.1 \times 0.1 \times 0.1 \times 0.1 \times 0.1 \times 0.1 \times 0.1 \times 0.1 = 0.000\ 000\ 01$$

$$10^{-9} = 0.1 \times 0.1 \times 0.1 \times 0.1 \times 0.1 \times 0.1 \times 0.1 \times 0.1 \times 0.1 = 0.000\ 000\ 001$$

I can also multiply $10^{-7} \times 0.1$ to get 10^{-8}.

I can also multiply $10^{-8} \times 0.1$ to get 10^{-9}.

Example 2: Change Standard Form to Exponential Form

Express 0.000 000 000 01 in exponential form.

How many times must I multiply one-tenth (0.1) to get 0.000 000 000 01?

Solution

$$0.000\ 000\ 000\ 01 = \frac{1}{100\ 000\ 000\ 000} = \frac{1}{10} \times \frac{1}{10} \times \frac{1}{10} \times \frac{1}{10} \times \frac{1}{10} \times \frac{1}{10} \times \frac{1}{10} \times \frac{1}{10} \times \frac{1}{10} \times \frac{1}{10} \times \frac{1}{10} = \frac{1}{10^{11}} = 10^{-11}$$

Did You Know?

Not all calculators are the same. The (aᵇ) key may be (y^x) or something else. Refer to your calculator manual for the correct key.

Example 3: Complete the Statement to Make It True

$10^{-3} \times ■ = 10^{-2}$

Solution

$10^{-3} \times ■ = 10^{-2}$
$0.001 \times ■ = 0.01$
$0.001 \times 10 = 0.01$

10^{-2} is 10 times greater than 10^{-3}.

Communicate the Key Ideas

1. Breanna and Jon are working with negative exponents. Jon thinks 10^{-3} equals -30. What should Breanna tell Jon to help him correct his thinking?

2. Shawna reasoned that 10^4 in standard form will have four zeroes, that is, 10 000. So she reasoned that 10^{-4} will also have four zeroes, so she wrote the value as .000 01. Describe the error in her thinking.

3. Alene and Tom are growing a tomato plant for a science project. After one week, Alene recorded the height of the plant as 10^{-2} m. Tom looked at her report and said, "That can't be right. You can't have a negative number for height." Alene disagreed. Who is correct and why?

Check Your Understanding

1. Which value is not equal to the other three? Explain.
   A 10^0 **B** 1
   C $\frac{10}{10}$ **D** 0

2. Which value is not equal to the other three? Explain.
   A $\frac{1}{1000}$ **B** 0.000 1
   C 10^{-3} **D** $\frac{1}{10^3}$

3. Write in exponential form.
   a) 0.000 000 01
   b) 0.000 000 000 000 01
   c) 0.000 01
   d) $\frac{1}{100}$

4. $\left(\frac{1}{10}\right)^1$ is a power with a positive exponent.
   Write each power using a positive exponent.
   a) 10^{-3} **b)** 10^{-10}
   c) 10^{-5} **d)** 10^{-9}

5. $\frac{1}{10^1}$ is a fraction with a power in the denominator.
   - Write each as a fraction with a power in the denominator.
   - Write each fraction as a power with a positive exponent.

   a) 10^{-2} **b)** 10^{-12}
   c) 10^{-4} **d)** 10^{-15}

6. Find the value of n.

   a) $10^n = \frac{1}{10^7}$
   b) $10^n = \frac{1}{1000}$
   c) $10^n = 0.000\ 01$
   d) $10^{-11} = \frac{1}{10^n}$

7. Copy and complete the statements using $\times$ or $\div$.

   a) 10^{-6} ■ $10 = 10^{-5}$
   b) 10^7 ■ $10 = 10^6$
   c) 10^0 ■ $10 = 10^{-1}$
   d) 10^{-9} ■ $10 = 10^{-10}$

8. **a)** Order the numbers from least to greatest.
   $\frac{1}{10^4}$, 10^{-7}, 1, 0.000 01, 100, $\frac{1}{10}$
   b) Express the numbers in exponential form using 10 as the base.

9. **a)** Find the reciprocal of each number. Then, write it and its reciprocal in exponential form.
   i) 10 **ii)** 100 **iii)** 1000
   b) What pattern do you see?

10. Write the power that when multiplied by each number equals 1.
    a) 10^1 **b)** 10^2 **c)** 10^{-3}

11. Write the number in each statement in exponential form using a base of 10. Show your work.
    a) The People's Republic of China has a population greater than 1 000 000 000 people.
    b) The mass of an electron is approximately 0.000 000 000 000 000 000 000 000 001 g.
    c) The distance of Venus from the sun is approximately 1 000 000 000 km.
    d) The diameter of an eyebrow hair is approximately 0.0001 m.

12. Explain why 10^{-4} written in standard form gives a positive number.

13. An *E. coli* bacterium has an approximate length of 10^{-6} m.

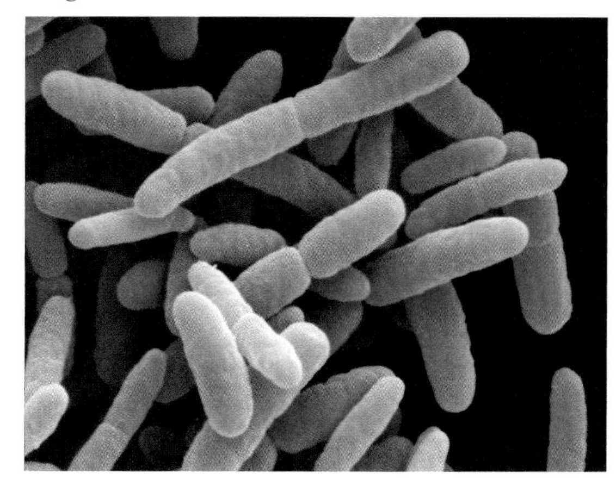

    a) Write this length using a power with a positive exponent.
    b) Write the length as a fraction with a power in the denominator.
    c) How many *E. coli* bacteria would fit end to end along a 1-cm line segment? Explain how you found your answer.

    1 cm

    d) Write your answer to part c) in exponential form.

Chapter Problem

14. Today you will make your first check on your stocks.

a) In your Portfolio Performance Record, record the new price for each of your stocks.

b) Above each price (in the same column), write the overall change since your first recording. If the price for a certain stock decreased by 20¢, you would write −0.20. If the price increased by 20¢, you would write 0.20.

c) How many of your stocks increased in value? How many decreased in value? Show your work.

15. For a research project in science class, Lesley wanted to compare the lengths of a bacterium, an atom, and a sugar crystal. A sugar crystal is approximately 0.001 m in length, a bacterium is approximately 0.000 001 m in length, and an atom is approximately 0.000 000 001 m in length.

a) Write the lengths in exponential form.

b) Lesley decided to draw her sugar crystal 1 cm in length. How long should the bacterium and the atom be in her scale drawing?

c) Lesley soon realized that she had a problem with her scale drawing. What is the problem? Explain.

16. A container is shaped so that half of the liquid it contains evaporates each hour. If 64 mL of liquid is placed in the container, how much will remain after 12 h? Use a table to help you. Express your answer as a fraction and as a fraction with a power in the denominator.

Extend

17. A power with a negative exponent is equal to the reciprocal of the base to the positive exponent.

$$10^{-2} = \left(\frac{1}{10}\right)^2 = \frac{1}{100} \qquad 6^{-2} = \left(\frac{1}{6}\right)^2 = \frac{1}{36}$$

$$(-6)^{-2} = \left(\frac{1}{-6}\right)^2 = \frac{1}{36} \qquad \left(\frac{2}{3}\right)^{-4} = \left(\frac{3}{2}\right)^4 = \frac{81}{16}$$

Evaluate. Express as a fraction with a positive exponent and as a fraction with no exponent.

a) 3^{-1}

b) 8^{-2}

c) 2^{-5}

d) 5^{-1}

e) $(-4)^{-2}$

f) $(-2)^{-3}$

g) $\left(\frac{3}{4}\right)^{-2}$

h) $\left(\frac{1}{2}\right)^{-4}$

18. a) Copy the table. Take a sheet of paper. Fold the sheet in half. How many layers thick is the paper? Fold the sheet in half again. Try folding the paper six times, and record the number of layers after each fold in your table. In the third column, record the answers as powers of 2.

Number of Folds	Number of Layers	Exponential Form
0	1	2^0
1		
2		
3		
4		
5		
6		

The increasing number of layers shows an exponential growth pattern. You were doubling the number of layers with each fold.

b) If it were possible to fold the paper 10 times, how many layers of paper would you have? How could you write the answer in exponential form?

19. a) Consider the sheet of paper in question 18 as one whole. With one sheet of paper, you have one whole region (2^0). When you folded the paper in half, you showed only half of the sheet. When you folded that sheet in half, one-fourth of the sheet showed. Copy and complete the table for four more folds. Complete the last column using negative exponents.

Number of Folds	Area of the Region Showing	Exponential Form
0	1	2^0
1	$\frac{1}{2}$	
2	$\frac{1}{4}$	
3		
4		
5		
6		

The decreasing area is called an exponential decay pattern. The area is halved with each fold.

b) If it were possible to fold the paper 10 times, what would be the area showing? Write this in exponential form.

Puzzler

How many kilograms of food will you eat in a lifetime?

Hint: An elephant has an average mass of 4.5 × 103 kg. In your lifetime, you will eat the equivalent mass of six elephants.

Making Connections

Exponential decay is a process where a quantity loses its mass over time at a proportional rate.

For example, the remains of every living organism contain a certain amount of carbon after death. As the mass decays, half of the carbon is changed over a specific period of time. Then, over the next period of time, half of the remaining carbon is changed, and so on. Scientists measure the amount of carbon in the mass to determine its age. This process is called carbon dating. Carbon dating is used to date fossils.

6.2 Scientific Notation

Focus on...
- writing any number in scientific notation
- changing numbers written in scientific notation to standard form

Scientists often use very large and very small numbers in their work. Scientific notation makes it easier to deal with these numbers. Numbers written in scientific notation are compact and easier to compare. What are some examples of the numbers that scientists use?

Discover the Math

Part A: How can numbers close to zero be written in scientific notation?

The common cold and the flu are two of many illnesses caused by viruses. Viruses are very small. The largest virus is approximately 0.000 000 25 m in length. How can you write this number in scientific notation?

Recall that a number in scientific notation is written as the product of: a number greater than or equal to 1 but less than 10, and a power of 10. For example, $186\ 000 = 1.86 \times 10^5$.

1. a) How many times do you need to multiply 2.5 by $\frac{1}{10}$ to get 0.25?

b) How many times do you need to divide 2.5 by 10 to get 0.25?

c) Copy and complete.
- $0.25 = 2.5 \div ■$
- $0.25 = 2.5 \times 10^{■}$

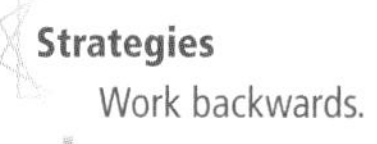
Strategies
Work backwards.

2. a) How many times do you need to multiply 2.5 by $\frac{1}{10}$ to get 0.025?

b) How many times do you need to divide 2.5 by 10 to get 0.025?

c) Copy and complete.
$0.025 = 2.5 \div ■ \div ■$

d) Write 0.025 in scientific notation. How can you tell it is scientific notation?

e) Compare your answer to part d) with a classmate's answer.

3. a) How many times do you need to multiply 2.5 by $\frac{1}{10}$ to get 0.0025?

b) How many times do you need to divide 2.5 by 10 to get 0.0025?

c) Copy and complete.
$0.0025 = 2.5 \div ■ \div ■ \div ■$

d) Write 0.0025 in scientific notation.

4. a) Copy the table. Complete the first three rows and look for patterns.

Standard Notation	Number Greater Than or Equal to 1 But Less Than 10	Number of Times Needed to Divide 2.5 by 10 to Get Standard Form	Scientific Notation
0.25	2.5		$2.5 \times 10^{■}$
0.025	2.5		$2.5 \times 10^{■}$
0.002 5	2.5		$2.5 \times 10^{■}$
0.000 000 25	2.5		$2.5 \times 10^{■}$

b) Predict the numbers that go in the fourth row of the table. Describe how you found your answers.

c) Compare your answers and method with a classmate's.

d) Check your work by converting the scientific notation to standard form.

5. Which values are equal to 0.0009?

A 9×10^{-4} **B** 9×0.0001

C $9 \times \frac{1}{10\ 000}$ **D** $9 \div 10\ 000$

6. **Reflect** Explain how you would decide which exponent to write in the equation.
$0.000\ 029 = 2.9 \times 10^{■}$

Part B: How can a number written in scientific notation, with a negative exponent, be written in standard form?

Materials
- calculator

1. a) The size of a red blood cell in scientific notation is 6.5×10^{-6} m. Just how large is that?
Consider the examples.

$$6.5 \times 10^{-1} = 6.5 \times \frac{1}{10}$$
$$= 6.5 \div 10$$
$$= 0.65$$

Multiplying by $\frac{1}{10}$ gives the same result as dividing by 10.

$$6.5 \times 10^{-2} = 6.5 \times \frac{1}{100}$$
$$= 6.5 \div 100$$
$$= 0.065$$

Multiplying by $\frac{1}{100}$ gives the same result as dividing by 100.

$$6.5 \times 10^{-3} = 6.5 \times \frac{1}{1000}$$
$$= 6.5 \div 1000$$
$$= 0.0065$$

Multiplying by $\frac{1}{1000}$ gives the same result as dividing by 1000.

Copy and complete the table. Use the examples to help you.

Scientific Notation Multiplying by a Power	Multiplying by Fractions	Division by Tens	Standard Form
6.5×10^{-1}			
6.5×10^{-2}			
6.5×10^{-3}			

Strategies
Organize your work in a table. Look for patterns.

b) Examine the table. Discuss with a partner any patterns you see.
c) Consider 6.5×10^{-4}. By what power of 10 do you need to divide 6.5 to get the standard form?
d) Write 6.5×10^{-4} in standard form.
e) Consider the size of the red blood cell. By what power of 10 do you need to divide 6.5 to get the standard form?
f) Write the size of the red blood cell in standard form.

2. a) Evaluate. What can be said about the answers?
5.3×10^{-1}
$5.3 \times \frac{1}{10}$
5.3×0.1
$5.3 \div 10$
Look closely at the operation.
b) How do the numbers in standard form compare with 5.3?

3. Consider 8×10^{-2}.

a) Copy and complete.

By multiplying by 10^{-2}, 8 will become $\frac{■}{■}$ times as large.

b) How many times do you need to divide 8 by 10 to get the standard form?

c) What is the standard form for 8×10^{-2}?

4. Consider 6.09×10^{-3}.

a) Copy and complete.

By multiplying by 10^{-3}, 6.09 will become $\frac{■}{■}$ times as large.

b) How many times do you need to divide 6.09 by 10 to get the standard form?

c) What is the standard form for 6.09×10^{-3}?

5. **Reflect** Explain in your own words what happens to a positive number when you multiply it by a power of 10 with a negative exponent.

Example 1: Change From Scientific Notation to Standard Form

Express in standard form.

a) 8.15×10^3

b) 1.3×10^{-2}

Solution

a)
$$\begin{aligned} 8.15 \times 10^3 &= 8.15 \times 10 \times 10 \times 10 \\ &= 8.15 \times 1000 \\ &= 8150 \end{aligned}$$

Since the exponent is positive, I know I have to multiply 8.15 by 10 three times. I am making 8.15 one thousand times as large.

b) ***Method 1: Divide by 10.***

$$\begin{aligned} 1.3 \times 10^{-2} &= 1.3 \div 10 \div 10 \\ &= 1.3 \div 100 \\ &= 0.013 \end{aligned}$$

Since the exponent is negative, I know I have to divide by 10 twice. I am making 1.3 one hundredth times as large.

Method 2: Multiply by $\frac{1}{10}$.

$$\begin{aligned} 1.3 \times 10^{-2} &= 1.3 \times \frac{1}{10} \times \frac{1}{10} \\ &= 1.3 \times 0.1 \times 0.1 \\ &= 1.3 \times 0.01 \\ &= 0.013 \end{aligned}$$

Example 2: Change a Number to Scientific Notation

Express in scientific notation.

a) 320 000 **b)** 0.000 002 7 **c)** 625×10^{-5}

Solution

a) $320\ 000 = 3.2 \times 10 \times 10 \times 10 \times 10 \times 10$
$= 3.2 \times 10^5$

I need to multiply 3.2 by 10 five times to equal 320 000. 320 000 is 100 000 times as large as 3.2. Multiplying by 100 000 is the same as multiplying by 10^5.

b) *Method 1: Divide by 10.*

$0.000\ 002\ 7 = 2.7 \div 10 \div 10 \div 10 \div 10 \div 10 \div 10$
$= 2.7 \times 10^{-6}$

I need to divide 2.7 by 10 six times to equal 0.000 002 7. So 0.000 002 7 is one millionth times the size of 2.7. Dividing by 1 000 000 is the same as multiplying by 0.000 001 or 10^{-6}.

Method 2: Multiply by $\frac{1}{10}$.

$$0.000\ 002\ 7 = 2.7 \times \frac{1}{10} \times \frac{1}{10} \times \frac{1}{10} \times \frac{1}{10} \times \frac{1}{10} \times \frac{1}{10}$$
$$= 2.7 \times 0.1 \times 0.1 \times 0.1 \times 0.1 \times 0.1 \times 0.1$$
$$= 2.7 \times 0.000001$$
$$= 2.7 \times 10^{-6}$$

c)
$$625 \times 10^{-5} = 6.25 \times 10 \times 10 \times \frac{1}{10} \times \frac{1}{10} \times \frac{1}{10} \times \frac{1}{10} \times \frac{1}{10}$$
$$= 6.25 \times \frac{1}{10} \times \frac{1}{10} \times \frac{1}{10}$$
$$= 6.25 \times 0.1 \times 0.1 \times 0.1$$
$$= 6.25 \times 0.001$$
$$= 6.25 \times 10^{-3}$$

$10 \times \frac{1}{10} = 1$

Communicate the Key Ideas

1. In his science report on the planets, Gustavo wrote that the closest planet to the sun is Mercury. He said that it is 5.8×10^{-7} km away from the sun. Is this distance reasonable? Why or why not?

2. In changing 0.000 000 000 7 to scientific notation, Gina wrote 7×10^{-9}. What do you think Gina might have done to get this answer? What is the correct answer?

3. There are about 1 670 000 000 000 000 000 000 molecules in one drop of water. Why is scientific notation useful in describing the number of molecules in one drop of water?

Check Your Understanding

1. Select the best answer to complete the sentence.

a) If you multiply 8000 by 0.0001, the answer is

A one thousand times as large.
B ten thousand as large.
C one thousandths times as large.
D ten thousandths times as large.

b) The value of 8.2×10^{-11} is

A greater than 1
B between 0 and 1
C equal to 0
D a negative number

2. Copy and complete the statements.

a) The number 830 is $\blacksquare$ times as large as 8.3.

b) The number 0.051 is $\frac{\blacksquare}{\blacksquare}$ times as large as 0.51.

c) The number 35 is $\blacksquare$ times as large as 0.35.

d) The number 0.000 074 is $\frac{\blacksquare}{\blacksquare}$ times as large as 7.4.

3. Which number is greater? How many times as great?

a) 10^5 or 10^3 **b)** 10^{-8} or 10^{-7}

4. Copy and complete the statements. Write the numbers in standard form.

a) $8 \times 10^3 = \blacksquare$ **b)** $8 \times 10^{-3} = \blacksquare$
c) $9.6 \times 10^7 = \blacksquare$ **d)** $6.1 \times 10^{-5} = \blacksquare$
e) $6.98 \times 10^0 = \blacksquare$ **f)** $124 \times 10^1 = \blacksquare$
g) $3567 \times 10^{-6} = \blacksquare$ **h)** $428 \times 10^2 = \blacksquare$

5. Rajiv said that question 4 was easy. He said that once he looked at the exponent, he knew exactly how to get the answer. What pattern did Rajiv apply?

6. Express in scientific notation.

a) 0.000 000 000 08
b) 0.000 006 5
c) 400 000 000 000 000
d) 76 800 000
e) 986×10^2
f) 125437×10^{-7}

7. Skylar was changing 0.085×10^8 to scientific notation. She could not decide whether the answer should be 8.5×10^6 or 8.5×10^{10}. Which answer is correct? Explain why.

8. Copy and complete the statements. Write the numbers in standard form.

a) $9800 \times 10^4 = \blacksquare$
b) $370\,000 \times 10^{-9} = \blacksquare$
c) $0.002 \times 10^6 = \blacksquare$
d) $0.94 \times 10^{-1} = \blacksquare$

9. Express in scientific notation.

a) 60×10^3 **b)** 0.007×10^8
c) 300×10^{-7} **d)** $0.000\,006\,5 \times 10^5$

10. Mrs. MacDonald has lived for approximately 2.903×10^9 s. Which birthday did she recently celebrate?

11. Some calculators show scientific notation like this: 7E12. Write this number in scientific notation.

12. On your calculator, enter 83 followed by a string of zeroes. How many zeroes can you add? Press ENTER. What does your calculator show? Write this number in scientific notation.

13. a) Write the number shown on the calculator display in scientific notation.

9.247E-12

b) Write the number in standard form.

14. Write the numbers in scientific notation.

a) The mass of a hydrogen atom is 0.000 000 000 000 000 000 000 001 675 g.

b) The width of a strand of DNA is about 0.000 000 002 m.

c) Some mitochondria—the organelles that transform energy inside a cell—measure 0.000 000 5 m in length.

d) About 9 000 000 t of rock were used to build the Canso Causeway. In imperial measure, this represents 10 000 000 tons.

15. Order the numbers from least to greatest.

a) 6.5×10^{-3}, 0.059, 1×10^{-4}, 6×10^{-2}, 0.091

b) 0.6×10^{-1}, 5.9×10^{-2}, 59.1×10^{-3}

16. a) Use a calculator to multiply 3 by 10, then press (=). Keep pressing (=). What happens each time you press (=)? Stop when your calculator no longer shows the answer in standard form. Record the number as it is displayed on the screen. Now write this number in scientific notation.

b) Clear your calculator screen. Divide 3 by 10 and press (=). Keep pressing (=). What happens each time you press (=)? Stop when your calculator converts the answer to scientific notation. How does your calculator show the answer? Write this number in proper scientific notation.

Did You Know?

Not all calculators have this constant feature. You may have to multiply your number by 10 and press (=) each time.

17. Robyn must find the quotient:
570 000 000 000 ÷ 190 000
She cannot enter the first number on her calculator because there are too many zeroes. What can Robyn do to the question so that she can use her calculator to find the answer? Write the answer in scientific notation. Show your work.

18.

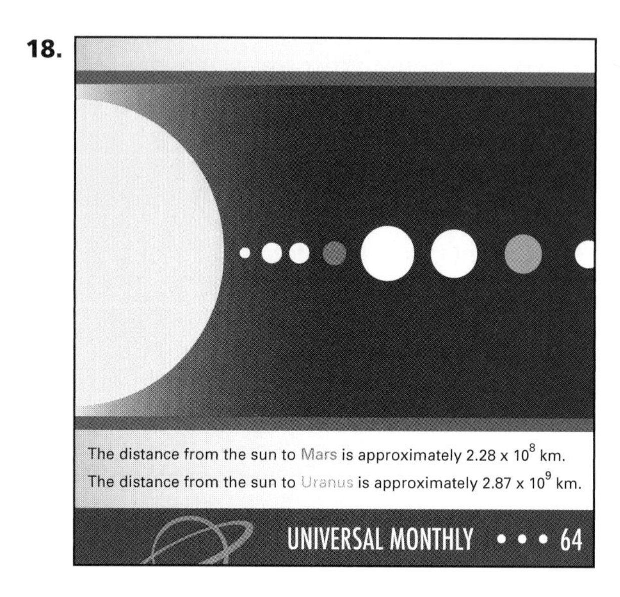

The distance from the sun to Mars is approximately 2.28 x 10^8 km.
The distance from the sun to Uranus is approximately 2.87 x 10^9 km.
UNIVERSAL MONTHLY • • • 64

Pierre was confused.

That's not much of a difference. I thought that Uranus was a lot farther from the sun than Mars!

What might be the cause of his confusion? Approximately how much farther is Uranus from the sun than Mars?

Internet Connect To learn more about the planets, go to **www.mcgrawhill.ca/links/math8NS** and follow the links.

Chapter Problem

19. a) Record the new price of each of your stocks in your Portfolio Performance Record.

b) Above each new price, write the overall change since your last recording.

c) Which stock is performing the best at this time? Which stock is performing the worst?

20. Scientists explain the behaviour of light using the **wave model of light**. The model shows light travelling as a wave. Just as with water waves, the distance from crest to crest or trough to trough of a wave of light as it travels is called a wavelength. These wavelengths are very small and are measured in nanometres. One nanometre measures 0.000 000 001 m.

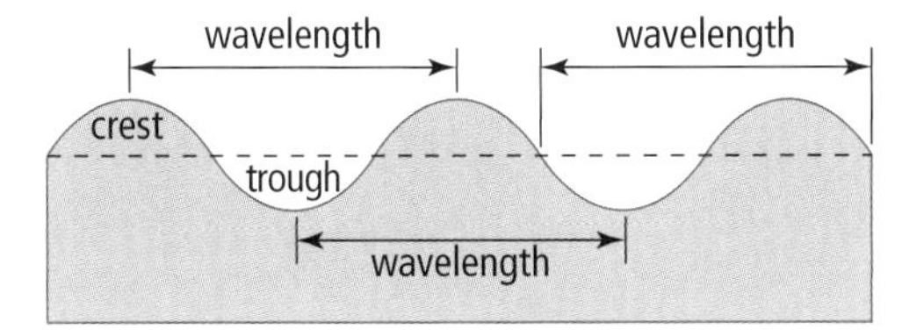

a) The shortest wavelength of light that the human eye can detect is about 3.8×10^{-7} m long. Write this in standard notation.

b) The longest wavelength of light that the human eye can detect is about 720 nm long. Write this length in scientific notation. Then write this length in metres in scientific notation.

Communicating Mathematically

The symbol for nanometre is nm.

21. Express in scientific notation.

a) 0.000 000 089

b) 0.000 000 000 652

c) 789 000

d) 0.456×10^{-3}

e) $0.000\ 259 \times 10^{-2}$

f) 146 987 327

Extend

22. a) A certain bacterium begins with a length of 6.25×10^{-5} m. It doubles its length after 2 h. What will be its length after 2 h? Write the length in scientific notation.

b) How many of these bacteria, arranged end to end, would it take to cover 1 cm?

23. Wendy and Rosalie were looking at the sunset. Wendy said, "Do you realize that the sun is already well below our horizon?"

"How can that be?" replied Rosalie.

"Well," Wendy said, "it takes time for the sun's light to reach us—we are looking at the sun where it was some time ago!"

a) The sun is approximately 1.5×10^8 km from Earth. Light travels at 3×10^5 km/s. How long does it take for the sun's light to reach us?

b) Wendy and Rosalie were watching the sunset in Yarmouth, NS, where sunsets can last from $2\frac{1}{2}$ to $3\frac{1}{2}$ min. Was Wendy correct?

6.3 Compare and Order Rational Numbers

Focus on...
- comparing and ordering rational numbers

When stock markets began in North America, in 1792, fractions were used to record changes in stock value. In 1996, Canada switched to the use of decimals.

The old system was a bit awkward. If a stock changed by for example, $1\frac{7}{16}$, it was not very easy to find the new value of the stock. Now, for example, $+0.23$ is a gain of 23¢ per share and -1.41 is a loss of \$1.41 per share. Decimals make the calculations much easier!

Discover the Math

Materials
- calculator

How do I compare and order rational numbers?

1. When Brittany was checking the newspaper to see how her stocks had performed the day before, she saw these changes.

$-0.02, -2.14, 0.95, 0.16, 1.09, -0.33, 0.00, -0.10$

a) How many of her stocks lost money?
b) How many of her stocks gained money?
c) What is the most money lost by one share of her stocks?
d) What is the most money gained by one share of her stocks?
e) Order the numbers from least to greatest.

2. Emma and Mariele were both shocked by the change in one of their stocks. Emma said that the change in her stock was -5 while Mariele said that the change in her stock was -4.8. Whose stock had the greater loss and by how much?

3. a) Copy the number line. Show the approximate locations for $+1.2$ and $+0.48$ on the number line.

−2 −1 0 +1 +2

b) Which number is farther to the right of zero?
c) Which number is greater?
d) Show the approximate locations of -1.2 and -0.48 on the number line.
e) Which number is farther to the left of zero? Which is greater?

4. Reflect Explain why -4.1 is greater than -4.11.

5. If Brittany had invested in stocks during the mid-1990s, she might have seen results recorded like this.

$-\frac{5}{8}, \frac{1}{2}, -1\frac{3}{16}, -\frac{3}{8}, \frac{3}{4}, -\frac{15}{16}, 1\frac{1}{4}, \frac{7}{8}$

a) How many of these stocks lost money?
b) How many of these stocks gained money?
c) Which number represents the greatest loss?
d) Which number represents the greatest gain?

Fractions and mixed numbers are examples of **rational numbers** . Numbers like $\sqrt{2}$ or π are not rational numbers. Their decimal forms are approximations and the digits do not terminate or repeat in any pattern.

rational numbers

- numbers that can be written as the ratio $a:b$ or fraction $\frac{a}{b}$ of two integers ($b \neq 0$)
  Examples: $-3 = -\frac{3}{1}$, $2\frac{3}{4} = \frac{11}{4}$
  When a is divided by b, the result is a terminating decimal, such as 1.52, or a non-terminating, repeating decimal, such as 0.727 272 … .

6. The numbers -1, $-\frac{1}{2}$, 0, $\frac{1}{2}$, and 1 are common benchmarks for rational numbers. Draw a number line using these benchmarks. Order the numbers in question 5 from least to greatest using these benchmarks.

7. Refer to question 6. Explain your reasoning.

a) There are four numbers that are located to the left of zero. Which number is farthest to the left?
b) Which number is closest to -1? Is this number less than or greater than -1?
c) How does $\frac{5}{8}$ compare with $\frac{3}{8}$? How does $-\frac{5}{8}$ compare with $-\frac{3}{8}$?
d) Which is greater: $\frac{3}{4}$ or $\frac{7}{8}$?

8. **a)** If Brittany wanted to see how much money each stock had lost or gained, she could change all the fractions to decimals.

$-\frac{5}{8}, \frac{1}{2}, -1\frac{3}{16}, -\frac{3}{8}, \frac{3}{4}, -\frac{15}{16}, 1\frac{1}{4}, \frac{7}{8}$

I can change a fraction to a decimal by dividing the numerator by the denominator.

Copy and complete the table.

Negative Fraction	Decimal Equivalent	Positive Fraction	Decimal Equivalent

b) Write the fractions in order from least to greatest.

c) Compare this order with your answer to question 6. Check your results with a classmate.

9. **Reflect**

a) Explain how you could use the benchmark $-\frac{1}{2}$ to compare the fractions $-\frac{3}{7}$ and $-\frac{5}{9}$.

b) On a number line, would you place -1.1 to the left or right of -1?

Example: Order Rational Numbers

Strategies

Visualize a number line.

Order the numbers from least to greatest.

$\frac{7}{5}, -0.95, \frac{3}{4}, 1.52, 0.777..., -\frac{11}{10}$

Solution

Method 1: Use benchmarks.

Use the benchmarks $-1, -\frac{1}{2}, 0, \frac{1}{2}$, and 1 to order the numbers.

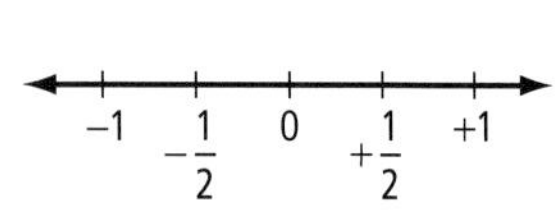

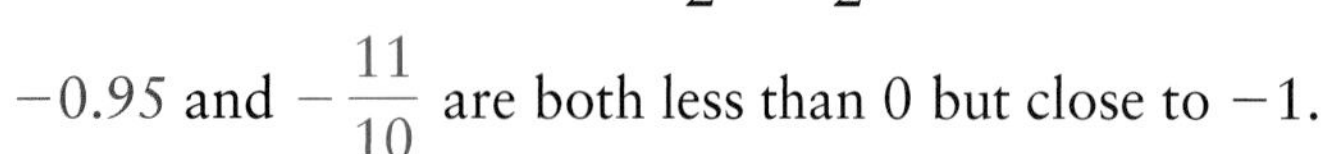

-0.95 and $-\frac{11}{10}$ are both less than 0 but close to -1.

$-0.95 > -1$ so it is placed to the right of -1.

$-\frac{11}{10} < -1$ so it is placed to the left of -1.

$\frac{3}{4}$ and 0.777... are both less than 1.

$\frac{3}{4}$ is halfway between $\frac{1}{2}$ and 1.

0.777... is almost 0.8, so it is closer to 1 than $\frac{3}{4}$.

$\frac{7}{5}$ and 1.52 are both greater than 1.

$\frac{7}{5} = 1\frac{2}{5}$ and is less than $1\frac{1}{2}$; 1.52 is greater than $1\frac{1}{2}$.

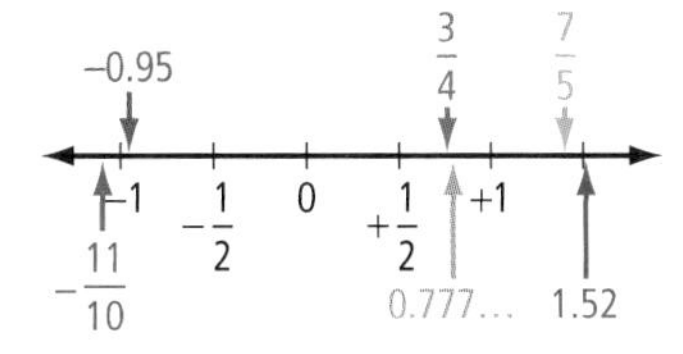

Method 2: Convert to decimals.

When you convert all numbers to the same form, it makes it easier to compare them.

-0.95 $\qquad \frac{7}{5} = 1.4$ $\qquad 1.52$

$-\frac{11}{10} = -1.1$ $\qquad \frac{3}{4} = 0.75$ $\qquad 0.777...$

The order from least to greatest is: $-\frac{11}{10}$, -0.95, $\frac{3}{4}$, $0.777...$, $\frac{7}{5}$, 1.52.

Communicate the Key Ideas

1. Yvette is ordering rational numbers. She came up with this method for comparing two numbers.

Is Yvette's method correct? Explain.

If I'm comparing −5.2 and −5.24, I think of the positive numbers 5.2 and 5.24. I know that 5.24 is greater than 5.2, but I know the opposite is true with −5.2 and −5.24. So I know that −5.2 is greater.

2. Which is greater: $-\frac{1}{99}$ or $-\frac{1}{100}$? Explain how you know.

Check Your Understanding

1. Which statement is true about $-3\frac{1}{3}$?

A It is equal to -3.3.
B It is a little greater than -3.
C It is a little less than -3.
D It is an integer.

2. Make a large copy of the number line.

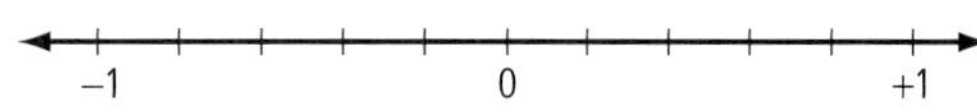

Place each pair of numbers on the number line.

- Which number is greater?
- Which number is farthest from zero and in which direction?

a) $\frac{4}{5}$ and $\frac{2}{5}$ $\qquad$ **b)** $-\frac{4}{5}$ and $-\frac{2}{5}$

3. Which is greater? Explain how you know.

a) $\frac{3}{8}$ or $\frac{1}{2}$ **b)** $-\frac{3}{8}$ or $-\frac{1}{2}$

c) 1 or 0.9 **d)** -1 or -0.9

4. Describe how you could determine which is greater.

a) $-\frac{3}{4}$ or $-\frac{3}{5}$ **b)** -2.95 or -2.9

5. Replace the ■ with a number to make each sentence true. Explain your choice.

a) $\frac{2}{\blacksquare} < \frac{2}{3}$ **b)** $-\frac{5}{\blacksquare} > -\frac{6}{8}$

c) $-0.457 < -0.45\blacksquare$ **d)** $-\frac{5}{\blacksquare} < -\frac{5}{7}$

6. Make a large copy of the number line.

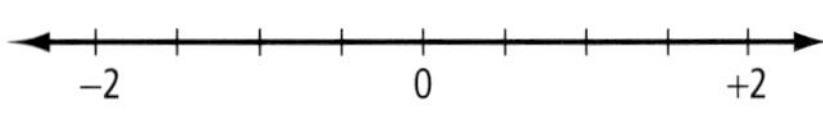

Show the approximate positions of these rational numbers on the number line.

a) -1.1 **b)** $\frac{4}{7}$

c) $1\frac{7}{8}$ **d)** 1.5×10^{-4}

e) $-\frac{11}{12}$ **f)** $-\frac{13}{9}$

7. Which is greater? How do you know?

a) $-\frac{1}{2}$ or $-\frac{1}{3}$ **b)** -0.7 or $-\frac{4}{5}$

c) $-\frac{5}{8}$ or $-\frac{5}{9}$ **d)** -0.3 or -0.31

e) $-4\frac{1}{4}$ or -4 **f)** $-\frac{1}{9}$ or $-\frac{1}{10}$

8. For each pair of numbers, find three numbers between them.

a) 0 and -1 **b)** $\frac{1}{4}$ and $\frac{1}{5}$

c) -2.7 and -2.8 **d)** $-\frac{1}{3}$ and -0.3

e) -3.2 and $-3\frac{1}{4}$ **f)** $-2\frac{3}{4}$ and -2.6

9. **a)** Find the next three numbers in each pattern.

b) Make a large copy of each number line. Show the approximate locations of all seven numbers on the number line. Label the points.

i) $-1, -\frac{1}{2}, -\frac{1}{4}, -\frac{1}{8}, \ldots$

−1 $-\frac{1}{2}$ 0

ii) 1.6, 1.1, 0.6, 0.1, …

−2 −1 0 +1 +2

iii) $-0.2, -\frac{1}{2}, -0.8, -1\frac{1}{10}, \ldots$

−2 −1 0 +1 +2

10. Is each statement sometimes true, always true, or never true? Justify your answer.

a) A natural number is a rational number.

b) A rational number is a whole number.

c) An integer is a rational number.

d) A fraction is a rational number.

e) All numbers are rational numbers.

11. Arrange the numbers from least to greatest. Explain your method.

$-2\frac{1}{3}, -\frac{26}{11}, -2.360\ 360\ 360 \ldots,$

$-2\frac{2}{5}, -2.36$

12. The low and high tides are shown for different locations in Nova Scotia on the same day.

Location	Low Tide (m)	High Tide (m)
Halifax	0.10	2.04
Yarmouth	0.19	4.69
Sydney	0.17	1.27
Pugwash	0.46	2.47
Horton Bluff	0.48	14.29

a) Arrange the low tides in order from lowest to highest.

b) The difference in height between high tide and low tide is known as the daily inequality. What is the daily inequality for Horton Bluff? for Sydney?

Internet Connect To learn more about tides, go to **www.mcgrawhill.ca/links/math8NS** and follow the links.

Did You Know?

The largest difference in tides ever recorded was at Burntcoat Head, NS. The difference was 16.27 m.

13. Estimate the location of each letter on the number line. Give your estimate as a fraction or a mixed number and as a decimal.

A B C D

−4 −3 −2 −1 0 +1 +2 +3 +4

Chapter Problem

14. Obtain a Portfolio Snapshot sheet from your teacher.

a) Record the new price of each of your stocks in your Portfolio Performance Record.

b) Record the change since your last recording.

c) Fill out your Portfolio Snapshot sheet.

d) Which stock is showing the greatest percent increase in value?

e) What is the overall value of your portfolio?

15. There are several ways to indicate the latitude of a city. The latitude of Bogota, Colombia, is 4°32′N and can be written as +4.5°. The table shows the latitudes for 10 cities using positive and negative values. A negative value indicates the city is south of the equator, while a positive value indicates it is north of the equator.

City	Latitude
Bogota, Colombia	+4.5°
Cape Town, South Africa	−33.9°
Edmonton, Alberta	+53.6°
Halifax, Nova Scotia	+44.6°
London, England	+51.5°
Melbourne, Australia	−37.8°
Quito, Ecuador	−0.2°
Reykjavik, Iceland	+64.1°
Rio de Janeiro, Brazil	−22.9°
Stockholm, Sweden	+59.3°

a) Which city is farthest north? Which city is farthest south? Which city is closest to the equator?

b) Make a table in your notebook similar to the one shown. List the cities in order from the farthest north to the farthest south.

c) The city of Montevideo has a latitude of −34.8°. The city of Buenos Aires has a latitude of −34.5°. Which city is farther north? Explain your reasoning.

d) City A has a latitude of $-22\frac{2}{3}°$. City B has a latitude of −22.6°. Which city is farther north? Explain your reasoning.

Extend

16. Which is greater?

a) 8^9 or 9^8

b) 2007^4 or 4^{2007}

17. What is the value of the ones digit in the sum $9^1 + 9^2 + 9^3 + 9^4 + \ldots + 9^{2008}$?

6.4 Operations With Rational Numbers

Focus on...
- working with positive and negative decimal numbers
- using estimation and mental math strategies to solve rational number problems
- solving and creating problems involving rational numbers

We use rational numbers as a part of everyday life. In carpentry, baking, managing a chequing account or investments, or estimating sale prices or the amount of tax—the ability to work with rational numbers is a very important life skill. Outside of the mathematics classroom, where else do you see and deal with rational numbers?

Discover the Math

Materials
- calculator
- ruler

How can you use your knowledge of integer operations to understand rational number operations?

Part A: Add Rational Numbers

Strategies
Use a diagram.

1. Jean owed Tanya \$3.50. She paid her back \$2. What is Jean's net worth how? The solution can be represented by the addition statement: $-3.5 + (+2)$. You can model this situation using a number line.

a) Make a large copy of the number line.

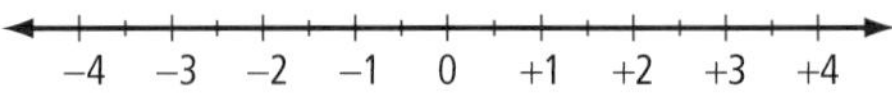

b) Start at zero, then draw an arrow above the number line to represent -3.5.

c) From -3.5, draw an arrow to represent $+2$. Where does this take you on the number line?

d) Copy and complete this number sentence: $-3.5 + (+2) = ■$ Why did your results make sense in terms of Jean's net worth?

Part B: Subtract Rational Numbers

Materials

- red and yellow coloured pencils

Optional:

- red and yellow construction paper

1. Jean owed her father \$5.50. Since she shovelled the front steps, her father decided to forgive \$3.00 of her debt. What is Jean's net worth now?

   You can model this situation using a diagram.

   The diagram represents the number -5.5, the amount Jean owed her father.

   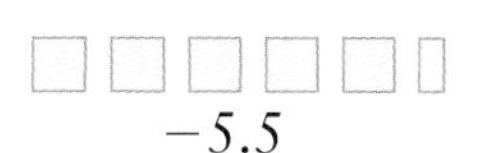

   a) What is the result of forgiving \$3 of her debt?

   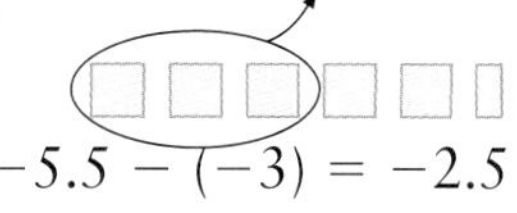

   b) Instead, suppose Jean's father gave her \$3.00. What is Jean's net worth now?

   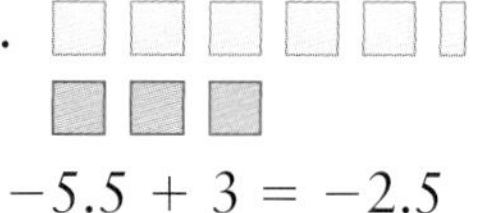

   c) When you subtract an integer, you get the same result as adding its opposite. For example, $5 - 3$ has the same answer as $5 + (-3)$. Do you think the same is true with rational numbers? Test your prediction by making up two more questions. Draw coloured squares (and half squares) to help you. Were the models you created the same for both expressions?

   d) Compare your results with several classmates' answers.

Strategies

Use a two-colour model.

Let one yellow square represent −1.
Let one red square represent +1.
Use the zero principle.

Part C: Multiply Rational Numbers

Materials

- BLM 6.4 Multiplication Table

In a card game, a penalty of -1.5 points is given if you end up with the last card. Lois ended up with the last card four times. She wondered what her score was at the end of playing the four hands.

1. **a)** Make a large copy of the number line.

   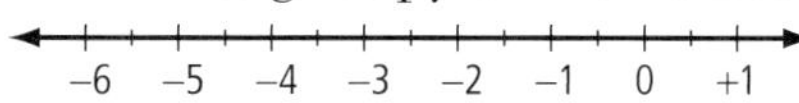

   b) Represent the multiplication, $1 \times (-1.5)$, on the number line by drawing a curved arrow from 0 to -1.5 as shown.

   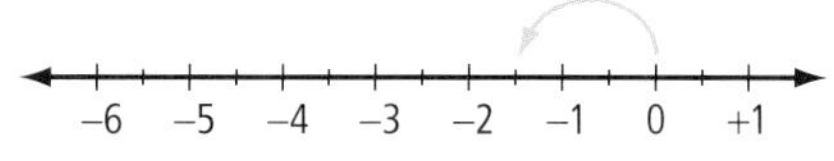

   c) What is $1 \times (-1.5)$?

   d) Represent $2 \times (-1.5)$ by drawing a second curved arrow on your number line from -1.5 to -3.

   e) What is $2 \times (-1.5)$?

   f) Use the number line to determine $3 \times (-1.5)$ and $4 \times (-1.5)$. Write the number sentences and record your answers.

2. a) Examine the patterns in the table. Copy and complete the table.

A	B	C	D
$+4 \times (-1) = ■$	$+4 \times (-1.5) = ■$	$+4 \times (+1) = ■$	$+4 \times (+1.5) = ■$
$+3 \times (-1) = ■$	$+3 \times (-1.5) = ■$	$+3 \times (+1) = ■$	$+3 \times (+1.5) = ■$
$+2 \times (-1) = ■$	$+2 \times (-1.5) = ■$	$+2 \times (+1) = ■$	$+2 \times (+1.5) = ■$
$+1 \times (-1) = ■$	$+1 \times (-1.5) = ■$	$+1 \times (+1) = ■$	$+1 \times (+1.5) = ■$
$0 \times (-1) = ■$	$0 \times (-1.5) = ■$	$0 \times (+1) = ■$	$0 \times (+1.5) = ■$
$-1 \times (-1) = ■$	$-1 \times (-1.5) = ■$	$-1 \times (+1) = ■$	$-1 \times (+1.5) = ■$
$-2 \times (-1) = ■$	$-2 \times (-1.5) = ■$	$-2 \times (+1) = ■$	$-2 \times (+1.5) = ■$
$-3 \times (-1) = ■$	$-3 \times (-1.5) = ■$	$-3 \times (+1) = ■$	$-3 \times (+1.5) = ■$
$-4 \times (-1) = ■$	$-4 \times (-1.5) = ■$	$-4 \times (+1) = ■$	$-4 \times (+1.5) = ■$

b) Look at your results for columns A and C. What patterns do you see? Copy and complete the statements using the words *positive* or *negative*.

- Multiplying two integers with the same sign gives a ■ product.
- Multiplying two integers with different signs gives a ■ product.

c) Look at your results for columns B and D. Would the two statements in part b) apply if you replaced the word *integers* with *rational numbers*? If so, write these new statements.

3. a) Write a situation to fit each expression.

- $5 \times (-1.5)$
- -1.5×5

b) Predict whether or not the product of $5 \times (-1.5)$ will be equal to -1.5×5. Explain your reasoning.

c) Verify your prediction by using a calculator.

d) Try other pairs of numbers and compare the results when you multiply both ways.

Part D: Divide Rational Numbers

Materials

- BLM 6.4 Division Table

A triangle can be used to illustrate the multiplication statement $(-5) \times (-4) = +20$ and the related division statements $+20 \div (-5) = -4$ and $+20 \div (-4) = -5$.

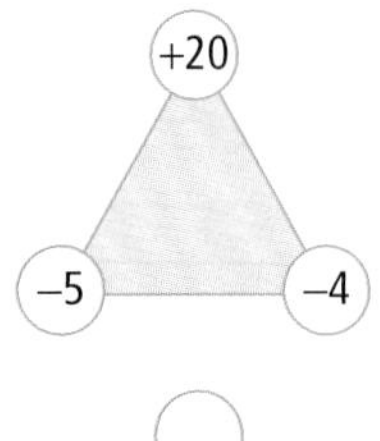

1. Copy the triangle. Add symbols and arrows to show the related multiplication and division statements.

2. a) Draw a similar triangle to illustrate $-4 \times (-2.5) = +10$.

b) Use the triangle to write the related multiplication and division statements.

c) What sign does each of your answers have?

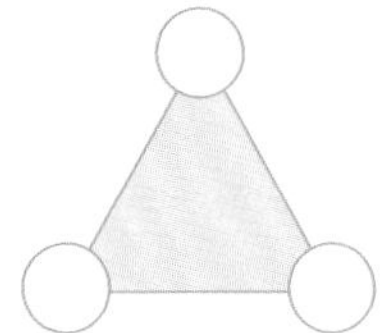

3. a) Refer to the table in Part C, question 2. Copy and complete the related division statements. The first cell of column A has been done for you.

A
$-4 \div (+4) = -1$
$-4 \div (-1) = +4$

b) Look at your results for columns A and C. What patterns do you see? Copy and complete the following statements using the words *positive* or *negative*.

- Dividing two integers with the same sign gives a ▬ quotient.
- Dividing two integers with different signs gives a ▬ quotient.

c) Look at your results for columns B and D. Would the two statements in part b) apply if you replaced the word *integers* with *rational numbers*? If so, write these new statements.

Remember: all integers are rational numbers, but not all rational numbers are integers.

4. Reflect Write a situation for each of the following.

a) $-6.5 + 4.3$

b) $-10 - (-2.47)$

c) $-6 \times \frac{1}{2}$

d) $\frac{-37.50}{3}$

Example 1: Use a Number Line to Represent a Sum

The temperature recorded earlier in the day was -2.8°C. Since that time, the temperature dropped by 2.5°.

a) Use the information above to make an addition problem.

b) Use a number line to show the solution.

c) Write the solution in a number sentence.

Solution

a) $-2.8 + (-2.5)$

b) −6 −5 −4 −3 −2 −1 0

c) $-2.8 + (-2.5) = -5.3$

When adding a negative number, I move left on the number line; so I go from 0 to –2.8. Then I go 2.5 farther left on the number line. This takes me to –5.3.

Example 2: Use Coloured Squares to Subtract

Subtract using squares of two different colours. $(-3.5) - (-4.25)$

Solution

Start with three and a half yellow squares to represent -3.5.

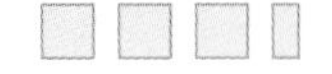

I don't have 4.25 squares to subtract. If I had $\frac{1}{2}$ and $\frac{1}{4}$ of a yellow square, I could remove 4.25 yellow squares.

I can use the zero principle. I'll add $\frac{3}{4}$ of a yellow square and $\frac{3}{4}$ of a red square. This still represents –3.5.

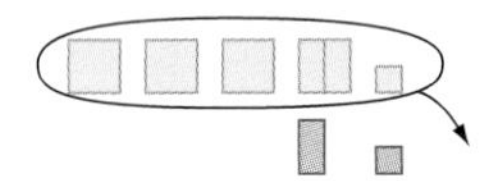

Now I can remove 4.25 yellow squares.

The difference is $\frac{3}{4}$ of a red square, or $+0.75$.

$(-3.5) - (-4.25) = +0.75$

Example 3: Estimate With Rational Numbers

Estimate.

a) $-2.95 + (-7.17) + (+3.1)$

b) $-1.5 \times (-11.9)$

c) $-11.5 \div 2.5$

d) $219.7 - (-97.8)$

Solution

a) $-2.95 + (-7.17) + (+3.1)$

Use near compatible numbers. –2.95 and +3.1 are near compatibles. Together their sum is approximately 0. So, the answer will be approximately –7.

b) $-1.5 \times (-11.9)$

Multiplying a negative number by a negative number will give a positive answer. I can use the "Double and Halve" strategy. The 11.9 is about 12. If I take half of 12, I get 6. If I double the 1.5, I get 3. So, 6 × 3 is 18.

c) $-11.5 \div 2.5$

A negative number divided by a positive number will give a negative answer. I can use the "Divide by Balancing" strategy. I can double both numbers to make an easier division problem. Doubling both numbers keeps the answer the same. I will then have –23 ÷ 5. I know the answer will be between –4 and –5, and a little closer to –5.

d) $219.7 - (-97.8)$

I can use the "Counting On" strategy for subtraction. If I start with –97.8, I can see what I must "count on" to reach 219.7.

–97.8 is almost –100 and 219.7 is almost 220. So if I start at –100, I must count on +100 to reach 0, and then another +220 to reach +220. Altogether I must "count on" 320. So my estimate is +320.

Communicate the Key Ideas

1. Alene was having trouble determining the difference: $-3.5 - 5.8$.

What confuses me is whether that sign in front of the 5.8 is a negative sign or subtraction symbol!

What explanation can you give Alene to help her understand?

2. Copy and complete each statement using the words *always*, *never*, or *sometimes*.

a) When multiplying or dividing two negative numbers, the result is ▬▬ positive.

b) When subtracting, you ▬▬ get less than what you start with.

c) When multiplying rational numbers, if the signs are the same your answer is ▬▬ positive.

d) When adding, the sum is ▬▬ greater than the addends.

e) When dividing, the dividend is ▬▬ greater than the quotient.

Check Your Understanding

1. Estimate. Select the best answer.

a) $-97.43 - (-205.4)$

A -300 **B** -100

C $+100$ **D** $+300$

b) -0.11×495

A -50 **B** -5

C $+5$ **D** $+50$

c) $-63.368 \div (-8.9)$

A 800 **B** 80

C 70 **D** 7

2. Make two large copies of the number line. Use the number line to help you evaluate.

−5 −4 −3 −2 −1 0

a) $-3.75 + (+1.5)$

b) $-2.4 + (-1.8)$

3. Use red and yellow squares to model each subtraction statement. Draw pictures to record the steps. What is the result?

a) $(-5.5) - (-3.5)$

b) $(-2.25) - (+1.5)$

4. Create a word problem for each subtraction statement in question 3. Exchange with a partner and solve each other's problem.

5. Copy and complete the statements.

a) $-5.5 + ■ = -3$
b) $-5.5 - ■ = -3$

6. Look for compatible numbers to help you solve this problem mentally.

$0.43 + (-1.8) + (+0.57) + (-0.2)$

7. $-21.27 + (+53.9) + (-142.8) + (-35.6)$

a) Estimate the sum. Explain how you got your answer.
b) Compare your method with a classmate's. Did you use the same strategies?

8. Estimate. Explain how you got your answer.

a) $-198.3 - (-407.2)$
b) $-6.4 + 17.9 + (-8.4)$
c) $1.5 \times (-4.41)$
d) $1.21 \div 0.2$

9. Estimate first, then calculate the answer using pencil and paper.

a) $-40.6 + (-73.4)$
b) $0.14 - (+2.3) - (-0.85)$
c) $-9 \times (-3.9)$
d) $5.1 \div (-0.03)$

10. Estimate to determine which answer is correct.

$0.4 \times (-2.89)$

A -1.734 **B** -10.86
C -1.156 **D** $+12.76$

11. Explain how you could solve these problems mentally. What is the result?

a) $-3.5 \times (-18)$
b) $4.2 \div \left(-\frac{1}{2}\right)$

12. Create a word problem for each number statement in question 11. Exchange with a partner and solve each other's problem.

13. Kentville, Nova Scotia, has a latitude of 45°. Buenos Aires, Argentina, has a latitude of −34.5°. What is the difference in latitude between the two cities?

14.

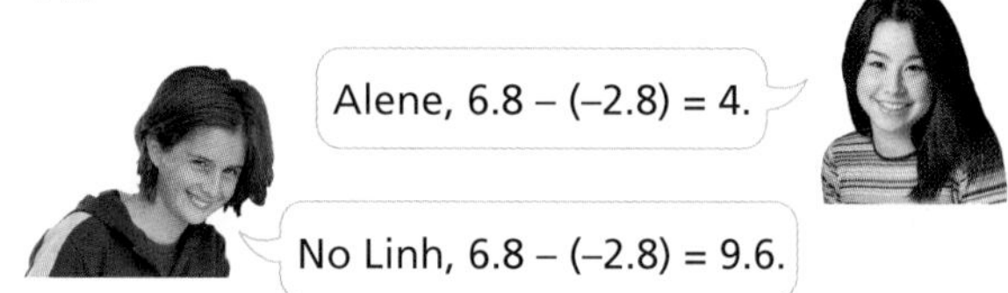

Alene, 6.8 – (–2.8) = 4.
No Linh, 6.8 – (–2.8) = 9.6.

That's impossible! How can you subtract and get more than what you started with?

Help Alene explain to Linh why the answer is 9.6. Include a diagram.

15. It was a bad week on the stock market for one of Ms. Daniels' stocks. It showed changes of -0.12, -0.3, $+0.14$, -1.29, and -0.8 during the week. How much did the stock lose during the week?

16. Mrs. MacEachern decided to invest in a company whose stock was priced at \$14.91.

a) She bought 120 shares of the stock. How much did she pay for her shares?
b) The next day she noticed that the price had dropped. The newspaper showed a change of -0.14. How much money did she lose on the stock for that day?

17. Mr. Aucoin saw that one of his stocks showed a change of -0.60. The total difference for his shares in the stock was −\$150. How many shares of the stock did he have?

Chapter Problem

18. a) Record the new price of each of your stocks in your Portfolio Performance Record.

b) Record the change since your last entry.

c) Suppose last summer, you earned \$2000 by cutting lawns. You decide that you want to invest it in stocks. Choose one of your stocks and buy another \$2000 worth. Record your purchase on a new line in your Portfolio Performance Record.

19. a) The Marianas Trench in the Pacific Ocean is the deepest place in the ocean. It has a depth of -10.9 km. Mount Everest in Nepal is the highest point on Earth. It has a height of $+8.85$ km. If Mount Everest had formed at the bottom of the Marianas Trench, would it have reached the water's surface?

b) Where would the tip of Mount Everest have been located?

c) Mauna Kea in Hawaii, U.S.A., rises to 4.2 km above sea level. The base of Mauna Kea is at -5.9 km–quite a distance below sea level! How high is Mauna Kea above the ocean floor?

d) A submersible used to explore the ocean can dive at a rate of -2.5 m every 4 s. How long would it take to reach a depth of -3.6 km?

20. Write a word problem that can be solved using the number sentence

$[-2.25 + (-1.5) + (0.8)] \div 3$

21. If two numbers are selected from the set of numbers and then multiplied, what is the least possible product? Explain your strategy.

$-6.5, -5, -2.5, 0, 3, 7, 9$

Extend

22. Examine the following sequence of numbers.

$-32, 16, -8, 4, -2, \ldots$

a) What are the next three numbers in the sequence?

b) Would the product of the first 10 numbers be positive, negative, or zero? Explain how you know.

c) Find the product of the first 11 numbers in the sequence.

23. Evaluate. Remember the order of operations.

a) $-\frac{1}{2} + \frac{3}{4}$

b) $-\frac{1}{2} + \frac{3}{4} \times 2.5$

c) $1.7 - 3.5 \div \left(-\frac{7}{10}\right)$

d) $10 + 2\frac{3}{5} \div \left(-\frac{2}{5}\right) \times \left(-\frac{1}{2}\right)$

24. Create a word problem for each number statement in question 23. Exchange with a partner and solve each other's problems.

Did You Know?

Mauna means mountain in the Hawaiian language. Mauna Kea means white mountain. Mauna Kea is usually snow capped during winter.

6.5 Properties of Operations

Focus on...

- working with decimal numbers
- understanding the properties of operations with rational numbers
- solving and creating problems involving rational numbers

Armand, Ravi, and Yvette want to order submarine sandwiches.

How much money do we have altogether?

$27

Do we have enough for 3 subs and 3 drinks?

The subs cost $5.89 each.

Three subs will cost a little less than $18.00.

We can get drinks for $1.90 each.

That's a little less than $6.00. So let's see ... that's less than $24.00 in total.

But the tax is 14%!

Well the tax can't be more than $3.60. So we're safe. The total cost will definitely be under $27!

Strategies
Estimate.

Is Yvette correct? How was she able to accurately estimate the total cost without using a calculator?

Discover the Math

Materials
Optional:
- calculator

How can the order of operations and properties of operations be useful when working with multi-step problems?

Order of Operations

B Brackets
E Exponents
D, M — Division and Multiplication, from left to right
A, S — Addition and Subtraction, from left to right

Commutative property: You can change the order of numbers when you add or multiply them and still get the same answer.

a) $1.5 + (-4.1) = (-4.1) + 1.5$
b) $2.5 \times (-6) = (-6) \times 2.5$

Associative property: You can regroup numbers when you add or multiply them and still get the same answer.

a) $[(-3.2) + (2.1)] + (-3.5) = (-3.2) + [(2.1) + (-3.5)]$
b) $[(-0.2) \times (2.1)] \times (-3.5) = (-0.2) \times [(2.1) \times (-3.5)]$

Distributive property: When multiplying two numbers, you can separate a product into two parts and multiply each part by the other to help you get the answer.

a) $4 \times 9.9 = 4 \times (9 + 0.9) = 4 \times 9 + 4 \times 0.9$
b) $0.6 \times 9.8 = 0.6 \times (10 - 0.2) = 0.6 \times 10 - 0.6 \times 0.2$

1. Mr. Barnes was making an outdoor rink for his children. He prepared a rectangular area 7.2 m wide and 12.0 m long. He drove to the hardware store to pick up plastic sheets to cover the area he had prepared. On his way, he determined the area of the rink.

a) First he multiplied 7 by 12. What is the product?
b) Then he multiplied 0.2 by 12. Why do you think he did this? What is the product?
c) What is the total area of the rink?

Mr. Barnes used the distributive property and separated 7.2 into two parts. He multiplied each part by 12. He then added the two products to get the total area.

2. Ms. Himmelman bought six items priced at \$6.95 each. She wanted to know the total cost.

a) First she multiplied 6 by 7 to get \$42. She then multiplied 6×0.05. Why do you think she did this? What is the product?
b) She then subtracted the second amount from \$42 to get her total. What was her total?

3. Ms. Himmelman had \$48 in cash. She wondered if she had enough money to pay for her purchases in question 2. She knows the tax is 14%. She decided to use an estimate of 15% for the tax and \$42 for her total purchase before taxes.

a) What is 10% of \$42?
b) What is 5% of \$42?
c) What is 15% of \$42?
d) What is her estimate for the total with tax?
e) Does Ms. Himmelman have enough money to make her purchase?
f) How much money did the extra 1% on the tax rate represent?

4. The distributive property can also be applied to this number statement.
-14×2.2
   a) Will the product be negative or positive? Explain how you know.
   b) What is 14×2?
   c) What is 14×0.2?
   d) What is -14×2.2?

The distributive property allows you to separate 2.2 into parts, 2 and 0.2. You can multiply –14 by both parts separately and then add the products.
-14×2.2
$= -14 \times 2 + (-14) \times 0.2$

5. **Reflect** Explain to a younger student how to use the distributive property to evaluate 12.1×8.

Example: Use the Properties of Operations to Evaluate Mentally

Evaluate.

a) $8.3 \times (-8)$

b) $-\frac{1}{4} \times 4.5 \times (-4) \times 2$

c) $-3.7 + (-6.3 + 12.1)$

d) $-\frac{3}{4} + \left(-\frac{7}{8}\right) + 1\frac{1}{4} + \frac{7}{8} + \frac{3}{8}$

Strategies
Look for ways to simplify the calculations. Apply procedures you learned earlier.

Solution

a)

I know the product is negative. I can express 8.3 as 8.0 + 0.3. I can then use the distributive property.

$$\begin{aligned} & 8.3 \times (-8) \\ &= (8 + 0.3) \times (-8) \\ &= 8 \times (-8) + 0.3 \times (-8) \\ &= -64 + (-2.4) \\ &= -66.4 \end{aligned}$$

b)

I know the product will be positive. There are two pairs of compatible numbers: $-\frac{1}{4}$ and 4, and 4.5 and 2. I can regroup these compatible numbers.

$$\begin{aligned} & -\frac{1}{4} \times 4.5 \times (-4) \times 2 \\ &= \left[-\frac{1}{4} \times (-4)\right] \times (4.5 \times 2) \\ &= 1 \times 9 \\ &= 9 \end{aligned}$$

You can use the commutative and associative properties. The commutative property allows you to change the order when multiplying or adding numbers. The associative property allows you to regroup the numbers when multiplying or adding.

c)

–3.7 and –6.3 are compatible numbers, because they add to a number that is easy to work with, –10.0.

$$\begin{aligned} & -3.7 + (-6.3 + 12.1) \\ &= -3.7 + (-6.3) + 12.1 \\ &= -10 + 12.1 \\ &= 2.1 \end{aligned}$$

d) $\left(-\frac{7}{8}\right) + \frac{7}{8} = 0$ because the sum of a pair of opposite numbers is zero.

$$-\frac{3}{4} + \left(-\frac{7}{8}\right) + 1\frac{1}{4} + \frac{7}{8} + \frac{3}{8}$$
$$= -\frac{3}{4} + 1\frac{1}{4} + \frac{3}{8}$$
$$= -\frac{3}{4} + 1 + \frac{1}{4} + \frac{3}{8}$$
$$= -\frac{2}{4} + 1 + \frac{3}{8}$$
$$= \frac{2}{4} + \frac{3}{8}$$
$$= \frac{4}{8} + \frac{3}{8}$$
$$= \frac{7}{8}$$

$1\frac{1}{4} = \frac{5}{4}$

$\frac{2}{4} = \frac{1}{2}$ or $\frac{4}{8}$

Communicate the Key Ideas

1. Tiffany and Shannon both answered the following problem using pencil and paper.

Tiffany's Solution

$\frac{1}{7} \times 2.9 \times (-14)$
$= (-2) \times 2.9$
$= -5.8$

Shannon's Solution

$\frac{1}{7} \times 2.9 \times (-14)$
$= 0.4 \times (-14)$
$= -5.6$

$\frac{1}{7} \times 2.9 = 2.9 \div 7$

```
   0.41
7 ) 2.9
    2.8
    0.10
    0.07
    0.030
```

When the work was checked in class, Tiffany's answer was correct. Shannon was surprised. When she checked her work again, she still got an answer of −5.6. Explain the error in Shannon's work. What properties of operations could she have used to make the work easier?

2. Andre has entered a contest to win a pen used by astronauts when travelling in space. First he must answer a skill-testing question. He has Armand check his solution for him before submitting his answer.

Andre's Solution

$-8.2 \times 5.4 + (-1.8) \times 5.4$
$= [-8.2 + (-1.8)] \times 5.4$
$= -10 \times 5.4$
$= -54$

Andre, you can't add –8.2 and –1.8 first! You've got to follow the order of operations. Use BEDMAS to help you.

Who is correct, Andre or Armand? Explain and provide the correct answer, if appropriate.

3. Beth was asked to evaluate this expression.

$5 \div (-2.5 \div 2)$

She thought she could just use the associative property as shown.

$5 \div (-2.5 \div 2)$
$= [5 \div (-2.5)] \div 2$
$= -2 \div 2$
$= -1$

However, the answer is incorrect. Explain where Beth went wrong in her thinking.

4. Give an example of each of the three properties of operations with rational numbers.

Check Your Understanding

1. Copy and complete the number sentences. Use the properties of operations to determine the missing numbers.

a) $-2.5 \times 13 \times 2 = ■ \times 13$

b) $1\frac{1}{2} \times (-2 \times 8) = ■ \times 8$

c) $-6 \times 2.8 = -6 \times 2 + (-6) \times ■$

d) $\frac{1}{3} \times 17 \times (-9) = ■ \times 17$

e) $13 \times (-1.4) - 3 \times (-1.4) = ■ \times (-1.4)$

2. Explain how the commutative property can be used to evaluate this product mentally. What is the result?

$\frac{2}{3} \times 25 \times \frac{3}{2}$

3. Use the commutative property to help you evaluate. Show your steps.

a) $\frac{1}{2} \times (-17) \times 20$

b) $-3.8 + (-7.32) + (-1.2)$

c) $0.1 \times (-7) \times 3 \times (-30)$

d) $2\frac{3}{8} + 6\frac{1}{3} + (-8) + 5\frac{5}{8} + 2\frac{2}{3}$

4. Create a word problem for each number statement in question 3. Exchange with a partner and solve each other's problem.

5. Copy and complete the number sentences. Insert = or ≠ to make each sentence true.

a) $7.5 \times (-0.28)$ ■ $(-0.28) \times 7.5$

b) $25\frac{1}{10} - (-21.9)$ ■ $(-21.9) - 25\frac{1}{10}$

c) $(-2.4) \div (-0.2)$ ■ $(-0.2) \div (-2.4)$

d) $-0.7 + \frac{1}{2}$ ■ $\frac{1}{2} + (-0.7)$

6. Refer to question 5.

a) By using the commutative property, the order of numbers can change but the answer remains the same. Give an example to illustrate when the commutative property is useful. Explain why it is useful in your example.

b) For which operations does the commutative property not apply?

7. Use the associative property to help you solve these problems mentally. Show your steps.

a) $1\frac{2}{5} + \left(3\frac{3}{5} + 1\frac{1}{4}\right)$

b) $\left(1.5 \times \frac{1}{4}\right) \times 24$

8. Evaluate mentally. Select the correct answer. Which property of operations did you use?

a) $(-0.2) \times (-0.27) \times (-5)$

A -0.027 B -0.27
C 0.27 D 2.7

b) $-4.1 + (-5.9 + 8.7)$

A -6.9 B -2.8
C -1.3 D -0.9

c) $7 \times (-3.6) + 3 \times (-3.6)$

A -36 B -30.6
C -14.4 D 31.2

d) $\frac{7}{8} \div \left(\frac{1}{2} + \frac{1}{4}\right)$

A $\frac{3}{8}$ B $1\frac{1}{6}$
C $1\frac{1}{8}$ D $\frac{3}{4}$

9. Select two number statements from question 8. Create a word problem for each statement. Exchange with a classmate and solve each other's problems.

10. Petra and Melina were given this problem to evaluate.

Petra's Solution

$-10 \times (-0.25) \times (-8)$
$= [-10 \times (-0.25)] \times (-8)$
$= (2.5) \times (-8)$

Melina's Solution

$-10 \times (-0.25) \times (-8)$
$= -10 \times [(-0.25) \times (-8)]$
$= -10 \times 2$

a) Petra and Melina used the associative property to group the numbers differently. What should be their answers?

b) Which grouping do you prefer? Why?

11. Mrs. Chandler was helping her daughter Kaitlyn with her math homework. Each solved the problem a different way.

Kaitlyn

$\frac{1}{5} \times 46 + \frac{1}{5} \times 54$
$= 9.2 + 10.8$
$= 20$

Mrs. Chandler

$\frac{1}{5} \times 46 + \frac{1}{5} \times 54$
$= \frac{1}{5} \times (46 + 54)$
$= \frac{1}{5} \times 100$
$= 20$

a) Are both solutions correct?

b) Explain the thinking involved in both solutions.

c) If you were trying to solve the problem mentally, whose method would you use?

d) Create two different problems that you can evaluate using Mrs. Chandler's method. One of the problems should be easy to evaluate mentally and the other should be difficult to evaluate mentally.

12. Use the distributive property to help you solve these problems mentally. Show your steps.

a) $\frac{3}{4} \times (-275) + \frac{1}{4} \times (-275)$

b) -2.1×35

c) $(-23.4) \times 17 - (-23.4) \times 16$

d) $66 \times (-2.9) + 34 \times (-2.9)$

13. Copy and complete the statement. Insert *always*, *never*, or *sometimes*. Explain your reasoning.

a) When I add two integers, the sum is ▬ another integer.

b) When I subtract two integers, the difference is ▬ another integer.

c) When I multiply two integers, the product is ▬ another integer.

d) When I divide two integers, the quotient is ▬ another integer.

14. Mr. Thimot bought three pens for \$2.25 each and another package of three pens for \$1.75.

a) Write a number sentence that represents Mr. Thimot's purchases.

b) Use one of the properties of operations to find the total cost of his purchases. Show your steps.

15. Mr. Sutherland had 250 shares of a particular stock that on Tuesday showed a change of -0.09 in value. On Wednesday the shares showed another change of -0.11.

a) Write a number sentence that represents the change in the value of his shares over the two days.

b) Use one of the properties of operations to evaluate the number sentence. Show your steps. Write a statement to tell how much he lost on those shares over the two days.

16. One of Nick's stocks showed changes in value of -0.67, $+1.12$, -0.33, and $+0.08$ over a four-day period.

a) Write a number sentence that shows the overall change in value for the four days.

b) Use the properties of operations to evaluate the overall change. Show your steps.

17. The depth of a submarine is changing at a rate of -3.8 m/min for $7\frac{1}{4}$ min. It stops to conduct a test for several minutes, then continues to travel at -3.8 m/min for another $2\frac{3}{4}$ min.

a) Write a number sentence that represents the overall change in depth.

b) Use one of the properties of operations to evaluate the overall change. Show your steps.

Chapter Problem

18. a) You decide to sell all the shares that you own in one company. Decide which company's shares you will sell.

b) Reinvest the money by purchasing stock from a company that is not already part of your portfolio. Enter this new stock in your Portfolio Performance Record on a new line. Be sure to record the original price that you paid for each share and how many shares you bought.

c) Fill in your Portfolio Performance Record.

d) Fill in your Portfolio Snapshot sheet.

e) Compare this snapshot with the one you completed in section 6.3. Have there been any significant changes? Have your investments grown? How much have you made (or lost) from your investments? (Remember, you have now invested \$12 000.)

f) Compare your portfolio with your classmates' portfolios. Are the more expensive stocks the better investment or is it the low-priced stock? Is there a relationship between stock performance and the price of the stock? Explain.

19. Decide whether each problem can be evaluated mentally or not. Solve each problem, showing your steps.

a) $-2 \times \left(-1\frac{1}{2} \times 9\right) \div \frac{2}{3}$

b) $\frac{1}{2} \times (-24) \times 0.9 \times 0 \times (-0.5)$

c) $(-0.9) \times (-0.31) + (-8.8) \times (-0.31)$

d) $(-4.8) + (-9.32) + (+3.8)$

e) $-5.2 \times (-11)$

f) $7.3 \times (-0.6) + 7.3 \times (-0.4)$

g) $(-119.7 + 43.62) + 26.08$

Extend

20. Copy and complete the number sentence. Use the distributive property to help you.

$1.7 \times 1.9 + 1.7 \times 2.9 - 1.7 \times ■ = 3.4$

21. a) Create a story problem for the number sentence.

$2.45 + (-6.88) + (-1.45) + (-0.12)$

b) Explain how to apply the properties of operations to solve the number sentence mentally.

22. a) Each stair in a staircase has a rise of 18.6 cm. Write a story problem involving the staircase that could be represented by the number sentence.

$-4 \times 18.6 + 14 \times 18.6$

b) Use the distributive property to simplify the expression.

c) What does the simplified expression represent in terms of the story?

Puzzler

A frog is sitting at the bottom of the stairs. It wants to get to the eighth step, so it leaps up 3 steps, and then back down 2. Then it leaps another 3 steps, and back 2. How many leaps will it have to take, if it follows this same pattern, until it reaches the eighth step?

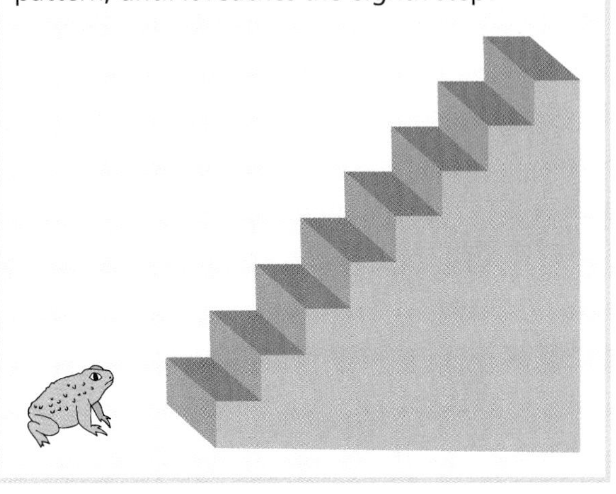

Did You Know?

Write Numbers Using Roman Numerals

Roman numerals are a numeral system that uses seven letters to represent all possible numbers by adding and subtracting the letters.

I	V	X	L	C	D	M
1	5	10	50	100	500	1000

Rules

- Write the numerals left to right, with the largest numeral first. Use the largest possible numeral for each part. So, 120 is CXX (100 + 10 + 10) not LLXX (50 + 50 + 10 + 10).
- A small numeral in front of a large numeral means it is subtracted from the large numeral. So, 192 is CXCII [100 + (100 – 10) + 1+ 1].
- You cannot use more than three of the same numeral in a row (except for M). So, 4 is IV (5 – 1) not IIII (1 + 1 + 1 + 1).

Try writing a power or number in scientific notation using Roman numerals. Now that is a number that is hard to work with!

CHAPTER 6 Review

1. Write each number as a power of 10.
   a) 10 000
   b) 0.001
   c) 0.000 001
   d) 0.1

2. Ashton missed the lesson on negative exponents. Write a summary about negative exponents for Ashton in point form.

3. Give the value of n.
   a) $10^n = \frac{1}{10^5}$
   b) $10^n = \frac{1}{10}$
   c) $10^n = 0.001$
   d) $10^{-6} = \left(\frac{1}{10}\right)^n$
   e) $10^n = 1$
   f) $\frac{1}{10^4} = 10^n$

4. An *E. coli* bacterium is approximately 10^{-6} m in length. A strand of DNA is approximately 10^{-7} m in length. Brandon drew a DNA strand to be 2 cm long. How long should Brandon draw an *E. coli* bacterium if he wants to keep the sizes proportional?

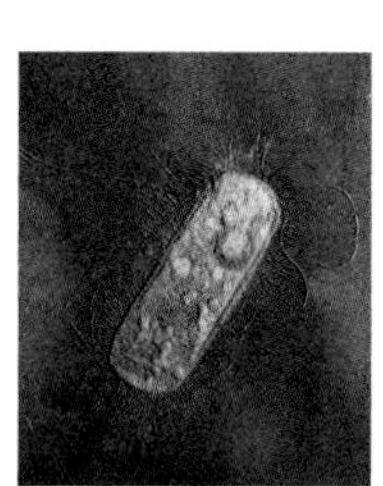

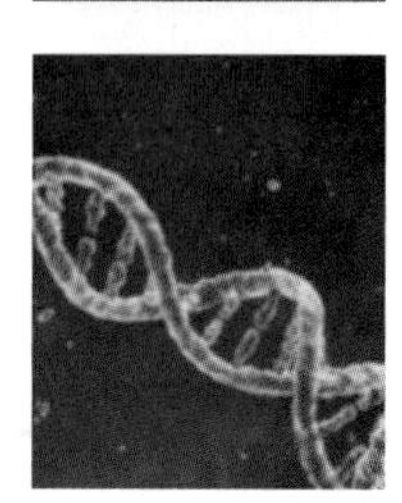

5. Is each number in scientific notation? If not, explain why and write the number in scientific notation.
   a) $6 + 10^{-4}$
   b) 10×10^7
   c) 4.21×10^{-8}
   d) 0.85×10^9
   e) 86×10^4
   f) 9351×10^{-4}

6. Given: 8×10^{-5}. How many times would you need to divide 8 by 10 to represent the number in standard form?

7. Write each number in standard form.
   a) 9×10^{-4}
   b) 7.1×10^{-7}
   c) 2500×10^{-6}
   d) 8.72×10^{-2}

8. Write each number in scientific notation.
   a) 0.000 000 07
   b) 0.000 000 000 098
   c) 0.003×10^{-3}
   d) 0.0041×10^{-2}

9. Copy the number line. Label the number line with the letters to show the approximate position of each number.

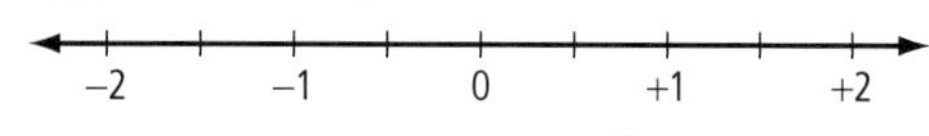

   A -1.42
   B $\frac{7}{16}$
   C -0.91
   D $\frac{8}{7}$
   E $-\frac{11}{8}$
   F 1.9×10^{-1}

10. How would you determine which number is greater?
   a) $-\frac{4}{5}$ or $-\frac{5}{6}$
   b) -0.91 or -0.9

11. Order the numbers from least to greatest.
   $-1.21, -1\frac{1}{6}, 1.23 \times 10^{-1}, -\frac{6}{5}, \frac{1}{10}$

12. Write the addition sentence that is represented by the diagram.

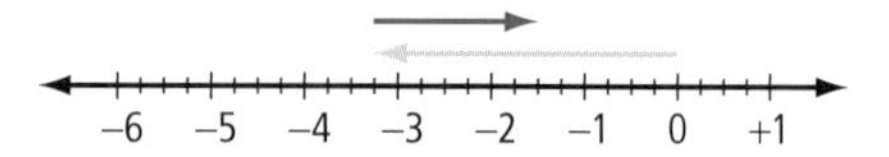

13. The diagram represents -2.5.
   Draw diagrams to represent $-2.5 - (+1.5)$. What is the result? Show your steps.

14. Estimate. Explain how you found your estimate.
   a) $-9.4 + (-39.2) + 22.1 + (-5.3)$
   b) $-212.9 - 91.45$
   c) -1.43×59.4
   d) $155.6 \div (-47.3)$

15. Evaluate. Show your work.
   a) $18.3 + (-4.8) + (-11)$
   b) $54.9 - 86.4$
   c) -8.6×5.2
   d) $-0.648 \div (-0.09)$

16. Evan calculated the mean of five numbers. As he was working, he smudged one of the numbers. What is the number?

$$\text{Mean} = \frac{-3.8 + 7.45 + (-19) + (-8.7) + \blacksquare}{5} = -5$$

17. Vicki evaluated an expression as shown.
$(-42.5) \times 18 - (-42.5) \times 16$
$= -765 - (-680)$
$= -85$
She could have saved time by using the properties of operations and evaluating the expression mentally. Explain what Vicki could have done.

18. Margaret MacLean helped to preserve Scottish culture in Nova Scotia by contributing over 45 songs to the Cape Breton Gaelic Folklore Collection housed at St. Francis Xavier University in Antigonish. Mrs. MacLean has lived for approximately 3.101×10^9 s. How old is she?

19. Explain how the properties of operations can help you to evaluate these expressions. Show your steps.
   a) $-7.8 + 14.9 + (-2.2)$
   b) $\frac{3}{8} \times (-32) + \frac{5}{8} \times (-32)$
   c) $3\frac{3}{4} + \left(3\frac{1}{4} + 6\frac{1}{2}\right)$
   d) $\frac{3}{4} \times (-16.5) \times \left(-\frac{4}{3}\right)$
   e) $45 \times (-2.2)$
   f) $-40 \times (-1.4) \times \frac{1}{5}$

20. When checking his stock portfolio, Brian found that the value of one stock changed by -2.38.
   a) If the same stock was worth \$16.09 the day before, what was the new price for that stock?
   b) If Brian had 80 shares of that stock, how much did he lose on that particular day?
   c) Mallory had the same stock and lost \$83.30. How many shares of that stock did she have?

21. Write a story problem for the expression $(+1.7) + (-0.24) + (-0.76) + (+2.3)$. Explain how you can solve it mentally. What properties of operations would you apply?

22. Order the numbers from least to greatest.
$0.000\ 000\ 000\ 076$, 8.9×10^{-12}, 0.694×10^{-9}

23. Write a story problem to fit the expression.
$8 \times (-1.5) + 2 \times (-1.5)$
Use properties of operations to simplify the calculation.

CHAPTER 6

Practice Test

Selected Response

Select the best answer.

1. The value of 10^{-2} is

A -20 **B** -100 **C** 0.01 **D** 20

2. The product of 2.4×10^{-4} is

A much greater than 1
B between 0 and 1
C a negative number
D equal to 2.4

3. The answer to -0.52×78.3 is

A approximately -40
B approximately -350
C approximately 400
D approximately 35

4. The answer to $-5 + 2.5 \div 0.5$ is

A -15 **B** -5 **C** 0 **D** $+5$

5. The answer to $0.5(-7.5)(4)$ is

A -15 **B** -37.5 **C** -7.5 **D** -14

Short Response

Provide a complete solution.

6. Write each number in exponential form.

a) 0.001 **b)** 0.000 000 000 01

c) 0.0001 **d)** $\frac{1}{1000}$

7. Write each number in question 6 as a power of $\frac{1}{10}$ with a positive exponent.

8. Write each number in standard form.

a) The wavelength of violet light is about 4.1×10^{-5} cm.
b) The diameter of a helium atom is 2.56×10^{-7} mm.
c) Carlos collected 12.5×10^{2} signatures for his petition.

9. Write each number in scientific notation.

a) A human hair is approximately 0.05 mm thick.
b) A particular virus measures 0.000 25 mm in diameter.
c) A snail can travel at a rate of 0.001 m/s.

10. Order the numbers from least to greatest. Show your work.

$0.9, -\frac{1}{3}, -\frac{1}{10}, 10^{-2}, -0.3$

11. Gregor was asked to represent 10^{-3} in standard form. He wrote -1000.

a) Explain to Gregor where he went wrong in his thinking.
b) What is the standard form for 10^{-3}?

12. Write the addition problem that is represented by each number line. Write the answer.

a)

-6 -5 -4 -3 -2 -1 0 $+1$

b)

-3 -2 -1 0 $+1$

13. Draw your own number line and use it to illustrate $4 \times \left(-\frac{3}{4}\right)$.

14. What subtraction statement is represented by the diagrams? Explain the steps. Write the answer.

Step 1: 

Step 2:

Step 3:

Step 4:

15. Use squares to represent $-4.5 - (+1.5)$. What is the result? Show your steps,

16. Estimate. Explain how you found your estimate.

a) $-14.82 + 29.7$
b) $65.2 - (-23.9)$
c) $7.1 \times (-88.4)$
d) $-23.6 \div (-1.85)$

17. Use the properties of operations to evaluate. Explain how you found your answer.

a) $\frac{2}{3} \times (-4.6) \times \frac{3}{2}$
b) $1.7 \times 5.8 + 1.7 \times 4.2$
c) $-9.3 + (-0.7 + 12)$
d) $-8 \times (-8.8)$

Extended Response

Provide a complete solution.

18. A common dust mite has an approximate length of 3×10^{-4} m.

a) Write the length using a power with a positive exponent.
b) Write the length as a fraction with a power in the denominator.
c) How many dust mites would fit end to end along a 3-cm line segment? Explain how you found your answer.
d) Write your answer to part c) in exponential form.

Chapter Problem Wrap-Up

a) Complete your Portfolio Performance Record.
b) How much did you gain or lose on the stock market? Suggest one change that would have increased the value of your investments more. Explain why.
c) What percent of your class saw the value of their investments increase?
d) Write a report telling grade 7 students what you learned from your experience with the stock market. Make some suggestions that may help the students when they make up their stock market portfolio next year.
e) Draw a graph that would best represent the changes in your stocks over time.

Chapters 4–6 Review

1. a) What percent of each entire figure is shaded? Explain how you found your answer. Round your answer to two decimal places, where necessary.
   i) ii)
   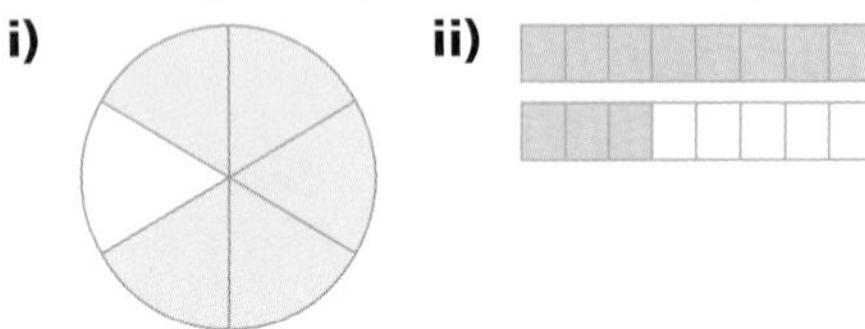
   b) Express each percent in part a) as a decimal to the nearest hundredth.

2. Jess scored 5 goals in 8 hockey games.
   a) Calculate his goals per game.
   b) If he scores at the same rate for each game, how many games will he have to play to score his twentieth goal?

3. The length of a model of the *Bluenose* is 35 cm. The scale used is 1:140. How long is the actual ship in metres?

4. Identify the sample and the population in each situation.
   a) A group of mechanics at an automotive store is asked what brand of engine oil they would recommend.
   b) Two plumbers are asked to provide an estimate for the cost of fixing the same shower.
   c) A group of teens is asked what brand of clothing they think most teenagers like to wear.

5. Find the missing number in each proportion.
   a) ■:60 = 3:4 b) 4:10 = 1:■
   c) 7.5:3 = 37.5:■ d) 0: 12 = ■: 144

6. Sandrine bought a pair of ice skates marked 25% off. She paid $112. What was the original price of the skates?

7. Suppose you roll two number cubes.
   a) What is the theoretical probability of rolling a sum greater than 10?
   b) What is the theoretical probability of rolling a sum of 5?
   c) What is the theoretical probability of not rolling a sum of 5?

8. The circle graph shows sample survey data about the summer sports that interest grade 8 students.

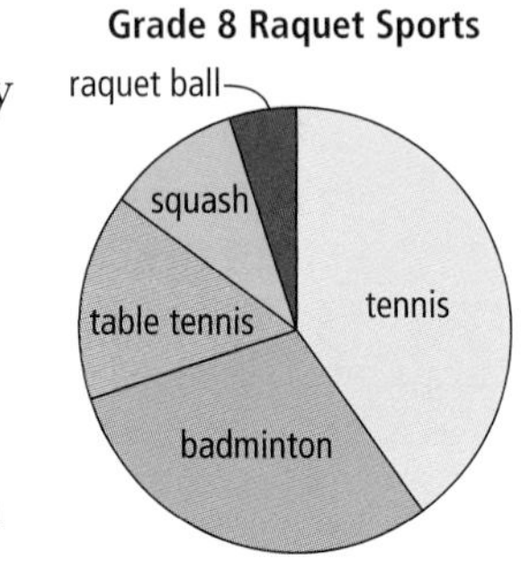

   a) Which racquet sport was chosen most often?
   b) There are 60 grade 8 students. How many like tennis best? racquet ball?
   c) According to the graph, what is the probability that a grade 8 student will prefer squash?

9. The Toronto Police kept track of the daily numbers of reported road accidents over three weeks.
   10 13 10 19 10 9 12 20 13 16 10
   18 14 15 14 13 20 15 12 9 16
   a) Calculate the mean, median, and mode.
   b) Which measure is most informative to a traffic safety officer? Explain.
   c) Multiply each value in the set of data from part b) by 0.5. Calculate the new mean, median, and mode.

10. Write in scientific notation.
   a) 0.00 003 b) 0.0 000 062
   c) 40 000 000 d) 76 800
   e) 715×10^2 f) $12\ 358 \times 10^{-7}$

11. Scores for a test marked out of 20 are shown.

14, 13, 15, 18, 13, 16, 17, 12, 20, 15, 11

a) Draw a box-and-whisker plot of the data.

b) Identify the median, the upper and lower quartiles, and the upper and lower extremes.

c) Identify the range.

d) Between which two marks (as percents) did the top 25% of students receive?

12. A group of teenagers is surveyed about their favourite type of TV show. The results are shown in the bar graph.

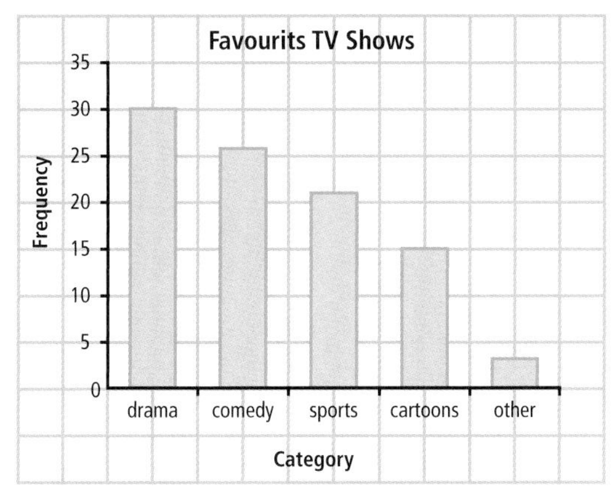

a) Draw a circle graph to display the information. Use percent labels.

b) If 200 teenagers were surveyed, how many would you expect to prefer sports?

c) If 150 teenagers were surveyed, how many would you expect to not prefer sports?

13. Seven 6-cm pea plants each were given a certain amount of water per day. The change in the height of the plant was recorded after 2 weeks.

Volume of Water (mL)	5	7	9	11	13	15	17
Change in Height of Plant (cm)	2	3	4	6	8	9	9

a) Draw a scatterplot of the data.

b) Is there a trend or relationship between the change in volume of water and the change in the height of the plant? If so, describe the relationship. If not, explain why not.

14. **a)** Arrange in order from least to greatest.

$\frac{1}{10^{-7}}$, $\frac{1}{10}$, 1, 0.00 001, 100, $\frac{1}{10^4}$

b) Express the numbers in exponential form using 10 as the base.

15. Arrange the rational numbers from least to greatest.

$4\frac{1}{6}$, 4.2p 271 …, 4.44, $-4\frac{3}{7}$, -4

16. Ms. Auberge saw that one of her stocks showed a change of −0.25. The total difference for her shares in that particular stock was −\$120. How many shares of that particular stock did she have?

17. Evaluate. Show your steps.

a) $-2 \times \left(-1\frac{1}{2} \times 9\right) \div \frac{1}{6}$

b) $(-1.5) \times 0.84 + (-7.7) \div (-1.1)$

c) $\frac{8}{5} \times 0.75 - (-6)$

d) $(-1.78) \times 22 - (-1.78) \times 24$

- add and subtract algebraic terms concretely, pictorially, and symbolically to solve simple algebraic expressions
- explore addition and subtraction of polynomial expressions concretely and pictorially
- multiply a polynomial by a scalar concretely, pictorially, and symbolically
- solve and verify simple linear equations algebraically

Key Words

variable

expression

equation

polynomial

numerical coefficient

term

CHAPTER 7

Algebraic Expressions and Solving Equations

How can a school kayaking trip involve algebra? You can use algebraic expressions to determine unknown quantities such as how much money you need for the trip, the length of time of the bus ride, or the amount of kayaking equipment needed for the whole class.

In this chapter, you will explore mathematical relationships. You will use your knowledge of algebra to model, solve, and create problems. You will develop new skills for solving equations.

Chapter Problem

Mr. Fisher's class is going on a trip. They will stay in cabins at a kayak camp, go kayaking, and search for fossils on the beach near the Joggins Fossil Cliffs. The class is deciding how many buses and kayaks they will need. There are 22 seats on one school bus. How can the students determine the number of buses needed for n students? Mr. Fisher plans to rent two-person kayaks for the trip. How can the students determine the number of kayaks needed for n students?

Get Ready

Order of Operations

To evaluate expressions, follow the order of operations.

B Brackets
E Exponents
D, M } Division and Multiplication, from *left* to *right*
A, S } Addition and Subtraction, from *left* to *right*

$3 + 4 \times 6 \div 3$ **Multiply and divide from left to right.**
$= 3 + 24 \div 3$
$= 3 + 8$
$= 11$

$6^2(3.7 - 2.2)$ **Brackets.**
$= 6^2(1.5)$ **Exponent.**
$= 36(1.5)$
$= 54$

1. Evaluate.
a) $15 + (7^2)(-3 - 2)$
b) $6(16.4 - 10.4) + 3^3$

2. Evaluate.
a) $6 + 5 \times 4 - 19$
b) $3 - 2(16.7 - 10.2) - 9 \times -2$

Represent Expressions Using Algebra Tiles

Algebra tiles are used for modelling algebraic expressions. These tiles represent **variables** because their value can change or vary.

These tiles represent **constants** because their values do not change.

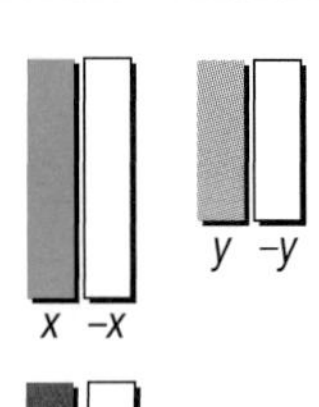

The model for the algebraic expression $2x - y - 1 + x + 3y + 2$ is shown.

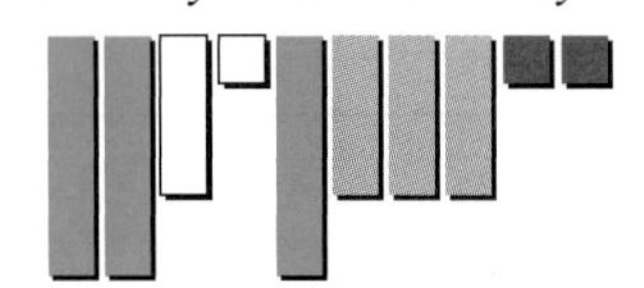

The algebraic expression is a **polynomial**.

This expression contains six **terms**. The terms are separated by the mathematical operators + and −.

A term is part of an expression or equation that consists of either a constant, a variable, or the product of a constant and a variable (or more than one variable).

Terms are separated by addition and subtraction symbols.

-1, x, and $3y$ are all examples of terms in the expression.

$2x$ and x are like terms.
$-y$ and $3y$ are like terms.
2 and -1 are like terms.

These are constants.

These are variable terms.

To collect like terms means to add or subtract like terms. When collecting like terms, you can apply the **zero principle** to **simplify** the expression.

$$
\begin{aligned}
& 2x - y - 1 + x + 3y + 2 \\
= {} & 2x + x - y + 3y - 1 + 2 \\
= {} & 3x + 2y + 1
\end{aligned}
$$

The sum of opposite quantities is equal to zero.

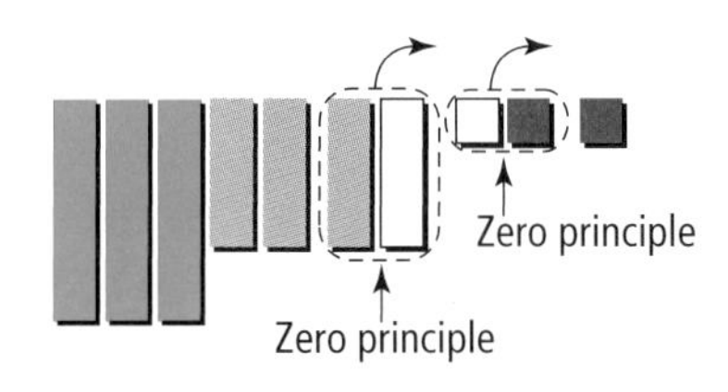

3. a) Model zero by using two x-tiles.
b) Model zero using eight unit tiles.
c) Can an odd number of tiles be used to model zero? Explain, using an example.

4. Identify the like terms in each set.
a) $3n$ $-4m$ 2 n -7
b) $-y$ $3y$ $4x$ x^2 $-7x$

5. Model each expression using algebra tiles. Record your models. What is the sum?
a) $2x + 3y - x + y$
b) $1 - x + 2y - 2x - 3y + 2$

6. Collect like terms.
a) $3 + 2x + 4x - 2y + 6y - 6$
b) $7x + 5y - 2x - 3y + 2x^2$
c) $3x + 7x + 2y + y$
d) $9m + 2n + 7 - 4m - 3n - 7$

Get Ready

Solve Equations by Inspection

Some equations can be easily solved mentally using inspection.
Look at the equation.
$x + 4 = 10$ **Ask: What number added to 4 is equal to 10?**
The number is 6, so $x = 6$.

7. Solve by inspection. What question could you ask when solving?

a) $x + 10 = 18$
b) $4 + x = 16$
c) $x + 15 = -10$
d) $12 - x = -2$
e) $4x = -24$
f) $\frac{x}{3} = 4$

Solve Equations Using a Model

An equation can be thought of as a balanced scale. A scale stays balanced when the same mass is added to or subtracted from each pan. The equal sign in an equation tells you that you have a balanced scale.

One green bag represents x and one red circle represents $+1$.

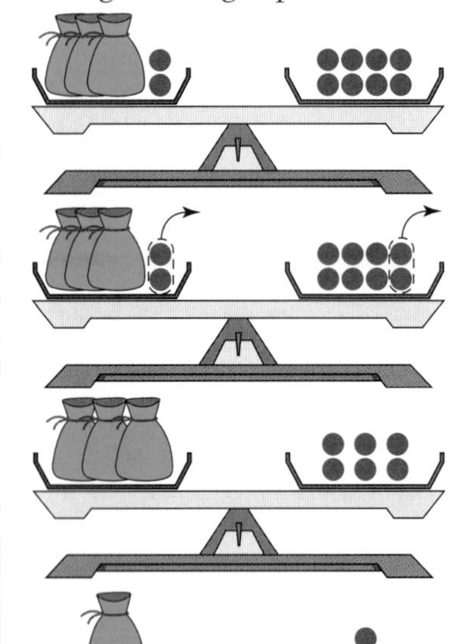

$$3x + 2 = 8$$

$$3x + 2 - 2 = 8 - 2$$

For the sides to be equal, the scale must balance. Remove 2 red circles from each side.

$$3x = 6$$

$$\frac{3x}{3} = \frac{6}{3}$$

$$x = 2$$

Divide the remaining 6 red circles into 3 equal groups.

8. The solution to an equation is shown. Write the symbols for each step and the solution.

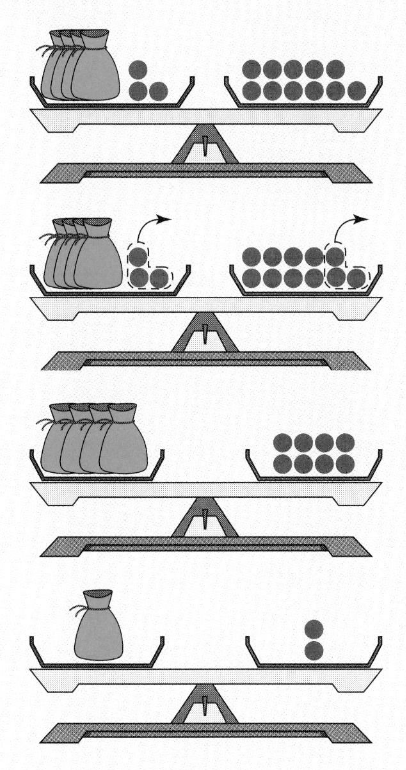

9. Write an algebraic expression for each of the following. Let x represent the number.

a) three times a number

b) a number increased by 10

c) one-half of a number

d) four less than a number

10. Solve by modelling. Record your diagrams. Write the symbols for each step.

a)

b)

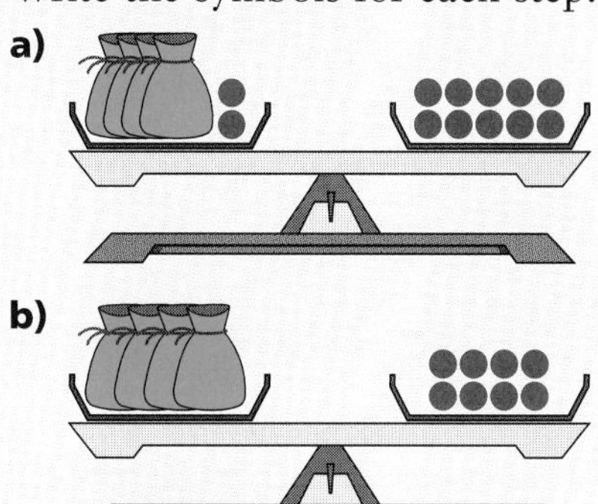

c) One orange bag represents y.

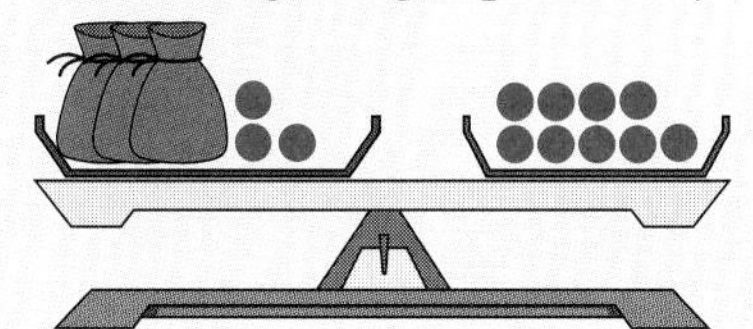

7.1 Add and Subtract Algebraic Expressions

Focus on...
- adding and subtracting algebraic terms concretely, pictorially and symbolically to solve simple algebraic problems
- adding and subtracting polynomial expressions concretely and pictorially

algebraic expression
- a mathematical statement that usually includes one or more variables, one or more constants, and at least one mathematical operation (+, −, ×, ÷)

Basketball courts in the National Basketball Association (NBA) are all a standard size. The courts for junior and senior high schools, colleges, and universities are also different standard sizes. Is there a relationship between the sizes of the courts? How can you model the size of each court using an **algebraic expression** ?

Discover the Math

Materials
- algebra tiles

How can you add and subtract polynomials?

Part A: Add and Subtract Polynomials

A community basketball court has a length that is 2 m more than twice its width.

1. Let x represent the width. Write an expression for the length that uses x as the variable.

2. Sketch the court. Label all four sides using the expressions.

3. a) Write an expression for the perimeter of the court.
b) Write another expression for the perimeter of the court.

4. Model your expression with unit tiles and x-tiles. Combine like terms to **simplify** the expression.

5. On Friday afternoons, a divider is put up that connects the midpoints of the lengths of the court, resulting in two half courts. Sketch the half-court and write an expression for its length. Label all four sides.

6. Wade usually walks two laps around the half-court. Write an expression for the distance Wade walks. Model the expression with tiles. Then, combine like terms to simplify the expression.

7. Is the distance Wade walks longer or shorter than the perimeter of the whole court? Write an expression for the difference. Simplify the expression.

8. On Saturday, a divider is put up that connects the midpoints of the two widths of the whole court. Sketch the new half court and write an expression for its width. Label all four sides. Write an expression for the perimeter of the half court. Model the expression with tiles. Simplify the expression. *Hint: You may need to make paper x-tiles that you can cut in half. Trace an x-tile to help you.*

9. Write an expression that shows the difference between the perimeters of the two different half-courts. Model the expression using tiles. Simplify the expression.

10. If Wade walks two laps of this half court on Saturday, how far does he walk?

11. On which day does Wade walk more and by how much?

Making Connections

To **simplify** an expression is to find an equivalent, simpler, and shorter expression.
Original:
$300 - 4x - 100 + 3x$
Simplified:
$200 - x$

Strategies

Model a situation using algebraic expressions and algebra tiles.

Part B: Area Model

For her 4-H club project, Adrienne decided to make a wall hanging in the shape of a square. She plans to put a border around her wall hanging. She needs to make a drawing of her project before she buys fabric.

1. Let x represent the side length of her wall hanging. Use the side of the x-tile as a ruler and draw a square that is x units long and x units wide. Write an algebraic term for the area of the square.

2. Select the tile from the algebra tile set that will exactly cover the square you drew.
   a) What is its side length as a variable?
   b) What is its area as an algebraic expression?

Did You Know?

The four Hs in 4-H stand for Head, Heart, Hands, and Health. The club promotes development of citizenship, leadership, and life skills of youth from ages 8 to 21.

3. Adrienne thought a border 1 unit wide along each edge of her wall hanging would look beautiful.

a) Let the width of an x-tile represent 1 unit. Use the width of the x-tile to draw the border. Cover the entire area of the wall hanging with tiles from the algebra tiles set. Which tiles did you use?

b) Write an algebraic expression for the entire area of the wall hanging.

c) Examine the terms of the expression you wrote in part b). Could you combine them to make a single term? Why or why not?

4. Look at the grey tile from the algebra tiles set.

a) Which tiles could be used as rulers to create this tile?

b) Write algebraic terms for its

i) length

ii) width

iii) area

c) What might the white side of this tile represent?

5. **Reflect** What kinds of situations could be modelled with the x^2-tile? the xy-tile?

Example 1: Add Like Terms

Callie's mother is building a walkway in her garden. Callie's job is to install the edging between the grass and the walkway. How much edging does she need if $x = 3$ m and $y = 2$ m? The width of each section of the walkway is 1 m.

Solution

Draw and label a diagram.

$x + x + x + x + y + y + y + y + y + y + 1 + 1 + 1 + 1$
$= 4x + 6y + 4$

The expression $4x + 6y + 4$ represents the perimeter of the walkway. Substitute $x = 3$ and $y = 2$ into the expression to determine the length of edging needed.

Length of edging $= 4x + 6y + 4$
$= 4(3) + 6(2) + 4$
$= 28$

Callie needs 28 m of edging.

Example 2: Subtract Polynomial Expressions I

Simplify.
$(3x + 3) - (x + 4)$

> **Strategies**
> Apply procedures you learned previously for integers.

Solution

Method 1: Use the take away method.

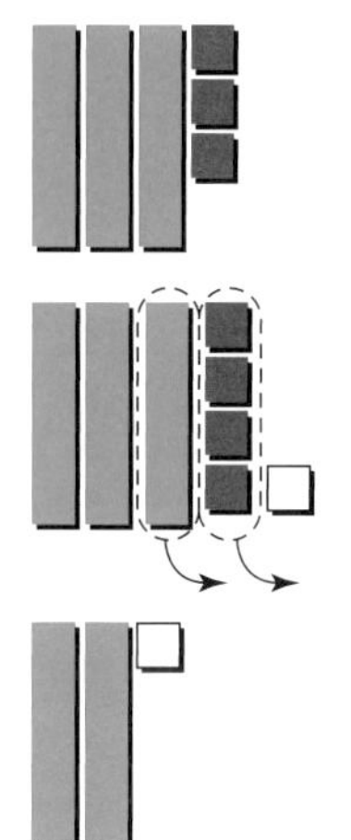

Model 3x + 3 using algebra tiles.
There are enough x-tiles to take away 1 x-tile. There are not enough unit tiles to take away 4 unit tiles. Add a pair of opposite unit tiles. Now you can take away 4 unit tiles.

The result is $2x - 1$.
$(3x + 3) - (x + 4) = 2x - 1$

Method 2: Add the opposite.
When you subtract an integer from another integer, it has the same result as adding the opposite integer.

With algebraic expressions, subtracting $(x + 4)$ has the same result as adding $(-x - 4)$.

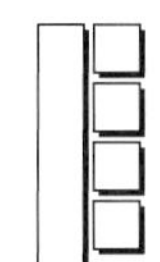
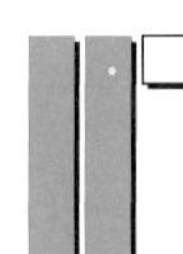

$(3x + 3) - (x + 4)$	
$= (3x + 3) + (-x - 4)$	**Add the opposite.**
$= 3x + 3 - x - 4$	**Remove brackets.**
$= 3x - x + 3 - 4$	**Group like terms.**
$= 2x - 1$	**Combine like terms.**

$(3x + 3) + (-x - 4) = 2x - 1$

Example 3: Subtract Polynomial Expressions II

Simplify.

$(3x + 3) - (x + 4)$

Solution

Method 1: Use comparison.

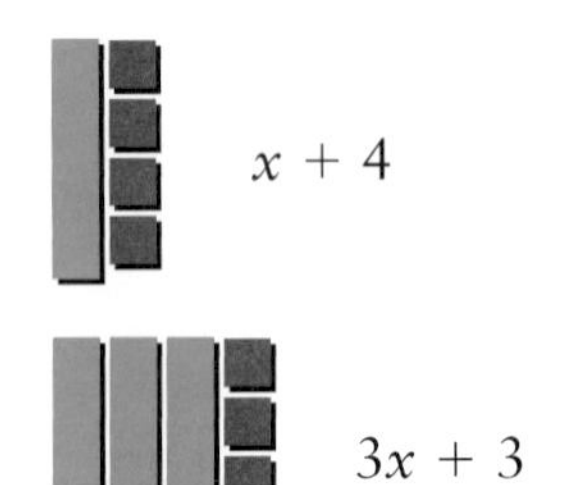

If I start with $x + 4$, I have to add two x-tiles to get three x-tiles and I have to take away one unit tile. The difference between the two tile models is $2x - 1$.

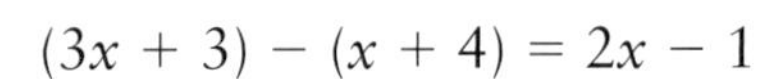

$(3x + 3) - (x + 4) = 2x - 1$

Method 2: Find the missing addend.

To simplify the expression using another method, you can find the missing addend.

Think addition: $(3x + 3) - (x + 4)$

What must be added to $x + 4$ to get $3x + 3$?

Add $2x$ to x to get $3x$.

Add -1 to 4 to get 3.

The missing addend is $2x - 1$.

So, $(3x + 3) = (x + 4) + (2x - 1)$

Example 4: Collect Like Terms

Simplify by collecting like terms.

$-y^2 + xy + x^2 + x^2 + x^2 + y^2 + y^2 - x^2 + xy + xy$

Solution

Model the original expression.

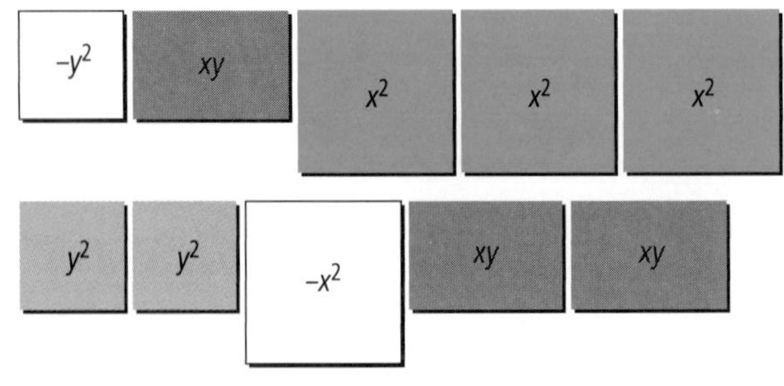

$-y^2 + xy + x^2 + x^2 + x^2 + y^2 + y^2 - x^2 + xy + xy$

Making Connections

The green tile is called x^2 because the dimensions of the tile are x by x, making the area x^2. Likewise for the orange tile, y^2. The grey tile is xy because one dimension is x and the other dimension is y, making the area xy. You can use any variable for the dimensions of a tile.

Rearrange the model of the original expression.
Apply the zero principle.

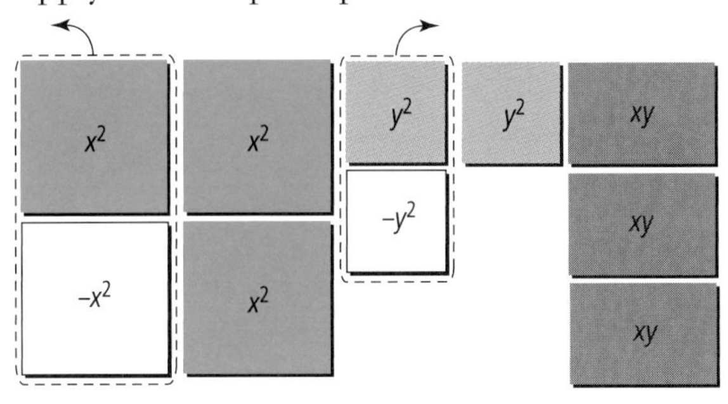

$= x^2 + x^2 + x^2 - x^2 + y^2 + y^2 - y^2 + xy + xy + xy$

I can apply the commutative property of addition to rearrange the terms and group like terms. Now the expression is easier to simplify.

Simplify.

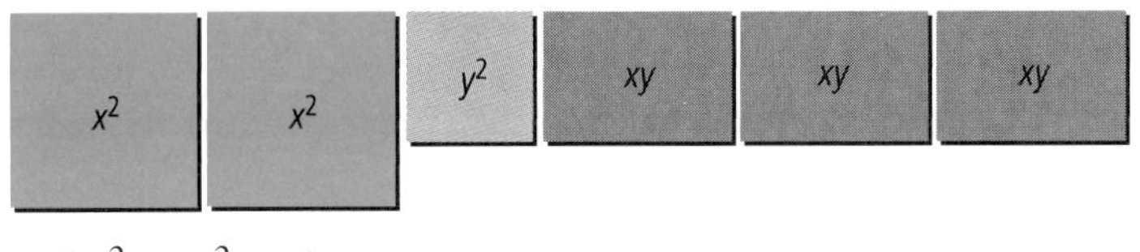

$= 2x^2 + y^2 + 3xy$

To rearrange the terms, order the variables alphabetically and in descending order of their powers. This is the usual practice.

Communicate the Key Ideas

1. Describe how to add one polynomial to another. Use an example and show the result using algebra tiles, pictures, and symbols.

2. a) Explain how you would simplify the expression $(5x + 2) - (3x + 4)$ using each method.
- the take away method
- adding the opposite
- finding the missing addend

b) Which method was easiest to use and why?

c) Write an expression similar to the one in part a) that could be simplified most easily by using the comparison method.

3. Explain why the grey tile is called the xy-tile.

4. Use algebra tiles to explain why unlike terms cannot be added or subtracted.

5. Explain why these algebra tile models represent equivalent expressions.

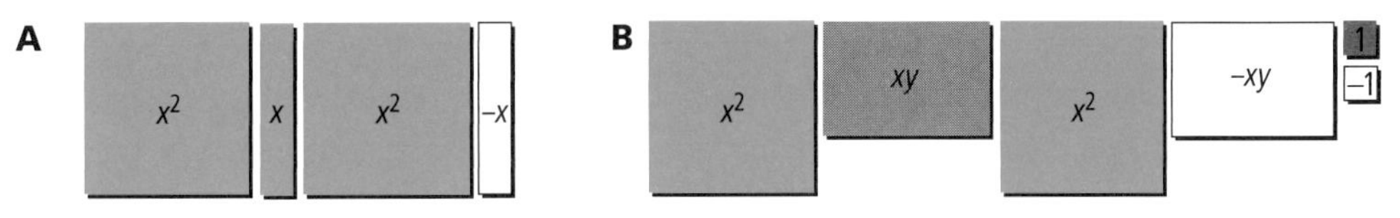

Check Your Understanding

1. How many terms are in each expression?
   a) $2b + 7$
   b) $6x$
   c) $x^2 + 3x + 5$
   d) -6

2. Simplify, if possible. If it is not possible, explain why.
   a) $3x + 5x$
   b) $2x^2 + 6x$
   c) $3x^2 + 4x^2 - 5x^2$
   d) $5xy - 7xy$

3. Copy and complete the table.

Expression	Model	Terms	Like/ Unlike Terms	Justification (Explain why terms can or cannot be combined.)
$3x + 4$		$3x$, 4	unlike terms	The second term does not have a variable.
$4x^2 - 3x^2$				
$2y^2 + 4y$				
$-3x^2 + 3x$				

4. Niall says the expressions $3x^2y + 5$ and $2xy + 7$ contain like terms and can be added together. Emily says they do not have like terms and cannot be added. Who is correct and why?

5. For each term, write two additional like terms.
   a) $3n^2$
   b) $-xy$
   c) $4t^2$
   d) $-7p^2$
   e) a^3
   f) x^4y

6. What would you need to add to each expression to get a sum of zero?
   a) $3x^2 - \frac{1}{2}x + 7$
   b) $4y^2 - 3y - \frac{4}{5}$
   c) $3n^2 + 2n + 4$
   d) $\frac{3}{2}m^3 - m^2 + 3m$
   e) $y^5 - 6n^2$
   f) $8w^4 - 0.8w^2 + 2$

7. Simplify by collecting like terms.
   a) $6x + 4 + 8x + 3$
   b) $2b - b + 7 - 8$
   c) $3r^2 + 2 - 6r + 9r - 3r^2$
   d) $0.5m - 4.5n + 0.7m + 4.5n - 1.5$
   e) $4p^5 - 6q^2 + p^4 - 2p^5 + 3q^2$
   f) $3xy + x^2y + y^2 + xy + 2x^2y + x^3y$
   g) $\frac{1}{2}a + \frac{2}{3}b + \frac{3}{4}a - \frac{1}{6}b$

8. a) Write an addition statement and a subtraction statement that can be simplified to $4x^2 - 2xy + 3$.
   b) Repeat part a) but include some fractions and/or decimals in each expression.

9. Use algebra tiles to model each expression. Record your models.
   a) $4x^2 + 5y - 3$
   b) $-3n^2 - mn + 2m^2 + 4$

10. Simplify the algebraic expression represented by each algebra tile model. Record your steps using pictures and symbols.
   a)

   b)
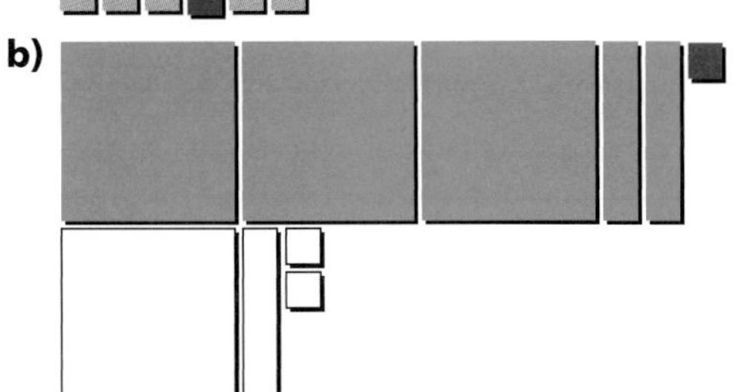

11. Use algebra tiles and symbols to simplify each expression. Record your steps.

a) $(5x - 3y + 2) + (3x - 7y + 5)$

b) $(2n^2 + 3n - 4) + (-4n^2 + 4n - 3)$

12. Subtract. Use the take away method. Use concrete materials, if needed.

a) $(13y + 5) - (7y + 4)$

b) $(3x - 5) - (x - 4)$

c) $(7m^2 + 2) - (3m^2 - 1)$

d) $(5p^3 - 4p^2) - (4p^3 + 2p^2)$

13. Subtract. Use the comparison method. Use concrete materials, if needed.

a) $(8s + 5) - (s + 5)$

b) $(6m + 4) - (2m + 1)$

c) $(4w^2 + 2y) - (5w^2 + 2y)$

d) $(6h^2 - 2xy) - (7h^2 + 2xy)$

14. For questions 12 and 13, find the difference by adding the opposite.

15. Simplify. Show your work using symbols. How are the simplified expressions similar? different?

a) $(4y^2 + 2y) + (-6y^2 + 3y)$

b) $(4y^2 + 2y) - (-6y^2 + 3y)$

16. Each solution contains an error. Describe the error and make the necessary corrections.

a) $(-6x^2 + 5xy) + (-4x^2 + 3xy)$
$= -6x^2 + 5xy - 4x^2 + 3xy$
$= -2x^2 + 8xy$

b) $(7x^2 - 4y) - (3x^2 - 5y)$
$= 7x^2 - 4y - 3x^2 - 5y$
$= 4x^2 - 9y$

Chapter Problem

17. At the beach near the Joggins Fossil Cliffs, Mr. Fisher divides the students into teams for a fossil hunt. He awards points for different types of fossils: tree trunks are 1 point, seed ferns are 2 points, worms are 3 points, and unknown species are 5 points. Let t represent the number of tree trunk fossils, f the number of seed fern fossils, w the number of worm fossils, and u the number of unknown species fossils.

a) Write an expression for the total number of possible points.

b) Khalid's team found 1 unknown species fossil, 3 seed fern fossils, 2 tree trunk fossils, and 5 worm fossils. Write an expression to represent his team's collection. Use the expression to find the total number of points they earned.

18. The cabins at the camp are of an unknown perimeter. Camp staff and teachers' cabins are smaller than the students' cabins.

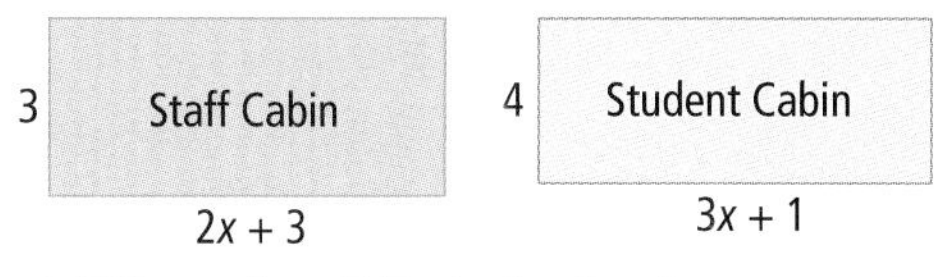

a) What simplified algebraic expression represents the perimeter of a student cabin?

b) What simplified algebraic expression represents the perimeter of a staff cabin?

c) Write an expression that can be used to find the difference in perimeter of the two cabins. Simplify your expression.

19. In basketball, a basket can be worth one, two, or three points. Let x represent the number of 1-point baskets scored, y the number of 2-point baskets scored, and z represent the number of 3-point baskets scored.

a) Write an expression to represent the total number of possible points.

b) Emma scored three 3-point baskets, four 2-point baskets, and three 1-point baskets. What were her total points scored in the game? Show your work.

20. a) Copy and complete the math path by combining like terms.

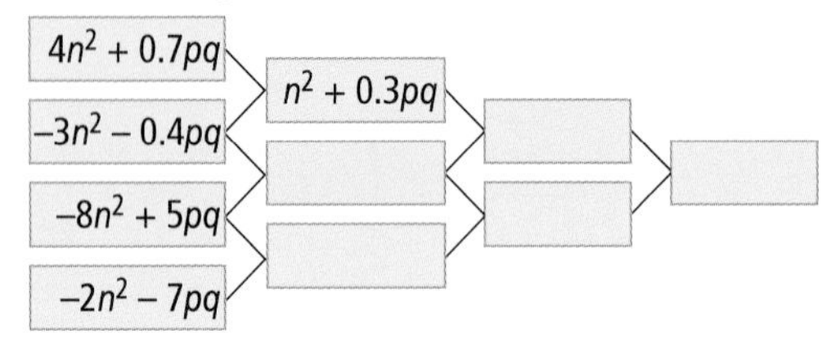

b) Make your own math path. Exchange problems with a classmate and verify each others' answers. Include at least one fraction or decimal in each expression.

21. Alex is painting on a canvas that is 9 cm longer than three times its width.

a) Draw a diagram of the painting and label each side length with an algebraic expression. Use the variable w to represent the width of the canvas.

b) Write an algebraic expression for the perimeter of the painting. Do not simplify.

c) Find the perimeter, if $w = 10$ cm.

d) Simplify your expression from part b). Evaluate the expression for $w = 10$ cm.

e) Are your answers to parts c) and d) the same? Which expression do you prefer to evaluate, the expression in part b) or part d)? Why?

f) Alex has a request to paint a new picture on a canvas with these dimensions.

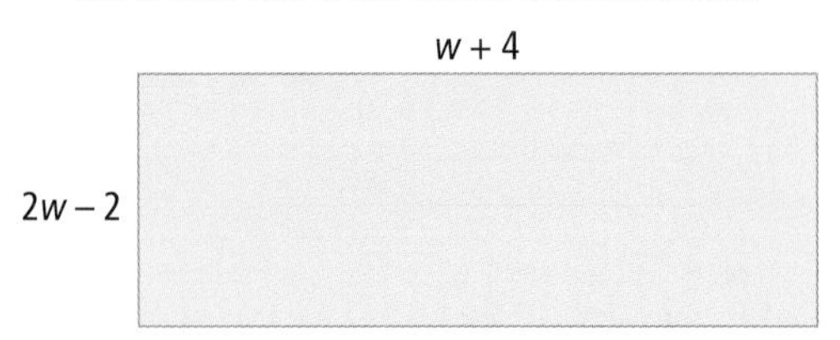

Write the simplified expression for the perimeter of the second painting.

g) Write an expression that shows the difference between the two perimeters. Simplify the expression by adding the opposite. If $w = 10$ cm, what is the difference between the perimeters?

22. The table shows the medal count for the four Atlantic Provinces at the 2007 Canada Winter Games in Whitehorse, NWT.

0 Gold	2 Gold	0 Gold	1 Gold
3 Silver	2 Silver	1 Silver	0 Silver
4 Bronze	8 Bronze	2 Bronze	3 Bronze

Write an expression for the total number of medals won by the Atlantic Provinces. Use g for gold, s for silver, and b for bronze.

Extend

23. The deer living on Big Tancook Island have no natural predators so the island has become overpopulated. The deer often destroy the islander's gardens. Jane and Sam have decided to build a fence around their garden to protect their vegetables.

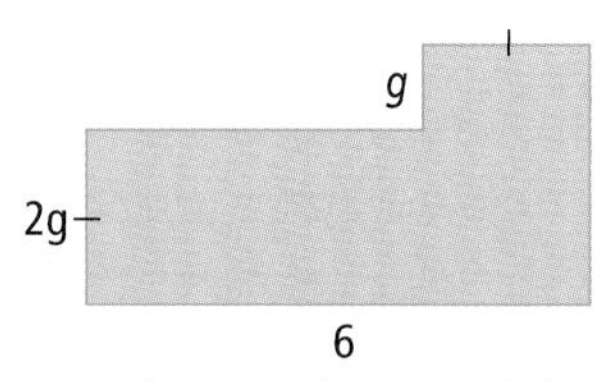

a) What is the simplified expression for the perimeter of the garden? Show your work. Include a diagram.

b) What is the perimeter of the garden if $g = 0.5$ m?

c) Jane and Sam estimate it will cost \$16/m to build the fence. This price includes materials and labour. Sales tax of 14% will be added to the cost. How much will the fence cost?

7.2 Multiply Polynomial Expressions

Focus on...

- multiplying a polynomial by a scalar concretely, pictorially, and symbolically.

The Cape Breton Highlands is a spectacular region of eastern Canada. Cape Breton Highlands Park was the first national park in Atlantic Canada. It covers an area of approximately 950 km^2. The park is known for its cycling, hiking, and skiing trails, as well as being a great place for kayaking.

The shape of the park approximates a parallelogram. A portion of the Cabot Trail runs through the park.

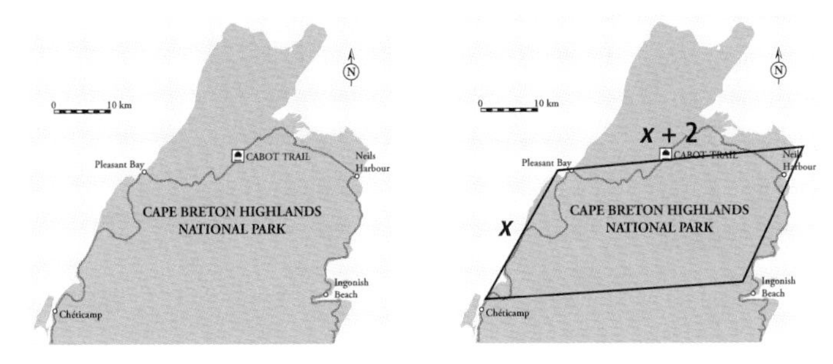

What expression could you write to represent the length of the trail that runs through the park?

Discover the Math

Materials

- algebra tiles

Optional

- trays

The Baddeck Hikers Club is having its annual outing on the hiking trail at North River. The members are making bags of trail mix for the hikers. Each bag contains equal masses of peanuts, almonds, raisins, dried currants, dried bananas, dried apples, and dried cranberries in grams. Logan brought sunflower seeds and added two scoops of sunflower seeds to each bag.

1. Model the mass of one bag of trail mix using one x-tile for the mass of each scoop of sunflower seeds and one unit tile for the mass of each of the other ingredients. Write an algebraic expression for the total mass of ingredients in one bag.

2. Sarah has more space in her backpack and offers to carry three bags of trail mix. Model this mass with your tiles. Write an algebraic expression to represent the mass of the three bags as
   a) a multiplication expression
   b) a repeated addition expression
   c) a simplified algebraic expression

3. **Reflect** How are the three algebraic expressions similar? different? Which expression do you prefer? Why?

Example 1: Multiply Polynomials by a Scalar

Simplify.

$4(3x + 2)$

Solution

Method 1: Use repeated addition.

4(3x + 2) means 4 sets or groups of 3x + 2. I can model 4 groups of (3x + 2).

$3x + 2 + 3x + 2 + 3x + 2 + 3x + 2$

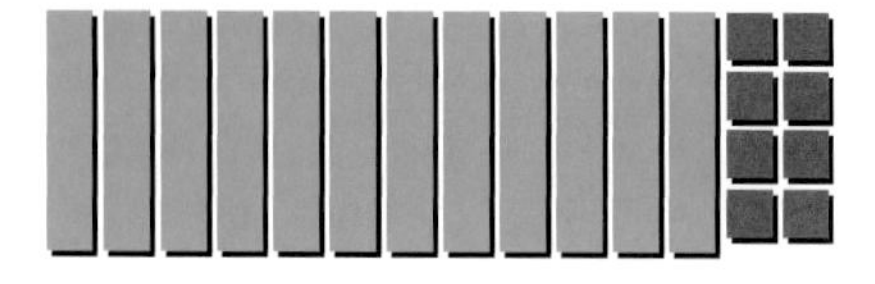

Now, I collect like terms.

$= 12x + 8$

Method 2: Use an area model.

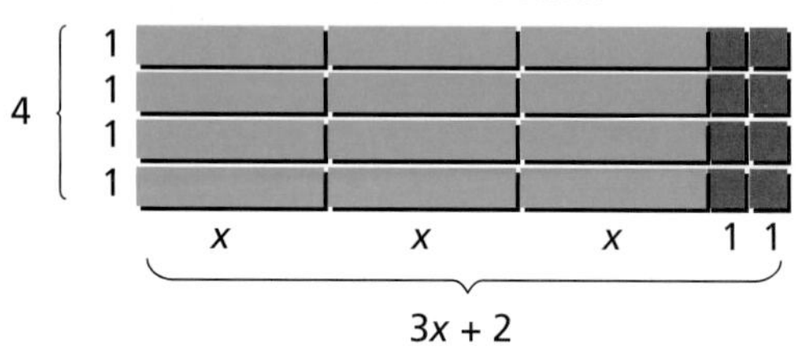

I can construct a rectangle with the dimensions 4 by (3x + 2).

$4(3x + 2) = 12x + 8$

Example 2: Volume of Solids

a) Find an expression for the volume of the square-based prism.

b) Find the volume of the prism, if $x = 3$ cm.

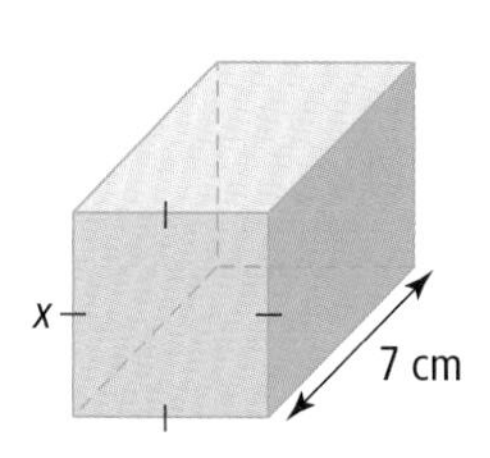

Solution

a) $\text{Volume}_{\text{prism}} = \text{area of base} \times \text{height}$
$= (x \times x) \times 7$
$= 7x^2$

> **Strategies**
> Use a formula.

b) $V_{\text{prism}} = 7x^2$
$= 7(3)^2$
$= 7 \times 9$
$= 63$

The volume of the square-based prism is 63 cm^3, when $x = 3$ cm.

Communicate the Key Ideas

1. Julien solved this problem as shown, without using tiles or drawings.
$5(x - 2) = 5x - 2$
Use tiles to show Julien the error in his thinking.

Check Your Understanding

1. Copy and complete the table for each model. The first one has been done for you.

Model	Width	Length	Area
a)	2	$2x - 1$	$2(2x - 1)$ $= 4x - 2$

b)

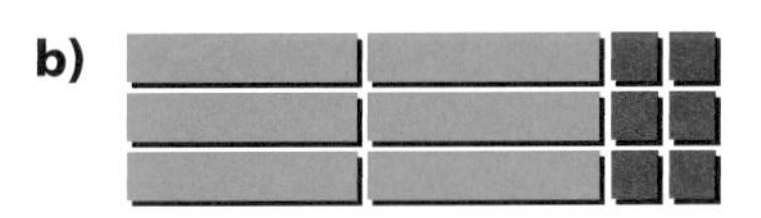

c)

d) Create your own model.

2. Copy and complete the table for each model. The first one has been done for you.

Model	Repeated Addition	Multiplication	Result
a)	$(x + 4) +$ $(x + 4)$	$2(x + 4)$	$2x + 8$

b)

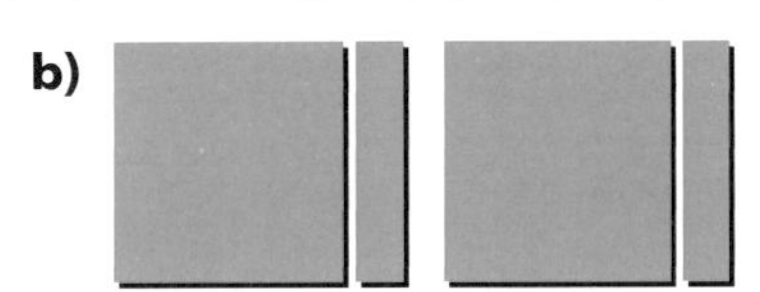

c)

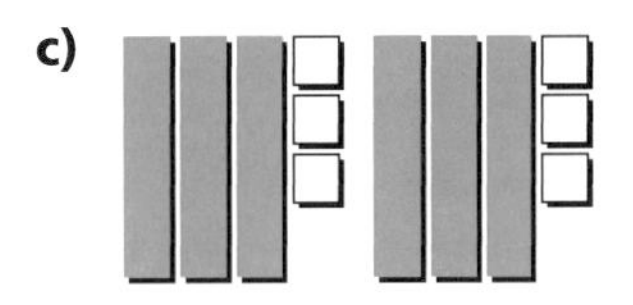

d) Create your own model.

3. Write an expression for each model as a repeated addition statement.

a) 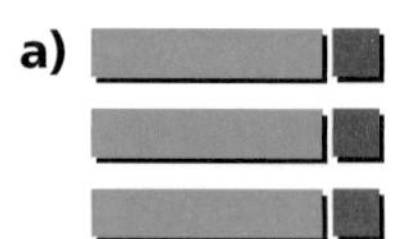

b) 

c) 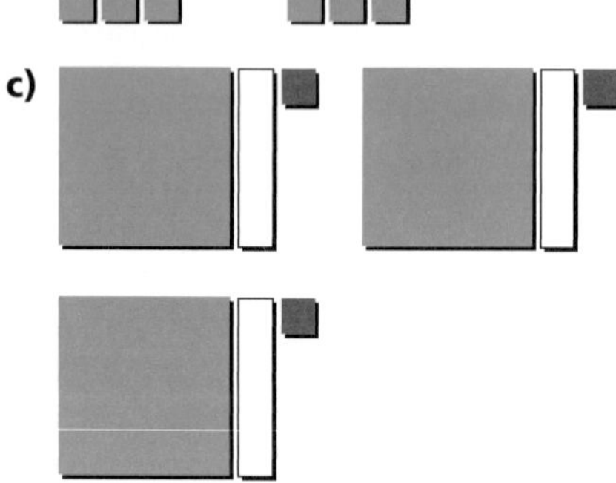

4. Write each addition expression in question 3 as an expression using multiplication.

5. Match each model with an expression.

a) b) 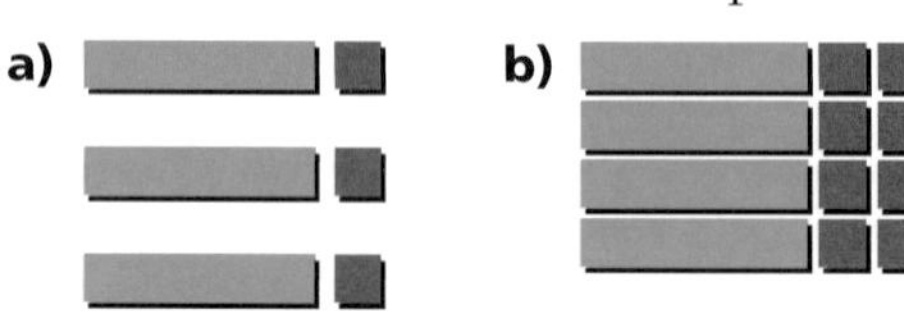

c) 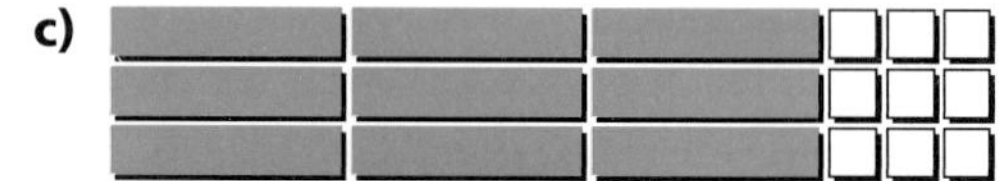

d) 

e) 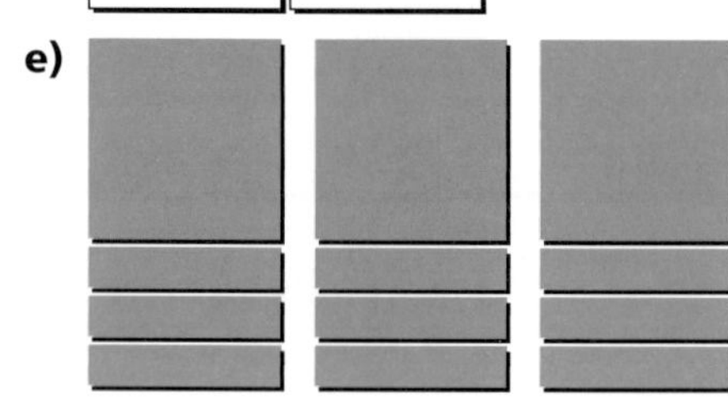

A $3(x + 1)$ **B** $2(x^2 - 2x)$
C $x(x - 4)$ **D** $4(x + 2)$
E $3(x^2 + 3x)$ **F** $3(3x - 3)$

6. Use algebra tiles to create a rectangle with the given dimensions. Draw the rectangle and label the width and length. Write the multiplication expression that could be used to find the area and give the result.

a) width 2, length $5x$
b) width 4, length $x + 1$
c) width 2, length $3x - 2$

7. For each rectangle, find the length and width then write the expression for the area.

a) b) 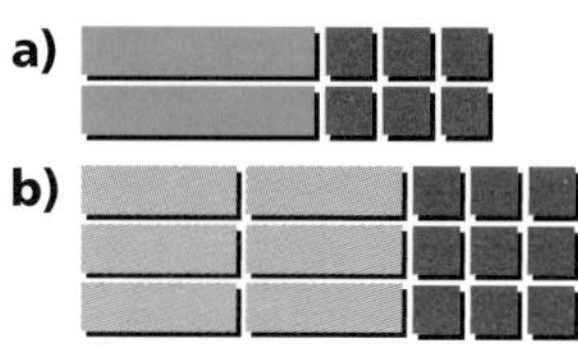

c) 

8. Write an expression for the area of each shaded region.

a) 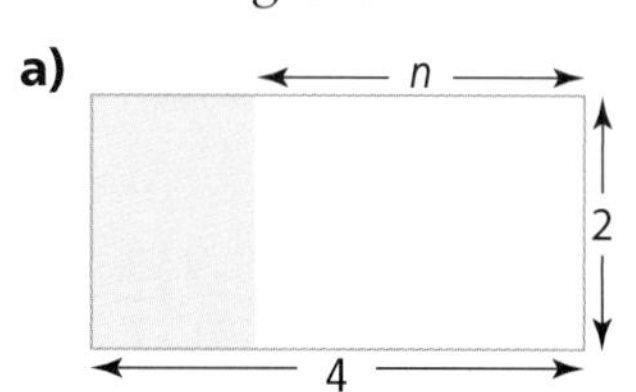

b) 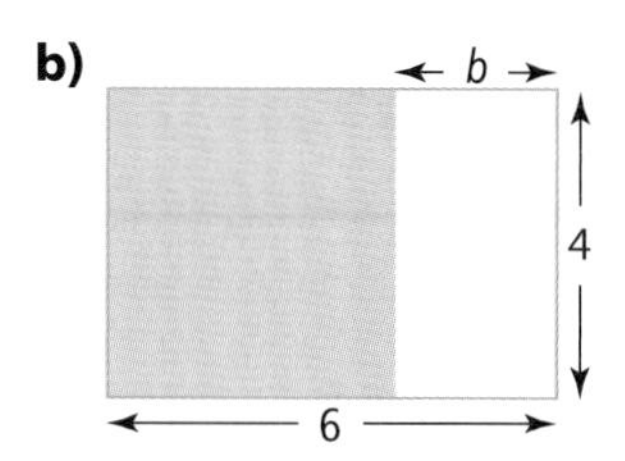

c) 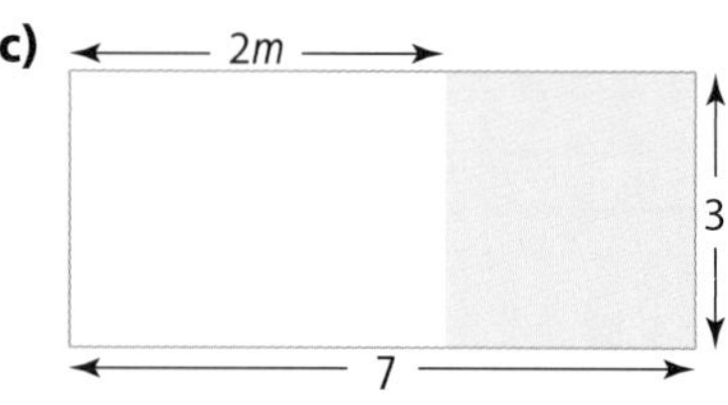

9. The diagrams show the backyards of two new houses being built on Xeno's street. Find the area of each backyard.

a)

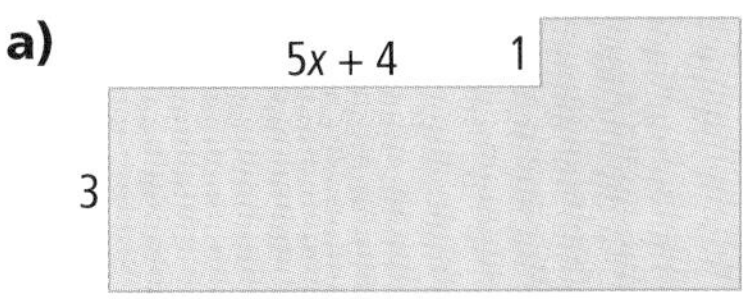

b)

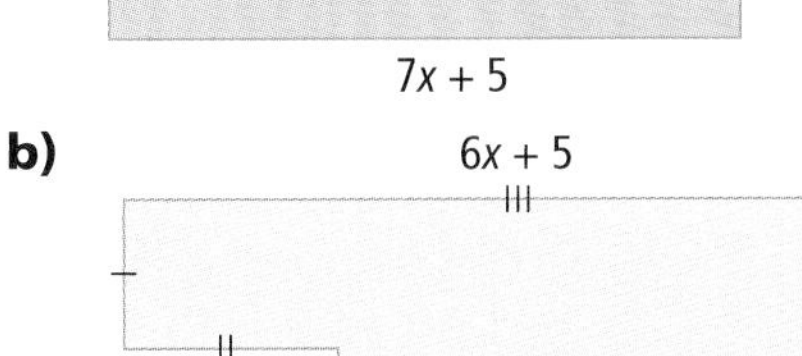

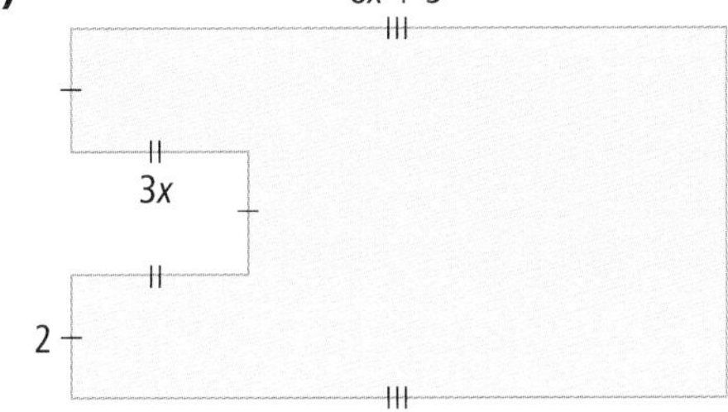

10. Which expression is equivalent to $5(x + 1)$? Support your answer with algebra tiles.

A $5x + 6$ **B** $5 + x + 1$
C $5x + 5$ **D** $5x + 1$

11. Which expression is equivalent to $\frac{1}{2}(4x - 8)$? Explain your reasoning with algebra tiles.

A $2x - 8$ **B** $4x - 4$
C $8x - 16$ **D** $2x - 4$

12. Model a rectangle with an area of $12x + 6$ using algebra tiles. Draw your model, showing the dimensions of the rectangle. Is there more than one model? Explain your answer.

13. Purpose Pools designs rectangular pools and decks. The standard width of a pool is 3 m but the length can vary and is modelled by the expression $(4x + 2)$, in metres. The standard width of a pool deck is 5 m but the length can vary and is modelled by the expression $(6x + 4)$, in metres.

a) Write an expression for the area of the pool and an expression for the area of the pool deck. Draw a labelled diagram to help you.

b) If $x = 2$, what is the area of the pool and the area of the pool deck?

Chapter Problem

14. Khalid notices the cabins at the camp vary in size but that the widths are standard measures. He wonders how much more area is covered by a student cabin than by a staff cabin. He drew these diagrams. All dimensions are in metres.

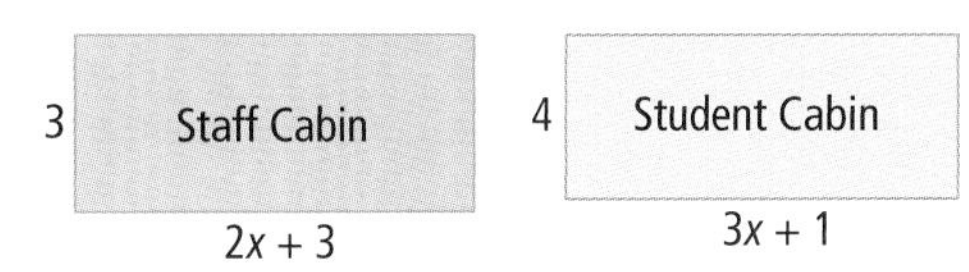

a) Write the algebraic expression for the area of a student cabin.

b) Write the algebraic expression for the area of a staff cabin.

c) Write the simplified expression for the difference between the areas of the two cabins.

d) If $x = 2$ m, what is the difference in area?

Extend

15. Simplify.

a) $4\left(\frac{1}{2}x + 3\right) + 5\left(3x + \frac{4}{5}\right)$

b) $3(4x - 0.2) - 2(x + 0.5)$

16. Tamara sells crafts. She packs her crafts in small boxes that have a standard height and width and a variable length, in cenitimetres, as shown.

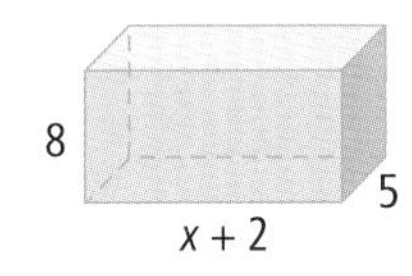

a) Write an expression for the surface area of the box.

b) Write an expression for the volume of the box.

c) What is the volume of the box, if $x = 10.8$ cm?

d) What would be the dimensions of a cube-shaped box having the same volume?

7.3 Solve Linear Equations

Focus on...
- solving and verifying simple linear equations algebraically

Some stores sell grab bags to reduce their inventory quickly. Elysia is having a party and needs small gifts for prizes so she buys two grab bags. She wonders how many items she has in each bag. The clerk tells her that there are an equal number of novelty pencils in each bag. Elysia decides to buy three erasers as well. The clerk tells her that she now has 17 items in total.

Discover the Math

Materials
- algebra tiles

How can you solve an equation for an unknown quantity?

How can Elysia determine how many pencils are in each grab bag, if each bag contains the same number of pencils?

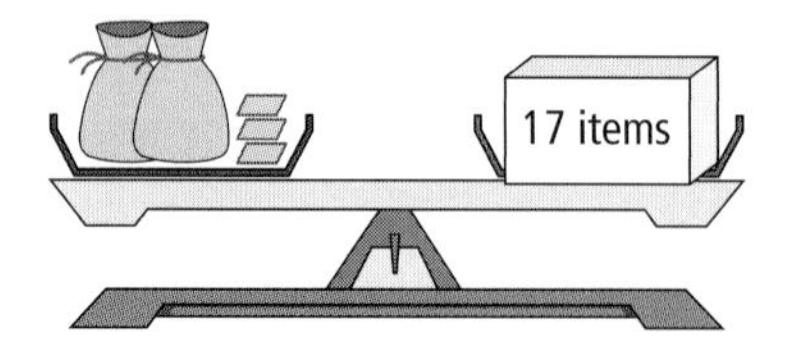

1. **a)** Represent Elysia's problem using a pan balance. Draw the steps you would use to balance the equation that represents the problem. *Hint: You know that three of the 17 items are erasers and the rest are pencils.*
   b) Write symbols beside each step of your solution.
   c) Explain in words how you know how many items are in each bag.

2. Jake modelled the solution using algebra tiles. Miriam solved the problem using symbols.

Strategies
Work backward.

Jake's Method

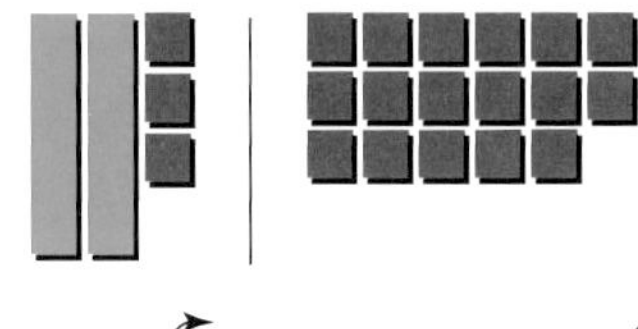

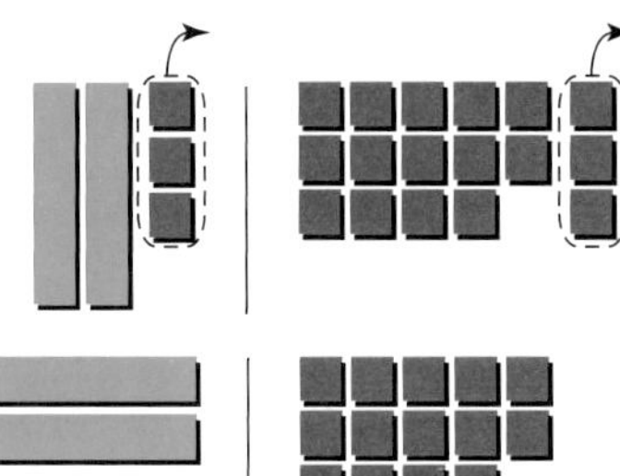

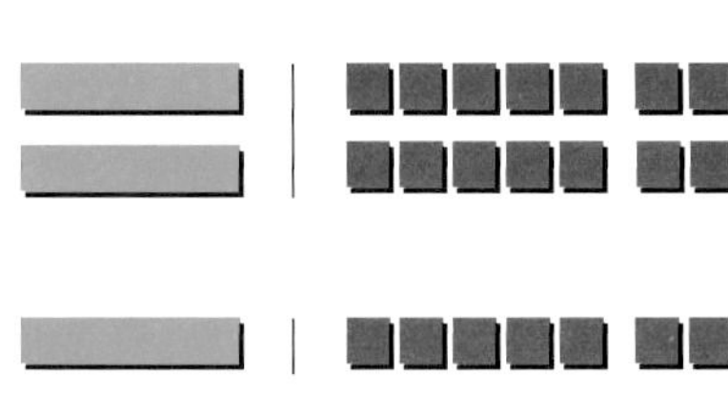

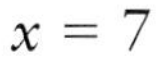

Miriam's Method

$$2x + 3 = 17$$

$$2x + 3 - 3 = 17 - 3$$

$$2x = 14$$

$$\frac{2x}{2} = \frac{14}{2}$$

$$x = 7$$

Check by working backward. The equation says double an unknown number x, and then add 3 to the result for a sum of 17. To work backward, I subtract 3 from 17 for a result of 14. Then I divide 14 by 2 for the unknown number 7. So $x = 7$.

a) Explain each step in Jake's solution.
b) Explain each step in Miriam's solution.
c) Compare the two methods to your answer in question 1, part c). How are the solutions similar?
d) How can you verify that the solution is 7 pencils?

3. Use algebra tiles to model and then solve each **algebraic equation**. Record each step using Jake and Miriam's methods.

a) $4x + 1 = 9$
b) $3n - 2 = 7$
c) $2x + 2 + x - 3 = 5$
d) $3(x + 1) = -6$
e) $2(y - 3) + 2(y + 1) = -4$
f) $3x + 5 - x - 3 = 5$

algebraic equation
- a mathematical statement that says two quantities or expressions are equal
- uses an equal sign

4. Reflect When solving an equation, why is it important to perform the same operation on both sides of the equation?

5. Reflect How is Miriam's algebraic method of solving an equation similar to the pan balance method you learned in grade 7?

Example 1: Solve One-Step Equations by Multiplying

Solve.

a) $\frac{x}{2} = 4$

b) $\frac{1}{3}y = -2$

c) $\frac{2}{3} = \frac{x}{15}$

Solution

a) ***Method 1: Use a balance.***

Let  represent x, then  represents $\frac{1}{2}x$ or $\frac{x}{2}$.

Model the equation.

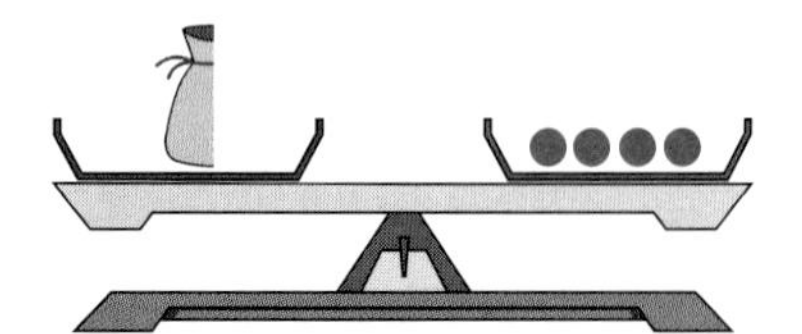

$$\frac{x}{2} = 4$$

To make one whole bag, double the quantity on the left side of the balance. To keep the pan balance balanced, you must also double the quantity on the right side of the balance.

$$2\left(\frac{x}{2}\right) = 2(4)$$

Simplify.

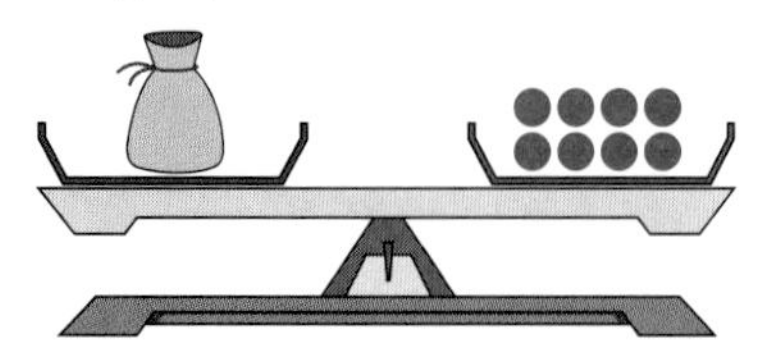

$$x = 8$$

Method 2: Use the cover-up method.

$$\frac{\quad}{2} = 4$$

I can cover up x and ask myself "What number divided by 2 equals 4?"

$$\frac{8}{2} = 4$$

$$x = 8$$

Since $8 \div 2 = 4$, $x = 8$.

Communicating Mathematically

When numbers are beside each other in brackets, it means to multiply. 4(2) means 4×2.

To verify that $x = 8$, substitute 8 for x into the equation.

$$\frac{x}{2} = 4$$
$$\frac{8}{2} = 4$$
$$4 = 4$$

Since the left side equals the right side, the equation is true when $x = 8$.

Strategies
Look back.

b) $\frac{1}{3}y = -2$

If ▭ represents y, then ▫ represents $\frac{1}{3}y$.

Model the equation. Use paper y-tiles to help you.

$$\frac{1}{3}y = -2$$

To make one whole y-tile, triple the quantity on the left side of the equation. To keep the equation balanced, you must also triple the quantity on the right side of the equation.

$$3\left(\frac{1}{3}y\right) = 3(-2)$$

Simplify.

$$y = -6$$

c) $\frac{2}{3} = \frac{x}{15}$

I can use mental math to solve for x.

$$\frac{2}{3} = \frac{x}{15} \quad (\times 5)$$

Think: What must I multiply by to get 15? The answer is 5.

So 2 multiplied by 5 gives what number? The answer is 10.

$$\frac{2}{3} = \frac{10}{15} \quad (\times 5)$$

Therefore, $x = 10$.

Example 2: Solve Multi-Step Equations I

$-2x + 7 = 13$

Solution

Method 1: Use the cover-up method.

$$\square + 7 = 13$$
$$-2\square = 6$$
$$x = -3$$

I can cover up –2x and ask myself "What number added to 7 equals 13?" Since 6 + 7 = 13, then –2*x* = 6.

I can cover up *x* and ask myself "What number, when multiplied by – 2 equals 6?" Since –2 × (–3) = 6, then *x* = –3.

To verify that $x = -3$, substitute -3 for x.

$$-2x + 7 = 13$$
$$-2(-3) + 7 = 13$$
$$6 + 7 = 13$$
$$13 = 13$$

Since the left side equals the right side, the equation is true when $x = -3$.

Strategies

Equations are solved by working backward. Think about the order of operations. Undo the last operation by applying the opposite operation to both sides.

Method 2: Use symbols.

$$-2x + 7 = 13$$
$$-2x + 7 - 7 = 13 - 7$$
$$-2x = 6$$
$$\frac{-2x}{-2} = \frac{6}{-2}$$
$$x = -3$$

To keep the equation balanced, subtract 7 from both sides.

Divide each side by the **numerical coefficient** of *x*, which is –2.

numerical coefficient
- a numeric factor of a term

Example 3: Solve Multi-Step Equations II

Solve.

$\frac{x}{2} + 5 = 8$

Solution

Method 1: Use algebra tiles.

If [tile] represents x, then [tile] represents $\frac{1}{2}x$.

Model the equation.

 | 

$$\frac{x}{2} + 5 = 8$$

Take away 5 from each side to keep the equation balanced.

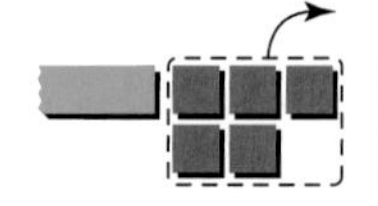 | 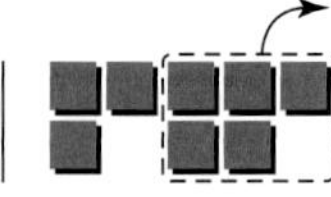

$$\frac{x}{2} + 5 - 5 = 8 - 5$$

Simplify.

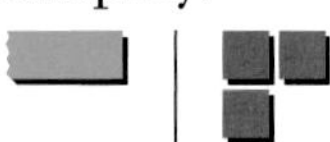

$$\frac{x}{2} = 3$$

Perform the inverse operation of division to "undo" the operation on the left side of the equation.

Strategies
Work backward.

To make one whole x-tile, double the quantity on the left side of the equation. To keep the equation balanced, you must double the quantity on the right side of the equation.

$$\frac{x}{2} \times 2 = 3 \times 2$$

Simplify.

$$x = 6$$

Method 2: Use the cover-up method.

$$\frac{x}{2} + 5 = 8$$

I can cover up $\frac{x}{2}$ and ask "What number added to 5 equals 8?" Since $3 + 5 = 8$, then $\frac{x}{2} = 3$.

$$\frac{x}{2} = 3$$

I can cover up x and ask "What number divided by 2 equals 3?" Since $6 \div 2 = 3$, then $x = 6$.

$$x = 6$$

To verify that $x = 6$ is the solution to this equation, substitute 6 for x.

$$\frac{x}{2} + 5 = 8$$

$$\frac{6}{2} + 5 = 8$$

$$3 + 5 = 8$$

$$8 = 8$$

Since the left side equals the right side, the equation is true when $x = 6$.

Example 4: Solve Multi-Step Equations III

Solve.

$$\frac{1}{2}x + 4 = -2$$

Solution

Method 1: Use symbols.

$$\frac{1}{2}x + 4 = -2$$

$$\frac{1}{2}x + 4 - 4 = -2 - 4$$

Subtract 4 from each side to keep the equation balanced.

$$\frac{1}{2}x = -6$$

$$(2)\frac{1}{2}x = -6(2)$$

Multiply each side by 2.

$$x = -12$$

Strategies
What happens if you first double each side of the equation?

Method 2: Use the cover-up method.

$\frac{1}{2}x + 4 = -2$ (with x covered up)

I can cover up $\frac{1}{2}x$ and ask "What number added to 4 equals –2?" Since 4 + (–6) = –2, then $\frac{1}{2}x = -6$.

$\frac{1}{2}x = -6$ (with x covered up)

I can cover up x and ask "What number multiplied by $\frac{1}{2}$ equals –6?" Since $\frac{1}{2} \times (-12) = -6$, $x = -12$.

$x = -12$

To verify that $x = -12$ is the solution to this equation, substitute -12 for x.

$$\frac{1}{2}x + 4 = -2$$
$$\frac{1}{2}(-12) + 4 = -2$$
$$(-6) + 4 = -2$$
$$-2 = -2$$

Since the left side equals the right side, the equation is true when $x = -12$.

Strategies
Can you solve this equation using systematic trial?

Example 5: Use Equations to Solve Problems I

In a triangle, the largest angle is three times the size of the smallest. The middle angle is 5 degrees greater than the smallest angle. Determine the size of each angle.

Solution

Let x represent the measure of the smallest angle in degrees.
The measure of the largest angle can be represented by $3x$.
The measure of the middle angle can be represented by $x + 5$.
The sum of the interior angles of a triangle is 180°.

The equation that models this problem is $3x + (x + 5) + x = 180$.

Simplify and solve for x.

$$3x + (x + 5) + x = 180$$
$$5x + 5 = 180$$
$$5x + 5 - 5 = 180 - 5$$
$$5x = 175$$
$$\frac{5x}{5} = \frac{175}{5}$$
$$x = 35$$

Middle angle $= x + 5$
$= 35 + 5$
$= 40$

Largest angle $= 3x$
$= 3(35)$
$= 105$

The angle measures are 35°, 40°, and 105°.

Example 6: Use Equations to Solve Problems II

The length of a rectangular garden is 4 units longer than twice its width. Find the dimensions of the garden, if its perimeter is 50 units.

Solution

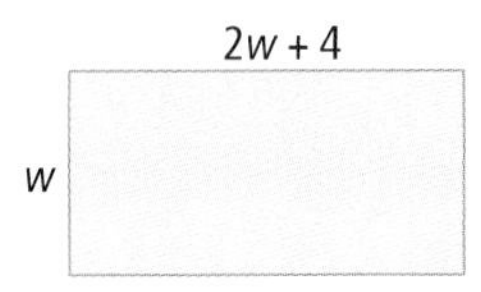

Strategies
Write an appropriate equation to model the problem. Solve the equation.

Let w represent the width.
Then $2w + 4$ represents the length.
The equation for the perimeter is $2(2w + 4) + 2w = 50$.

$2(2w + 4) + 2w = 50$

$4w + 8 + 2w = 50$ **Simplify the left side of the equation.**

$6w + 8 = 50$

$6w + 8 - 8 = 50 - 8$ **Subtract 8 from both sides of the equation to isolate 6*w*.**

$6w = 42$ **Simplify.**

$\frac{6w}{6} = \frac{42}{6}$ **To find the value of *w*, divide both sides by 6.**

$w = 7$ **Simplify.**

The width of the garden is 7 units.

Substitute $w = 7$ into the expression for the length.

$$\begin{aligned} \text{Length} &= 2w + 4 \\ &= 2(7) + 4 \\ &= 14 + 4 \\ &= 18 \end{aligned}$$

The length of the garden is 18 units.

Communicate the Key Ideas

1. What equation is modelled by the diagram? Explain how you could solve the equation using
- a pan balance
- algebra tiles
- algebraic symbols
- the cover-up method

2. Each solution is incorrect. Indicate where the error first occurred, then state how to correct the error. What is the correct solution?

a)
$$\begin{aligned} 16 - 2y &= 24 \\ 16 - 16 - 2y &= 24 - 16 \\ 2y &= 8 \\ \frac{2y}{2} &= \frac{8}{2} \\ y &= 4 \end{aligned}$$

b)
$$\begin{aligned} \frac{n}{4} - 12 &= -20 \\ \frac{n}{4} - 12 + 12 &= -20 + 12 \\ \frac{n}{4} &= -8 \\ \frac{n}{4} &= \frac{-8}{4} \\ n &= -2 \end{aligned}$$

c)
$$\frac{x}{21} = \frac{5}{-7}$$
$\div (-3)$
$$\frac{x}{21} = \frac{5}{-7}$$
$\div (-3)$
$$x = 15$$

3. Simplify. State the operations you would use to undo parts of the equation.
$$\frac{y}{2} + 1 = -5$$

4. Brandon wants to buy a new cell phone in 12 weeks. The phone costs \$240 and he has already saved \$60. He wrote the equation $12x + 60 = 240$ to model his cell phone problem. Explain Brandon's equation. How much money will he need to save each week?

Check Your Understanding

1. The solution to an equation is modelled using algebra tiles. Write the equation. Show each step in the solution using symbols.

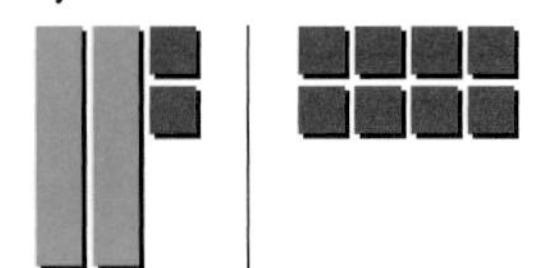

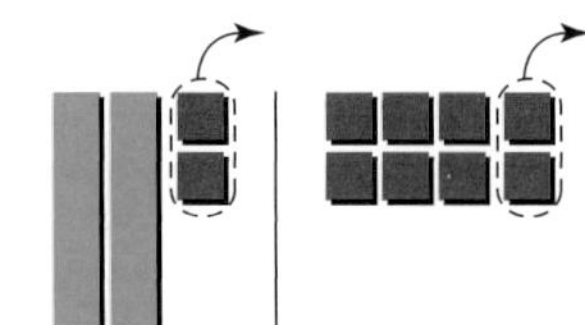

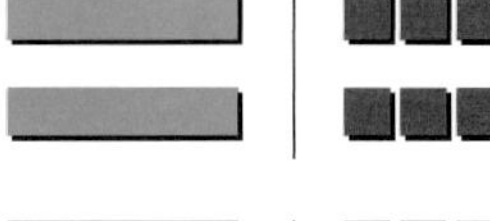

2. Model each equation using algebra tiles and then solve. Record each step with tile models and with symbols.

a) $2y - 3 = 9$
b) $4x + 5 = -3$

3. By what number would you multiply each side of the equation to solve for the variable?

a) $\frac{x}{2} = -12$

b) $\frac{1}{2}x = 10$

c) $\frac{m}{4} = -8$

d) $\frac{y}{-7} = -3$

4. Find the length of TM, to the nearest tenth of a metre. *Hint: Use the Pythagorean relationship.*

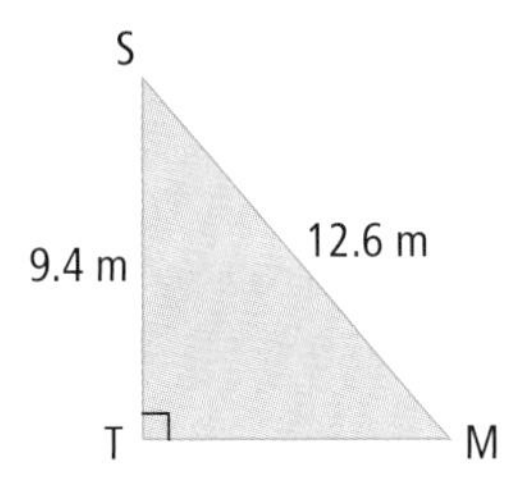

5. Solve each equation. Verify your solution.

a) $8 - 2x = 4$

b) $\frac{n}{2} - 1 = -8$

c) $\frac{1}{4}m + 5 = 9$

d) $\frac{x}{20} = \frac{3}{5}$

6. Describe a situation that can be modelled with each equation

a) $7y + 4 = 25$

b) $3(x - 4) = 18$

c) $\frac{x}{3} = 10$

7. Copy each equation.
For each:

- Circle the first operation you should undo.
- Underline the second operation you should undo.

a) $-2n - 4 = -8$

b) $0.8y - 7 = 9.8$

c) $\frac{b}{5} - 7 = -2$

8. Solve each equation in question 7. Verify your solution.

9. How is the equal sign the same as a balance scale? Use an example and a diagram to explain.

10. HTAM radio holds a Guess-the-Band contest. The radio station gives away three CDs for every correct answer, plus one CD just for calling in. Linh got 10 CDs.

a) Write an equation to model Linh's CDs.

b) Solve your equation to find how many questions she answered correctly.

11. Ravi is saving for a ski vacation that costs \$500. If he triples his savings, he will still need \$35. This situation can be modelled as $3s + 35 = 500$, where s represents his savings.

a) Explain how the equation $3s + 35 = 500$ models the problem.

b) How much money has Ravi saved so far?

c) What other strategy can you use to find Ravi's savings?

12. The length of a rectangular garden is 6 m longer than twice its width. Find the dimensions of the garden if its perimeter is 36 m. Write an equation to model the problem. Solve.

13. An isosceles triangle has congruent sides NE and EF. Find the value of t in centimetres. Write an equation to model the problem. Solve.

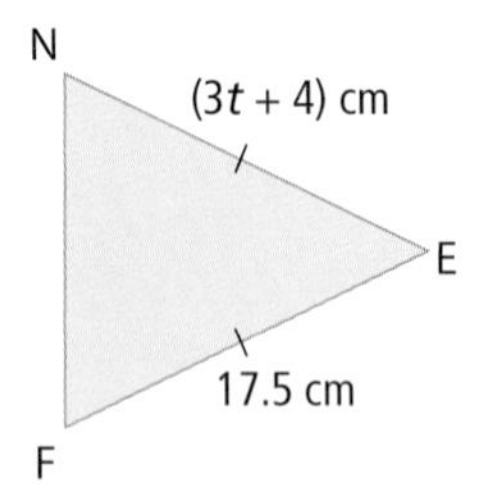

14. The width of a rectangular pool is 1 m less than half its length. Find the dimensions of the pool if its perimeter is 36 m. Write an equation to model the problem. Solve.

15. Hilda wants to order packages of seeds for her garden. Each package of seeds costs $3.75 including tax. Hilda wants to spend exactly $58. Shipping and handling is $5.50. How many packages of seeds can she buy? Write an equation to model the problem. Solve.

16. Brandon bought his new cell phone. His airtime plan costs $29.95 per month and includes 400 min during peak times. Each additional minute during peak time costs 45¢/min. His bill was $144.70 before tax. How many additional minutes did Brandon use? Write an equation to model the problem. Solve.

17. The school dance committee borrowed $1100 from the dance fund to pay for the year-end fundraising dance. They have also promised to donate $600 to a local food bank. If each ticket for the dance costs $10, how many tickets must they sell in order to meet their financial commitments? Write an equation to model the problem. Solve.

Chapter Problem

18. The camp charges schools $100 per day to use the equipment plus $25 per day for each student.

a) Write an equation to represent the cost for one day. Explain what the variables in your equation represent.

b) If 30 students want to go on the trip, how much will it cost per day?

c) The school raised $300 for a one-day trip. How many students can go?

19. a) Write an expression for the perimeter of the kite.

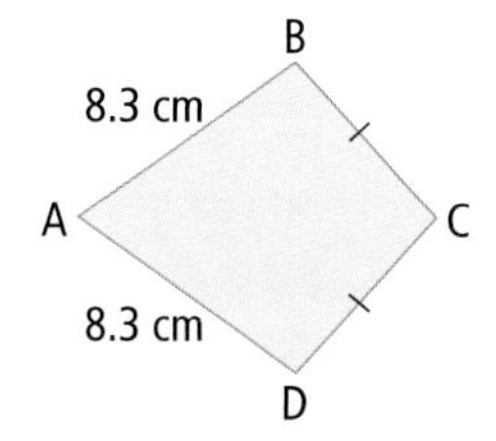

b) If the perimeter of the kite is 31.4 cm, what is the length of side CD? Use an equation to find your answer.

c) If the perimeter of the kite is 49.8 cm, what is the relationship among the side lengths? Model your answer using an equation.

20. Write an equation that you could use to find the value of x. Simplify your equation. Solve for x and find the measures of $\angle$A, $\angle$B, and $\angle$C.

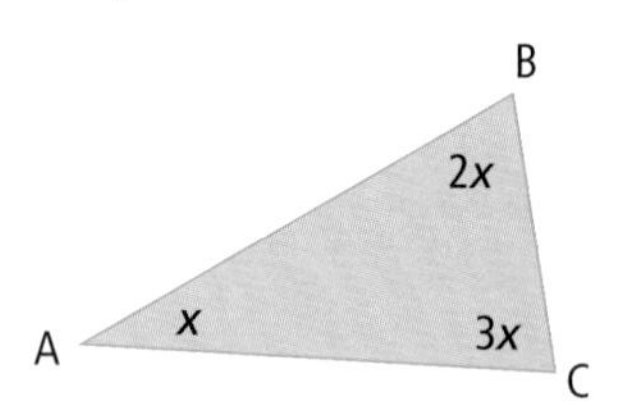

Extend

21. Yvette measures the circumference of each wheel.

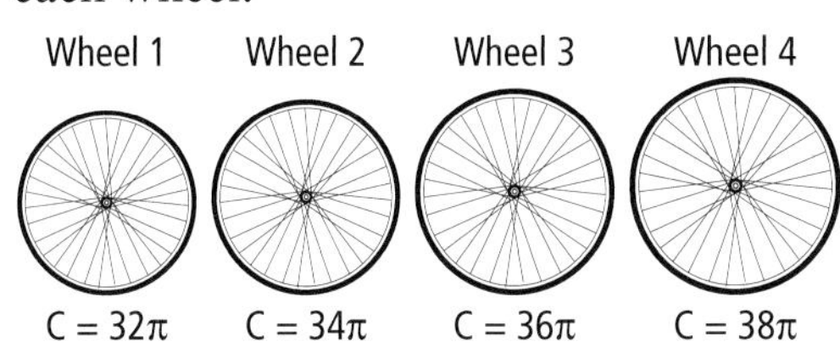

a) Copy and complete the table.

Wheel	Circumference, *C* (cm)	Pattern
1	32π	$2\pi(15 + 1)$
2		
3		
4		

b) The circumference of a wheel is 60π cm. Use an equation to find the wheel number.

22. a) Write an algebraic equation for the perimeter (P) of the triangle.

b) Write an algebraic equation for the area (A) of the triangle.

c) Determine P and A if $x = 2.5$ cm.

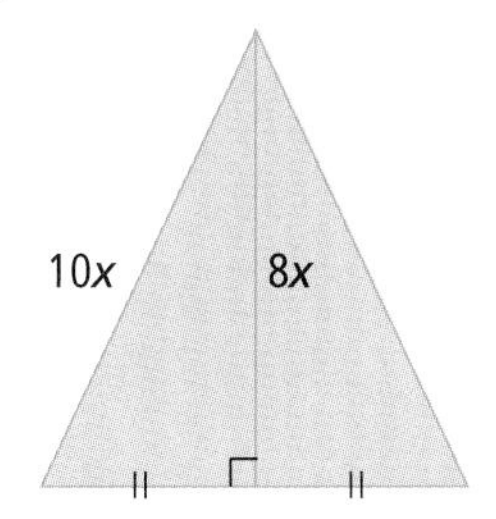

Puzzler

a) Draw an equilateral triangle. At each vertex, write a different number.

b) In the centre of the triangle, write another but different number.

c) Add the centre number to each number at a vertex. Write the sum outside of the opposite side. An example is shown.

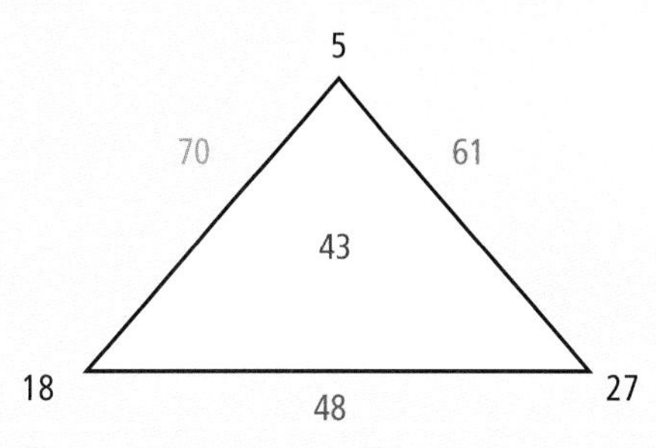

d) For each side, add the three numbers. What do you notice about the sums?

e) Try to show why this works using algebra.

f) Does this work for other numbers? Negative numbers? Try it!

Puzzler

You have a 5 × 5 geoboard.

How many squares can you make?

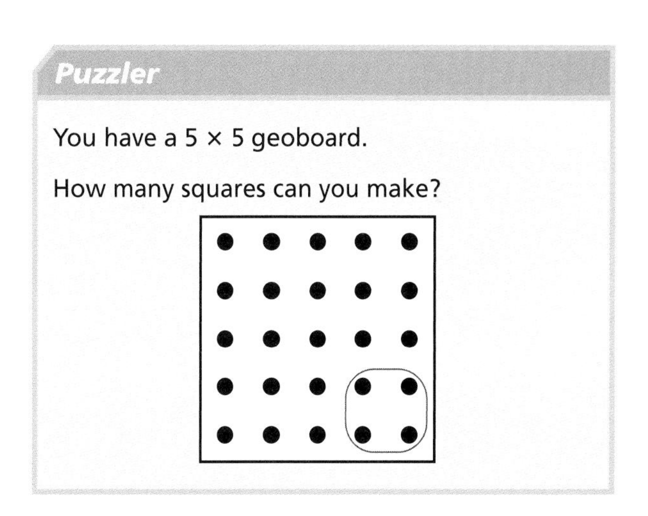

CHAPTER 7 Review

1. Match each description with the correct term.

a) Have identical variables or are constants.
b) x in the expression $3x + 1$
c) $5y - 7 = 23$ is an ■.
d) In $-3z + 5$, -3 is the ■.
e) In $-3z + 5$, 5 is the ■.
f) $2t + 13$ is an ■.
g) $3x + 2y + 6$ is a ■.

A variable
B expression
C equation
D like terms
E constant
F polynomial
G numerical coefficient

2. Which expression is equivalent to $3(x - 2)$?

A $3x - 2$ **B** $x - 6$
C $3x + 2$ **D** $3x - 6$

3. What number would you multiply each side of the equation $\frac{1}{3}x = -4$ by to solve for x?

A 12 **B** -12
C 3 **D** -4

4. Jenny says that $2x$ and $4y$ can be simplified to $6xy$ by combining like terms. Show the error in her thinking by drawing algebra tile models and providing explanations.

5.

Expression	Model	Like/Unlike Terms	Justification
$2x + 4$		unlike terms	The second term does not have a variable.
$3x^2 - 4x^2$			
$4y^2 + 4y$			
$-2x^2 + 3x$			

Copy and complete the table.

6. Simplify by collecting like terms.

a) $2r + 4 + 4t - t + r$
b) $4\frac{2}{3} + 3c^2 + d^2 + c^2 - 2\frac{1}{2}$
c) $-3.2 + 3x + (-3y) + 5.7 + x$
d) $p^3 + q^4 + 2p^3 - 4q^4 + 7 + pq$
e) $\frac{2}{3}b^4 - 5b^2 + 8 - 7 - 2b^3 - 2b^2$
f) $5y^5 - 2y^5 - 2xy + 3 - 3xy - y^5$

7. Write the area expression represented by each set of tiles. Simplify your area expression.

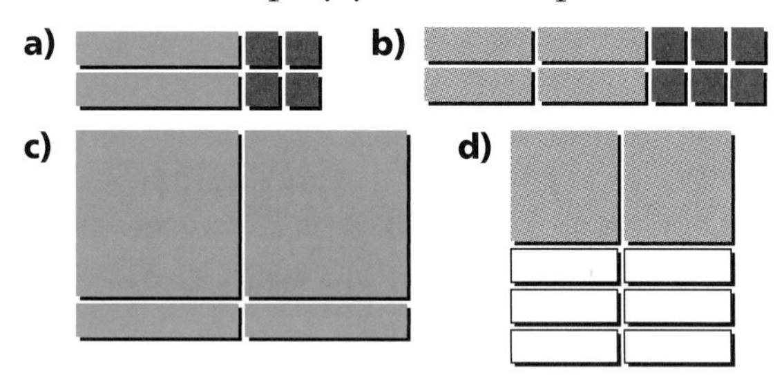

8. Write an expression to represent each tile model as

i) a repeated addition expression
ii) a multiplication expression

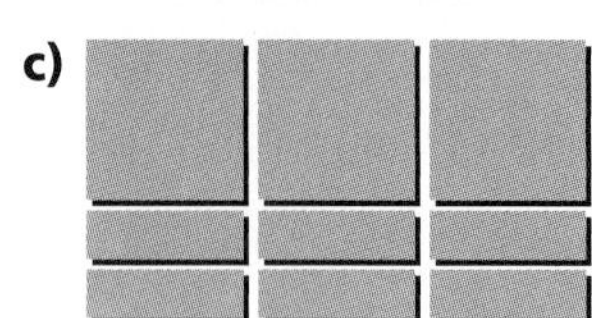

9. Anna's hockey team had a closing banquet for the league at a local hotel. The rental cost for the banquet room was \$125 plus \$11.25 per person for the meal. The total cost for the banquet was \$1812.50. Write an equation and use it to determine how many people attended the banquet.

10. Simplify and solve.

$$3(2x + 5 - 4x) + \frac{1}{3}(6x + 3) = 52$$

11. Add. Write the simplified expression. Use tiles to show your steps.

a)

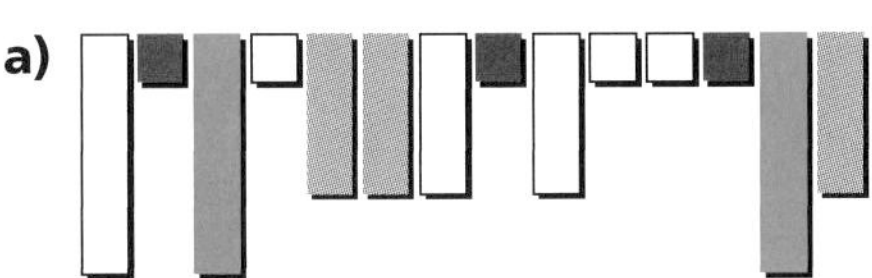

b)

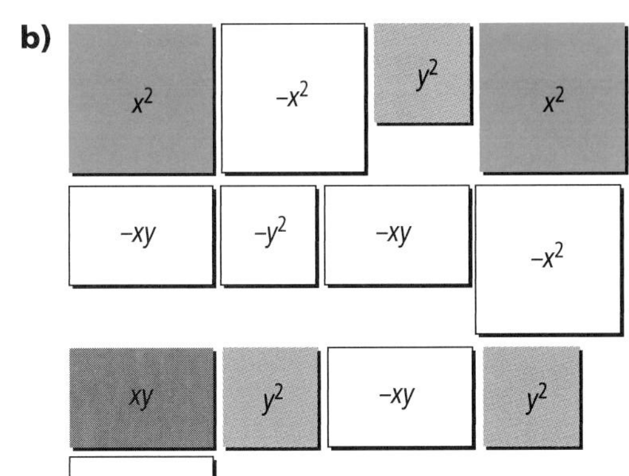

12. Write the subtraction expression if
- ii) is subtracted from i)
- i) is subtracted from ii)

Simplify the expressions. Show your steps.

a) i)

ii)

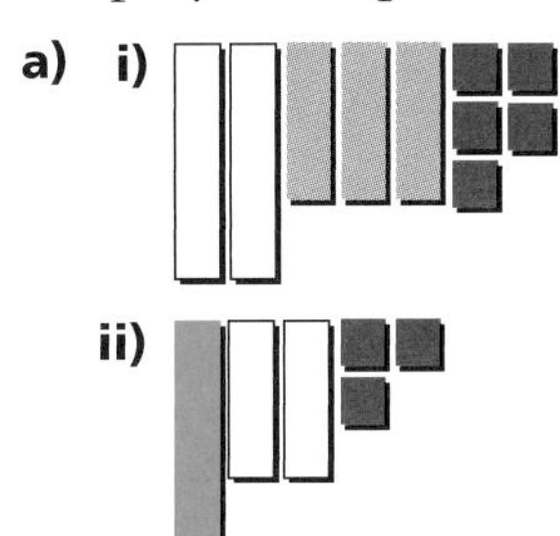

b) i)

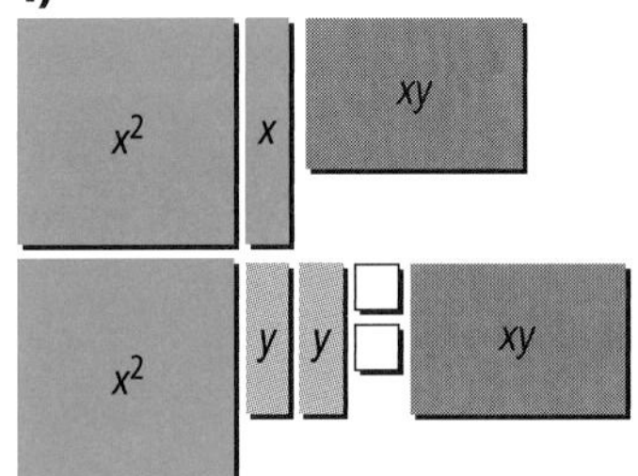

ii)

13. Solve. Verify your answer algebraically.

a) $5 - x = 11$

b) $7t + 3t = 30$

c) $-10 = \frac{x}{5}$

d) $\frac{x}{3} - 4 = 1$

e) $\frac{1}{4}x + 7 = 3$

f) $3x + 1.2 = 3.9$

g) $\frac{y}{27} = \frac{2}{3}$

h) $2(3 + 2w) + 1 = -5$

14. In a video hockey tournament, teams are awarded 3 points for a win, 2 points for an overtime win, and 1 point for an overtime loss.

a) Choose variables and write an expression for the total number of points a team might earn.

b) Use your expression to find the number of points earned by a team that has 4 wins, 3 overtime wins, and 2 overtime losses.

CHAPTER 7

Practice Test

Selected Response

Select the best answer.

1. There are 88 keys on a piano. Thirty-six keys are black. The rest are white. Which equation models this?

A $w + 52 = 88$ **B** $88 + 33 = w$
C $w + 36 = 88$ **D** $w - 88 = 36$

2. Which is a like term for $3t^2$?

A 3 **B** t **C** $3t$ **D** $-4t^2$

3. Which expression is equivalent to $-2(x - 3)$?

A $-2x + 6$ **B** $x + 6$
C $-3x - 2$ **D** $-2x - 6$

4. Which is the solution to $5.6 + 10b = 35.6$?

A $b = 20.6$ **B** $b = 3$
C $b = 4$ **D** $b = 15$

Short Response

Provide a complete solution.

5. Identify any like and unlike terms in each set.

a) $2p$, $3q$, -2, p, $3q^2$
b) $5x^2$, $5x$, x, $-5x^2$, -5, $3x^2$
c) $4y^4$, $5xy$, $6y^3$, $2xy$, y^4

6. A rectangular doorframe has dimensions expressed by $4c$ and $3c - 5$. Find a simplified expression for its perimeter.

7. Molly is solving the equation $2x - 4 = 10$.

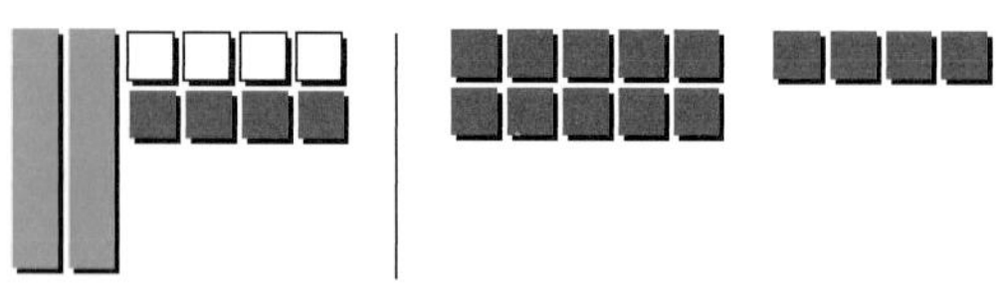

a) What is Molly modelling?
b) What will be her next step? Write an equation that represents this step.

8. Add. Write the simplified expression. Use tiles to show your steps.

a)

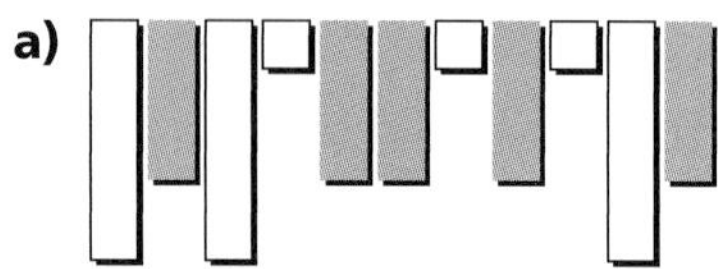

b)

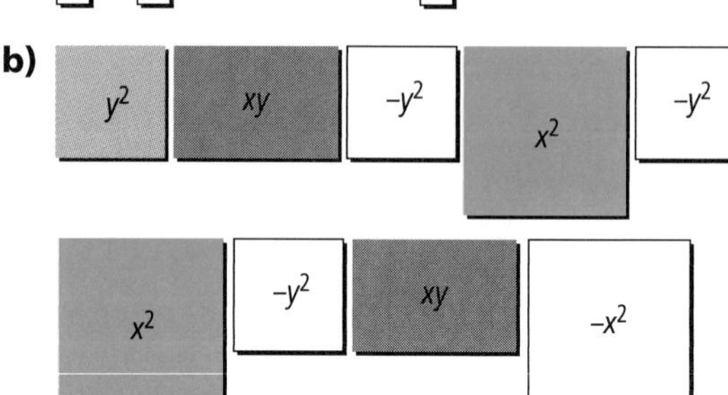

Extended Response

Provide a complete solution.

9. Write the subtraction expression if
- ii) is subtracted from i)
- i) is subtracted from ii)

Simplify the expressions. Show your steps.

a) i) **ii)**

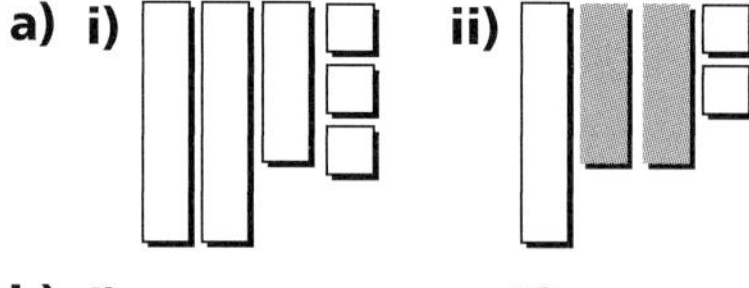

b) i) **ii)**

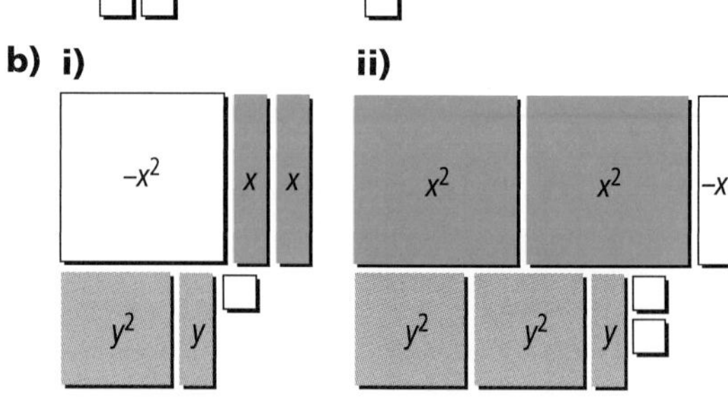

10. David bought a pair of jeans that were on sale at 25% off. He saved \$15. Write an equation for this situation. (Do not include the HST.) What was the regular price of the jeans?

11. When solving the equation $0.4n = 20$, Evan multiplies both sides by 10, but Jerod divides both sides by 0.4. Explain the reasoning behind each student's method.

12. PhotoCopiers R Us charges a flat fee of \$2 plus 3¢ per copy. A recent photocopying bill for a large company was \$101. Write an equation to represent this situation. How many photocopies were made for the company?

13. Solve.

a) $20 = \frac{x}{5}$

b) $26 + q = 27 + 6$

c) $12 = 11r - 10$

d) $2n + 1.3 = 7.8$

e) $\frac{x}{3} - 5 = -2$

f) $\frac{1}{4}x + 2 = 4$

g) $\frac{y}{48} = \frac{3}{4}$

h) $3(2 + 4w) + 2 = -16$

14. The school dance committee ordered six cases of pop and the principal ordered eight cases of pop. The total cost of the combined purchase was \$56. Write an equation to represent this situation. Use this equation to determine the price of each case of pop.

Chapter Problem Wrap-Up

1. Cal's grade 8 class is going on a four-day school trip to a different camp.
   a) It costs \$100 to rent the bus plus 25¢/km. What equation models the cost of renting the bus?
   b) This camp charges \$20 per person per day. How much money does Cal's class need to fund this trip? Use equations to model your solution and identify what the variables represent.

2. After kayaking, students hang their lifejackets on pegs to dry. The first peg has two lifejackets hanging on it. Every other peg has three lifejackets.

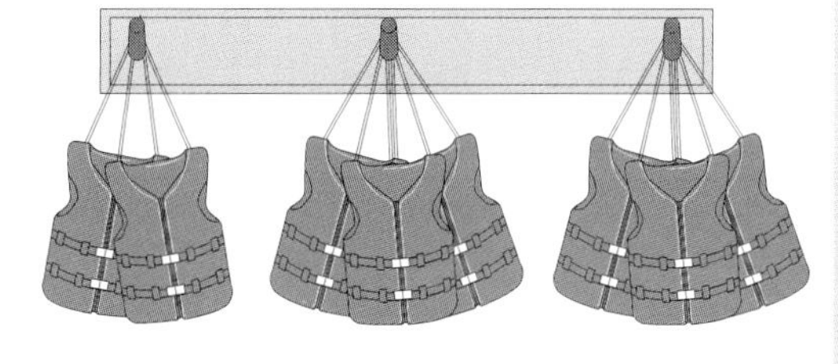

   a) Make a table showing the total number of lifejackets hanging from one to five pegs.
   b) Describe, in words, the relationship between the number of pegs and the total number of lifejackets.
   c) Write an equation for the total number of lifejackets, L, that are hanging on p pegs.
   d) Show how to use your equation to find the number of pegs needed to hang 35 lifejackets

- represent patterns and relationships in a variety of formats and use these representations to predict unknown values
- solve and verify simple linear equations algebraically
- construct and analyse tables and graphs to describe how change in one quantity affects a related quantity
- link visual characteristics of slope with its numerical value by comparing vertical change with horizontal change
- solve problems involving the intersection of two lines on a graph
- interpret graphs that represent linear and non-linear data

Key Words

linear relationship
non-linear relationship
rise
run
slope
intersection point
line of best fit
constant

CHAPTER 8

Patterns and Relations

There are patterns all around us. Look at what is happening in the photograph. What patterns can you see that involve:

- natural things?
- man-made things?
- human behaviour?

Are there other patterns you can see?

The understanding of patterns and how things are related is important both inside and outside the world of mathematics. Where else do you encounter patterns in everyday life?

Chapter Problem

Suppose you are the manager of a ski resort. Your main responsibility is to ensure that your customers have a safe and enjoyable time. How can an understanding of patterns and relations be useful in solving various day-to-day problems in running the resort?

Get Ready

Recognize and Describe Patterns

A **pattern** can appear in many forms.

Each number or diagram in the patterns shown is called a **term**.

i) 3, 6, 9, 12, ...

ii)  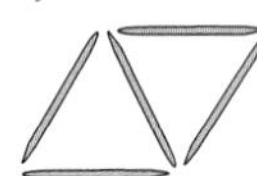 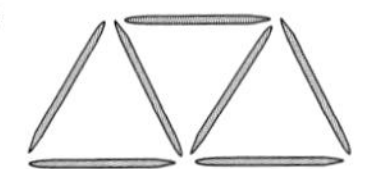

iii) 20, 15, 10, ...

For the number pattern: 20, 15, 10, ...
The first term is 20. The next term, 15, is 5 less than the term before.
The pattern continues for the third term, 10.
If this pattern continues, the next three terms will be 5, 0, and −5.

1. Refer to the patterns above.
a) Describe each pattern in words.
b) Write, draw, or model the next three terms in each pattern.

2. Write a pattern of numbers that:
a) increases by 6 each term
b) doubles each term

Tables, Graphs, Algebraic Expressions, and Equations

A pattern or **relation** can be represented in a table, a graph, words, and as an **algebraic expression** or **equation**.

Term Number, t	Value of the Term, v
1	5
2	9
3	13
4	17
⋮	⋮
t	v

+4 +4 +4

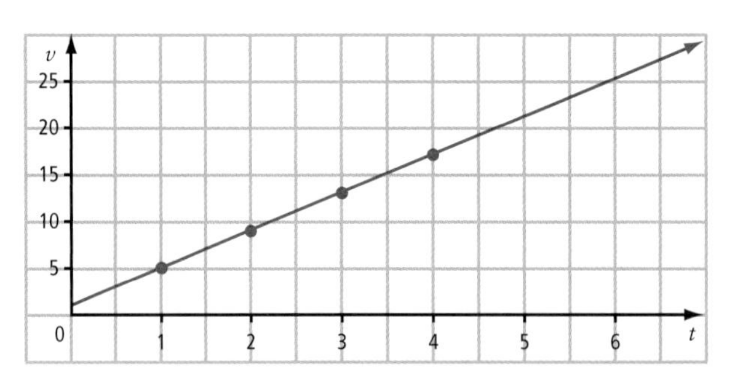

You can use words to describe the pattern 5, 9, 13, 17, ... in the table:

As the term number increases by 1, the value of the term increases by 4. To find the next term, add 4 to the value of the previous term.

For the tenth term, 4 will have been added many times. Instead of repeatedly adding 4, you could take a shortcut and multiply by 4.

A line was used to join the points to show a trend in the data.

Consider the first term.
$4 \times 1 + ■ = 5$

The unknown number must be 1. To check if this rule works, consider the second term.

Second term.
$4 \times 2 + 1 = 9$

You can check this rule for all the terms. Each value of the term is 4 times the term number increased by 1.

The algebraic expression for the value of the term is $4t + 1$.

The algebraic **equation** for the value, v of the term is $v = 4t + 1$.

You can use the equation to find the value of the tenth term.

$v = 4 \times 10 + 1$
$= 41$

3. a) Draw the next three terms of the pattern.

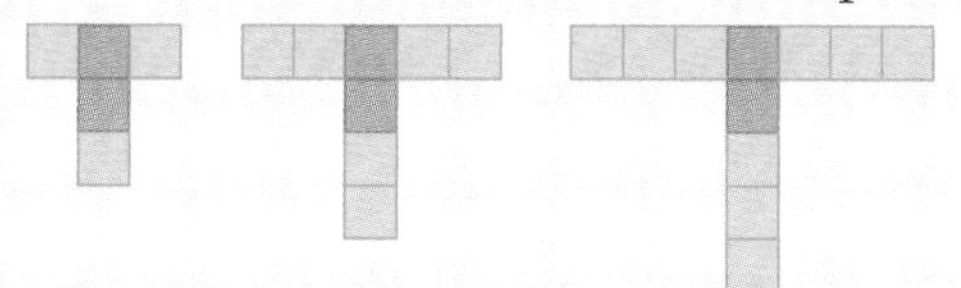

b) Copy and complete the table for the first six terms.

Term Number, t	Number of Squares, s
1	5
⋮	⋮
t	s

c) Graph the data in the table.

d) Describe the pattern in words. Copy and complete the statement.
If I know the term number, t, I can find the number of squares, s, by ■.

e) Write an algebraic equation that describes the number of squares, s in relation to the term number, t.

f) Explain how you could use the equation to find the number of squares in the eighth term. Calculate the number of squares in the eighth term.

g) Check your answer to part f) by extending the table and graph.

h) Use the equation to find the value of s when:
- $t = 10$
- $t = 15$

4. Determine the value of s when $t = 25$. Explain how you found your answer.

8.1 Explore Patterns and Relations

Focus on...
- representing patterns and relationships in a variety of formats
- using the representations to predict unknown values

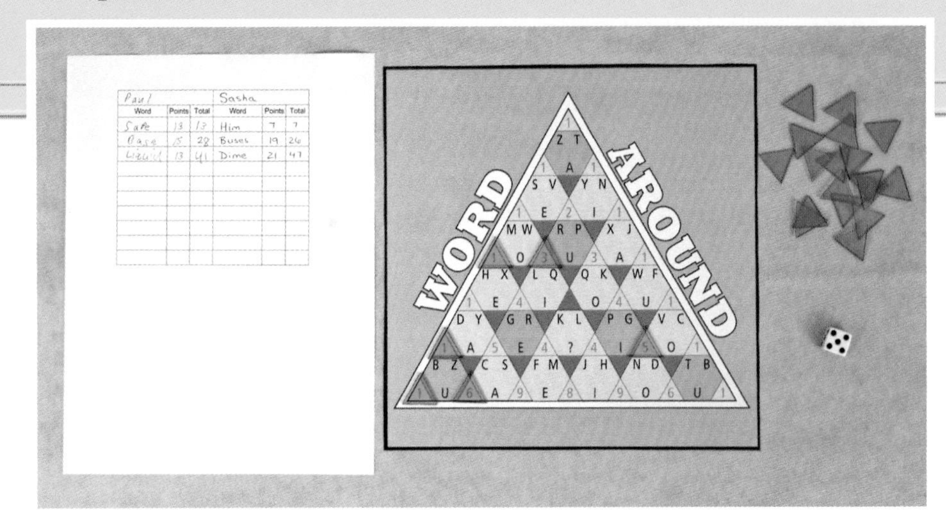

Look at the game board. Describe any patterns you can see that use:
- geometric shapes
- numbers
- words or letters

What other patterns can you see?

You can describe patterns in mathematics in various ways. In this section, you will explore how patterns can be represented using:
- concrete materials
- tables
- graphs
- words
- equations
- context

In addition, you will learn how to change one type of representation to another. You will discover how each representation can tell you something different about the pattern.

Materials
- toothpicks
- graph paper
- small manipulatives (colour tiles, linking cubes, pattern blocks)

Optional:
- BLM 8.1 DTM Pattern Table

Discover the Math

How can you identify a pattern and describe it in various ways?

Part A: Square Pattern

Examine the pattern of squares.

Term 1

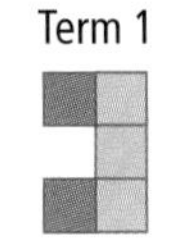

Term 2

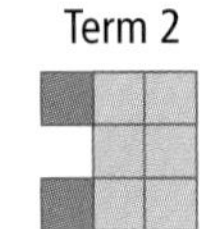

Term 3

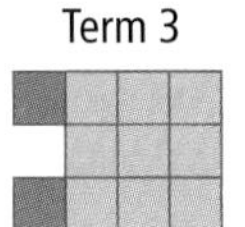

1. a) Copy and complete the table.

Term, n	Number of Red Squares	Number of Blue Squares	Total Number of Squares
1	2	3	5
2			8
3			
4			
5			
⋮	⋮	⋮	⋮
10			
⋮	⋮	⋮	⋮
20			
⋮	⋮	⋮	⋮
			92
⋮	⋮	⋮	⋮
			122
⋮	⋮	⋮	⋮
100			
⋮	⋮	⋮	⋮
n			

Strategies

Organize your work in a table. Find a pattern. Use algebra to describe the pattern.

Describe how each number in the table changes from one term to the next.

b) If you know the term number, describe in words how to find the number of red squares, blue squares, and total number of squares.

c) Write an equation that relates the total number of squares, s, to the term, n. $s =$ ▇

d) Use the equation to find the total number of squares in the 12th term.

e) Is it possible for a term in this pattern to have a total of 27 squares? Explain.

2. The data in the table can be written as ordered pairs. The first ordered pair that relates the term, n, to the total number of squares, s, is (1, 5). Write the next four ordered pairs relating n and s.

3. Graph the pattern by plotting the ordered pairs. Plot the term, n, on the horizontal axis and the total number of squares, s, on the vertical axis.

4. a) Describe how you can move from one point to the next, beginning at point (1, 5) and moving to the right. Use words like *left*, *right*, *up*, and *down* and tell the number of squares to move. What pattern develops from the words and numbers? Explain how the "right" and "up" numbers are used in the table, and how they are visible in the pictures.

b) Extend the graph to find the value of s when $n = 6$.

c) Use the graph to find the value of s when $n = 0$. How is this value related to the growing pattern?

d) Does it make sense to join these points to form a straight line? Explain why or why not. *Hint: Consider what a fractional value of either coordinate would mean.*

Part B: Pattern in a Graph

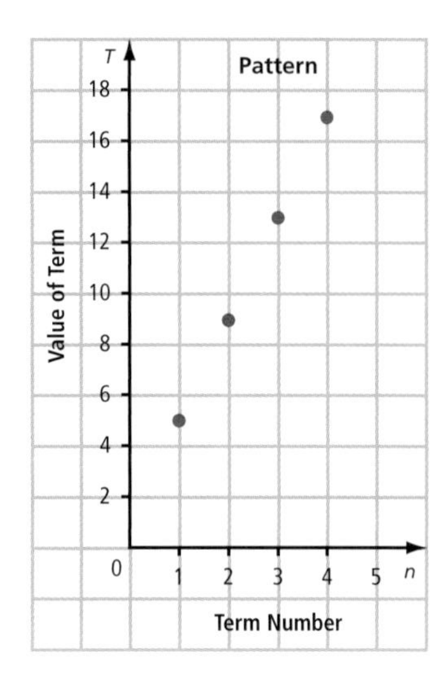

The graph shows the relationship between the value of the term, T and the term, n, in a pattern.

1. a) Write the coordinates of each point as a set of ordered pairs.

b) Describe how you can move from one point to the next, beginning at the left-most point.

c) Create a table of values. Find the differences in the values of the terms and compare this number with the "up" movement of the points on the graph.

2. a) Explain in words how you can determine the value of the term if you know the term number.

b) Write an equation relating T and n.

c) Use the equation to predict the value of the sixth term. Check using the graph.

3. a) Create a model of this pattern, using concrete materials of your choice.

b) Draw diagrams to represent your pattern.

c) Explain how your pattern relates to the graph.

d) Describe a problem or situation that this pattern could represent.

4. Reflect

a) List as many ways as you can think of to represent a pattern.

b) List two advantages and two disadvantages of each type of representation.

When the graph of the relationship between two quantities or variables is a straight line, the relationship is called **linear**.

linear relationship

- as one variable changes by a constant amount, the other variable changes by a constant amount
- a graph of the ordered pairs are in a straight line

Example 1: Graph and Analyse a Linear Relation

The cost of a taxi trip is a flat rate of $3.00 plus $2.50 per kilometre.

a) Find the cost (c) of a 1-km, 2-km, and 3-km trip (k). Write your answers as ordered pairs (k, c).

b) Plot the ordered pairs on a graph. Extend the graph to find the cost of a 5-km trip.

c) Describe the pattern in words. Write an equation that relates c and k. Check that your equation is true.

d) Find the cost of a 1.5-km trip, assuming the taxi will charge for partial kilometres.

Solution

a) The cost of a taxi trip depends on the number of kilometres travelled. Start with \$3.00, and add \$2.50 for each kilometre.

Strategies
Make an organized list, table, or chart.

Distance, k (km)	Cost, c (\$)	(k, c)
1	$3.00 + 2.50 \times 1 = 5.50$	(1, 5.5)
2	$3.00 + 2.50 \times 2 = 8.00$	(2, 8.0)
3	$3.00 + 2.50 \times 3 = 10.50$	(3, 10.5)

b) Graph the relation.

Taxi Ride
c
12
10
8
6
4
2
0
1 2 3 4 k
Cost (\$)
Distance (km)

Distance is the independent variable because you can choose the distance to travel. Distance is graphed along the horizontal axis. Cost is the dependent variable because the cost of the taxi ride depends on the distance travelled. Cost is graphed along the vertical axis.

Distance and cost are continuous data because you can have partial values for distance and cost. I can join the points with a line to show the trend in the data.

The increase in the cost of the taxi trips is constant (\$2.50) just as the increase in distance is constant (1 km). When the points are joined, they produce a straight line. This suggests that the relationship is linear.

Notice that as you move right from one point to the next you move
- right 1 km (the run)
- up \$2.50 (the rise)

Both the rise and the run are constant, which creates the straight line.

Strategies
A graph is a diagram of a relation. Does it help you see the pattern?

You can continue this pattern to find the cost of a 5-km trip.

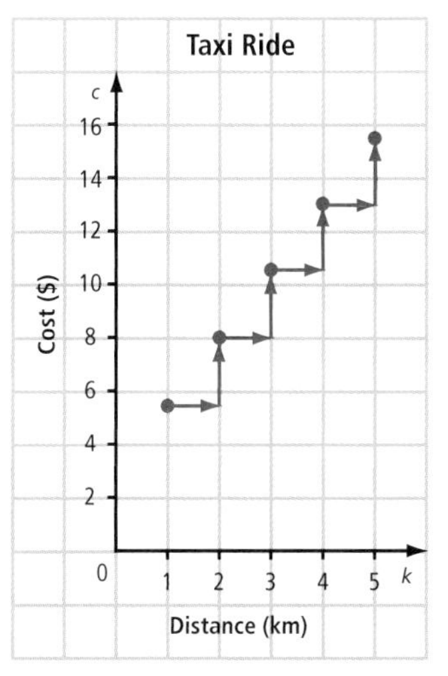

The graph shows that a 5-km trip would cost \$15.50.

c) The cost of a taxi trip is a flat rate of \$3.00 plus \$2.50 per kilometre. This can be written as an equation: $c = 3.0 + 2.5k$

This means that the cost of a trip can be found by multiplying the number of kilometres by \$2.50, and adding \$3.00.

I can check the equation by substituting ordered pairs to verify that the equation holds true. For example, (1, 5.5) means that for $k = 1$, $c = 5.5$.
Substitute.
$5.5 = 3.0 + 2.5(1)$
$5.5 = 3.0 + 2.5$
$5.5 = 5.5$
The left side equals the right side. The equation holds true for (1, 5.5). Try the other ordered pairs.

Communicating Mathematically

$2k$ means the same as $2 \times k$.

d) You can use the equation or the graph to find unknown values in a relation.

Find the cost of a 1.5-km trip.

Method 1: Use the equation.

Substitute $k = 1.5$ into the equation and solve for c.

$c = 3.0 + 2.5k$
$= 3.0 + 2.5(1.5)$ **Follow the correct order of operations when simplifying.**
$= 3.0 + 3.75$
$= 6.75$

The cost of 1.5-km taxi ride will be \$6.75.

Method 2: Use the graph.

You can find values that occur between the given points by using interpolation. Find the cost when the distance is 1.5 km.

The cost of 1.5-km taxi ride will be \$6.75.

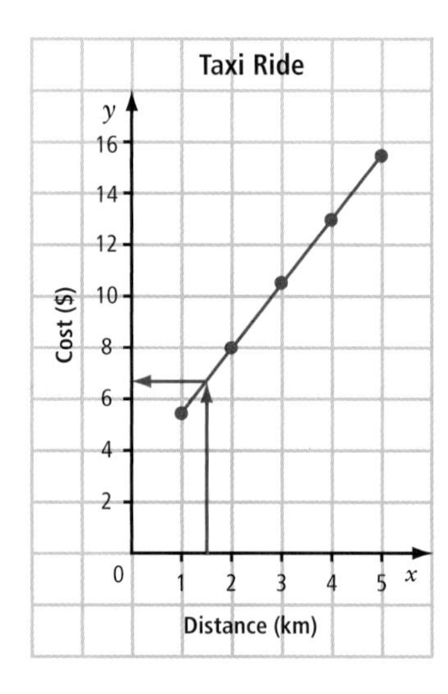

Example 2: Represent a Linear Relation in Various Ways

The graph of a pattern of squares is shown. Represent this relation using:

a) a table of values

b) words

c) an equation

d) a concrete model

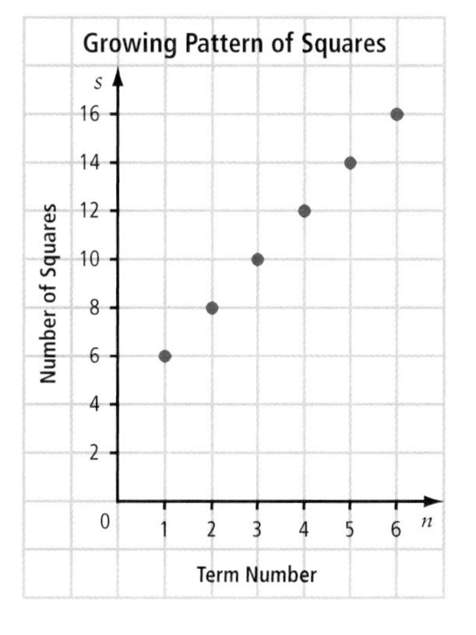

Solution

a) To represent this relation as a table of values, read the coordinates of several points on the line.

Write the ordered pairs as lists of values for the term, n, and the total number of squares, s.

Term, n	Total Number of Squares, s
1	6
2	8
3	10
4	12

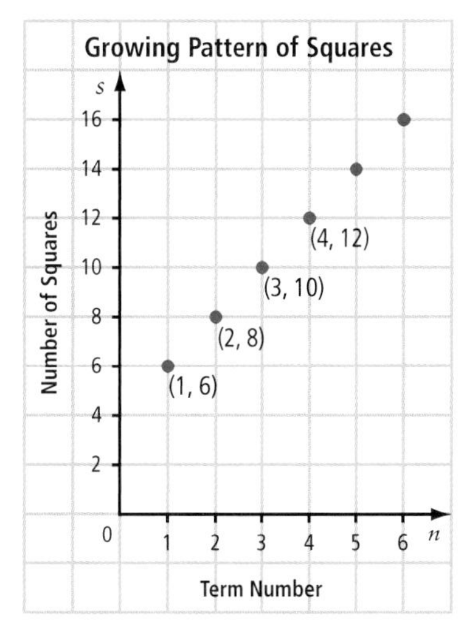

Strategies

Make an organized list, table, or chart.

b) To find the number of squares, double the term number and add 4.

c) Use the table to find the pattern.

Term, n	Total Number of Squares, s
1	$4 + 2 \times 1$
2	$4 + 2 \times 2$
3	$4 + 2 \times 3$
4	$4 + 2 \times 4$
n	$4 + 2 \times n$

The equation is $s = 4 + 2n$.

I can check this equation by substituting known values to verify that the equation holds true.

$n = 2$ Total number of squares = 2(2) + 4 = 8

(2, 8) is correct.

$n = 3$ Total number of squares = 2(3) + 4 = 10

(3, 10) is correct.

d) To develop a concrete model, consider the equation.

Total number of squares $= 4 + 2n$

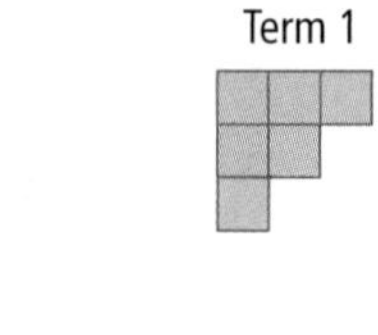

Build a pattern of squares that has 4 squares that never change plus 2 squares times the term number.

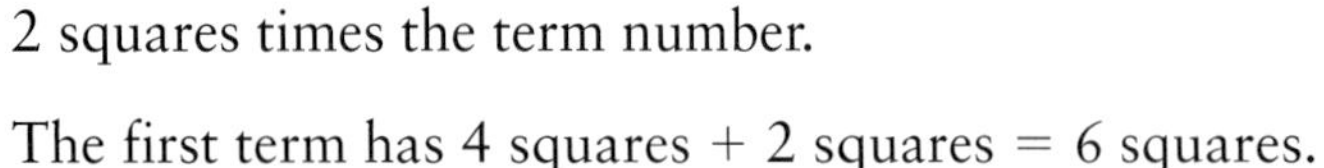

The first term has 4 squares + 2 squares = 6 squares.

Add two more blue squares to each following term.

Term 1: $4 + 1(2)$ $= 6$ squares

Term 2: $4 + 2(2)$ $= 8$ squares

Term 3: $4 + 3(2)$ $= 10$ squares

Communicate the Key Ideas

Use this information to answer questions 1 to 3.

The graph shows the cost of renting a snow blower. Assume that customers will be charged for partial hours.

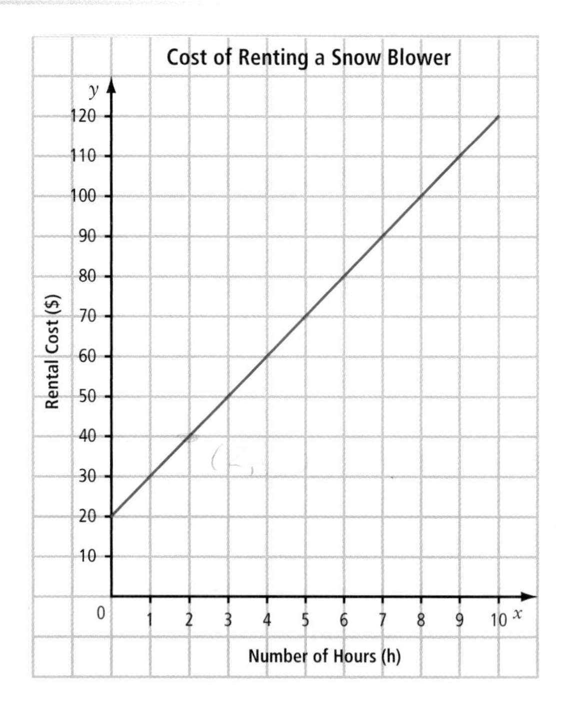

1. a) Is this relation linear? Explain why or why not.

b) Identify two points on the graph and explain what they mean.

c) Explain how you could produce a table of values for this relation.

d) Describe another method you could use to represent this relation.

2. a) Describe two methods you could use to find the cost of renting a snow blower for 8 h.

b) A variable represents a quantity that has changing values. What are the variables in this situation? Choose two different letters to represent the variables.

c) Write an equation that relates the cost to the number of rental hours.

d) Explain how you found the equation.

3. a) List all the different methods that you can use to represent a linear relation.

b) Describe two advantages and two disadvantages of each method.

4. Explain how you know if a relation is linear or not.

Check Your Understanding

Use this information to answer questions 1 to 3.

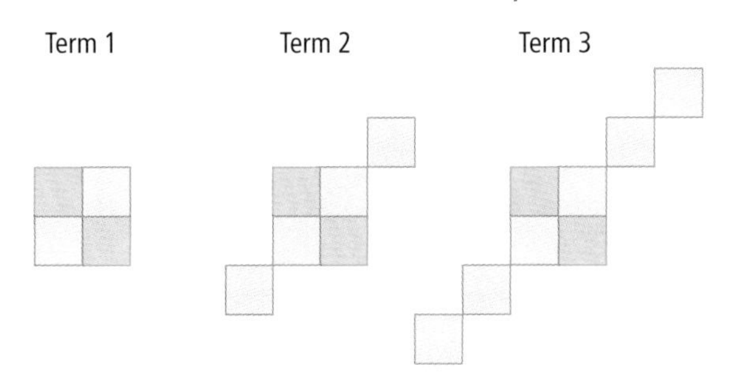

1. a) Represent this pattern in a table of values that shows the number of squares for each term. Continue the pattern up to the fifth term.
   b) By how many squares is the pattern increasing from one term to the next?

2. a) Use the data in the table of values to graph the relation. Plot term on the horizontal axis and number of squares on the vertical axis. Should you join the points with a line? Explain.
   b) Is this a linear relation? Explain.

3. a) Describe in words how you can determine the number of squares if you know the term.
   b) Write an equation that relates the number of squares, s, and the term, n.
   c) Use the equation to determine how many squares will there be in the:
      - 8th term?
      - 25th term?
   d) Use the equation to determine which term has 82 squares.

Use this information to answer questions 4 to 5.

A trapezoid-shaped meeting table can seat five people as shown.

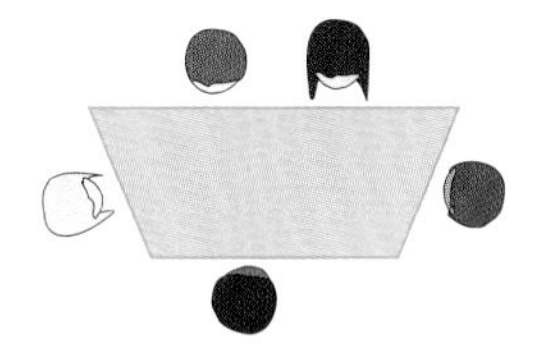

4. a) How many people can be seated at each arrangement of tables?

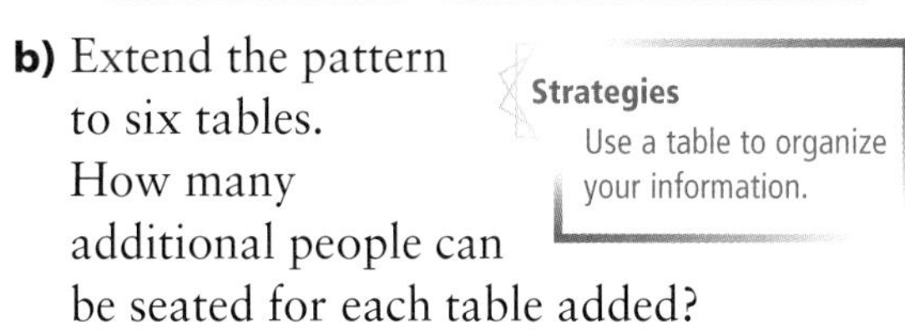

   b) Extend the pattern to six tables. How many additional people can be seated for each table added?

Strategies
Use a table to organize your information.

   c) Use your data to graph the relation. Plot number of tables on the horizontal axis and number of people on the vertical axis.
   d) Use the graph to find the number of people that can sit at nine tables.

5. a) Describe in words how you can determine the number of people if you know the number of tables.
   b) Write an equation that relates the number of people, p to the number of tables, t.
   c) Use the equation to check your answer to question 4, part d).
   d) Use the equation to find the number of people that can sit at 30 tables. Can you check your answer using the graph? Explain why or why not.

Use this table to answer questions 6 to 8.

Term, n	Number of Toothpicks, t
1	3
2	5
3	7
4	9

6. a) By how much is the number of toothpicks increasing from one term to the next?

b) Use the data in the table of values to graph the relation. Plot term on the horizontal axis and number of toothpicks on the vertical axis. Should you join the points with a line? Explain.

c) Describe in words how you can determine the number of toothpicks if you know the term.

d) Write an equation that describes this relation.

7. a) How many toothpicks will the sixth term have? Explain how you know.

b) How many toothpicks will the 75th term have? Explain how you know.

c) Is it possible for one of the terms in this pattern to have 50 toothpicks? Explain why or why not.

8. a) Build a model to represent this relation, using toothpicks. Draw your model.

b) Look at your model. Describe the pattern relating the perimeter and term number using:
- a table
- a graph
- an equation

c) Is this the same relation as the one in the table? Explain.

Use this information to answer questions 9 and 10.

The table shows a plumber's earnings, which are made up of a fixed rate and a variable rate.

Time, t (h)	Earnings, e ($)
1	60
2	85
3	110
4	135

9. a) By how much are the plumber's earnings increasing from one hour to the next?

b) Use the data in the table to graph the relation. Should you join the points with a line? Explain.

c) Describe in words how you can determine the earnings if you know the time worked.

d) Write an equation that describes this relation.

10. a) How much will the plumber earn in 6 h? Explain how you know.

b) How much will the plumber earn in 75 h? Explain how you know.

c) Is it possible for the plumber to earn $150? Explain why or why not.

Use this graph to answer questions 11 and 12.

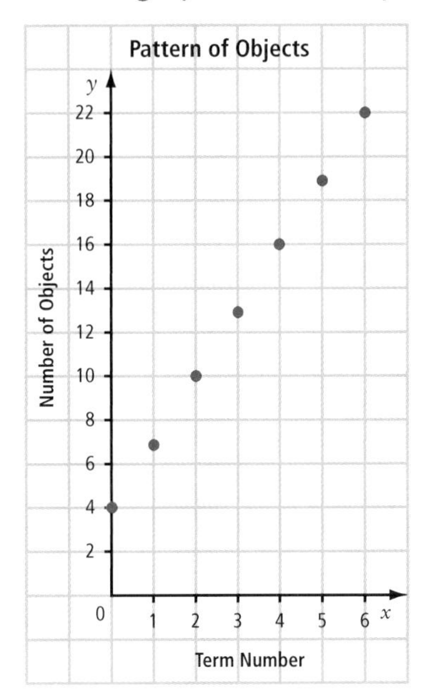

11. a) Is this relation linear? Explain.

b) Represent this relation using a table of values.

c) By how much is the number of objects increasing from one term to the next?

d) Write an equation that describes this relation.

12. a) Build a model to represent this relation, using materials of your choice. Draw your model.

b) Compare your model with those of some classmates. Is there more than one possible model for this relation? Explain.

c) Write a problem situation that this relation could represent.

13. a) Create a pattern that decreases by three objects each term. Start with the fifth term. Draw your pattern.

b) Construct a table of values for the first five terms of your pattern.

c) Graph the relation. Should the points be joined? Explain. Is the relation linear? Explain how you know.

d) Write an equation that relates the number of objects to the term.

e) Write a problem situation that this relation could represent.

Chapter Problem

14. You do not need Mother Nature to create snow. As long as it is cold enough, you can make your own!

The table shows the amount of snow created over time by a snowmaking machine.

Time, t (h)	Amount of Snow, V (m^3)
0	0
1	20
2	40
3	60
4	80

a) Graph the data. Should you join the points? Explain why or why not.

b) Is this a linear relation? Explain two ways how you know.

c) How much snow will be produced in 10 h?

d) How long will it take to produce 300 m^3 of snow? Discuss any assumptions you must make.

e) Describe how the graph would change if you doubled the rate of snow production. Justify your answer.

f) How can you use proportions to answer part c)? Show your work to confirm your answer.

g) Can you use proportions to answer parts d) and e)? Explain.

Extend

15. Each regular polygon has a side length of 1 cm.

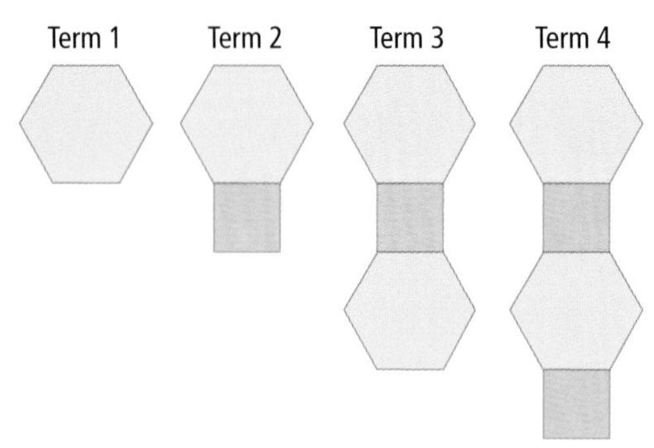

a) What is the perimeter for each term in the pattern? Copy and complete the table.

Term, n	Perimeter, P (cm)
1	6
2	8
3	
4	

b) Extend the pattern for three more terms. Add the values to your table.

c) Graph the data. Plot the term on the horizontal axis and the perimeter on the vertical axis. Is this a linear relationship? Explain.

d) What will be the perimeter of the
- 20th term?
- 50th term?

Explain how you found your answers.

e) Is it possible for a term to have a perimeter of
- 100 cm?
- 150 cm?

If it is possible, give the term number. If it is not possible, explain why not.

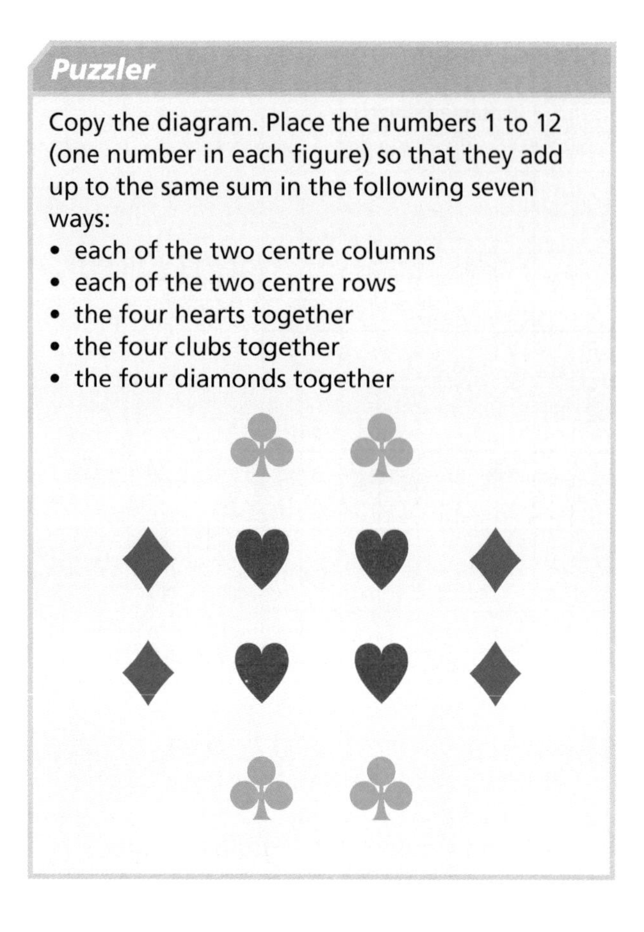

Puzzler
Copy the diagram. Place the numbers 1 to 12 (one number in each figure) so that they add up to the same sum in the following seven ways:
• each of the two centre columns
• each of the two centre rows
• the four hearts together
• the four clubs together
• the four diamonds together

8.2 Linear and Non-Linear Relations

Focus on...
- interpreting graphs that represent linear and non-linear data
- constructing and analysing tables and graphs

Mathematical relations are used to compare how two variables change with respect to each other. Look at the parachutist in the photograph. How do you think the distance she has fallen changes over time? Do you think her speed is changing as well? How will these relations change after she opens the parachute?

Some mathematical relations are simple and produce graphs of straight lines. Some are more complicated and produce graphs of different shapes, such as curves.

Discover the Math

Materials
- colour tiles or linking cubes
- graph paper
- ruler

How do the characteristics of linear and non-linear data compare?

1. Start with a single square that measures 1 cm by 1 cm.

a) What is its perimeter?

b) What is its area?

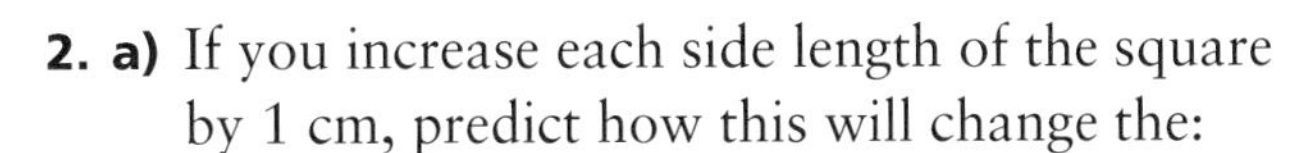

2. a) If you increase each side length of the square by 1 cm, predict how this will change the:
- perimeter
- area

Give reasons for your answers.

b) Build the square, as shown, and check your predictions.

3. Suppose you keep increasing each side length of the square by 1 cm.

a) Predict how this would change the perimeter each time. Would it change by the same amount? Explain why or why not.

b) Repeat part a) for the area.

4. a) Build squares having side lengths up to 10 cm and check your predictions. Copy and complete the table.

Square Side Length (cm)	Perimeter (cm)	Change in Perimeter (cm)	Area (cm^2)	Change in Area (cm^2)
1	4	—	1	—
2	8	$8 - 4 = 4$	4	$4 - 1 = 3$
3	12	$12 - 8 = 4$	9	$9 - 4 = 5$
4				
⋮				
10				

b) Examine how the perimeter changes each time. Compare this with your prediction in question 3, part a).

c) Examine how the area changes each time. Compare this with your prediction in question 3, part b).

5. a) What do you think a graph of side length and perimeter would look like? Explain your prediction.

b) Graph the data. Plot side length, l, on the horizontal axis and perimeter, P, on the vertical axis.

c) Compare the shape of the graph to the prediction you made in part a). Was your prediction correct? Explain.

6. Repeat question 5 for side length, l, and area, A.

7. Compare the two graphs you have constructed. How are they

a) similar?

b) different?

8. Reflect Look at the two graphs. Which of these relationships is linear? Which is non-linear? Explain in your own words.

Example 1: Graph and Analyse a Non-Linear Relationship

non-linear relationship

- as one variable changes by a constant amount, the other variable does not change by a constant amount
- a graph of the ordered pairs does not form a straight line.

A parachutist jumps from an airplane. The table shows the distance she has fallen over time.

Time (s)	Distance Fallen (m)
0	0
0.5	1.25
1	5
2	20
3	45
4	80
5	125

a) How is the distance fallen changing as time changes? Is the relationship linear or non-linear?

b) Graph the relationship. Plot time on the horizontal axis and distance fallen on the vertical axis. Should the points be joined? Explain.

c) How far did the parachutist fall after 3.5 s?

Solution

a) Add a column to the table. To find the change in distance fallen, subtract each distance from the previous whole number value.

Time (s)	Distance Fallen (m)	Change in Distance Fallen (m)
0	0	—
0.5	1.25	—
1	5	$5 - 0 = 5$
2	20	$20 - 5 = 15$
3	45	$45 - 20 = 25$
4	80	$80 - 45 = 35$
5	125	$125 - 80 = 45$

In a non-linear relation, as one variable changes by a constant amount, the other variable does not change by a constant amount. In this example, after every additional second, the change in the distance fallen increases, so the relationship is not linear.

Falling objects pick up speed the longer they fall. So as time passes at a constant rate, the distance fallen each second keeps increasing. Therefore, this relationship is non-linear.

b) Graph the relationship between time and distance fallen by plotting the ordered pairs in the table. The points are joined because distance and time are continuous data.

The path of the points is a smooth curve. This confirms the relationship is non-linear.

Since time and distance can be expressed as fractions, it makes sense to connect these points with a smooth curve.

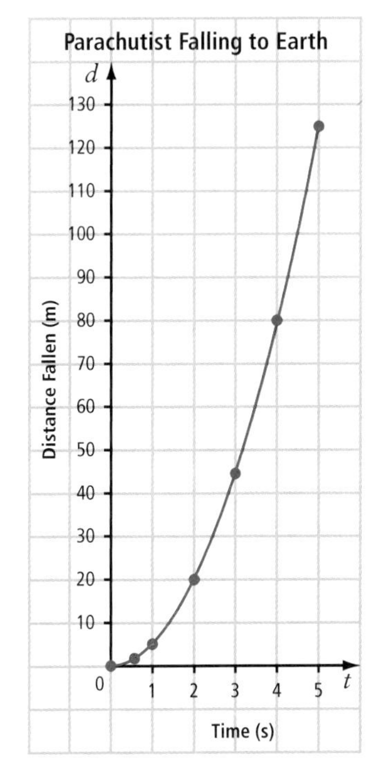

Strategies

Choose your scales for the graph carefully, so that all the points will fit.

c) Use interpolation to find how far the parachutist fell after 3.5 s.
After 3.5 s, the parachutist fell approximately 60 m.

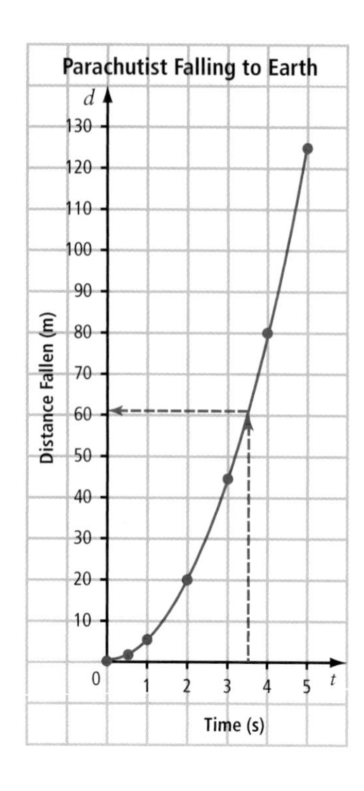

Communicate the Key Ideas

1. Explain how you can determine if a relation is linear or non-linear from its

a) written description
b) model
c) graph
d) table of values

2. Give an example of a situation that could be represented by a
a) linear relation
b) non-linear relation
Choose situations not already discussed in this section. For each situation, explain why you would expect its relation to be either linear or non-linear.

3. a) What does the graph tell you about how temperature is changing over time? Explain.
b) Describe a situation that this graph could represent.

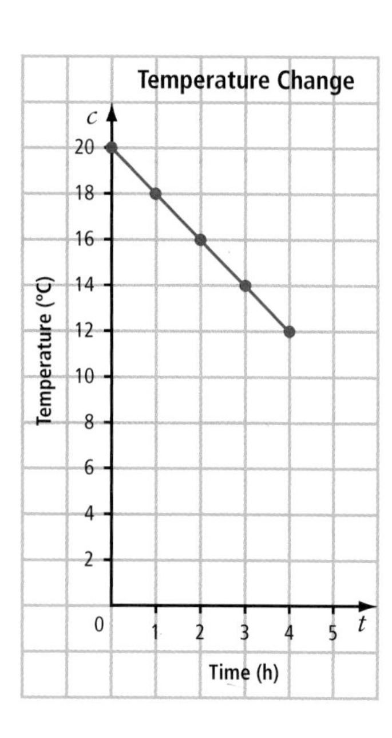

Check Your Understanding

1. a) Classify each relation as linear or non-linear. Justify your choice.

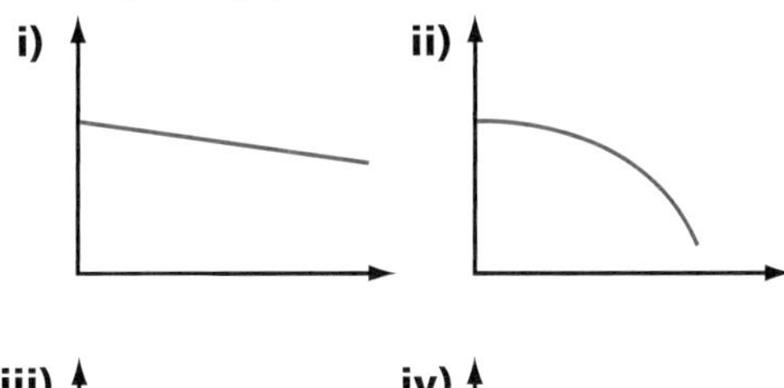

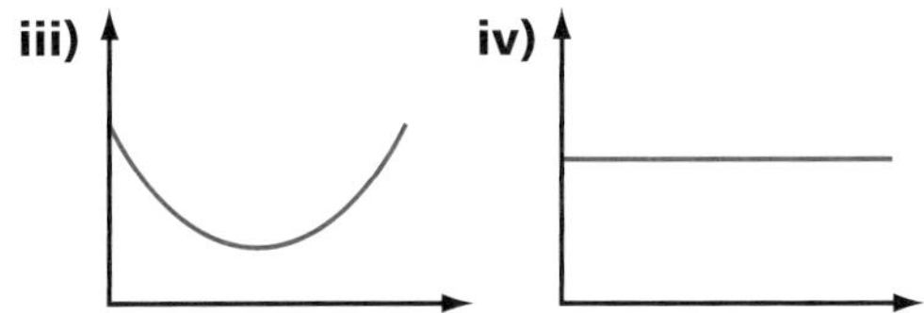

   b) Suggest a situation that each graph could represent. Identify the variables in each case.

2. Classify each relation as linear or non-linear. Justify your choice.

a)

x	y
0	3
1	8
2	13
3	18

b)

n	S
1	2
2	4
3	8
4	16

3. Refer to question 2.
   a) Graph each relation.
   b) Do the shapes of the graphs confirm your answers to question 2? Explain.

4. The cost for renting a banquet hall is a flat rate of $50 plus $10 per person attending the banquet.
   a) Write an equation that relates the cost, c, to the number of people, n.
   b) Find the cost of renting the banquet hall for $n = 0$, $n = 10$, $n = 20$, and $n = 30$. Organize your answers in a table.
   c) Graph the data. Should you join the points? Explain. Is the relationship linear or non-linear? Explain.
   d) Is there another way of determining whether the relationship is linear or non-linear without graphing the data? Explain.
   e) Describe how the change in number of people affects the cost.

5. A car is cruising at a steady speed of 50 km/h when it reaches a highway on-ramp. As the car enters the highway, it gradually picks up speed as shown in the table.

Time (s)	Speed (km/h)
0	50
5	60
10	70
15	80

   a) Describe how the car's speed is changing over time.
   b) Is this relationship linear or non-linear? How can you tell from looking at the table?
   c) Graph the data. Should you join the points? Explain. Does the shape of your graph confirm your answer to part b)? Explain.
   d) Suppose the car is coming onto a major Nova Scotia highway. Do you think the trend in the graph will continue? Extend the graph and show approximately when and how its shape will change.

Communicating Mathematically

Speed limits are posted on large signs located on the sides of roads and highways.

Most major highways in Nova Scotia have speed limits of 100 km/h or 110 km/h. Driving within the speed limit is an important part of driving safely!

6. The table shows the area of lawn infested by a weed over time.

Time (weeks)	Infested Area (m^2)
0	3
1	5
2	9
3	15

a) Graph this relation. Plot time on the horizontal axis and infested area on the vertical axis. Is the graph linear or non-linear?

b) Estimate the area infested by the weed after:
- 4 weeks
- 2 months

Hint: look for a pattern in how the infested area is changing over time.

c) Explain how you found your estimates. Discuss any assumptions you made.

7. The table shows the falling speed of a parachutist after jumping out of an airplane.

Time (s)	Falling Speed (m/s)
1	5
2	15
3	25
4	35
5	45
6	10
7	10
8	10

a) Graph only the first five ordered pairs. Is this relation linear or non-linear? Explain.

b) Estimate the falling speed of the parachutist after 2.5 s.

c) About how long was the parachutist falling when she reached a speed of 40 m/s?

d) Add the next three ordered pairs to your graph. Do these points follow the same pattern as the first five points? Explain.

8. Two friends, Nicole and Tanya, are running a race. One of the runners has a head start. The table shows each runner's distance from the finish line over time.

Time (min)	Nicole's Distance from Finish Line (m)	Tanya's Distance from Finish Line (m)
0	800	1000
1	700	920
2	600	830
3	500	730
4	400	620

a) Describe how each runner's distance from the finish line is changing over time.

b) Graph both sets of data. Are the relations linear or non-linear? Justify your answers.

c) Which runner had a head start? How much of a head start? How do you know?

d) Assume the trends in the table continue for both runners. Who will win the race? By how many metres?

e) How would your answer to part d) change if the head start were reduced by 100 m? Explain your thinking.

9. How does a child's height change over time?

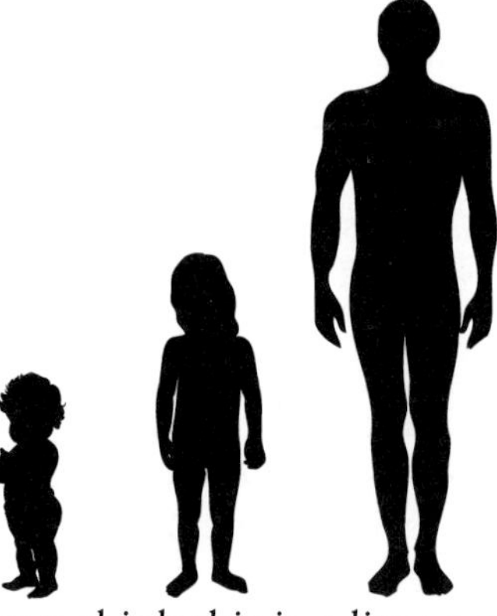

Do you think this is a linear or non-linear relation? Does the relationship change over time or remain the same? Use words and diagrams to explain your thinking.

Chapter Problem

10. One of the most exciting events at your ski resort is the Alpine Ski Jumping competition.

The table shows distance travelled by a ski jumper accelerating down a ramp.

Time (s)	Distance (m)
0	0
1	4
2	16
3	36
4	64

a) Is this relation linear or non-linear? Explain how you know.

b) Graph the relation. Does the shape of your graph confirm your answer to part a)? Explain.

c) Estimate how far the ski jumper travelled down the ramp after 2.5 s.

d) Estimate the speed of the skier over time by subtracting consecutive distance values. Copy and complete the table.

Time (s)	Distance (m)	Speed (m/s)
0	0	—
1	4	$4 - 0 = 4$
2	16	
3	36	
4	64	

e) Look at the relationship between speed and time. Is it linear or non-linear? Explain how you know.

f) Graph the relation. Compare this graph with the time and distance graph. How are the graphs similar? different?

Making Connections

The speed of the skier is the change in distance over time. You can estimate the average speed during each second by finding the difference in distance between the beginning and end of that second.

Extend

11. Refer to the data for Nicole in question 8. If a relation has at least one constant variable, you can find the first and second differences of the relation. The first differences are the differences between consecutive values of the second variable. The second differences are the differences between consecutive first difference values.

a) Copy and complete the table.

Time (min)	Nicole's Distance from Finish Line (m)	First Differences	Second Differences
0	800		
1	700		
2	600		
3	500		
4	400		

b) What do you notice about the first differences of this relation? What type of relation is this?

c) Calculate the first and second differences of Tanya's data. Describe what you notice.

12. Predict which relations are linear. Check by making a table of values and graphing each equation.

a) $t = 2n$ **b)** $t = 2^n$

c) $t = 2n - 2$ **d)** $t = n^2$

e) $t = \frac{2}{n}$ **f)** $t = n + 2$

8.3 Slope

Focus on...
- link visual characteristics of slope with its numerical value by comparing vertical change with horizontal change

Ski Cape Smokey, at Ingonish Bay, NS advertises a vertical lift of 305 m. On their trail map, symbols are used to rate the difficulty of each run. In general, the steeper the hill, the more difficult it is to ski. It is very important not to ski or snowboard on runs beyond your level of ability.

Symbol	Level of Difficulty	Description
●	easiest	for beginner skiers
■	difficult	for intermediate skiers
◆	very difficult	for advanced skiers
◆◆	most difficult	for expert skiers dangerous

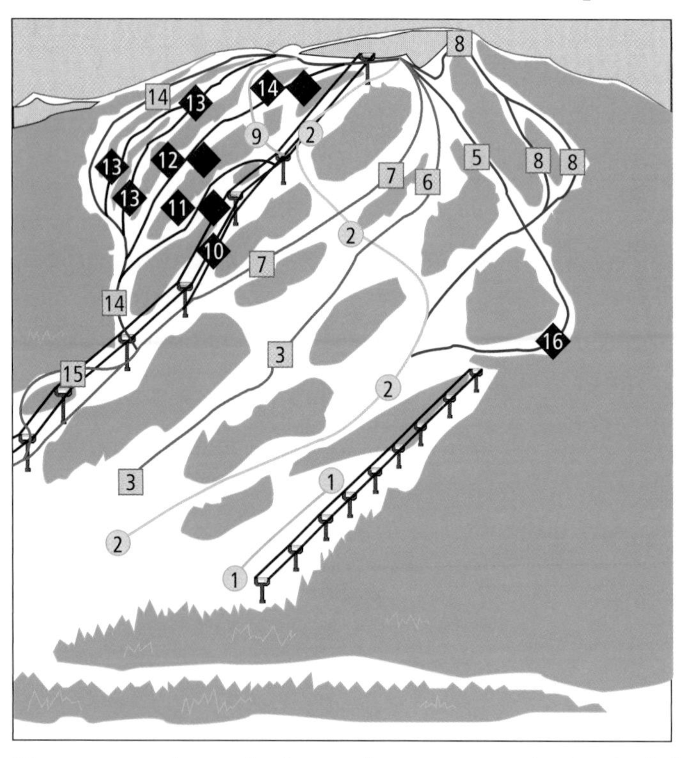

How can mathematics be used to measure the level of difficulty of a ski hill?

Discover the Math

Materials
- linking cubes
- graph paper
- ruler

How can you measure slope?

Part A: Measure Steepness and Slope

1. a) How are the staircases similar? different?

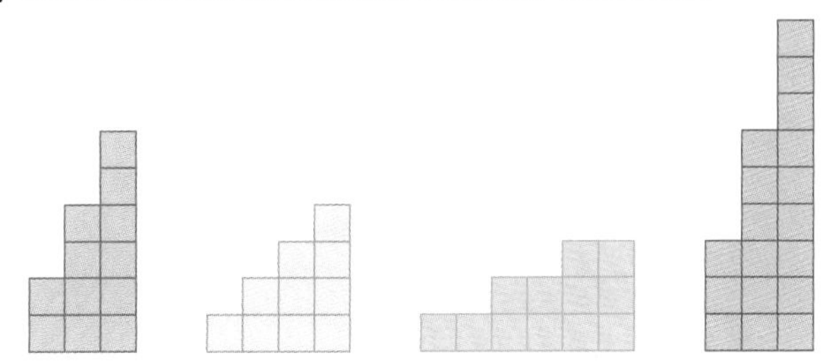

b) Arrange the staircases in order of steepness from most steep to least steep.

2. Construct or draw a staircase that is:

a) steeper than the steepest staircase
b) less steep than the least steep staircase

3. a) Determine a method to measure the steepness of the staircases. Describe your method.
b) Compare methods with some of your classmates. Is there more than one way to measure steepness? Explain.

4. Reflect Identify at least two different situations where it might be important to measure the steepness of an object or a surface.

One way to measure the steepness of a staircase is to compare its **rise** to its **run** using a ratio.

For example, the ratio of rise to run is called the **slope**.

$$\text{Slope} = \frac{\text{rise}}{\text{run}}$$

The slope of the blue staircase is:

$$\begin{aligned}\text{Slope} &= \frac{\text{rise}}{\text{run}}\\ &= \frac{6}{3} \quad \textbf{Simplify the fraction.}\\ &= \frac{2}{1}\\ &= 2\end{aligned}$$

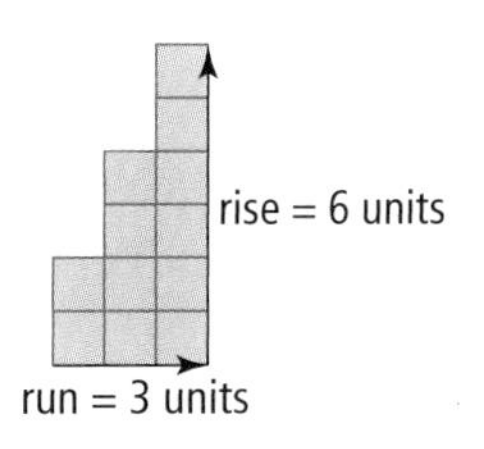

rise
- the vertical distance between two points on a line or line segment

run
- the horizontal distance between two points on a line or line segment

slope
- a measure of the steepness of a line
- Slope = $\frac{\text{rise}}{\text{run}}$

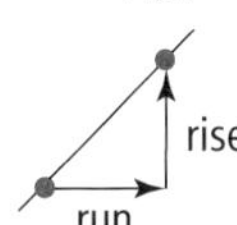

- written as a fraction or a decimal
- represents a **constant** rate of change

5. Determine the slope of the yellow staircase.

a) What is the rise?

b) What is the run?

c) Calculate the slope.

6. Determine the slope of the

a) red staircase

b) green staircase

7. Reflect

a) Arrange the slopes of all four staircases in order from least to greatest.

b) How is the steepness of a staircase related to its slope?

The slope of any object can be measured by comparing its rise to its run.

Example 1: Determining Slope

Determine the slope of each object.

a) ski hill

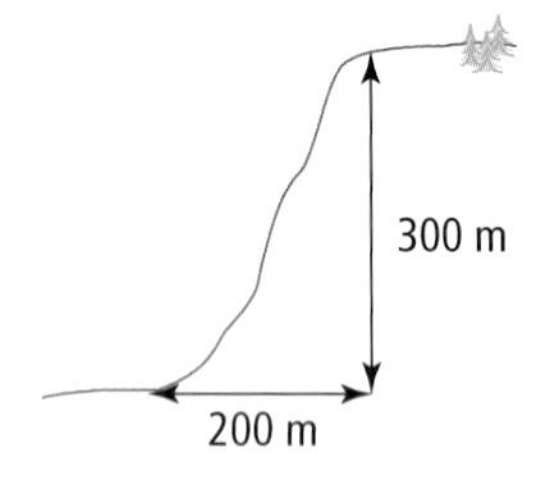

b) ramp

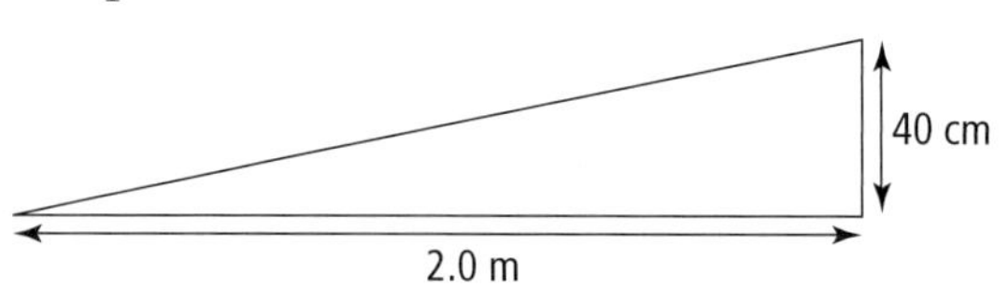

Solution

a) The slope of the ski hill is the ratio of its rise to its run.

$$\text{Slope} = \frac{\text{rise}}{\text{run}}$$

$$= \frac{300 \cancel{m}}{200 \cancel{m}}$$ **Simplify. Divide the numerator and denominator by 100.**

$$= \frac{3}{2} \text{ or } 1.5$$

To convert a fraction to a decimal, divide the numerator by the denominator: $3 \div 2 = 1.5$

The slope of the ski hill is $\frac{3}{2}$ or 1.5.

Notice that the units of measure (metres) are not applied after simplifying. Slope is a numerical value that has no units of measure.

Strategies
Apply previously learned skills with fractions.

b) When both variables represent the same kind of measure (e.g. length), then the slope is a ratio that compares measures having common units. When the measures are not in the same units, you must first convert one of the measures.

Method 1: Express rise and run in centimetres.
To convert the run to centimetres, multiply by 100.

$$\begin{aligned}\text{Run} &= 2.0 \times 100\\ &= 200\end{aligned}$$

The run is 200 cm.

Now find the slope.

$$\begin{aligned}\text{Slope} &= \frac{\text{rise}}{\text{run}}\\ &= \frac{40\ \cancel{\text{cm}}}{200\ \cancel{\text{cm}}}\\ &= \frac{1}{5} \text{ or } 0.2\end{aligned}$$

Simplify by dividing the numerator and denominator by 40.

Slope can be written either as a fraction in simplest form or as a decimal.

The slope of the ramp is $\frac{1}{5}$ or 0.2.

Method 2: Express rise and run in metres.
To convert the rise to metres, divide by 100.

$$\begin{aligned}\text{Rise} &= 40 \div 100\\ &= 0.4\end{aligned}$$

The rise is 0.4 m.

Now determine the slope.

$$\begin{aligned}\text{Slope} &= \frac{\text{rise}}{\text{run}}\\ &= \frac{0.4\ \cancel{\text{m}}}{2\ \cancel{\text{m}}}\\ &= 0.2\end{aligned}$$

Convert to a decimal by dividing the numerator by the denominator.

The slope of the ramp is $\frac{1}{5}$ or 0.2.

Did You Know?

A ramp with this slope could be used for loading furniture into a truck.

To measure the slope of a line or line segment, use two points on the line and draw a right triangle. Then use this to measure the rise and run.

Label the **rise** and **run**.

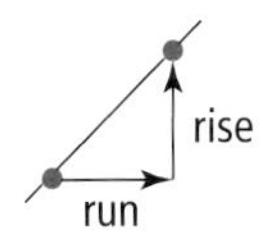

Example 2: Slope of a Line

Determine the slope of the line.

x	y
0	3.0
1	3.5
2	4.0
3	4.5
4	5.0
5	5.5
6	6.0
7	6.5
8	7.0
9	7.5
10	8.0

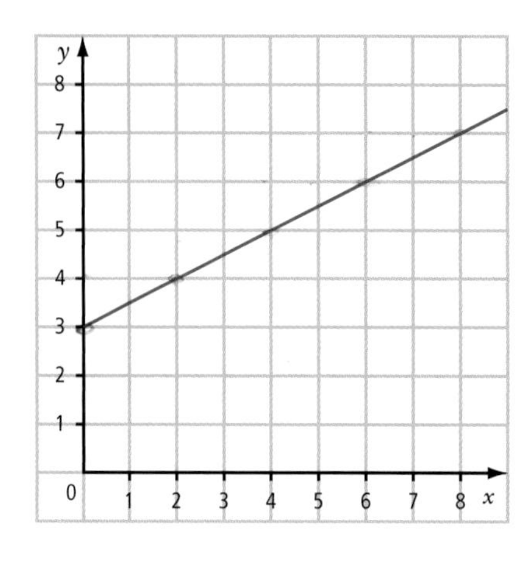

Solution

To determine the slope, identify two points on the line. Draw a right triangle and measure the rise and run.

$$\text{Slope} = \frac{\text{rise}}{\text{run}} = \frac{2}{4} = \frac{1}{2} \text{ or } 0.5$$

The slope of this line is $\frac{1}{2}$ or 0.5.

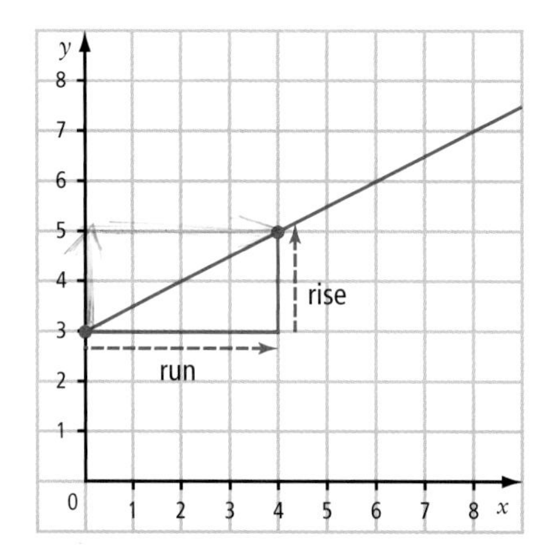

You could choose any two points on the line and the slope would be the same.

$$\text{Slope} = \frac{\text{rise}}{\text{run}} = \frac{3}{6} = \frac{1}{2} \text{ or } 0.5$$

The slope of this line is $\frac{1}{2}$ or 0.5.

The slope can be confirmed by looking at the table of values.

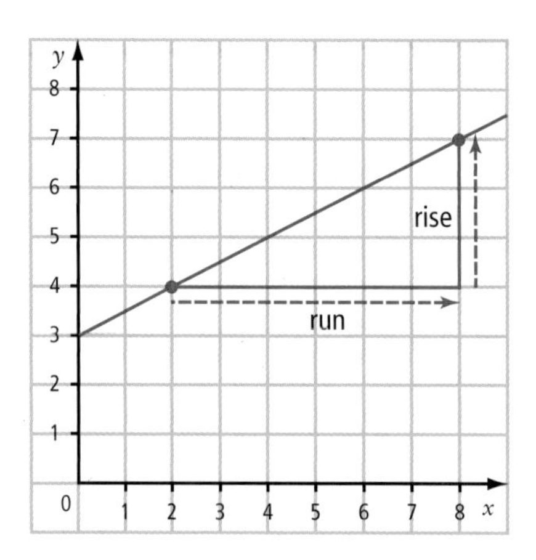

x	y
0	3.0
1	3.5
2	4.0
3	4.5
4	5.0
5	5.5
6	6.0
7	6.5
8	7.0
9	7.5
10	8.0

+ 0.5
+ 0.5
+ 0.5
+ 0.5
+ 0.5
+ 0.5
+ 0.5
+ 0.5
+ 0.5
+ 0.5

The difference between consecutive y-values is 0.5, which means that as the x-values change by 1, the y-values change by 0.5. The change in x-values is the run and the change in y-values is the rise and these can be used to determine the slope.

The difference between consecutive y-values in the table is constant, and is the same value as the slope.

Discover the Math

Part B: Direction of Slope

1. How are the graphs similar? different?

a)

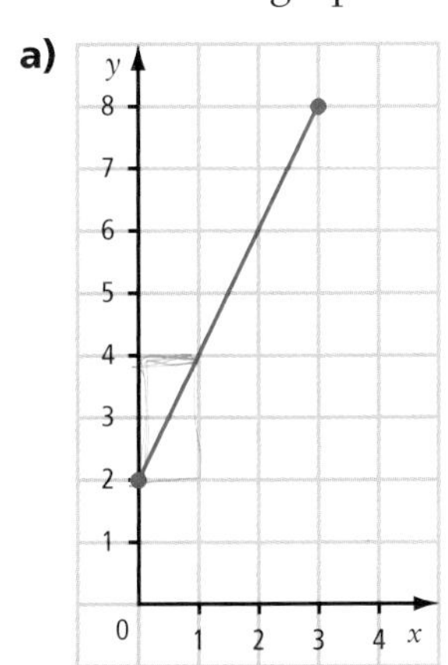

b)

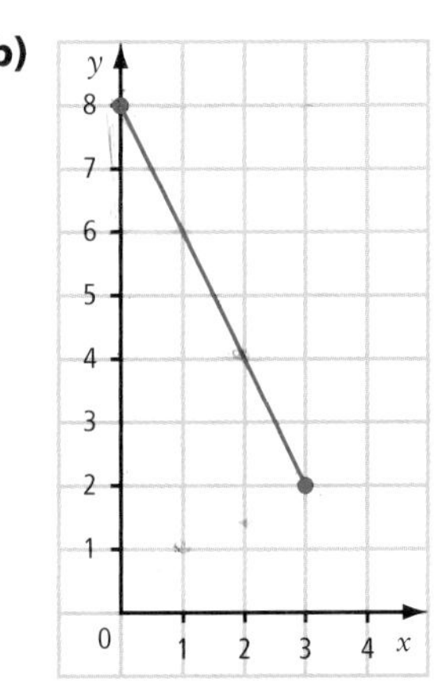

2. Describe in your own words how the slopes differ. Give reasons.

Example 3: Different Directions

Refer to the graphs in Discover the Math, Part B.
Determine the slope of each graph.

Solution

$\text{Slope} = \frac{\text{rise}}{\text{run}}$

The slope of the graph a) is:

$$\begin{aligned}\text{Slope} &= \frac{\text{rise}}{\text{run}} \\ &= \frac{6}{3} \\ &= \frac{2}{1} \\ &= 2\end{aligned}$$

Simplify the fraction.

The slope of the graph b) is:

$$\begin{aligned}\text{Slope} &= \frac{\text{rise}}{\text{run}} \\ &= \frac{-6}{3} \\ &= \frac{-2}{1} \\ &= -2\end{aligned}$$

Each graph has the same steepness of 2 but the direction of the line is different.
Graph a) rises to the right and is positive.
Graph b) falls to the right and is negative.

The slope of a line on a graph is a measure of how one variable changes with respect to another.

- If one variable *increases* as the other variable *increases*, the slope is *positive*.
- If one variable *decreases* as the other variable *increases*, the slope is *negative*.

Positive Slope

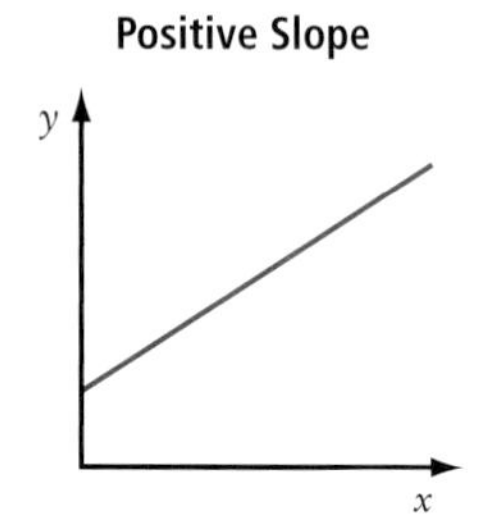

As x increases, y increases.

A line with a *positive* slope *rises* to the right.

Negative Slope

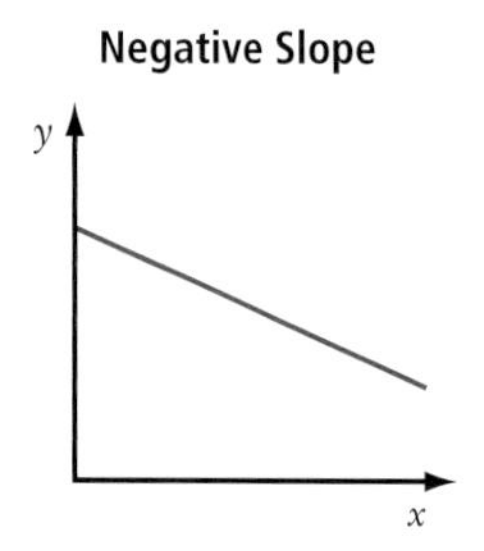

As x increases, y decreases.

A line with a *negative* slope *falls* to the right.

Strategies
Can you explain this difference between positive slope and negative slope in another way?

Example 4: Slope of a Linear Relation

Find the slope of each line and explain what it means.

a)

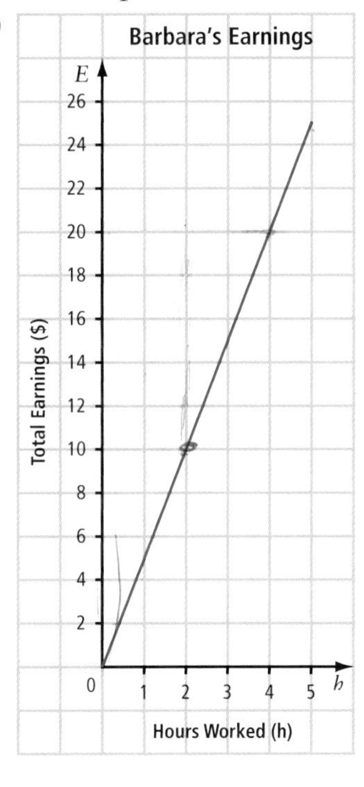

b)

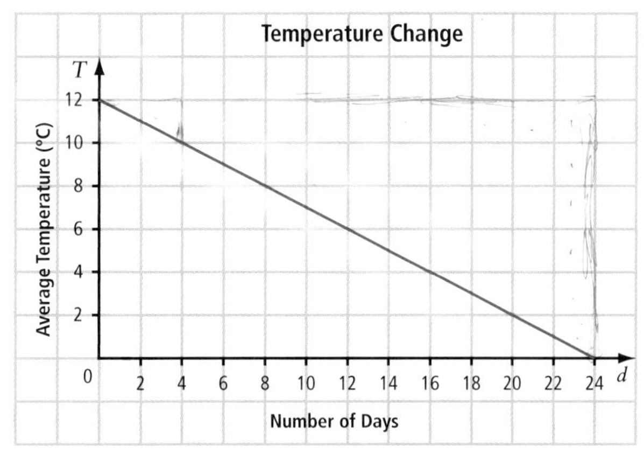

Solution

a) Notice that the graph *rises* to the right. As the number of hours increases, earnings also increase, which means the slope is *positive*.

To find the value of the slope, use two points on the line, and measure the rise and run.

$$\text{Slope} = \frac{\text{rise}}{\text{run}}$$
$$= \frac{5}{1}$$
$$= 5$$

The slope of the line is 5. This means that for every additional hour worked, Barbara's earnings increase by \$5. The rate of earnings is \$5/h.

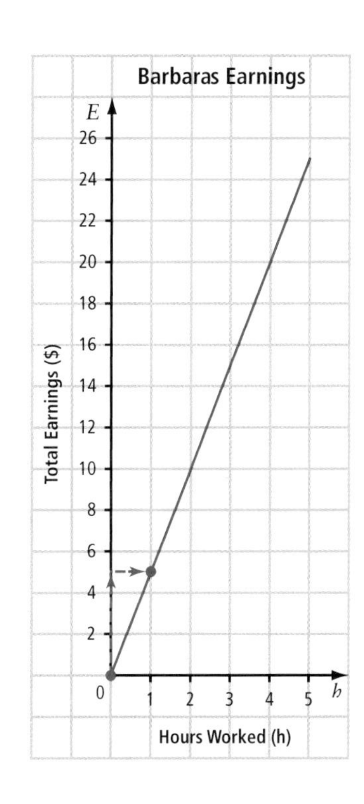

Communicating Mathematically

The slope of a linear relation is a measure of the change in the value of the variable measured on the vertical or *y*-axis divided by the change in the value of the variable measured on the horizontal or *x*-axis.

Slope = $\frac{5}{1}$ Rate = \$5/h

Earnings are measured in dollars. **Time** is measured in hours.

Slope is written as a unit rate, without units of measure. When describing the slope in a problem situation, the units of measure are included and it is a rate.

b) Notice that the graph *falls* to the right. As the number of days increases, temperature decreases, which means the slope is *negative*.

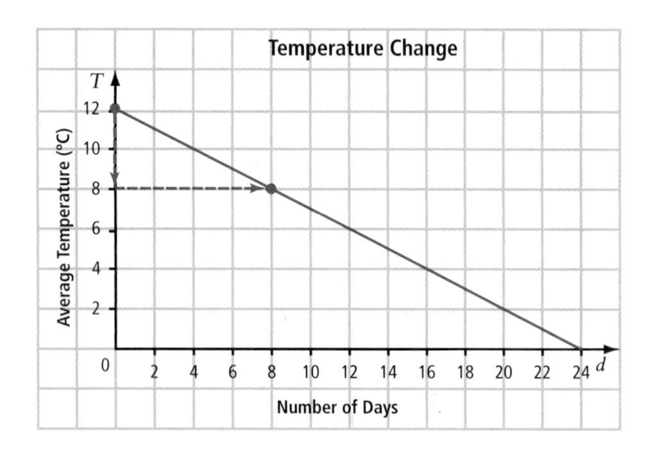

To determine the value of the slope, use two points on the line, and measure the rise and run.

$$\text{Slope} = \frac{\text{rise}}{\text{run}}$$

$$= \frac{-4}{8}$$ **Express in simplest form.**

$$= -\frac{1}{2}$$

The slope of the line is $-\frac{1}{2}$. This means that each day that goes by the temperature drops by half a degree. The rate of temperature change is $-\frac{1}{2}$°C per day.

Communicate the Key Ideas

1. a) What does the slope of a line measure?

b) Explain how you can determine the slope of line. Give an example to support your explanation.

c) Use your example to show how slope can be written as:
- a fraction in simplest form
- a decimal

2. A line passes through three points.

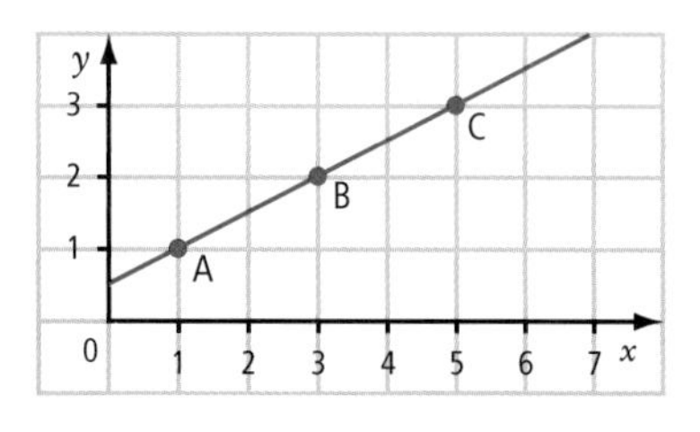

a) Does it matter which two points you use to determine the slope of the line? Explain.

b) Provide calculations to support your answer.

3. Explain the difference between positive and negative slope. Use words and diagrams.

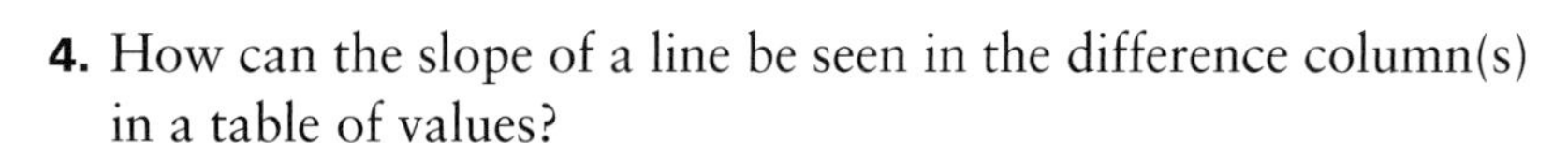

4. How can the slope of a line be seen in the difference column(s) in a table of values?

5. The graph shows the depth of water in a large swimming pool as it drains.

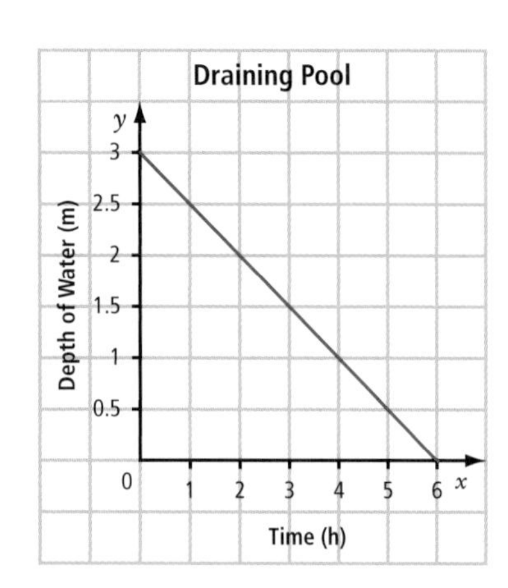

a) What is the slope of the line?

b) What does the slope mean in this case?

Check Your Understanding

1. Determine the slope of each object.

   a) staircase

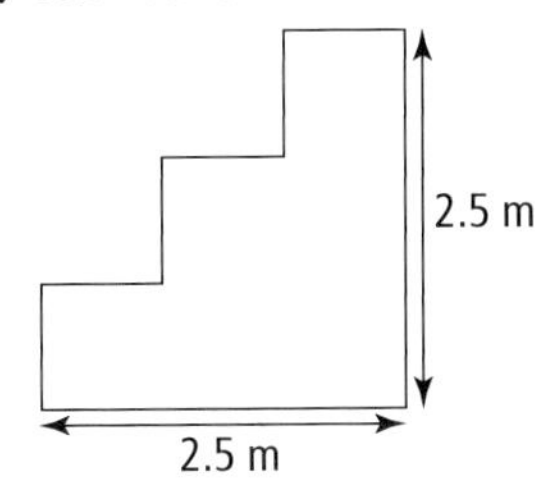

   b) roof

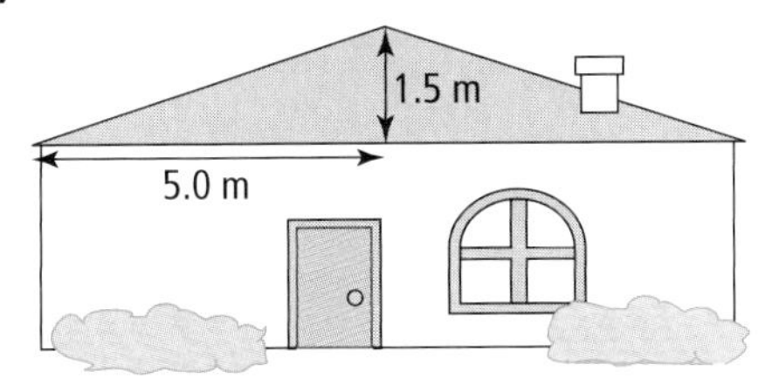

   c) mountain

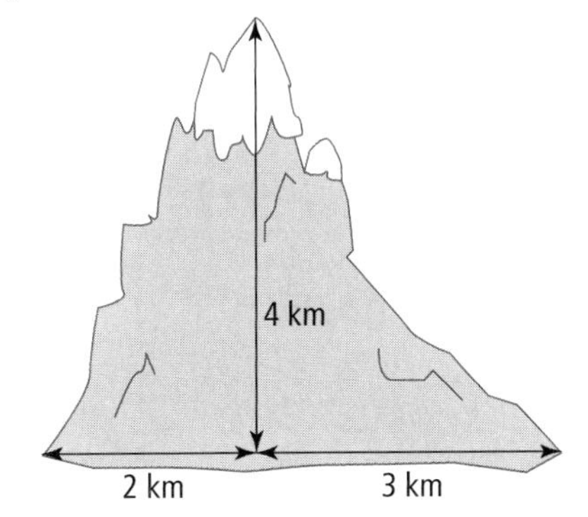

2. Draw an object that has a slope of $\frac{1}{3}$. Indicate the measures of the rise and run.

3. What is the slope of the line that is represented in the table of values?

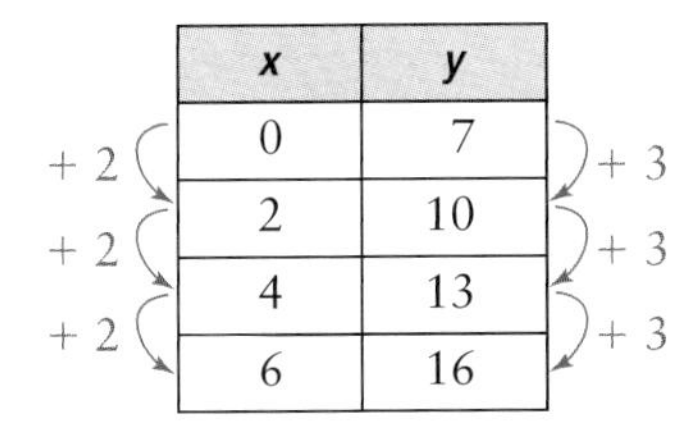

x	y
0	7
2	10
4	13
6	16

4. Match each line with its slope. Justify your answers.

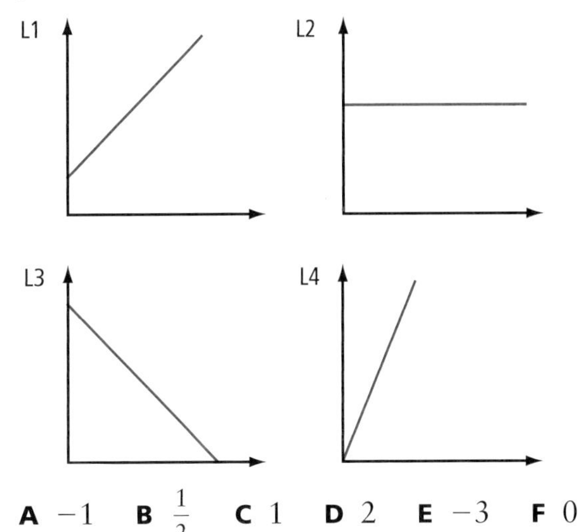

A -1 **B** $\frac{1}{2}$ **C** 1 **D** 2 **E** -3 **F** 0

5. Determine the slope of each line segment. Express your answer as a fraction in simplest form.

   a)

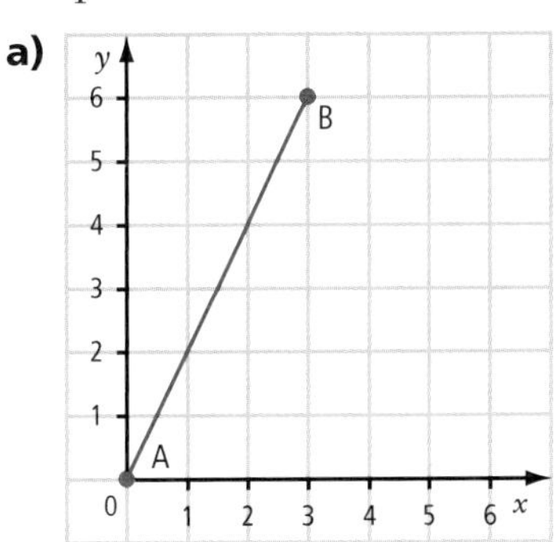

   b)

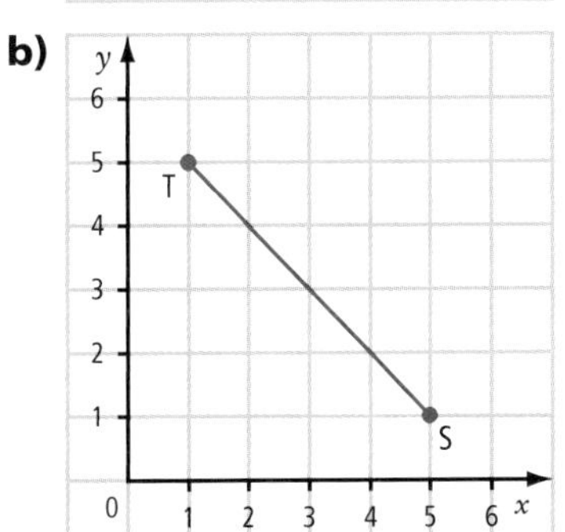

c)

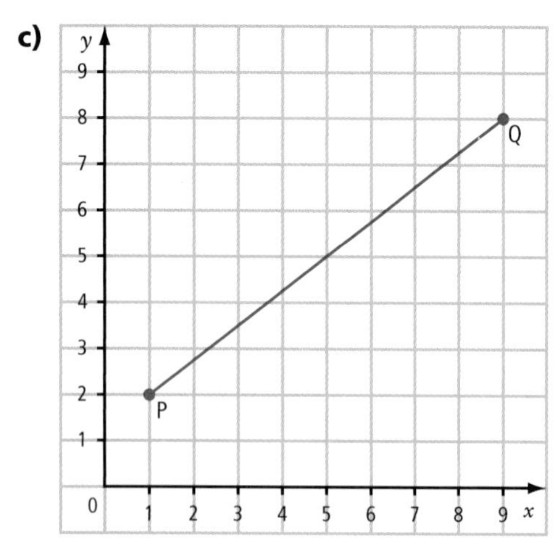

d)

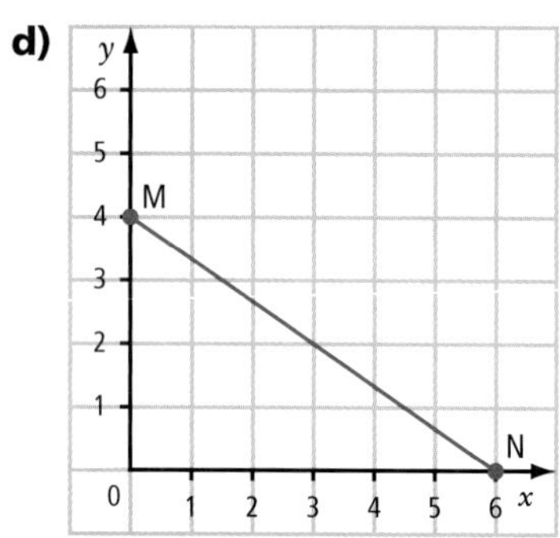

6. Determine the slope of each object. Express your answer as a decimal.

a) banister

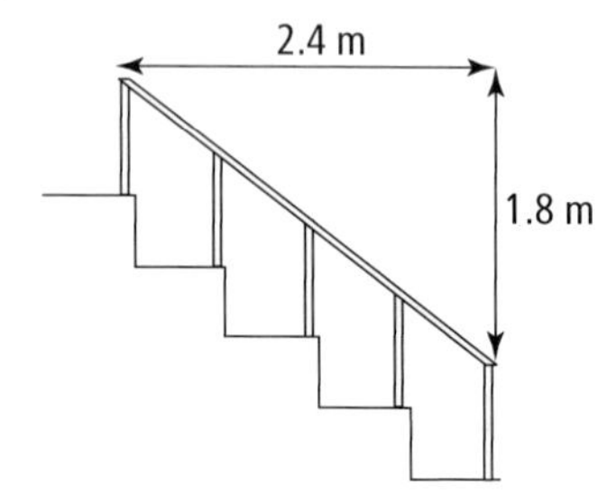

b) road ramp

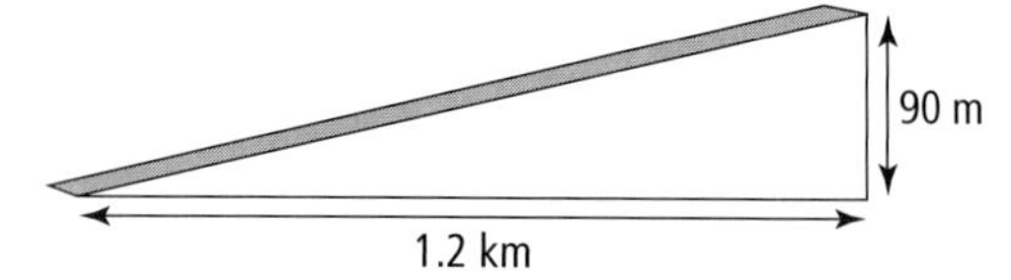

7. a) Draw a line segment that has a slope of $\frac{1}{4}$. Identify the coordinates of two points on the line segment.

b) Draw a line segment that has a slope of -3. Identify the coordinates of two points on the line segment.

8. The graph shows the distance from the starting blocks in a swimming pool for three swimmers.

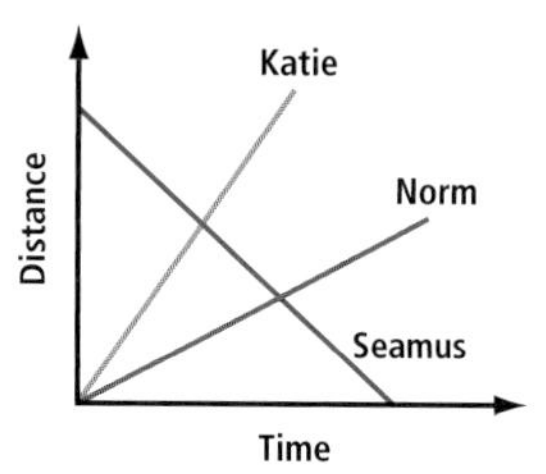

a) Who is swimming
- in the same direction?
- in the opposite direction?

Explain how you know.

b) Who is travelling
- the fastest?
- the slowest?

Explain how you know.

9. The graph shows how Rodney's savings grow over time.

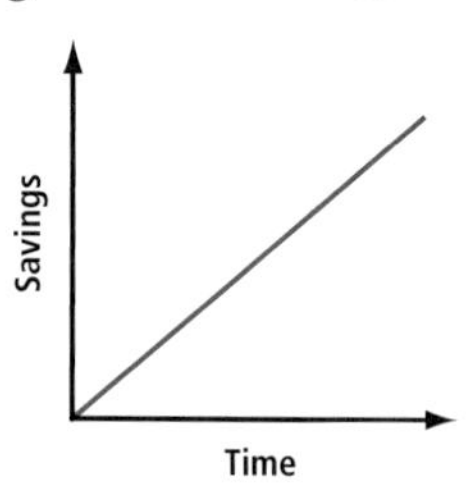

a) Copy the graph into your notebook.

b) With two different colours, sketch how the graph would change if Rodney were to
- double the amount he saves
- halve the amount he saves

10. Refer to question 9. Suppose Rodney saved \$500 over the summer to spend at a constant rate throughout the school year.

a) What is the maximum rate of spending to last Rodney from September to June? Discuss any assumptions you make.

b) Sketch a graph of savings over time for the school year if Rodney spends at this rate.

c) Show how the graph would change if Rodney were to spend his savings at a
- faster rate
- slower rate

Describe what would happen in each situation.

Use this information to answer questions 11 and 12.

Sarah pays \$30 per month for her cellular phone. The money is taken from her savings account each month.

11. a) Copy and complete the table.

Months, *m*	Savings, *s* (\$)
0	\$300
1	\$270
2	
3	
4	

b) Graph the relation.

c) Determine the slope of the line. What does the slope mean?

12. a) Predict how the graph would change if Sarah had to pay \$35 per month for her cellular phone.

b) Make a table of values for the new rate. Graph the line on the same set of axes using a different colour. How does this line compare to the original line?

c) Suppose Sarah found a new cellular phone company who charges \$25 per month. Predict how the original graph would change. Make a table of values. Graph the line on the same set of axes using a different colour to check your prediction.

13. Explain what happens to a line when its slope is
- increased
- decreased

Include diagrams to support your explanations.

14. The graphs show the distance travelled by two koalas over time.

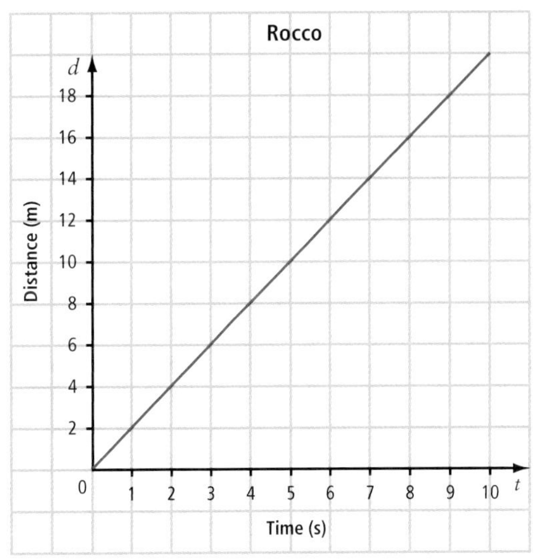

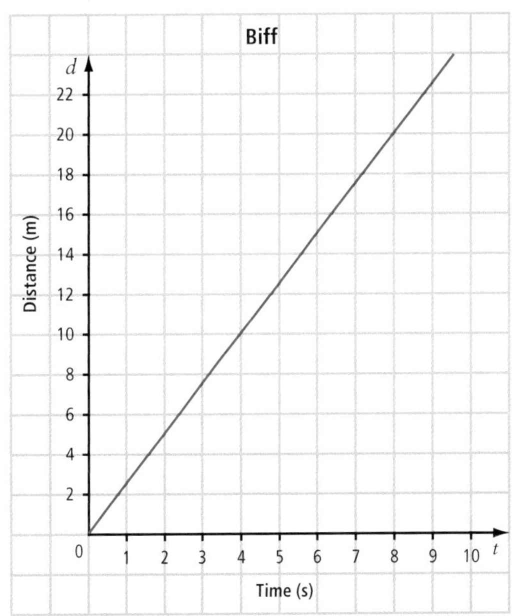

a) Which koala do you think is running faster? Justify your answer.

b) Measure the slope of each line. Explain what the slope means. Do the slopes confirm your prediction? Explain.

15. Beth starts out 90 m from her front door. She jogs toward her front door at a speed of 2 m/s.

a) Construct a table of values that shows Beth's distance from the door after each of the first 10 s.

b) Graph the relation. Plot time on the horizontal axis and distance on the vertical axis.

c) Calculate the slope.

d) Explain how you can see the value of the slope in

- the table of values
- the graph
- the written description

Puzzler

Copy the magic square. Place one of the integers –8, –6, –4, –2, 0, 2, 4, 6, 8, and in each of the smaller squares. The sum of the three integers in any row, column or diagonal must be the same.

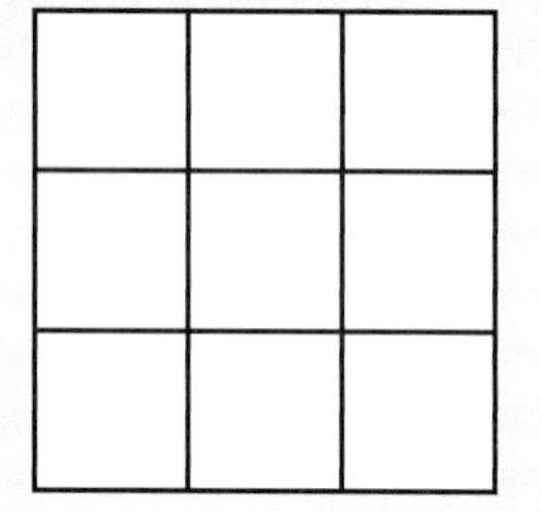

Use the information in the table to answer questions 16 to 18.

Symbol	Difficulty	Slope
●	green	0.1 – 0.4
■	blue	0.4 – 0.8
◆	black diamond	0.8 – 1
◆◆	double black diamond	> 1

Chapter Problem

16. As manager of the ski resort, you are proud to announce the opening of five new hills.

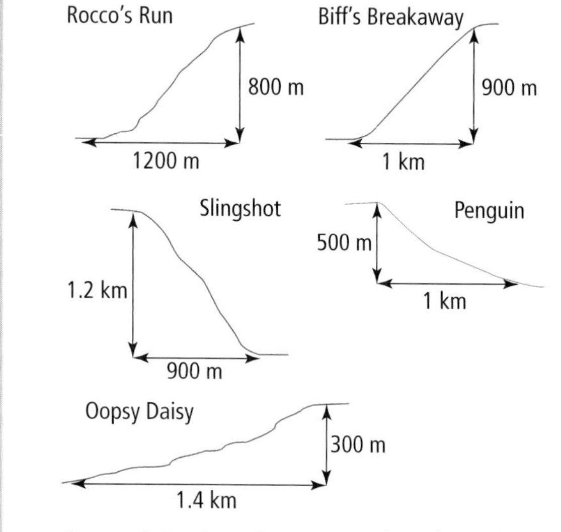

One of the key factors in deciding the difficulty level of a ski hill is its slope.

a) Use the table to rate each hill.

b) What other factors should you consider when assigning a difficulty rating to a ski hill?

17. Sidewinder is a black diamond rated hill. Its run is 750 m. Estimate its rise.

Extend

Journal

18. What is the slope of a horizontal line? What is the slope of a vertical line? Carry out an investigation and write a report of your findings. Include graphs in your explanation.

8.4 Intersection of Two Lines on a Graph

Focus on…
- solving problems involving the intersection of two lines on a graph

Recreational watercraft can be great fun in the summer. Suppose you want to rent a jet ski. You have two companies to choose from but they charge different rental rates.

How can you decide which company offers the best deal? Sometimes it is useful to examine more than one mathematical relationship at a time. This allows you to make better comparisons and wiser decisions.

Discover the Math

Materials
- linking cubes or colour tiles
- graph paper
- coloured pencils

What does the intersection point of two lines on a graph tell you?

Alma works in a greenhouse. She is studying the growth patterns of two types of tomato plants. The heights of the plants when she begins her observations are shown.

Plant A — 3 cm
Plant B — 5 cm

Alma notices that Plant A grows at a constant rate of 3 cm per week and that Plant B grows at a constant rate of 2 cm per week.

1. Are these plants growing at the same rate? Explain.

2. a) Copy and complete the table.

Time (weeks)	Height of Plant A (cm)	Height of Plant B (cm)
0	3	5
1	6	
2		
3		
4		
5		

b) Is the relationship between time and the height of Plant A linear or non-linear? Explain how you know.

c) Repeat part b) for Plant B.

3. a) Graph the data for Plant A. Use a coloured pencil to connect the points with a straight line and label the line Plant A.

b) On the same set of axes, graph the data for Plant B. Use a different coloured pencil to connect the points with a straight line and label the line Plant B.

c) Explain why it makes sense to join the points for each relation.

4. Identify the coordinates of the point where the two lines cross. This is called the **intersection point**. What do the values of these coordinates mean?

5. **Reflect** Look at each section of the graph. Determine which plant is taller:
- to the *left* of the intersection point
- to the *right* of the intersection point
- *at* the intersection point

When two or more linear relationships are shown on the same graph, a **linear system** is formed. Such graphs are useful for comparing two or more relationships.

Did You Know?

The mayflower is Nova Scotia's provincial flower. It bloms in the forest in early spring.

Strategies
Use a diagram.

intersection point
- the point on a graph where two lines meet

Example: Compare Cost Relationships

Cody wants to rent a jet ski.

Assume that customers will be charged for partial hours.

Which company should Cody choose? Will the number of hours he rents the jet ski affect his choice?

Solution

Find the cost of renting a jet ski from each company for up to 6 h. Fast Freddie charges $22.50/h. To find the rental cost, multiply the number of hours by 22.50.

Strategies
Use a table to organize the information. Look for a pattern.

Rental Time (h)	Fast Freddie's Cost ($)
0.5	11.25
1	22.50
2	45.00
3	67.50
4	90.00
5	112.50
6	135.00

Aquatic Adventures charges a $50 flat fee, plus $10/h. To find the rental cost, multiply the number of hours by 10 and add 50.

Rental Time (h)	Aquatic Adventures Cost ($)
0.5	$0.5 \times 10 + 50 = 55$
1	$1 \times 10 + 50 = 60$
2	$2 \times 10 + 50 = 70$
3	$3 \times 10 + 50 = 80$
4	$4 \times 10 + 50 = 90$
5	$5 \times 10 + 50 = 100$
6	$6 \times 10 + 50 = 110$

Technology Tip

- You can use a spreadsheet or graphing calculator to perform repeated calculations.

You can compare the costs by combining the information in a single table.

Rental Time (h)	Fast Freddie's Cost ($)	Aquatic Adventures Cost ($)
0.5	11.25	55
1	22.50	60
2	45.00	70
3	67.50	80
4	90.00	90
5	112.50	100
6	135.00	110

If Cody rents the jet ski for less than 4 h, Fast Freddie's Watercrafts offers the best deal. If he rents for more than 4 h, Aquatic Adventures offers the best deal. If he rents for 4 h, the cost will be the same with either company.

This information can be illustrated by graphing both relations on the same set of axes.

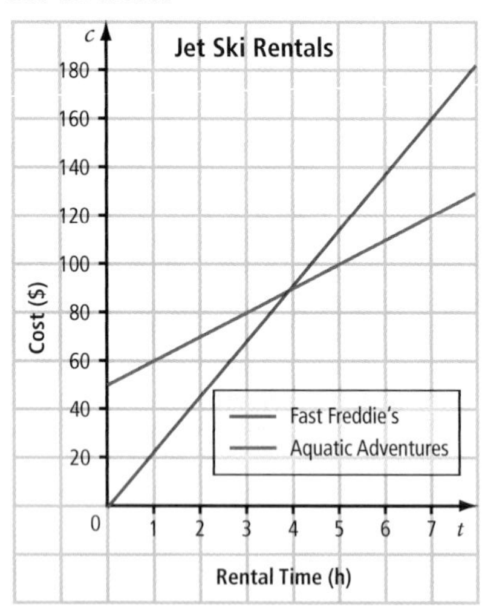

Look at both linear relations.

- To the left of the intersection point, Fast Freddie's cost (in red) is *lower* than Aquatic Adventures' cost (in blue).
- To the right of the intersection point, Fast Freddie's cost is *higher* than Aquatic Adventures' cost.
- At the intersection point, the two costs are equal. The coordinates of this point are (4, 90). This means that it costs $90 to rent the jet ski for 4 h from either company.

Rental Time	< 4 h	4 h	> 4 h
Who is less expensive?	Fast Freddie's	Equal cost	Aquatic Adventures

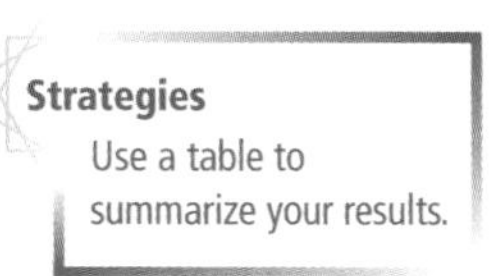

Communicate the Key Ideas

1. Give an example of a situation in which you would want to examine a pair of linear relationships. Choose a situation not already discussed in this section.

2. a) What is the intersection point of a linear system?
b) What does this point mean?

3. The graph shows the growth of two kittens that were born at the same time on the same day.
a) Which kitten was longer at birth? Explain how you know.
b) Which kitten was longer at 3 weeks? Explain how you know.
c) What is the intersection point? What does it mean?

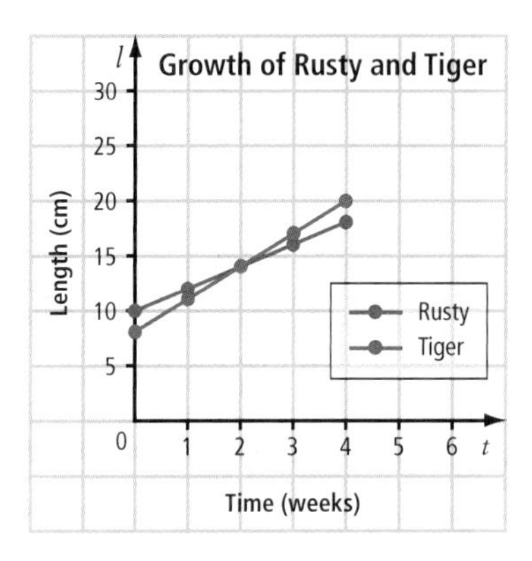

4. What advantages are there in comparing two linear relations on the same set of axes?

Check Your Understanding

1. The graph shows the cost of a ride from two rickshaw drivers.

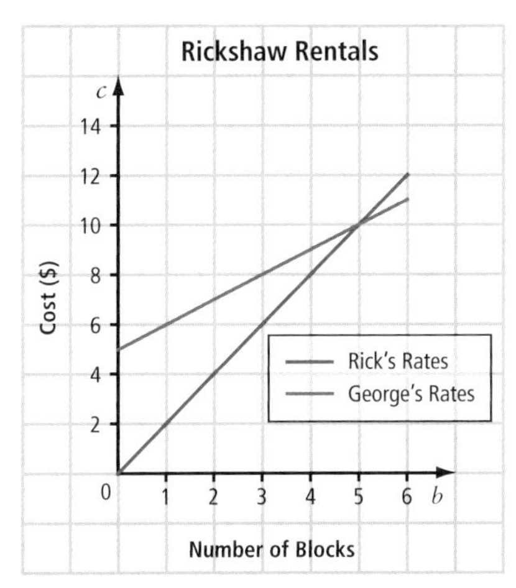

a) Which driver charges less for a 4-block ride? How much less?
b) Which driver charges less for a 6-block ride? How much less?
c) For what distance do the drivers charge the same amount? What is this amount?
d) Which driver would you choose and why?

Use this information to answer questions 2 and 3.

Membership rates for two fitness clubs are shown.

2. a) Copy and complete the table for up to 10 visits per month.

Visits per Month	Sally's Fitness Cost ($)	Gary's Gym Cost ($)
0	40	10
1	40	15
2	40	20
3		
⋮	⋮	⋮

b) When does Sally's Fitness offer the better deal?

c) When does Gary's Gym offer the better deal?
d) When do both clubs offer the same deal?

3. a) Graph the data for both gyms on the same set of axes. Plot visits per month on the horizontal axis and cost on the vertical axis. Use a different colour for each set of points and label each set.
b) Identify the coordinates of the intersection point. Explain what each coordinate means.
c) Look to the *left* of the intersection point. What does this part of the graph tell you?
d) Look to the *right* of the intersection point. What does this part of the graph tell you?

Use this information to answer questions 4 and 5.

Two cellular phone companies calculate their customers' monthly cost (C), in dollars, for talking m minutes, as shown. Customers will be charged for partial minutes of use.

Quick Talk: $C = 0.1m$
Chatters: $C = 10 + 0.05m$

4. a) Describe how each company calculates their customers' monthly cost, in words.
b) Copy and complete the table.

Number of Minutes	Quick Talk Cost ($)	Chatters Cost ($)
0	$0.1(0) = 0$	$10 + 0.05(0)$ $= 10 + 0$ $= 10$
100	$0.1(100) = 10$	$10 + 0.05(100)$ $= 10 + 5$ $= 15$
200		
300		
400		
500		

c) Graph the data for both companies on the same set of axes. Plot number of minutes on the horizontal axis and cost on the vertical axis. Use a different colour to join the points on each line and label the lines.
d) Identify the coordinates of the intersection point. Explain what each coordinate means.
e) Does one company offer a better deal than the other in all circumstances? Explain.

5. Suppose that Quick Talk dropped their rates to 7¢/min. How would this change your answer to question 4, part d)?

6. a) Create a problem in which two companies charge different rate structures. Describe the rate structures in words.
b) Graph the relations on the same set of axes. Plot time or number of people on the horizontal axis and cost on the vertical axis. Use a different colour to join the points on each line and label the lines.
c) Is there an intersection point? If yes, explain what it means. If no, change one of the relations so that there is an intersection point and explain what it means.

7. a) Create a problem involving two linear relations whose intersection point is (5, 20).
b) Pose two questions about the problem.
c) Answer the questions.
d) Trade problems with a classmate and answer each other's problems. Check your answers.

8. Carol works at a greenhouse. Her boss has told her to stack two types of flowerpots on a shelf.
- The brown pots are 24 cm tall. When stacked, the height of the stack increases by 6 cm for each pot.
- The green pots are 16 cm tall. When stacked, the height of the stack increases by 8 cm for each pot.

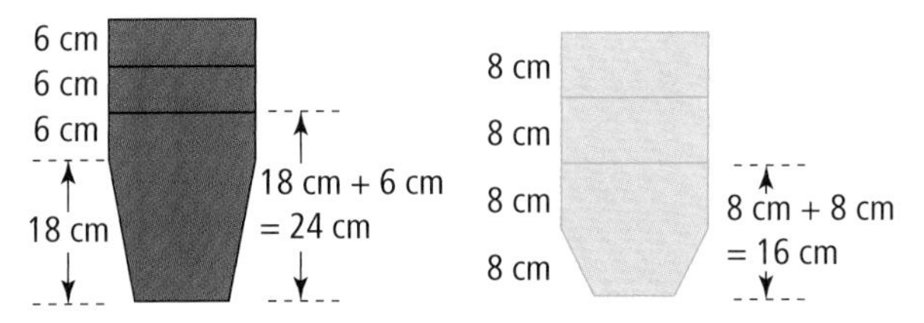

a) Copy and complete the table.

Number of Pots in Stack	Height of Brown Pots (cm)	Height of Green Pots (cm)
1	24	16
2	30	
3		
4		
5		

b) Graph the data for both stacks of pots on the same set of axes. Use a different colour to join the points on each line and label the lines.

c) For each stack, write an equation relating height and number of pots.

d) The brown pots and the green pots have the same circumference. Describe how Carol should arrange the pots on the display shelf shown below, to maximize the shelf space. Explain your solution. Discuss any assumptions you must make.

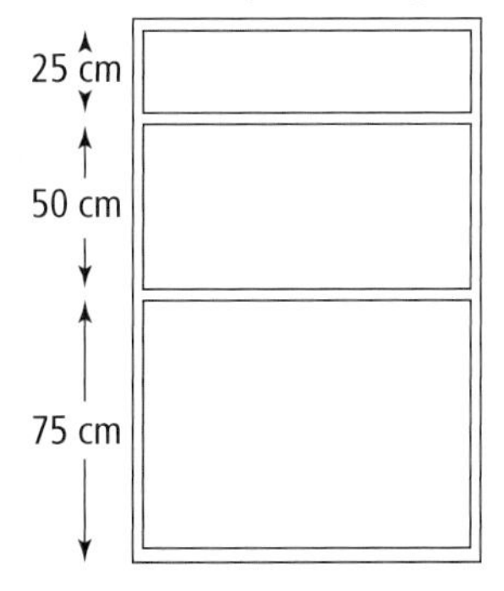

Journal

9. Search the Internet, newspapers, magazines, or non-math related books for a graph of a linear system. Explain what the graph is illustrating. Identify the coordinates of the intersection point and explain what the coordinates mean. Describe at least three things you learned from looking at the graph.

Extend

10. Scott and Donna cycled the 31.5 km stretch from Middle Sackville, NS, to Newport, NS. Donna started from Newport and Scott started from Middle Sackville.

The distance each person is from Newport is shown.

Time (min)	Donna's Distance (km)	Scott's Distance (km)
0	0	31.5
20	3	25.5
40	6	19.5
60	9	13.5
80	12	7.5
100	15	1.5

a) Graph the data for both cyclists on the same set of axes. Plot time on the horizontal axis and distance on the vertical axis. Use a different colour to join the points on each line and label the lines.

b) Determine the average speed of each cyclist. Show how you found your answer.

c) Is the relationship between distance and time linear or non-linear? Explain.

d) How long did it take each cyclist to reach the other's starting point? Explain how you found your answer.

e) When did the cyclists meet on the path? Explain your answer.

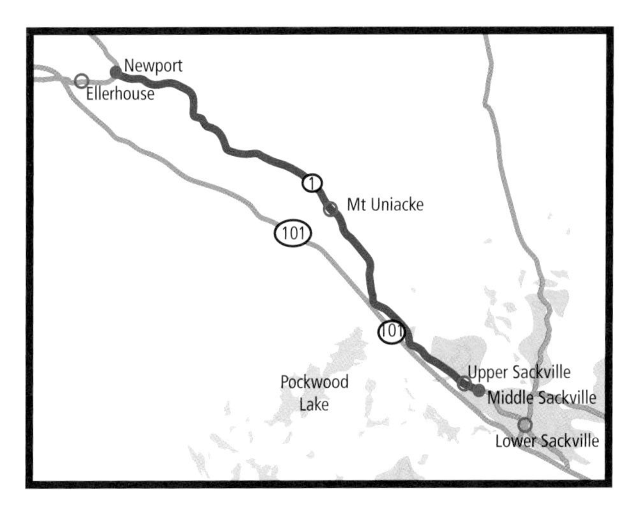

8.5 Analyse Relations

Focus on...
- constructing and analysing tables and graphs to describe how change in one quantity affects a related quantity
- interpret graphs that represent linear and non-linear data

Have you ever heard the expression "A picture is worth a thousand words"? Whoever coined this statement must have understood the value of graphs in mathematics.

Imagine pouring water into a tall, narrow glass. How quickly will the glass fill up? How would this change if the glass was wider? What if the glass had an unusual shape? What would these relationships look like in a graph?

The more complicated a relationship is, the more useful a graph can be to illustrate it.

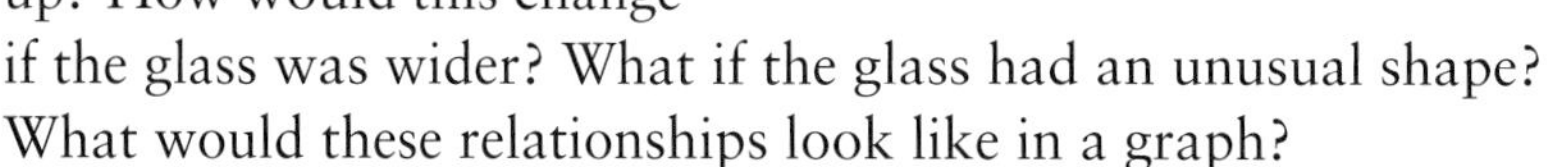

Discover the Math

Materials
- beakers (e.g, 500 mL) or graduated cylinders
- measuring scoops (e.g., 50 mL)
- Erlenmeyer flasks
- water or rice
- rulers

How can I construct and analyse graphs of linear and non-linear relations?

1. Work in a small group. Select a medium-sized beaker (e.g., 500 mL).

2. a) Fill the measuring scoop with water (or rice) and carefully pour the contents into the beaker.

b) Measure the height of the water with a ruler. Copy the table and enter your results.

Volume (mL)	Height (cm)

c) Repeat until the beaker is almost full. You should record at least 6 to 10 ordered pairs.

3. a) Graph the ordered pairs. Plot volume on the horizontal axis and height on the vertical axis.

b) Does the pattern appear to be a linear relation?

4. a) Draw a **line of best fit** through the points.

b) Find the slope of the line. Explain what it means.

line of best fit

- the straight line that passes through, or as near as possible to, the points on a scatterplot

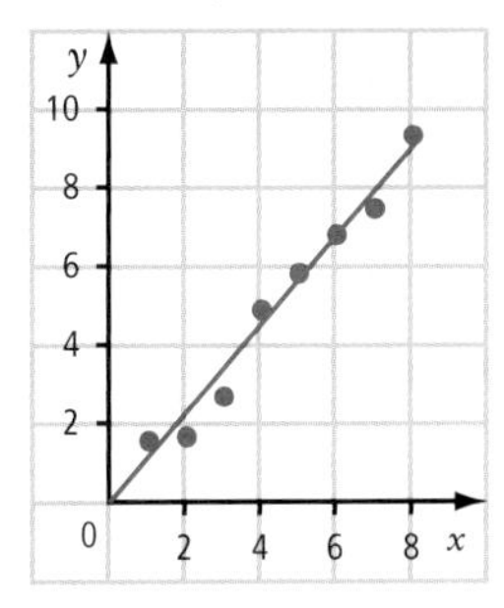

5. What happens if you change the diameter of the beaker?

a) Predict what would happen to the graph if you used a beaker that is
- wider
- narrower

b) Sketch graphs to support your answers.

c) Carry out an experiment to test your predictions. Compare your results to your predictions.

6. Select an Erlenmeyer flask.

Use only the tapered part of the flask.

Tapered part

a) Suppose you filled the flask. Predict the shape of the graph comparing volume to height. Sketch your graph.

b) Repeat steps 2, 3, and 5 using the flask. Was your prediction correct? Explain.

7. Reflect

a) What factors determined the slope of the graphs you created in this activity?

b) In which cases was the slope constant?

c) In which cases did the slope change?

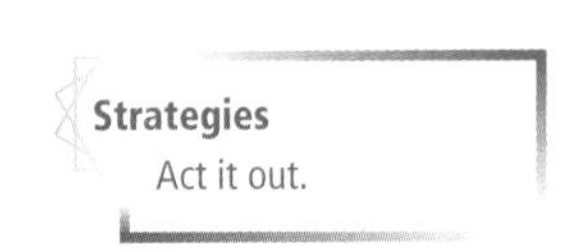

Example 1: Analyse a Broken Line Graph

The graph shows how temperature changes over a 6-h period, beginning at noon.

Find the slope of each segment of the graph and explain what it means.

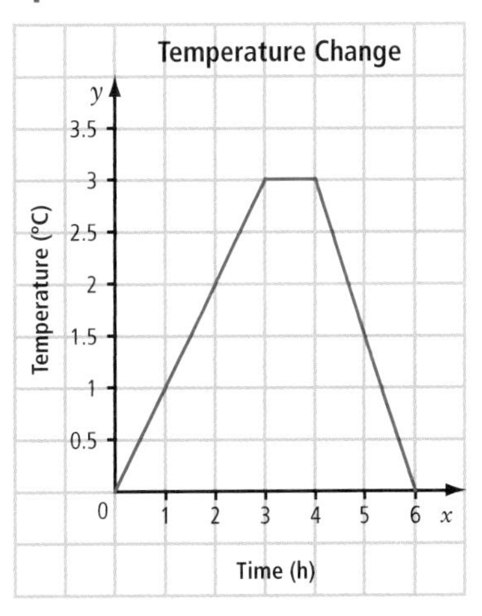

Solution

The first segment of the graph has a positive slope.

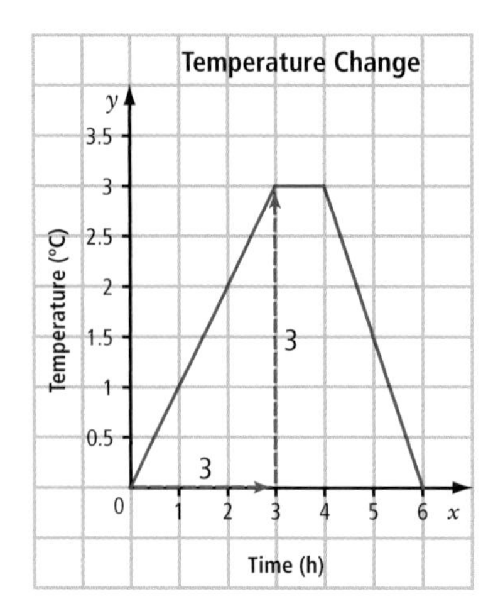

$$\text{Slope} = \frac{\text{rise}}{\text{run}}$$
$$= \frac{3}{3}$$
$$= 1$$

The slope of this segment is 1, which means that during this time interval the temperature increased by 1°C per hour. Therefore, the rate of change in temperature is 1°C/h.

The second segment of the graph has a flat slope.

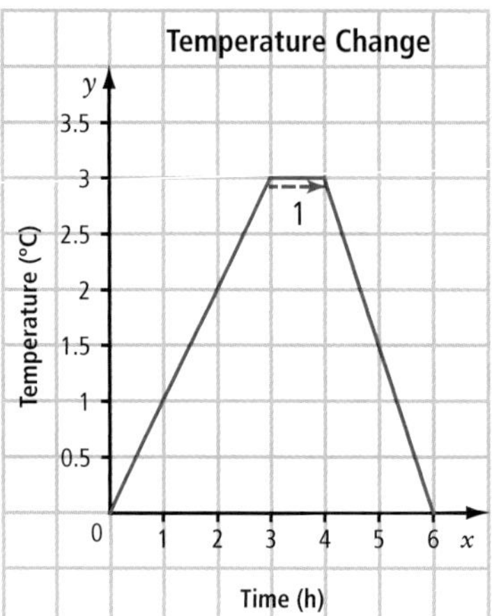

You can measure the run, but the rise is zero.

$$\text{Slope} = \frac{\text{rise}}{\text{run}}$$
$$= \frac{0}{1}$$
$$= 0$$

The slope of this line segment is zero. This means that during this time period the temperature did not change. Therefore, the rate of change in temperature is 0°C/h.

A horizontal line has zero rise.

rise = 0

$$\text{Slope} = \frac{\text{rise}}{\text{run}}$$
$$= \frac{0}{\text{run}}$$
$$= 0$$

Zero divided by any value is zero.

Therefore the slope of any horizontal line or line segment is zero.

The third part of the graph has a negative slope.

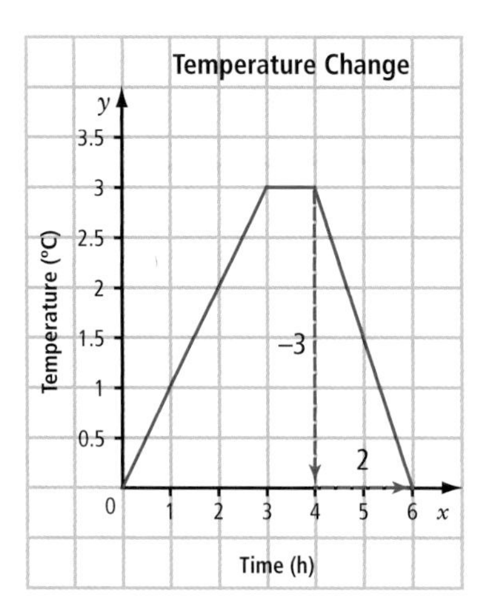

$$\text{Slope} = \frac{\text{rise}}{\text{run}}$$
$$= \frac{-3}{2}$$

The rise is negative, because the line segment falls to the right.

$$= -1.5$$

The slope of this segment is -1.5. This means that during this time period the temperature decreased by 1.5°C per hour. Therefore, the rate of change in temperature is -1.5°C/h.

Example 2: Analyse the Graph of a Non-Linear Relation

The graph shows the height of water in the flask over time as it is filled. The rate at which the water is added is constant.

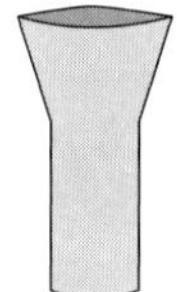

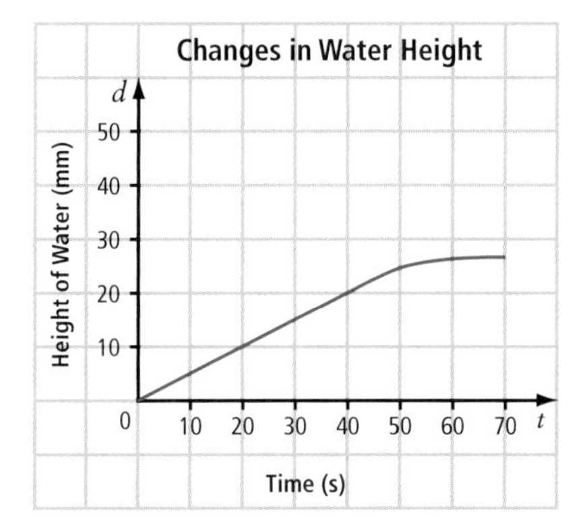

Describe how the height of the water in the flask is changing over time.

Solution

From 0 to 40 s, the height of the water changes from 0 to 20 mm. The slope of this segment is 0.5. This means the rate of change in the height of the water is 0.5 mm/s.

Since the line is straight, the change in the height of the water is constant. The vertical portion of the flask must be filled from 0 to 40 s.

From the 40 s to 70 s, the height of the water changes gradually from 20 mm to 30 mm.

The average change in the height of the water is 0.33 mm/s.

Since the line is curved and flattens out, the trapezoidal portion of the flask must be filled from 40 s to 70 s.

Example 3: Interpret a Time-Distance Graph

The graph shows the distance Marita travelled as she walked home.

Describe what Marita was doing in each segment of the graph in terms of distance, time, and speed.

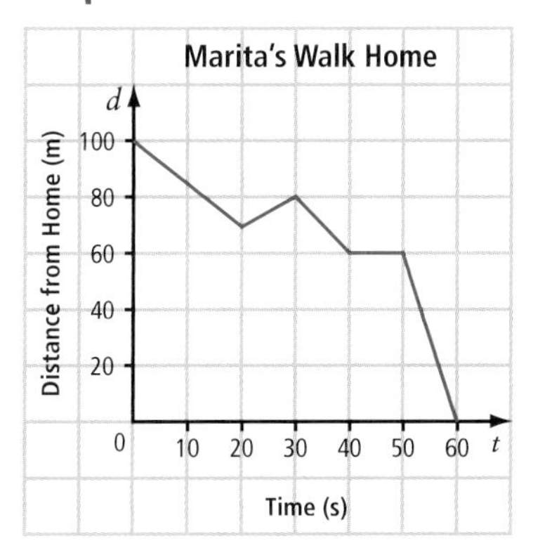

Solution

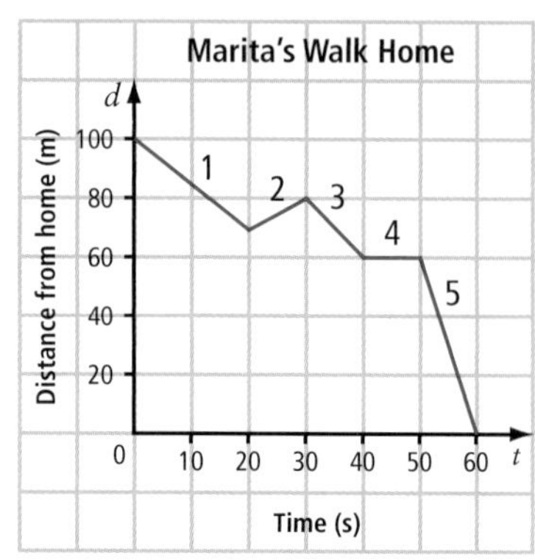

For segment 1, Marita walked a distance of 30 m in 20 s. Her speed was 1.5 m/s.

For segment 2, she walked a distance of 10 m in 10 s. Her speed was 1 m/s, so she slowed down. She was also farther from home at 30 s. She could have taken a detour.

For segment 3, she walked a distance of 20 m in 10 s. Her speed was 2 m/s, so she was speeding up.

For segment 4, the distance from home remained the same at 60 m. Her speed was 0 m/min, so she probably stopped to rest for 10 s.

For segment 5, she walked a distance of 60 m in 10 s. Her speed was 6 m/s, so she was probably running for the last segment.

The speed of each segment is the same as the slope of the line for each segment.

Communicate the Key Ideas

1. a) Sketch a time-distance graph for a person walking at a constant speed of 1 m/s. Plot time on the horizontal axis and distance on the vertical axis.

b) On the same set of axes, sketch a graph of
- a person walking at a slower constant speed
- a person walking at a faster constant speed
- a person standing still
- a person whose speed is constantly increasing

Use this information to answer questions 2 and 3.

The graph shows the distance of a person walking over time.

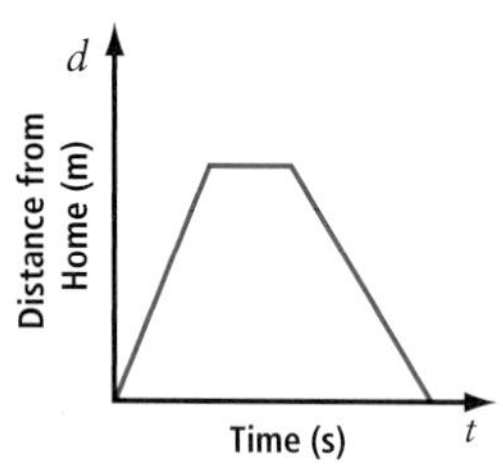

2. Describe three things you can tell about the person walking from the graph.

3. Write a story to explain the shape of the graph. In your story, be sure to mention distance, time, and speed.

Check Your Understanding

Use this information to answer questions 1 and 2.

The original volume of water in the can is 500 mL.

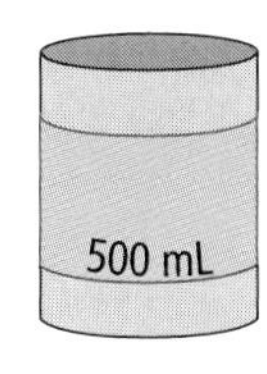

When a small hole is made in the bottom of the can, the water begins to drain out. The table shows the amount of water remaining in the can over time.

Time (s)	Amount of Water Remaining (mL)
0	500
1	450
2	400
3	350
4	300

1. a) Graph the data. Plot time on the horizontal axis and amount of water remaining on the vertical axis.

b) Find the slope and explain what it means.

c) Is the slope positive or negative? Explain what this tells you.

2. How do you think the graph would change if the hole in the bottom of the can was:

- made larger?
- made smaller?

Explain your reasoning.

3. Three bathtubs are filled at a constant rate.

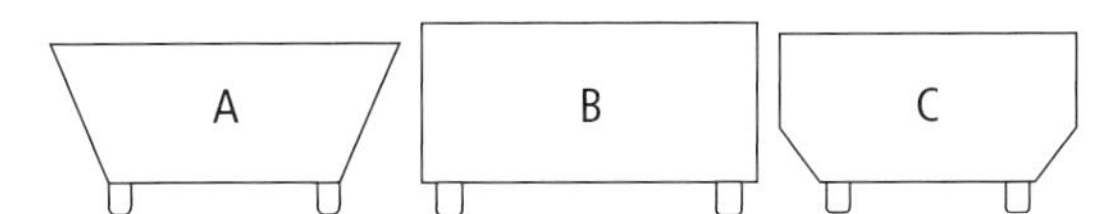

Sketch a graph of the height of the water over time for each tub. Plot time on the horizontal axis and height of water on the vertical axis. Explain the shapes of your graphs.

4. The graph shows the distance travelled as Billy walks to school one morning.

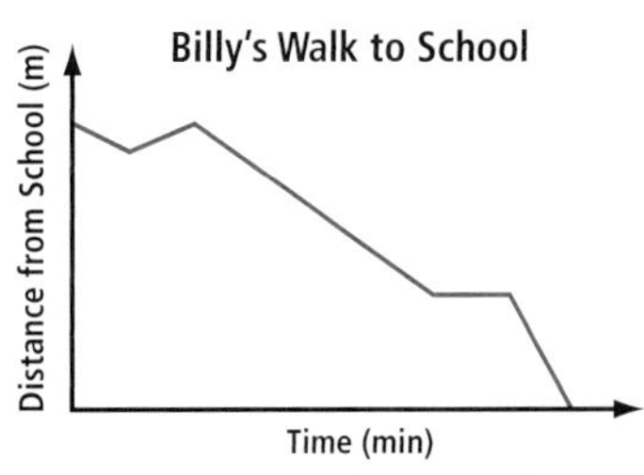

Write a story about Billy's walk that can be represented by the graph. Be sure to mention Billy's distance, time, and speed.

Use this information to answer questions 5 to 8.

The graph shows the distance Matilda hiked over time.

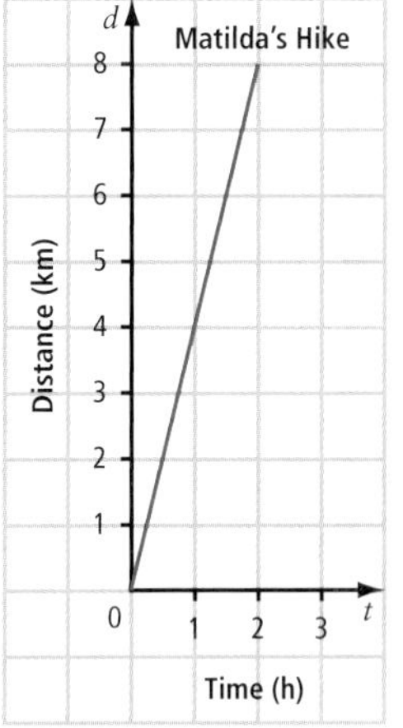

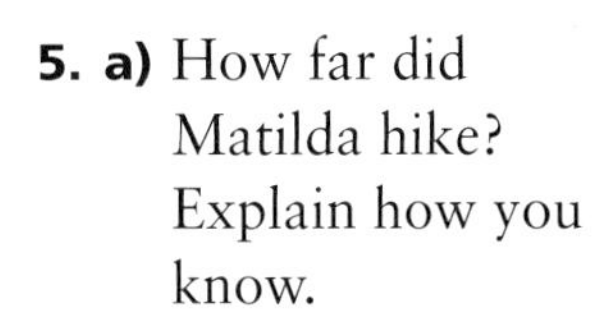

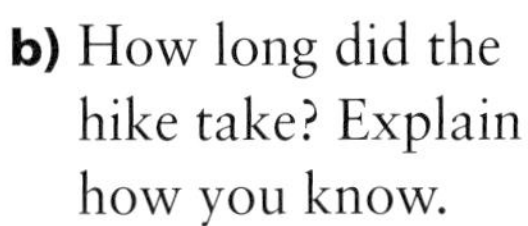

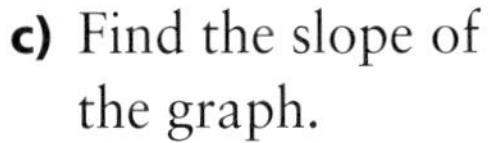

5. a) How far did Matilda hike? Explain how you know.

b) How long did the hike take? Explain how you know.

c) Find the slope of the graph.

d) What was Matilda's average speed? How is this related to the slope?

6. a) Carefully copy the graph.

b) On the same set of axes, draw a graph of a hiker who does the hike in

- half the time
- twice the time

c) Discuss how the slopes of these graphs are related to the slope of the original graph.

7. a) Draw a graph of a hiker who starts 0.5 km along the path and takes the same amount of time to reach the end as Matilda.

b) What is the slope of this graph? What is the speed of the hiker?

c) How does this speed compare to Matilda's speed?

8. Draw a graph of a hiker who starts at the same point and walks twice as fast as Matilda, but only half as far. Explain your answer.

9. Two koalas, Rocco and Biff, raced home from the stream where they were playing. The time-distance graph is shown.

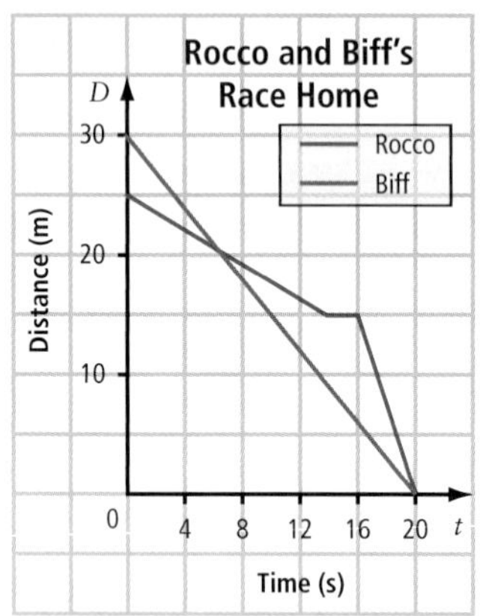

Rocco started from the stream, which is 30 m from home. Biff was several metres from Rocco when he started. Describe this race. In your description, be sure to mention the koalas' distance, time, and speed.

10. The graph shows the distance over time for two brothers who are walking home. Josh is coming from the library and Elijah is coming from the mall. They leave for home at the same time.

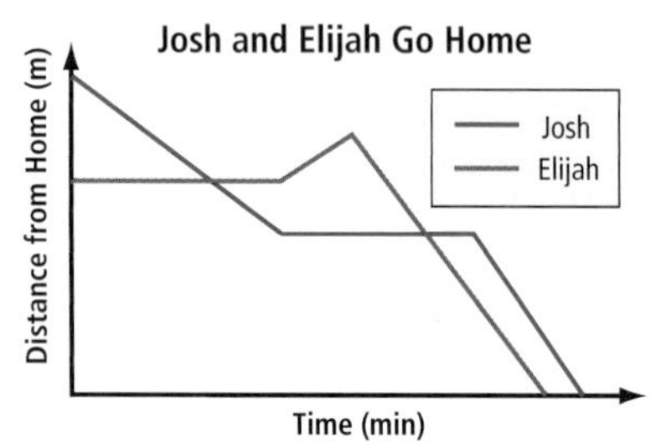

a) Which is farther from the brothers' home: the library or the mall? Explain how you know.

b) Who arrived at home first? Explain how you know.

c) Identify the point or points where the brothers were the same distance from home at the same time. Explain your answer.

d) Write a story that describes each brother's walk home.

11. Suppose the flask is filled with water at a constant rate.

a) Sketch a graph of the height of the water in the flask over time. Plot time on the horizontal axis and height of water on the vertical axis.

b) Describe what is happening in your graph by referring to the height of the water and time.

Extend

12. The graph shows how temperatures in degrees Fahrenheit are related to temperature in degrees Celsius.

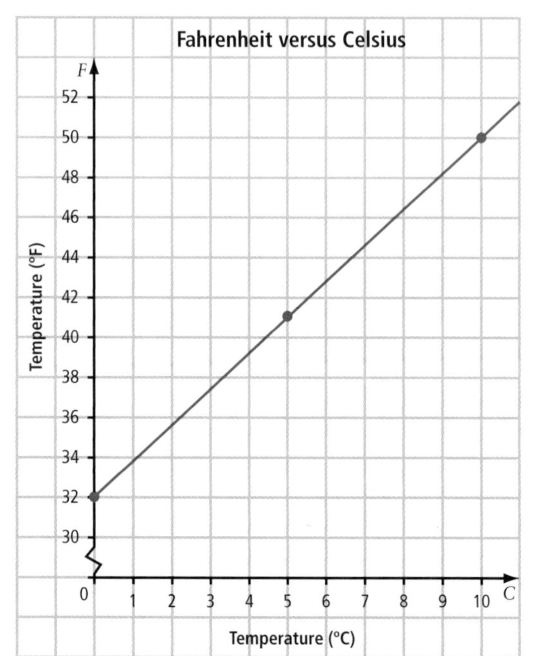

Suppose you changed the graph so that temperatures in degrees Fahrenheit were on the horizontal axis and temperatures in degrees Celsius were on the vertical axis.

a) Predict what you think would be the slope of the line.
b) Record the coordinates of several points on the given line. Reverse their coordinates and plot the new points. Graph the new line.
c) Find the slope and explain what it means. Compare this result with your prediction. Were you correct? Explain.

13. Match each scenario with its graph. Justify your choices.

Scenario

a) One person tells two friends a secret. Every day, each new friend who is told the secret tells two more friends.
b) Every minute, Julio moves halfway closer to the wall.
c) A video rental company charges a fixed amount to rent a movie for five days. After this, a small additional charge is applied for each day late.
d) An internet café charges a fixed amount for the first ten minutes. Then an additional charge is applied after each 5-min interval.

Graph

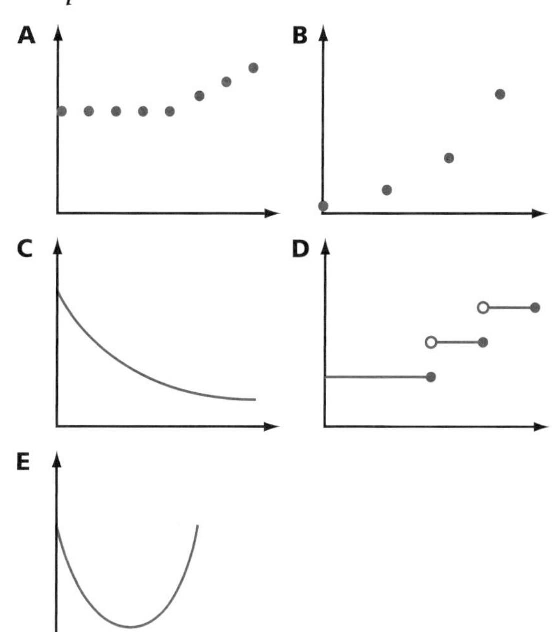

14. A new species of fish is introduced into a pond. Initially, the fish population grows faster and faster as more fish are born. Eventually, factors such as limited food supply, overcrowding, and disease interfere with population growth, causing it to slow down. Then the fish population stabilizes to a steady number, called the 'carrying capacity' of the living environment. The graph illustrates this growth pattern.

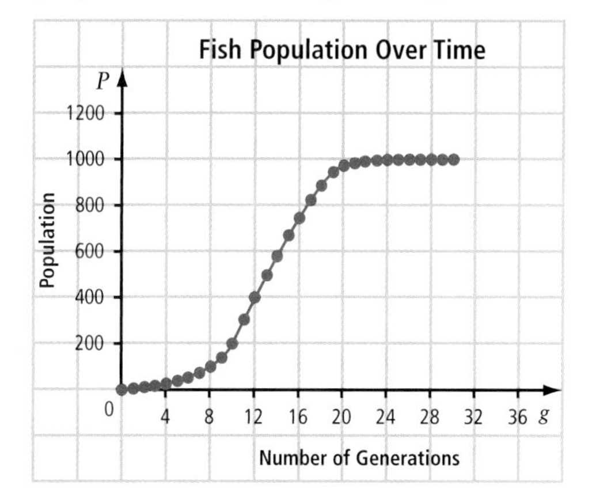

Internet Connect To see this fish population growth, go to **www.mcgrawhill.ca/links/math8NS** and follow the links.

Explain how each of the environmental factors would impact the shape of the graph. Give reasons for your answers using words and sketches. Consider only one factor at a time.

Environmental Factors

a) the initial population of fish was 100
b) the carrying capacity was 500
c) the food supply was reduced
d) several disease threats were eliminated

Did You Know?

Several organisms follow this type of growth pattern, which is called a **logistical curve**. You will learn more about growth patterns of animals if you study biology.

CHAPTER 8

Review

Use this information to answer questions 1 and 2.

The table shows the distance travelled by a person walking at a constant speed.

Time (s)	Distance (m)
0	0
1	3
2	6
3	9

1. a) Describe in words how distance changes as time changes.

b) Is this relation linear or non-linear? Explain how you know.

c) Graph the relation. How far will the person have walked in

- 5 s?
- 10 s?

Explain how you found your answers.

2. a) Write an equation that relates the person's distance and time.

b) Verify your equation by substituting two ordered pairs from the table.

c) Use your equation to find the distance walked after

- 25 s
- 1 min

d) How long will it take for the person to walk 1 km at this pace?

3. Find the slope determined by each object. Write your answer as a decimal, rounded to the nearest tenth.

a)

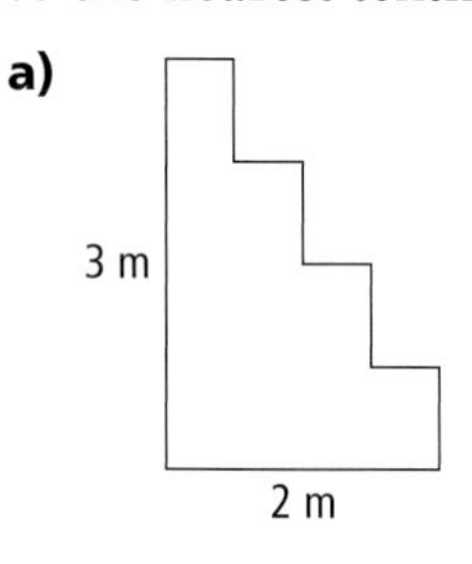

b)

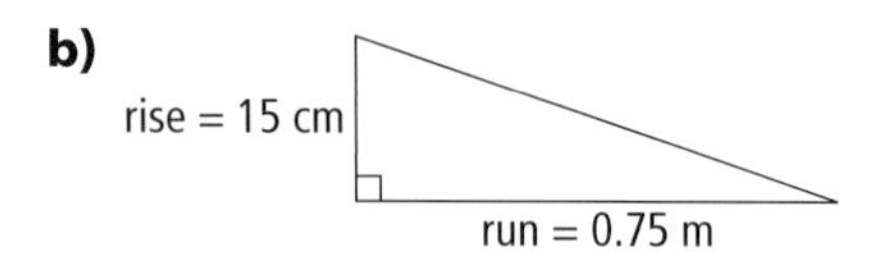

4. The graph shows the daily cost of renting a car from two companies.

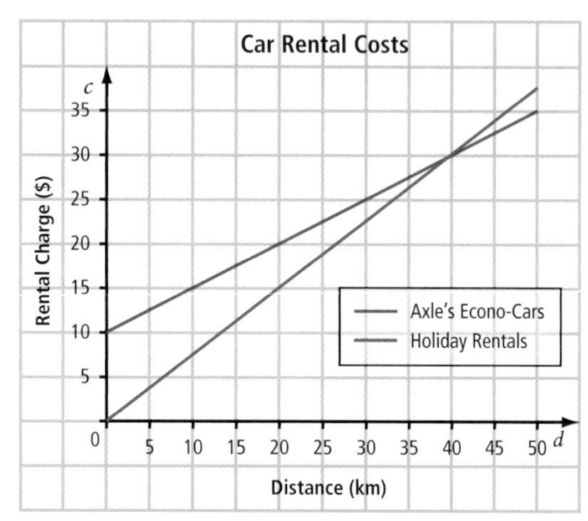

a) Which company would you rent from if you planned to drive 20 km? Explain your answer.

b) Would you change your mind if you planned to drive 50 km? Explain.

c) What are the coordinates of the intersection point? What does each coordinate mean?

d) Predict what would happen to the intersection point if Holiday Rentals increased their rental charge from \$0.75/km to \$0.85/km.

5. a) Draw a line or line segment that has

- a positive slope
- a negative slope
- zero slope

b) Determine the slope of the first two lines by measuring the rise and run with a ruler.

c) Explain why the third line has a zero slope.

6. A pattern is shown.

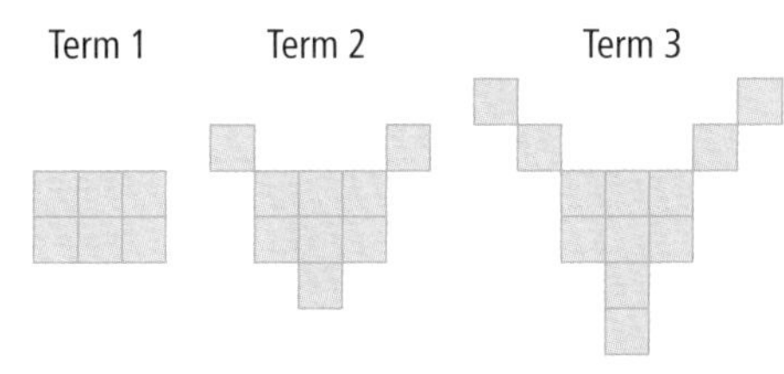

a) Construct a table of values that relates term number, n, to the total number of squares, s.

b) Graph the relation. Use the graph to predict the number of squares in the
- 5th term
- 11th term

c) Describe the relation using words.

d) Write an equation that relates s and n: $s = $ ▬

e) Use your equation to verify your answers to part b).

7. Susie took a bath. The graph below shows the water level in the tub over time.

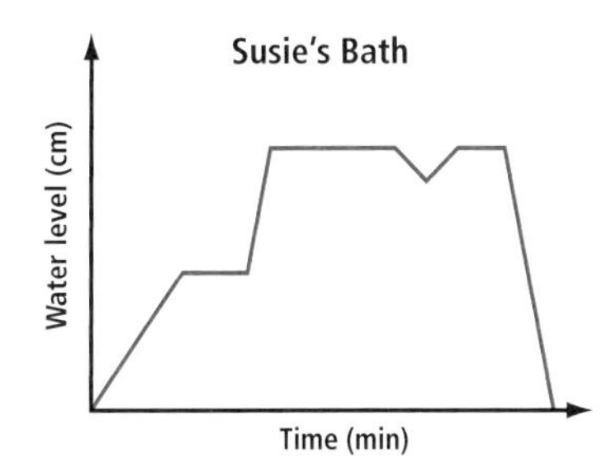

Write a story about Susie's bath that would explain the shape of this graph. Refer to the water level and time.

8. Juice is poured into three glasses at the same constant rate.

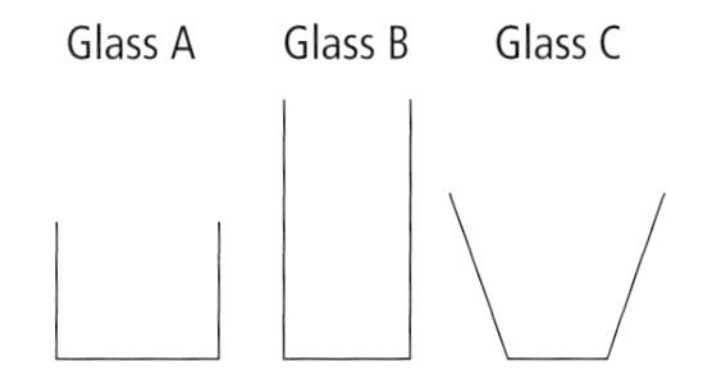

On the same set of axes, sketch a graph of the height of the juice over time for each glass.

9. Mitosis is the process of cell division, where one cell becomes two identical cells. Consider one type of cell that splits into two over the course of a 24-h period, so that the number of cells doubles every day. Suppose you start with one cell. How fast do you think the cell culture will grow?

a) Copy and complete the table.

Day	Number of Cells
0	1
1	$2 \times 1 = 2$
2	$2 \times 2 = 4$
3	
4	
5	

b) Is this relation linear or non-linear? Explain how you know.

c) Graph the data. Does the shape of the graph confirm your answer to part b)? Explain.

Making Connections

You will learn more about mitosis and other cellular functions in science class and in high school biology.

CHAPTER 8

Practice Test

Selected Response

Select the best answer.

1. What is the slope of the ramp?

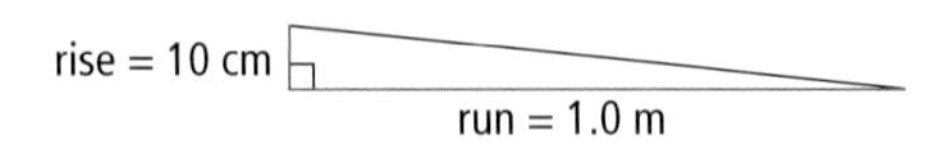

A $\frac{1}{100}$ **B** $\frac{1}{10}$

C 1 **D** 10

2. Which expression correctly gives the number of squares in each term of the growing pattern?

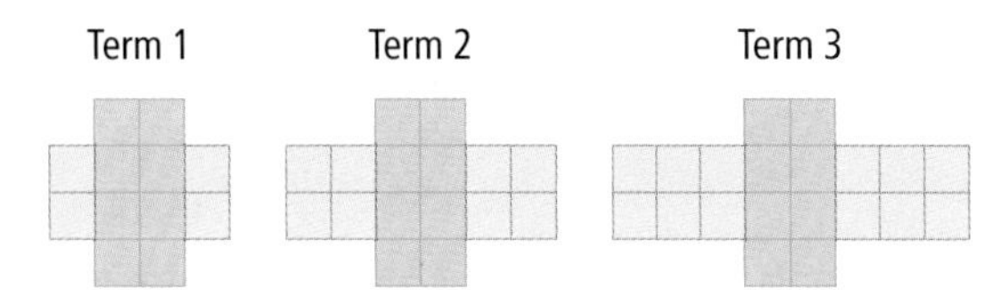

A $6n + 4$ **B** $8 + 2n$

C $4n + 8$ **D** $8n + 4$

3. Which relation is non-linear?

A

x	y
0	1
1	3
2	5
3	7

B

x	y
0	−3
1	−2
2	−1
3	0

C

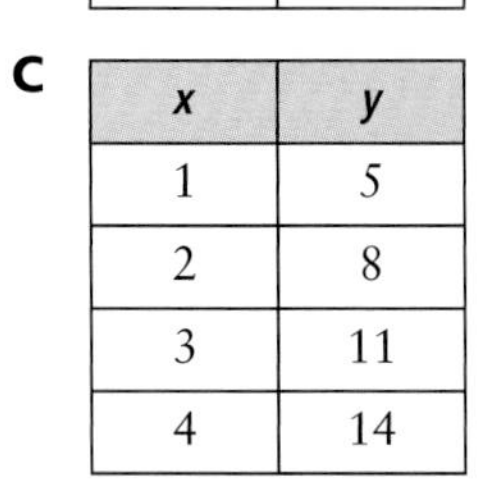

x	y
1	5
2	8
3	11
4	14

D

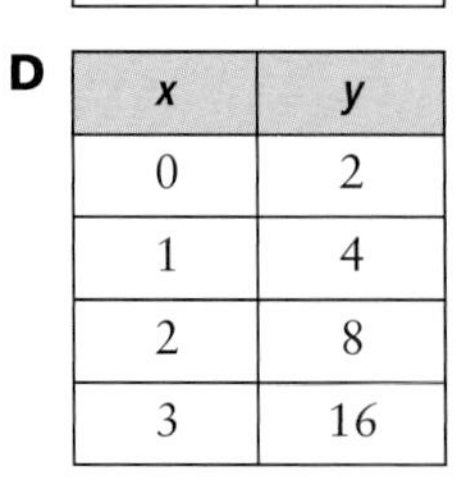

x	y
0	2
1	4
2	8
3	16

4. Water is poured into a cone-shaped container at a constant rate. Which graph best describes the water level compared to volume?

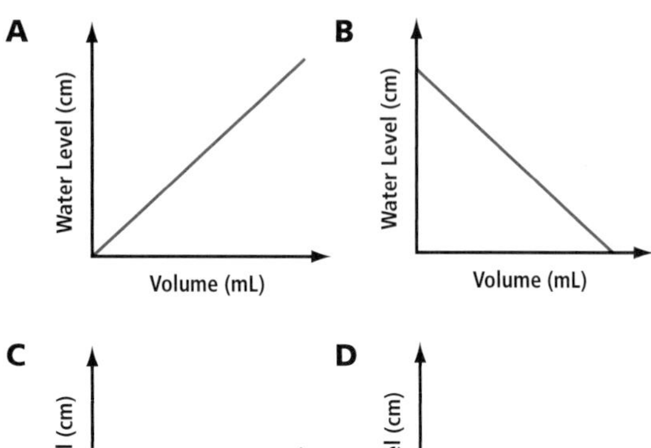

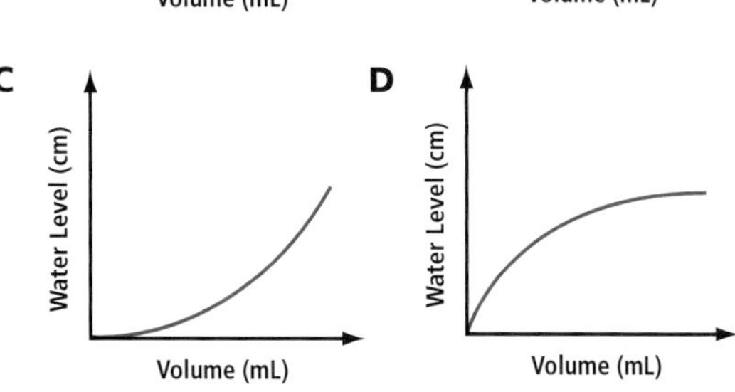

Short Response

Provide a complete solution.

5.

a) Determine the number of squares on the surface of each cube. Copy and complete the table.

Term Number, n	Number of Squares on Surface, s
1	6
2	
3	

b) Extend the pattern for the next two terms.

c) Graph the relation. Should you join the points? Explain.

d) Is this relation linear or non-linear? Explain two ways that you know.

6. The graph shows the cost of a taxi ride.

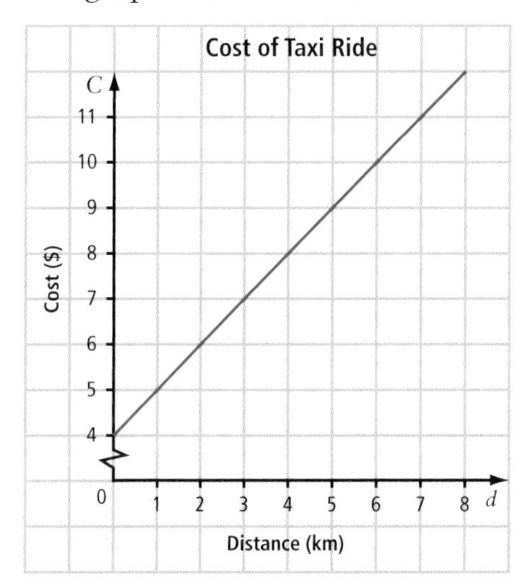

a) Copy and complete the table of values.

Distance, d (km)	Cost, C ($)
0	
1	
⋮	
8	

b) Describe the relation using words.

c) Write an equation that relates cost, C, to distance, d.

d) Use your equation to determine the cost of a 12-km trip.

Extended Response

Provide a complete solution.

7. Samantha has two choices for renting a guitar amplifier.

F-Sharp Music Supplies	*Super-Amps*
$20 insurance fee plus $6/h	$8/h

Which company should she choose? Justify your answer using at least two of the following:
- words
- equations
- a table
- a graph

8. Tommy and Rajiv are competing in a 1500-m race. Tommy's running speed is 2 m/s and Rajiv's speed is 3 m/s. To make this a fair race, Tommy is given a head start of 300 m.

a) Graph each runner's distance over time on the same set of axes. Use a different colour for each runner.

b) When will both runners have covered the same distance?

c) Who will finish the race first? Explain how you know.

Chapter Problem Wrap-Up

Another big event at your ski resort is the annual cross-country ski race. The time-distance graphs for the top two skiers are shown. Use the graph to write a play-by-play account of this race. Describe the race in terms of the skiers' time, distance, and speed. Add a reasonable scale to the axes.

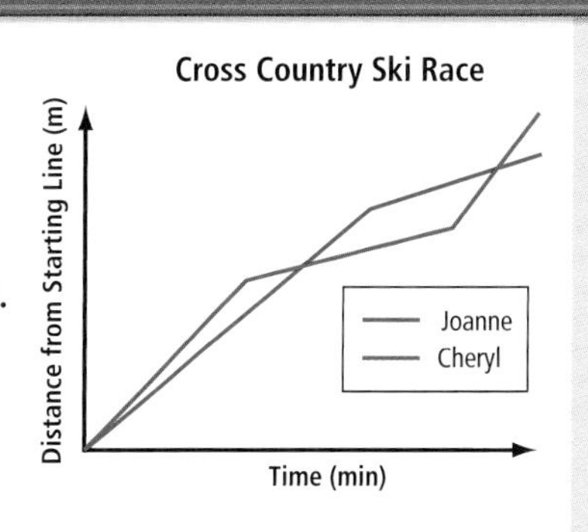

Be sure to mention:
- who skied the faster during the first leg of the race
- when both skiers were the same distance from the starting line
- who won the race

Making Connections

Equation Puzzle

Copy this three-by-three grid onto a piece of paper or thin cardboard. Make sure each section is square. Then, cut out the nine squares.

Rearrange the squares so that each equation is next to its solution. There will be extra equations and numbers around the outside edge that do not match up.

Hint: The piece from the top left corner slides, with no turn, to become the centre piece.

Materials

- scissors

Optional:

- BLM 8 Task Equation Puzzle

Square	Top edge	Right edge	Bottom edge	Left edge
Top left	7	13	$6m + 7 = 37$	$4k + 3 = 7$
Top centre	$m - 6 = 7$	2	7	$4w + 6 = 10$
Top right	4	$3c - 4 = 5$	8	$2b - 4 = 10$
Middle left	$c + 6 = 10$	$7w + 3 = 24$	3	1
Centre	$2k + 1 = 11$	$7x - 3 = 4$	2	2
Middle right	$4k + 1 = 9$	$y - 5 = 3$	6	4
Bottom left	9	$2x + 5 = 9$	1	$5x + 4 = 24$
Bottom centre	5	$5x = 15$	$8n = 24$	3
Bottom right	$2m - 6 = 8$	$3k - 1 = 5$	13	8

Magic Squares

1. Neeta and Sam found this card in the class math centre. They were not sure how to solve it. Help them find the missing numbers.

Magic Square

2		
	5	1
		8

What numbers belong in the empty squares? Remember that the rows, columns, and diagonals of a magic square all add to the same total.

2. Neeta made up this magic square where you must solve for x. Sam says that it is not a magic square. Who is correct? Explain why.

$2 \times 10^x = 0.002$	$1.3 \times 10^x = 130$	$7.1 \times 10^x = 71$
$8 \times 10^x = 80\ 000$	$5 \times 10^x = 5$	$40 \times 10^x = 0.004$
$7.7 \times 10^x = 0.77$	$6.1 \times 10^x = 0.061$	$0.9 \times 10^x = 900$

3. Sam found another card with these two magic squares. He thinks they must be related. Solve each magic square and explain how they are related.

		5 + 1
5 + 4	5	
5 − 1		

$\frac{1}{3}x - 1$	$\frac{1}{3}x - \frac{2}{3}$	
	$\frac{x}{3}$	
$\frac{1}{3}x - \frac{1}{3}$		

4. Create at least one original magic square. Use some positive and some negative rational numbers. Try to develop another using a variable or scientific notation.

- determine and apply the properties of dilatations
- determine and apply the properties of similar 2-D figures
- recognize, name, and describe prisms, pyramids, cones, and cylinders
- develop and apply the properties of prisms, pyramids, cones, and cylinders
- make and interpret isometric, orthographic, and perspective drawings

Key Words

dilatation

cone

oblique

cylinder

isometric drawing

orthographic drawing

perspective drawing

CHAPTER 9

Geometry II

Architects use geometry in their building designs. They combine geometric figures and repeated patterns of shapes to make buildings that are both functional and beautiful. The photograph of The Louvre Museum in Paris, France, shows part of the older palace and the famous glass pyramid. What geometric figures do you see? Do you see any repeated designs? Do you see any similar figures?

Architects need to record their designs on paper so builders can construct the buildings as the architects intended. These architectural drawings may include isometric, orthographic, and perspective drawings. All these drawings are drawn to scale so they are smaller, accurate descriptions of the buildings.

In this chapter, you will examine and investigate the properties of dilatations, similar figures, and some common three-dimensional objects. You will also learn several drawing techniques. Architects use these geometric ideas and techniques in their work.

Chapter Problem

Imagine you are an architect. You have been hired to design a new public building—it could be a museum, a sports arena, a concert hall, a train station, or a shopping centre.
Your client wants something original and eye catching. What type of building will you choose? You will learn about various geometric shapes you can use in your design and how to record your design using different types of scale drawings.

Get Ready

Congruent and Similar Figures

Congruent figures have the same shape and size. All corresponding sides are equal in length and all corresponding angles are equal in measure. These polygons are congruent.

Similar figures have the same shape, but may have different sizes. All corresponding angles are equal in measure and all corresponding sides are proportional. One figure is an enlargement or reduction of the other. These polygons are similar.

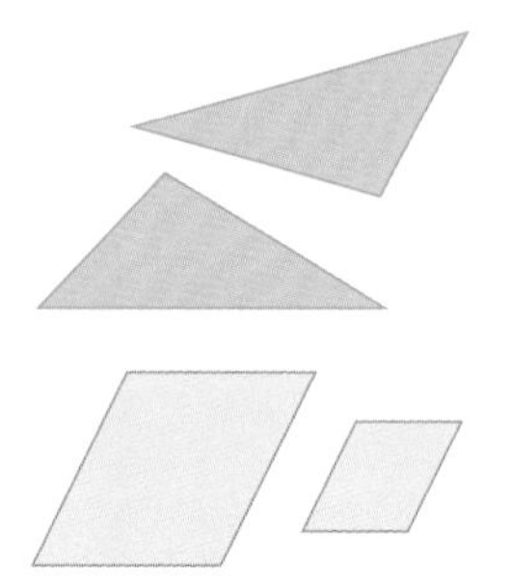

1. Classify each pair of figures as:
- congruent
- similar
- neither

a) 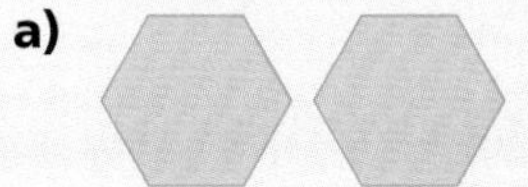

b) 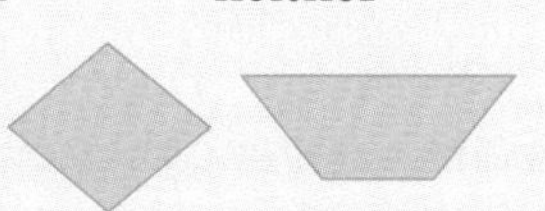

c) 

d) 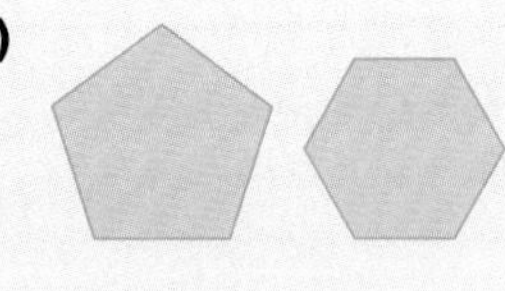

Proportional Length

If the lengths of two pairs of line segments can be compared using the same ratio, the pairs of segments are said to be **proportional**. Consider two pairs of line segments: 4 cm and 12 cm long, and 6 cm and 18 cm long. Compare the lengths of each pair of line segments as a ratio, 4:12 and 6:18, and you can see they are both equivalent to the ratio 1:3. Therefore, the 4-cm and 12-cm line segments are proportional to the 6-cm and 18-cm line segments. Each pair of proportional line segments will always be multiples of each other.

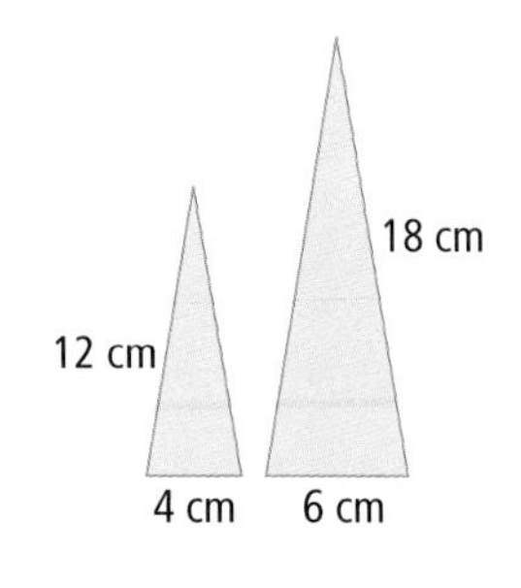

2. Which pairs of lengths are proportional to 2.5 cm and 10.0 cm? Explain how you know.

a) 4 cm and 16 cm

b) 5.0 cm and 12.5 cm

c) 12.5 m and 50.0 m

d) 10 cm and 40 cm

3. Give one pair of lengths that would be proportional to each pair.

a) 8.0 m and 1.6 m

b) 6 cm and 9 cm

c) 0.70 m and 0.35 m

d) 4 cm and 10 cm

4. Nathan said that 4 cm and 6 cm lengths are proportional to 5 cm and 7 cm lengths because 6 is 2 more than 4 and 7 is 2 more than 5. Is Nathan correct? Explain your decision.

Ratios and Proportions

Equal ratios form a proportion. If you know three numbers in a proportion, you can solve for the fourth.

$$\frac{3}{2} = \frac{18}{■}$$

× 6

$$\frac{3}{2} = \frac{18}{■}$$

× 6

$$\frac{3}{2} = \frac{18}{12}$$

5. What is the missing number in each proportion?

a) ■:30 = 2:3

b) 5:15 = 1:■

c) 10:■ = 120:12

d) 0:13 = ■:39

e) 2:■ = 5:30

f) 1:3 = ■:4.5

g) ■:12 = 2:3

h) 2:■ = 3:7.5

i) 6:5 = ■:12.5

6. Find the missing number in each proportion.

a) $\frac{1}{8} = \frac{■}{24}$

b) $\frac{3}{■} = \frac{90}{240}$

c) $\frac{5}{60} = \frac{1}{■}$

d) $\frac{■}{24} = \frac{8}{6}$

e) $\frac{64}{12} = \frac{48}{■}$

Get Ready

Name Pyramids and Prisms

Pyramids and **prisms** are special **polyhedra** and are named for the shape of their **polygon** bases.

This is a square pyramid or square-based pyramid.

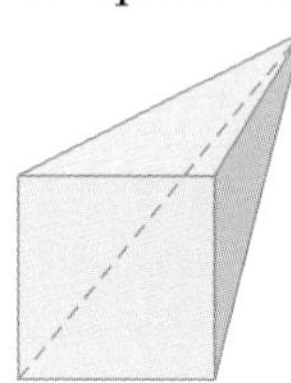

This is a rectangular prism or a rectangle-based prism.

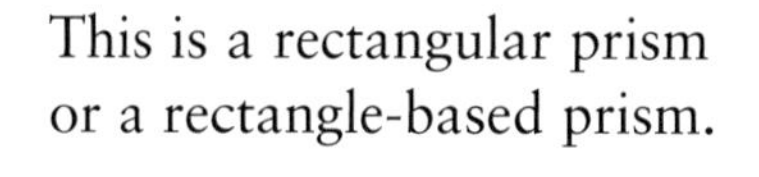

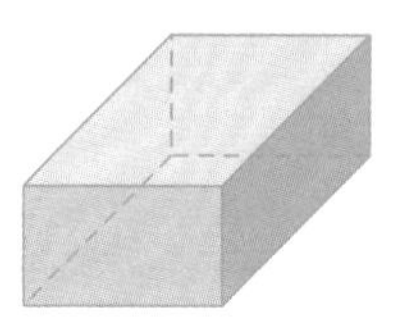

Polygons with five or more sides are named using Greek or Latin prefixes: penta for 5, hexa for 6, hepta (or septa) for 7, octa for 8, nona for 9, deca for 10, hendeca (or undeca) for 11, and dodeca for 12.

Therefore, a prism with nonagons as bases would be a nonagonal prism or nonagon-based prism.

7. Name each polyhedron.

a) 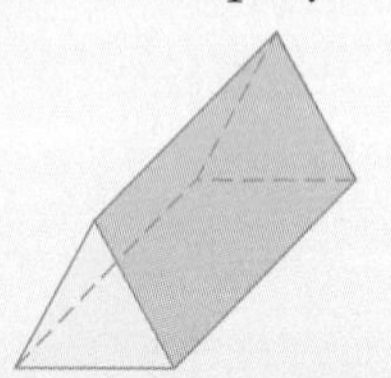

b) 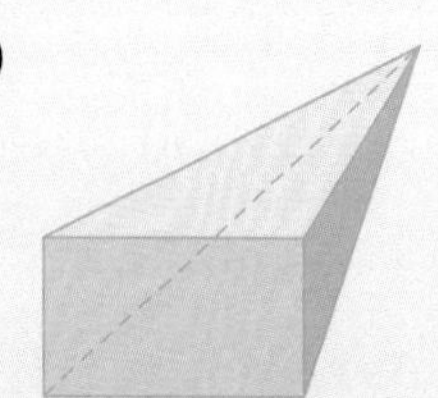

c) 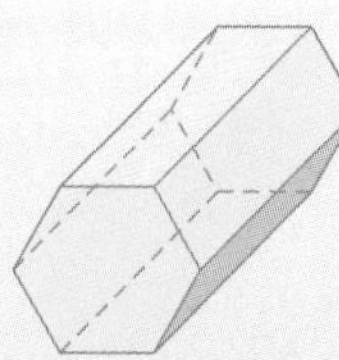

d) 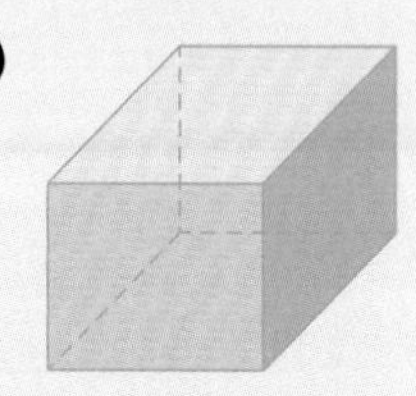

e) 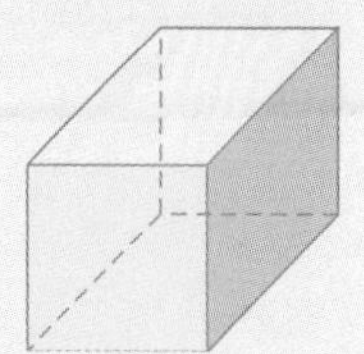

f) 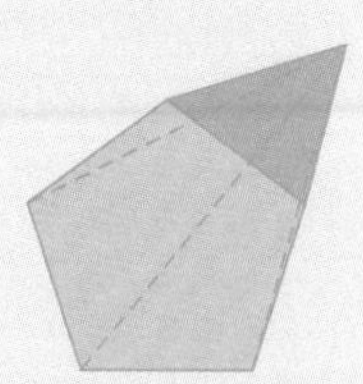

8. How many vertices, edges, and faces would each polyhedron have?

a) heptagonal pyramid **b)** octagonal prism

c) triangular pyramid **d)** decagonal prism

Name All Polyhedra

Any polyhedron can be named by referring to the number of faces using the same prefixes that are used for polygons. For example, a square pyramid (five faces) and a triangular prism (five faces) could both be called a pentahedron. Since the polyhedron names only indicate the number of faces, the prism and pyramid names are more useful because they tell you more about the objects.

The polyhedron names are used only for three-dimensional objects that lack more descriptive names.

9. The **Platonic solids** or **regular polyhedra** have faces that are all congruent regular polygons. What are the polyhedron names for the five Platonic solids?

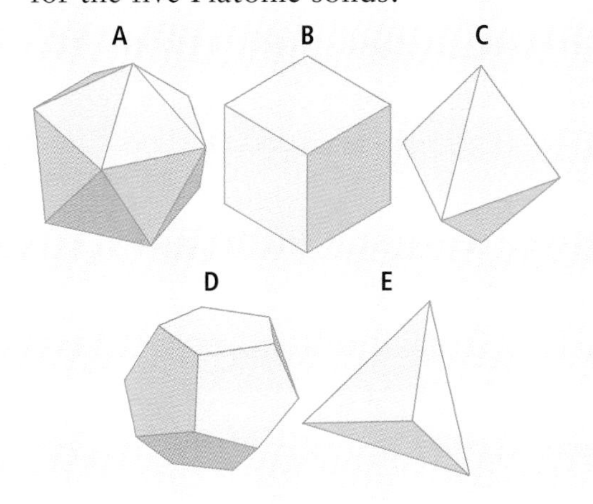

10. a) Yvette has a heptahedron. What do you know about this object?

b) Ravi has a hexagonal pyramid. What do you know about this object?

c) Could a hexagonal pyramid also be called a heptahedron? Explain.

d) Why is a hexagonal pyramid a better choice of name than heptahedron for this polyhedron?

11. a) What polyhedron name could be used for a decagonal prism?

b) Why is the prism name a better choice for this polyhedron?

12. Explain why a regular tetrahedron is a better name than triangular pyramid for this polyhedron.

9.1 Dilatations

Focus on...

- discovering and applying the properties of dilatations

Computer images and photographs can be enlarged or reduced, producing images that are similar to the original but that have a different size. Translations, reflections, and rotations are all transformations that produce images that are congruent to their pre-images. What kind of transformations can describe reductions and enlargements?

Discover the Math

Materials

- Pattern Blocks or Power Polygons®
- square dot paper
- isometric dot paper
- coloured pencils

Optional:

- BLM 9.2 DTM Dilatation Table 1 and 2

What are the properties of dilatations? How can you recognize and define dilatations?

dilatation

- a transformation that enlarges or reduces a figure by a scale factor

Work with a partner.

1. a) Start with one equilateral triangle block or polygon piece. Use congruent triangles to build an equilateral triangle with sides twice the length of the original triangle. How many triangles did you use?

b) Repeat part a) to build equilateral triangles with sides three and four times the length of the original. How many triangles did you use for each triangle?

c) Predict how many triangles you would need to construct an equilateral triangle with sides five times the length of the original. Build the triangle to test your prediction.

d) Draw an equilateral triangle with side lengths of 5 units on isometric dot paper.

All the triangles you built can be found within this triangle, sharing a common vertex and overlapping each other. Locate each of the other triangles and use different coloured pencils to highlight each one.

e) Label the common vertex A, and the other vertices of the smallest triangle B and C.

f) The larger triangles are dilatation images of △ABC. Label the corresponding vertices of the triangles using prime notation: B′, C′, B″, C″, B‴, C‴, B⁗, and C⁗.

2. a) Let the side length of the original triangle be one unit, and the triangle itself be one unit of area. Copy and complete the table using the results from question 1 and by finding patterns.

Equilateral Triangle	Length of One Side (units)	Perimeter (units)	Area (triangles)
△ABC	1	3	1
△AB′C′			
△AB″C″			
△AB‴C‴			
△AB⁗C⁗			
△AUV	10		
△AXY	n		

b) A dilatation is defined by a centre point, called the **dilatation centre**, and a **scale factor**. In the picture from question 1, part d), the centre of dilatation is the point A, and the scale factor is the ratio of the lengths of corresponding sides in two triangles. For example, △AB″C″ is the image of △ABC under the dilatation with A as the dilatation centre and a scale factor of 3, since AB″:AB is 3:1.

- Describe △AB‴C‴ as a dilatation image of △ABC.
- Describe △AB‴C‴ as a dilatation image of △AB′C′.
- Describe △ABC as a dilatation image of △AB′C′.

scale factor

- a ratio or number that represents the amount by which a figure is enlarged or reduced

c) Copy and complete the table.

Pre-Image	Image	Scale Factor (Pre-image to Image)	Ratio of The Areas
△ABC	△AB″C″		
△ABC	△AB‴C‴		
△ABC	△AB⁗C⁗		
△AB′C′	△AB‴C‴		
△AB′C′	△AB″C″		
△AB″C″	△AB‴C‴		
△AB⁗C⁗	△AB″C″		
△AB′C′	△ABC		
△AB‴C‴	△AB′C′		

d) Look for and describe the pattern between the scale factors and the ratios of the areas.

e) Predict the relationship between the scale factor of a dilatation and the ratio of the perimeters of the two triangles under that dilatation. Verify your prediction.

3. a) Copy the similar triangles on square dot paper.

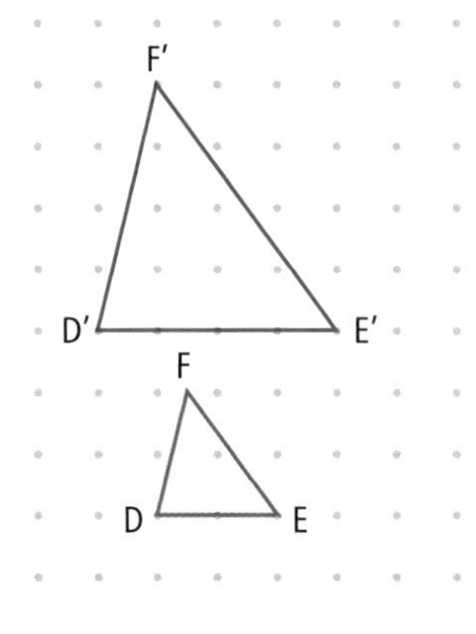

b) Describe two relationships between the corresponding sides of the triangles.

c) Describe the relationship between the corresponding angles of the triangles.

d) Join D′ to D and extend the line beyond D. In the same way, join E′ to E and F′ to F. What do you notice about these three lines? This point of intersection is the centre of dilatation. Label the point P.

e) Measure and compare PD′ and PD, PE′ and PE, and PF′ and PF. What do you notice about these three comparisons? How do these comparisons relate to the comparisons of the corresponding sides of the two triangles? Explain.

f) How do the areas of the two triangles compare?

4. a) Trace the quadrilateral and point C onto the middle, left side of a sheet of paper.

b) Join CJ, CK, CL, and CM.

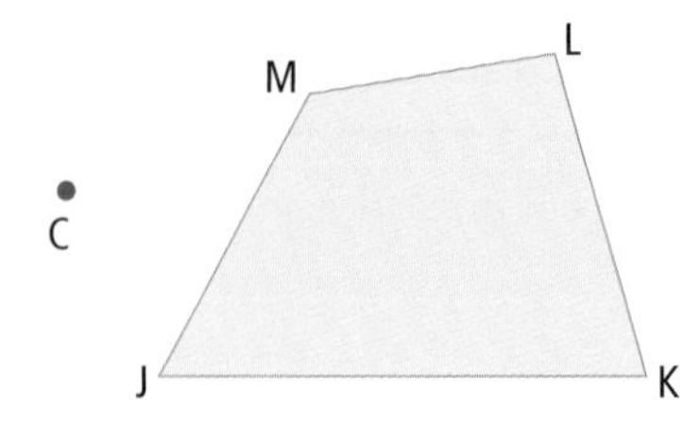

c) Measure CJ and extend CJ four times its length beyond J, labelling the endpoint J′. Repeat the process for CK, CL, and CM. Join the new points to form another quadrilateral.
d) Describe the larger quadrilateral as a dilatation image of the original.
e) Are the corresponding angles of the two quadrilaterals congruent? Explain how you know.
f) Are the corresponding sides of the two quadrilaterals parallel? Explain how you know.
g) Are the corresponding sides of the two quadrilaterals proportional? Explain how you know.
h) Are the two quadrilaterals similar? Explain how you know.
i) Are the line segments joining the centre of dilatation and corresponding points of the two quadrilaterals proportional? Explain how you know.
j) How do the areas of the quadrilaterals compare? Divide the larger quadrilateral into multiple copies of the smaller quadrilateral to illustrate this area relationship.

5. Reflect

a) Explain what you need to know to define a dilatation.
b) If a polygon and its dilatation image are a distance apart on a sheet of paper, how could you find the centre and scale factor of this dilatation?

Communicate the Key Ideas

1. If a polygon is an image of another polygon under a dilatation, list all the things you know about the relationships between the angles, sides, perimeters, and areas of these two polygons.

2. If you are given △PQR and a point J outside the triangle explain the steps you would follow to draw the dilatation image of △PQR with centre J and a scale factor of 3.

Example 1: Use Properties to Confirm a Dilatation

a) Is the larger rhombus a dilatation image of PQRS? Explain how you know.

b) If it is a dilatation image, describe the dilatation.

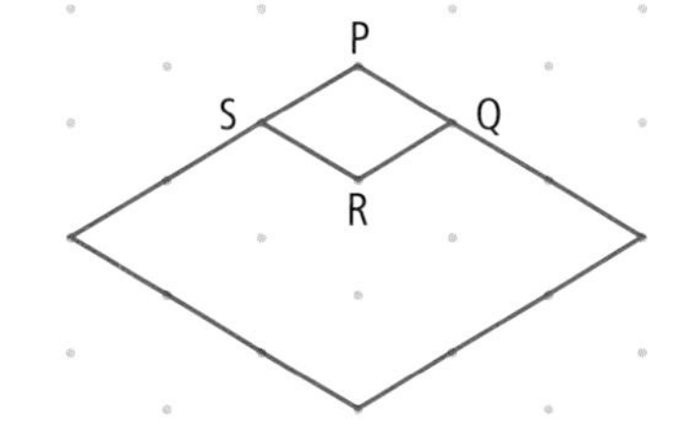

Solution

a) Yes, the larger rhombus is a dilatation image of PQRS. PS, PQ, and PR are all extended three times their length, and SR and QR are parallel to the corresponding sides of the larger rhombus.

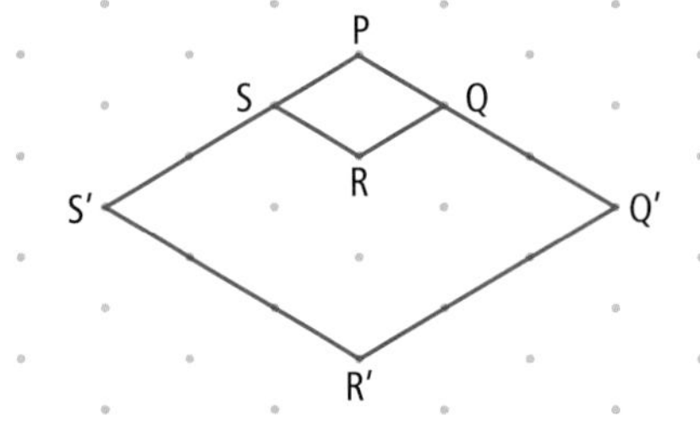

b) Label the corresponding vertices of the larger rhombus as Q′, R′, and S′. PQ′R′S′ is the image of PQRS under a dilatation with centre P and a scale factor of 3.

Communicating Mathematically

You can describe a dilatation as the ratio of the size of the image to pre-image or the pre-image to image.

Example 2: Use the Area Relationship of Dilatation Images

$\triangle A'BC'$ is a dilatation image of $\triangle ABC$. If the area of $\triangle ABC$ is 16 times the area of $\triangle A'BC'$, describe this dilatation.

Solution

Compare $\triangle A'BC'$ to $\triangle ABC$ for this dilatation.

- Area $\triangle A'BC'$:Area $\triangle ABC = 1{:}16$
- $A'B{:}AB = \frac{1}{4}$ because the ratio of areas is the square of the ratio of the side lengths. The scale factor must be $\frac{1}{4}$ because the ratio of the side lengths is the same as the scale factor of the dilatation.
- Since point B did not change under the dilatation, it must be the centre of the dilatation.
- Therefore, $\triangle A'BC'$ is the dilatation image of $\triangle ABC$ under a dilatation with centre B and a scale factor of $\frac{1}{4}$.

Did You Know?

Dilatations are used to re-size computer images. The point opposite the one selected is the centre of dilatation and a variable scale factor is applied horizontally and vertically depending on how the point is dragged.

Check Your Understanding

1. a) Copy rectangle ABCD onto the bottom right corner of a sheet of square dot paper.

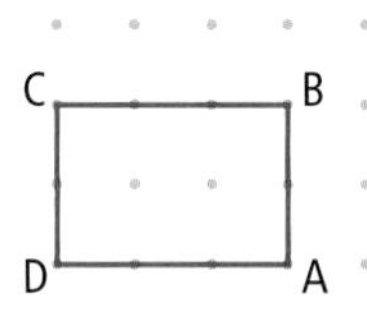

b) Draw the image of this rectangle under a dilatation with centre A and scale factor 2. Label the image.

c) Draw the image of ABCD under a dilatation with centre A and scale factor 3. Label the image.

d) Draw diagonal AC and extend it to C″. What do you notice about this diagonal in the pre-image and images?

2. a) Copy $\triangle JKL$ and point P onto the bottom left corner of a sheet of square dot paper.

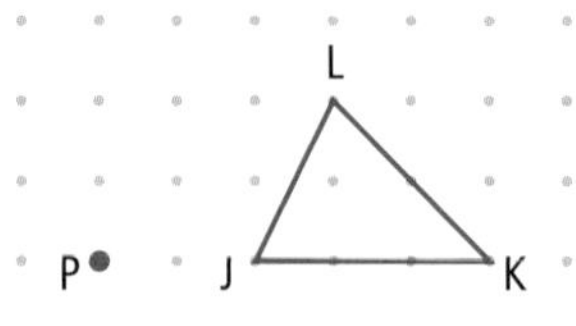

b) Draw the image of $\triangle JKL$ under a dilatation with centre P and scale factor 4. Label the image.

c) List everything you know about the corresponding angles and sides of the triangles.

d) Is JK parallel to J′K′? Use a mathematics dictionary to help you explain.

e) What is the ratio of the areas of the triangles? Divide △J′K′L′ into multiple copies of △JKL to illustrate this area relationship.

f) What is the ratio of the perimeters of the triangles? Explain how you know.

g) Suppose that △JKL was the image of △J′K′L′ under a dilatation. Describe this dilatation.

3. a) Is the smaller hexagon a dilatation image of the larger hexagon? Explain.

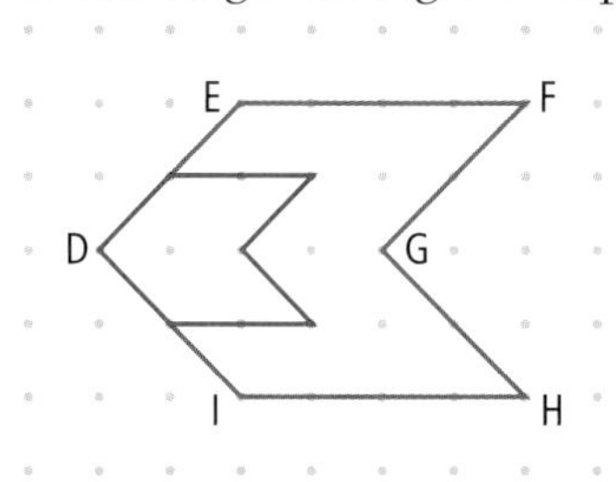

b) If it is a dilatation image, suggest appropriate labels and describe the dilatation.

c) Find the areas of the hexagons and the ratio of these areas. Is this ratio what you expected? Explain.

d) You cannot divide the larger hexagon into multiple copies of the smaller hexagon. Why do you think this is not possible?

4. a) Copy △PQR onto the middle of a sheet of square dot paper.

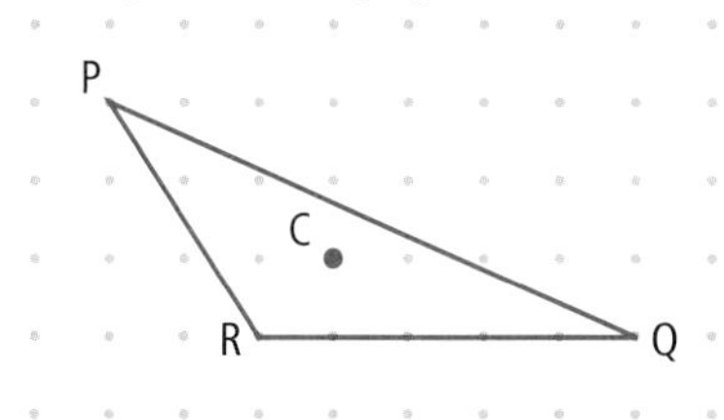

b) Draw the image of △PQR under a dilatation with centre C and scale factor 2.

c) List all the relationships between the angles and sides of the two triangles.

d) Draw the image of △PQR under a dilatation with centre C and scale factor 0.5.

e) Sharon says these triangles are *nested*. Explain what she means by this, and what caused them to be nested.

5. a) On square dot paper, draw a rectangle that has an area of 16 cm^2.

b) Draw a rectangle with an area of 4 cm^2 that shares a vertex with the original rectangle and is its dilatation image.

c) Describe the 4 cm^2 rectangle as a dilatation image of the original.

6. a) Copy the quadrilaterals on a coordinate grid.

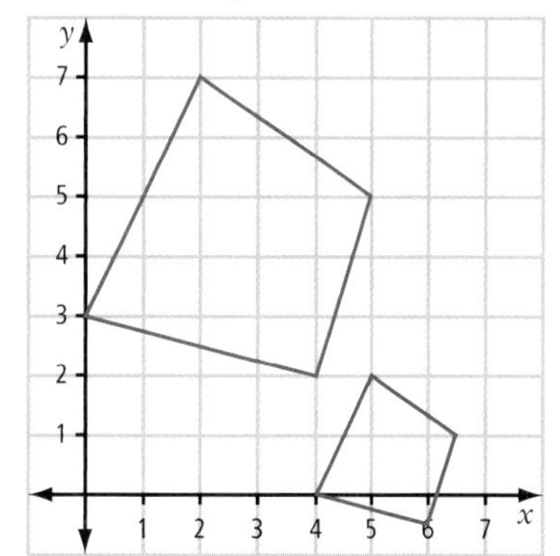

b) How does their appearance suggest that these quadrilaterals may be dilatation images of one another?

c) To check if they are dilatation images, locate a dilatation centre and determine a scale factor.

d) Label the larger quadrilateral KLMN and the smaller quadrilateral with primes.

e) Describe the smaller quadrilateral as a dilatation image of the larger one.

7. a) Use a pattern block to draw a regular hexagon in the middle of a sheet of paper and locate its centre.

b) Use the centre of the hexagon as the centre of dilatation. Draw the images of the hexagon with scale factors 0.5, 2, and 3.

c) Are the corresponding sides of the images parallel to the sides of the original? Measure the sides and find the ratios of the corresponding sides. Are the ratios of the sides the same as the scale factors of the dilatations? Explain.

d) Explain how the areas of the images compare to the area of the original.

e) Draw lines from the centre through the six vertices of the original hexagon and extend these lines. What do you notice about these lines? If you were to draw other dilatation images of the original hexagon with the same centre but with different scale factors, where would the vertices lie? Explain.

8. a) Copy the polygons on square dot paper.

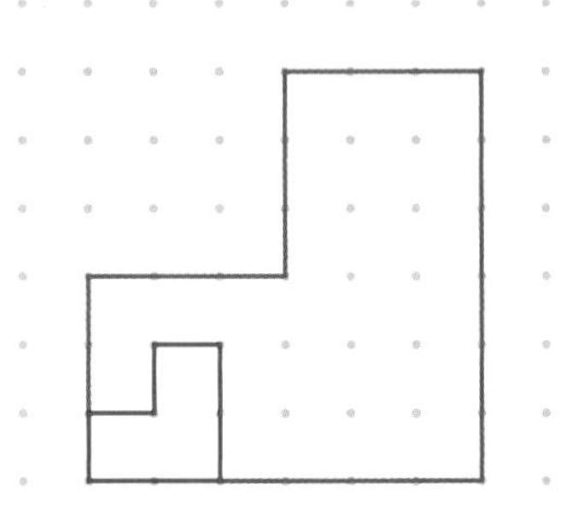

b) Divide the large polygon into as many copies of the small polygon as you can. How many times greater is the large polygon than the small polygon?

c) Is the large polygon a dilatation image of the small polygon? How do you know?

d) If the large polygon is a dilatation image of the small polygon, what is the scale factor?

e) Label the small polygon using letters and the large polygon using prime notation.

f) Describe the dilatation of the small polygon that results in the large polygon.

9. a) On a coordinate grid, draw $\triangle$LMN with vertices L(4, 4), M(6, −2), and N(10, 2).

b) Draw the image of $\triangle$LMN under a dilatation with centre (0, 0) and scale factor 1.5. Label the image.

c) List all the relationships between the sides and angles of the image and $\triangle$LMN.

Extend

10. a) On a coordinate grid, draw parallelogram JKLM with vertices J(1, 2), K(5, 2), L(6, 3), and M(2, 3).

b) Multiply the x- and y-coordinates of each given point by −2 to get a new parallelogram with vertices J′, K′, L′, and M′. Draw the parallelogram on the same grid.

c) The parallelograms are dilatation images. How are they different from other dilatation images you have seen? Do they have the same properties as other dilatations? Explain.

d) What is the centre of dilatation? How is the location of this centre different from other dilatations you have seen?

e) Draw the image of JKLM under a dilatation with centre (0, 0) and scale factor 2. Label the image using double prime notation.

f) Compare J′K′L′M′ and J″K″L″M″. How are the images the same? different? What do you think has caused the differences? Explain.

g) Describe J′K′L′M′ as a dilatation image of JKLM.

9.2 Properties of Similar Figures

Focus on...
- discovering and applying the properties of similar 2-D figures

An overhead projector, a projection television, and the zoom feature on a photocopier all produce larger or smaller images that are identical to the original. The original and its image are described in mathematics as being **similar**. You might say two people are similar even if they are not identical but in mathematics the word has a very precise meaning and is only used when certain conditions are met. What conditions must exist to call two figures similar? How can you apply the properties of similar figures to solve problems?

Discover the Math

Materials
- protractor
- ruler

Part A: What makes two figures similar?

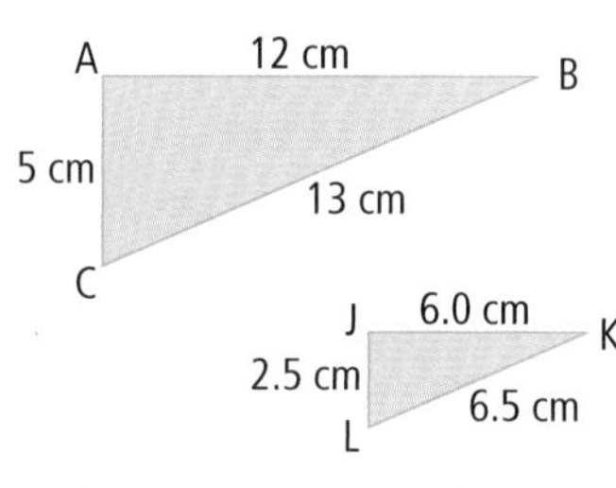

1. a) Without comparing the measurements, does △JKL look as if it could be a reduced image of △ABC? Explain.
   b) Compare AB in △ABC to JK in △JKL by determining the ratio AB:JK.
   c) Determine the ratios AC:JL and BC:KL.
   d) What do you notice about the three ratios? What is the name for this relationship between the corresponding sides?
   e) Compare the measures of ∠A and ∠J, ∠B and ∠K, and ∠C and ∠L. What do you notice about the angles?
   f) Mathematicians would say that △ABC and △JKL are similar, or that △ABC ~ △JKL. What conditions must exist for two figures to be called similar?

2. Rectangle PTUV was created on a computer by highlighting rectangle PQRS, and dragging point R down and to the right.

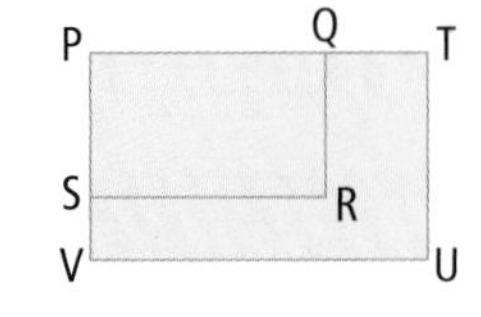

a) How do the measures of the corresponding angles of the two rectangles compare?

b) Measure and find the ratios of the corresponding sides.

c) How do the ratios of the corresponding sides compare?

d) Are the two rectangles similar? Explain.

e) Calculate the areas of the two rectangles and find the ratio of these areas. How does the ratio of the areas compare to the ratio of the sides?

3. a) Sonjay says the corresponding angles of these two quadrilaterals are congruent. Use a protractor or tracing paper to compare the pairs of corresponding angles. Is Sonjay correct? Explain.

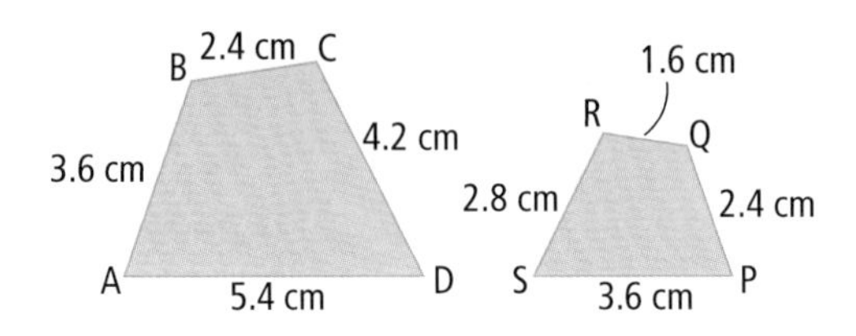

b) Sonjay also says the corresponding sides of these quadrilaterals are proportional. Compare the side measures and explain if Sonjay is correct.

c) Would Sonjay be correct in saying that the two quadrilaterals are similar? Explain.

d) What are the two possible scale factors for the two quadrilaterals? Describe the relationship between the two quadrilaterals. Use the scale factors to help you explain.

e) What would you expect to be the ratio of the areas of the quadrilaterals? Why?

4. a) Draw a line segment of any length and label it DE. To create $\triangle DEF$, draw an angle of 50° at D and extend the arm of the angle. At E, draw an angle of 70° and extend the arm until it intersects the other extended arm at F.

b) Draw another line segment PQ of a different length than DE. To create $\triangle PQR$, draw an angle of 50° at P and extend the arm of the angle. At Q, draw an angle of 70° and extend the arm until it intersects the other extended arm at R.

c) Are the three corresponding angles of both triangles congruent? Explain.

d) Measure the lengths of the three sides of both triangles. Calculate the ratios of corresponding sides. Are the corresponding sides of the two triangles proportional? Explain.

e) Is $\triangle DEF \sim \triangle PQR$? Explain.

Communicating Mathematically

The symbol $\sim$ means that two figures are similar. $\triangle ABC \sim \triangle RST$ means "triangle ABC is similar to triangle RST." The convention is to name the two similar figures so the letters of the corresponding vertices are in the same order. Compare this symbol to $\cong$, the symbol for congruency that uses a combination of the equality (=) and similarity ($\sim$) symbols.

5. **a)** Refer to question 4. Which parts of the triangles did you know the actual measurements for after you drew them?
   b) Based on your experience, do you agree or disagree with Pierre? Explain.
   c) Draw two quadrilaterals that show Pierre's statement only applies to triangles.

If a pair of triangles each have two corresponding angles of equal measure then the triangles are similar.

6. **Reflect** If someone tells you that pentagon ABCDE is similar to pentagon PQRST, list everything you would know about the relationships between the parts of the two pentagons.

Part B: How can you use the properties of similar figures to solve problems?

1. Rectangles ABCD and AEFG are similar.
   a) Determine the ratio FG:CD.
   b) What should be the ratio AG:AD? Why?
   c) Use what you know from part b) to determine the length of AG. Show your work.

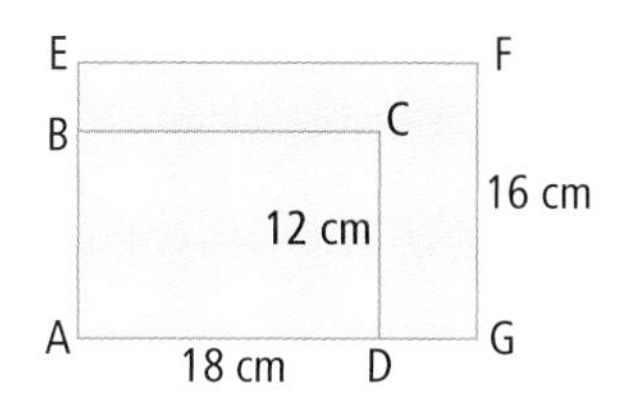

2. $\triangle RST \sim \triangle JKL$.
   a) Determine the measures of the angles in both triangles.
   b) Determine the ratio RS:JK.
   c) What should be the ratios RT:JL and ST:KL? Why?
   d) Determine the lengths of ST and RT. Show your work.
   e) If the area of $\triangle JKL$ is 177 cm^2, what is the area of $\triangle RST$? Explain how you know.

L
26.4 cm
14.9 cm
82°
34°
J
24 cm
K
T
S
12 cm
R

3. Mary and Sue wanted to find the height of a tree but have no ladder.

   Their teacher explained that they could use what they know about similar triangles to find the tree's height.

   "Mark the end of the shadow

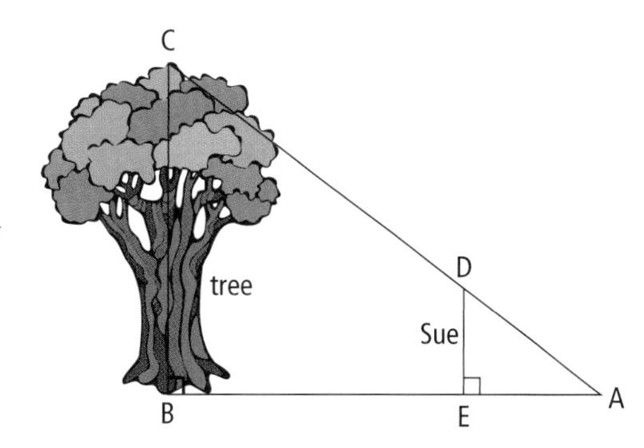

being cast by the tree as point A. Sue should stand on line segment AB so the end of her shadow is also on point A. Mary should measure the length of the tree's shadow (AB) and the length of Sue's shadow (AE)."

a) Are △ABC and △AED similar? How do you know?
b) Sue is 1.6 m tall and is standing 2.8 m from point A. The length of the tree's shadow is 8.4 m. Determine the ratio AB:AE.
c) What should be the ratio BC:DE? Why?
d) Use this ratio to determine BC, the height of the tree.

4. The next day, Sue and Mary try to determine the height of another tree. They find the length of that tree's shadow is 7.2 m and the length of Sue's shadow is 1.2 m. What is the height of the tree?

5. Reflect What property of similar figures enables you to determine the lengths of the unknown sides? Explain. How much information must you know about the similar figures to use this property?

Communicate the Key Ideas

1. What does Andre mean? Explain.

The corresponding sides of similar figures are proportional.

2. What do you need to measure and calculate to show that two polygons are similar? Explain how triangles are an exception.

3. Suppose you have two triangles, one on a chalkboard and one in your notebook. You suspect they are similar but have no ruler to take measurements.

a) How could you use a protractor to verify your suspicion? Why would this work?
b) If the triangles are similar, how many sides would you need to measure to know the ratio of their areas?

4. If the ratio of two corresponding sides in two similar quadrilaterals is 4:3, what might be the lengths of the sides in each quadrilateral?

Example 1: Show Two Figures are Similar

In the diagram, $AG \parallel BC \parallel EF$ and $AE \parallel CD \parallel FG$.
Is $ABCD \sim AEFG$? Explain.

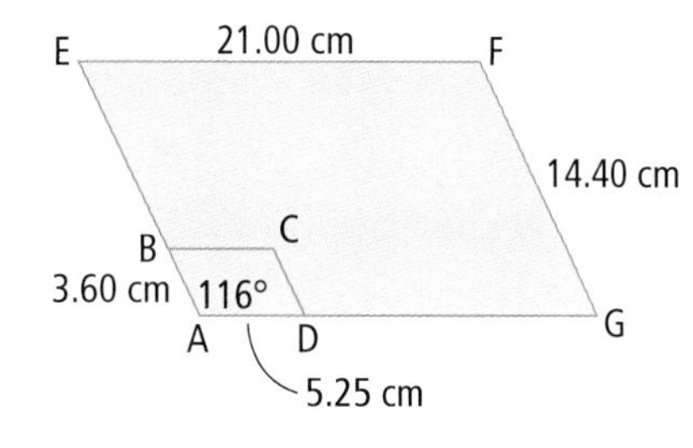

Solution

ABCD and AEFG are parallelograms because the opposite sides are parallel.

$\angle A$, $\angle C$, and $\angle F$ all equal 116° because they are opposite angles of the parallelograms and $\angle A = 116°$.

$\angle ABC$, $\angle ADC$, $\angle E$, and $\angle G$ are all equal 64° because adjacent angles in parallelograms are supplementary.

AE = 14.40 cm, AG = 21.00 cm, BC = 5.25 cm, and CD = 3.60 cm because opposite sides of a parallelogram are equal.

AB:AE = 3.60:14.40 or 1:4 and DC:FG = 1:4.

AD:AG = 5.25:21.00 or 1:4 and BC:EF = 1:4.

Therefore, $ABCD \sim AEFG$ because corresponding angles are congruent and corresponding sides are proportional (have the same ratio).

Example 2: Use Properties of Similar Figures to Find Missing Parts

If AE = 11.7 cm, what is the length of DE?

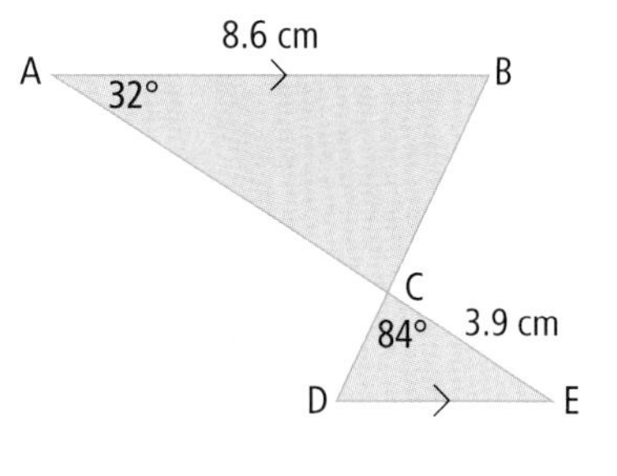

Solution

If the two triangles are similar, then use the ratio of the side lengths.

$\angle E = 32°$ because $AB \parallel DE$ and $\angle A = 32°$ (alternate interior angles).

$\angle ACB = \angle ECD$ because they are opposite angles.

$\triangle ABC \sim \triangle EDC$ because triangles are similar if two pairs of angles are equal.

Since AE = 11.7 cm and CE = 3.9 cm, then AC = 7.8 cm.

The ratio of the sides in the similar triangles is 3.9:7.8 = 1:2.

DE:8.6 must also be 1:2 because corresponding sides of similar triangles are proportional.

DE = 4.3 cm because 4.3 is one half of 8.6 cm.

Making Connections

Architects use scale drawings of the floor plan for a house.

Check Your Understanding

1. Show that each pair of figures is similar.

a)

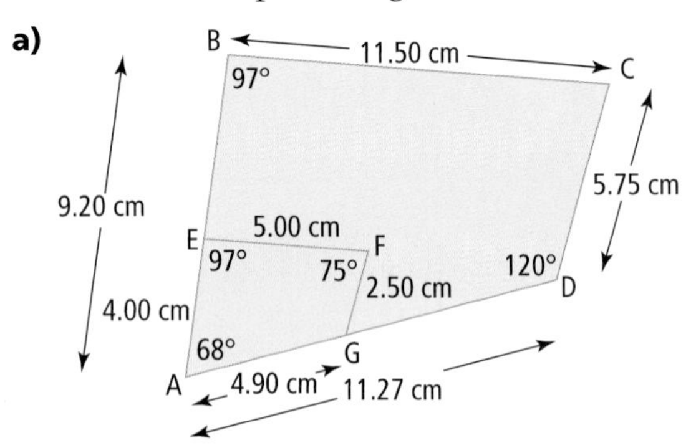

b)

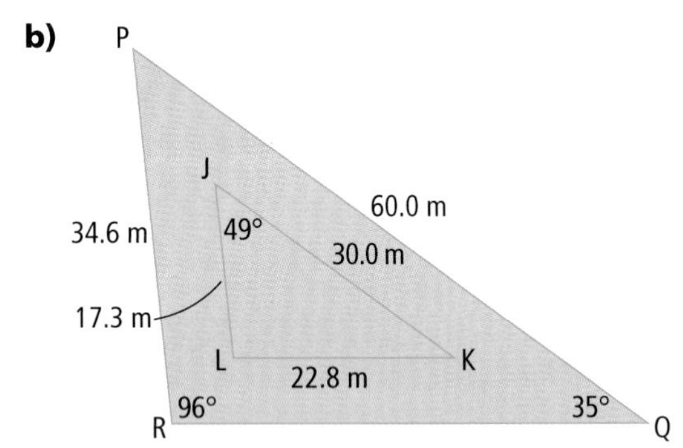

2. a) Draw each figure on 1-cm square dot paper.
   i) a 6 cm by 15 cm rectangle
   ii) a rectangle similar to the rectangle in part a) with a scale factor of one-third
   iii) a parallelogram that has a base of 15 cm and a height of 6 cm
   iv) a parallelogram similar to the one part c) with a scale factor of one-third

   b) Was it easier to draw the similar rectangle or the similar parallelogram? Why?

3. Which rectangles are similar? Justify your choices.

Rectangle	Width	Length
ABCD	6 m	8 m
EFGH	4.6 cm	9.8 cm
IJKL	12.6 cm	16.8 cm
MNOP	2 m	4 m
QRST	18 cm	24 cm
UVWX	23 cm	49 cm

4. Mr. Reynolds has a rectangular building lot that is 40 m wide and 120 m long. Mr. Jones has a similar lot that is 144 m long. How many metres wide is Mr. Jones's lot?

5. PQRS ~ PTUV.

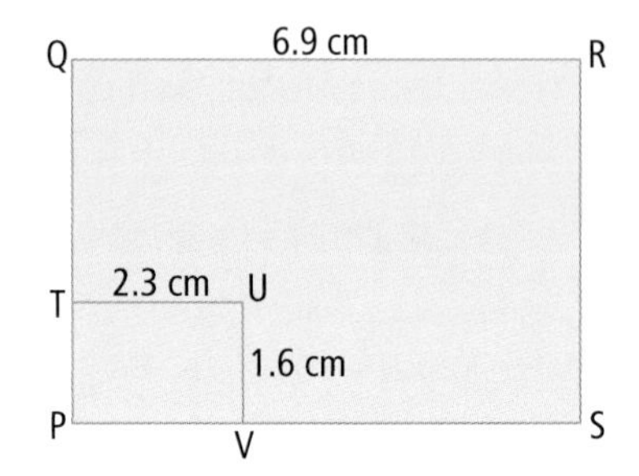

What is the area of the hexagon QRSVUT?

6. Kyle measured the length of side LM in △LMN that was drawn on the chalkboard. He found it was 28 cm long. He also measured ∠L and ∠M before returning to his seat.

   a) Kyle wanted to know if △PQR in his notebook is similar to △LMN. Explain how Kyle could check.

   b) Kyle measured the side in △PQR that corresponded to LM and found it was 7 cm. Explain how Kyle could determine how the areas of the two triangles compared.

7. Rob of Johnston's Photos explained that when he develops film, the 4 in. by 6 in. prints contained everything that was on the negatives. If someone chose the 3.5 in. by 5 in. prints, he had to crop the negatives, so those prints would not quite be similar to the 4 in. by 6 in. prints. Does Rob crop the width or the length to fit the 3.5 in. by 5 in. prints? Show your calculations and explain your decision.

8. Create two triangles: one using an R1, a B1, and a Y1 Geostrip® and one using an R2, a B2, and a Y2 Geostrip®.

a) Are the two triangles similar? Explain how you know.

b) Leon claims that if you draw two triangles that have sides that are proportional, the corresponding angles will always be congruent, and the two triangles will be similar. Based on your experience with the Geostrip® triangles, do you agree with Leon? Explain.

c) Leon says that this only applies to triangles. To prove his point, he drew two quadrilaterals with proportional sides that were not similar. Create two parallelograms: one using two R1 and two B1 strips and one using two R2 and two B2 strips. Arrange the strips so the parallelograms are not similar. How can you re-arrange the strips so the parallelograms are similar? What did you have to watch carefully to make them similar?

9. Each pair of figures is similar. Determine the measure of all the missing sides.

a)

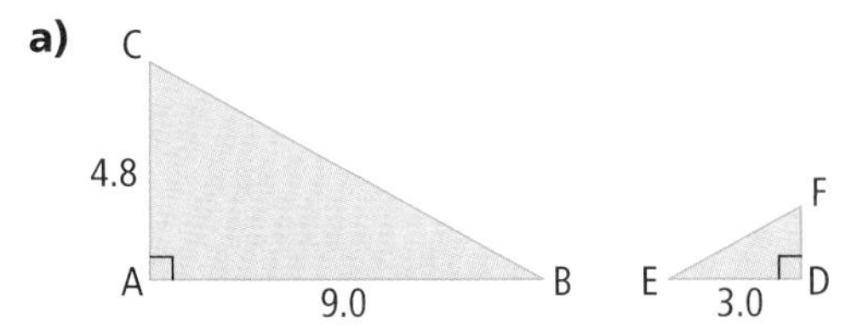

b)

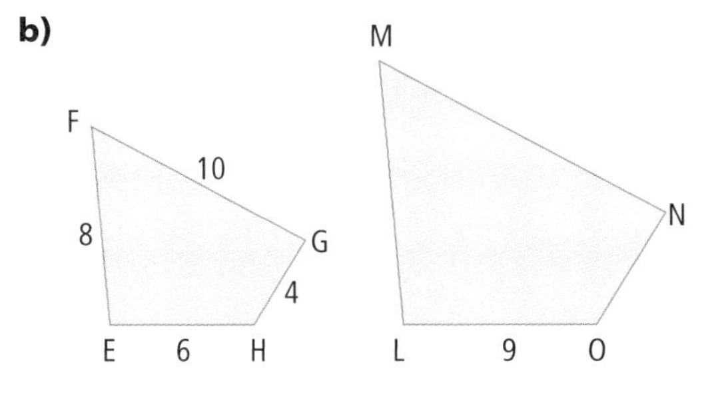

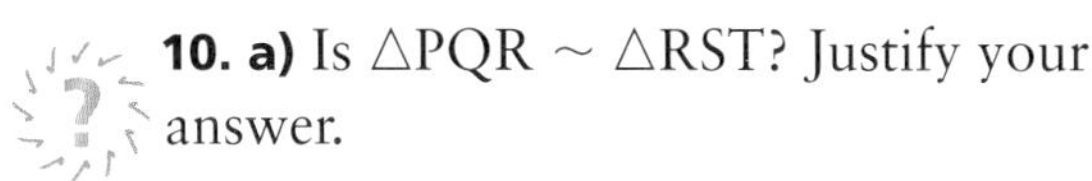

10. a) Is $\triangle PQR \sim \triangle RST$? Justify your answer.

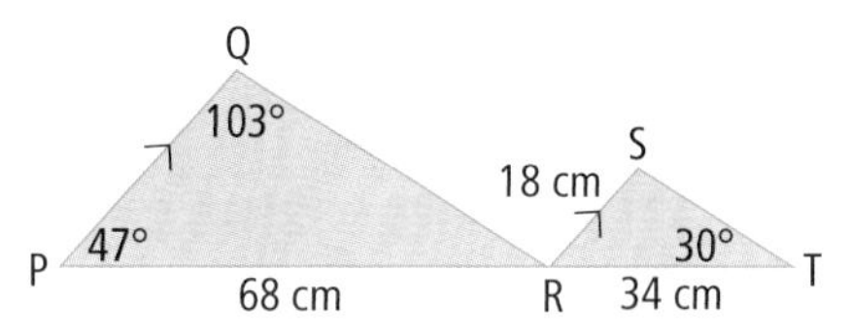

b) Determine the length of PQ. Show your work.

11. a) Draw a quadrilateral with all sides of different lengths.

b) Draw another quadrilateral that is similar but not congruent to the original quadrilateral. Use a scale factor of your choice. Explain your steps.

c) Label the angle and side measures of each quadrilateral.

12. a) Explain why Yvette is able to make this claim.

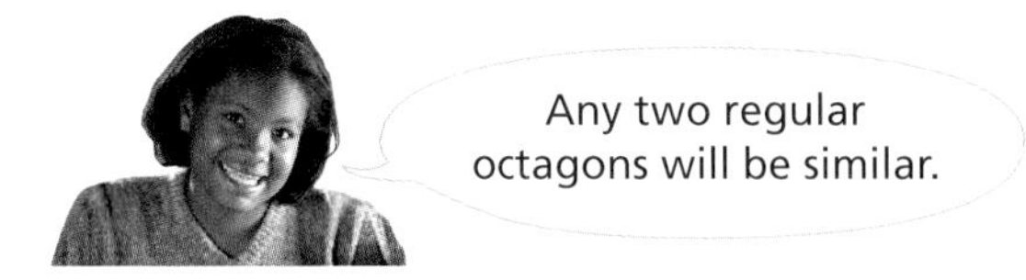

b) Is this true for any two equilateral triangles or any two squares? Explain.

13. Tricia measured the perimeter of the living room and her bedroom in her family's new house. The living room is 4.5 m by 6.5 m. Her bedroom is 3.5 m by 4.0 m. The living room has a 2.5 m wide doorway in the middle of one long wall and her bedroom has a 0.8 m wide doorway in the middle of one shorter wall.

a) On 1-cm square dot paper, make a scale drawing of the floors of both rooms. Include the doorways. Use the scale of 0.5 cm = 1 m.

b) Is your drawing similar to the floors of the actual rooms? Explain.

14. Jake and Leanne wondered how tall the flagpole in the schoolyard was. They measured a distance of 15 m out from the base of the flagpole and marked it with a stake. Jake lay down so his eyes were at ground level by the stake, with the flagpole directly in his line of sight. Leanne walked toward the flagpole in Jake's line of sight and stopped when Jake could just see the top of the flagpole over her head. Jake then measured how far Leanne was from the stake and found it was 3 m. Leanne is 1.7 m tall.

   a) Draw a diagram that represents this situation.

   b) Explain how Jake and Leanne could use the measurements they have to find the height of the flagpole.

   c) What is the height of the flagpole?

Chapter Problem

15. One wall in your building will have a door and a window that are the same size as a door and window in your home. Choose a door and a window in your home and make all the necessary measurements. On grid paper, make a scale drawing that shows where you have placed the door and the window on the wall. Choose a scale that allows you to fit the entire wall on the paper. Label your drawing with the scale and the measurements.

Extend

16. Suppose you have a set of right triangles that have legs with lengths $5x$ and $12x$, where $x > 0$. Write a convincing argument that all triangles in this set are similar to one another.

17. AB = 12.6 cm, JK = 4.2 cm, and AB $\parallel$ JK. AK and BJ intersect at L. Write a convincing argument that AL:KL = BL:JL.

Did You Know?

Three natural numbers that satisfy the Pythagorean relationship are called a Pythagorean triple.

Puzzler

Are all right triangles that have side lengths that are Pythagorean triples similar?

Provide examples to help you explain.

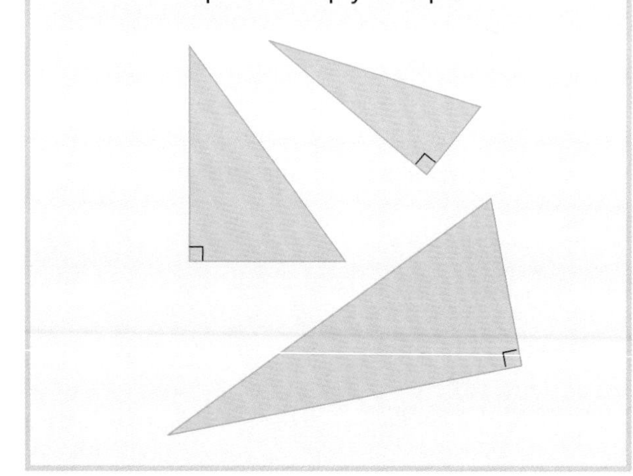

9.3 Cones and Cylinders

Focus on...

- recognizing, naming, and describing prisms, pyramids, cones, and cylinders
- developing and applying the properties of prisms, pyramids, cones, and cylinders

Materials

- sets of 3-D objects
- Polydron® pieces
- Geoblocks
- Geometric Solids

Optional:

- BLM 9.3 DTM Pyramid Table

Architects use a variety of three-dimensional figures, and combinations or modifications of these figures in their designs. Some everyday objects, such as cardboard boxes and tin cans, are easy to recognize and name. It is not as easy to recognize and name other objects, especially when only parts of them are used, or when they are used in combinations. Examine the picture of Province House in Halifax, NS, the home of the Nova Scotia Legislature. How many three-dimensional figures can you find? Sketch and name as many as you can.

cone

- a three-dimensional object with a base that is a flat surface and a point not on the base called an apex. All points on the base are connected to the apex by a line segment.
- a right circular cones has a circular base and a line drawn through the apex and the centre of the base is perpendicular to the base

Discover the Math

Part A: Cones and Pyramids

Which 3-D objects can be classified as cones or pyramids in mathematics?

Work in pairs.

Some of these three-dimensional objects are **cones** .

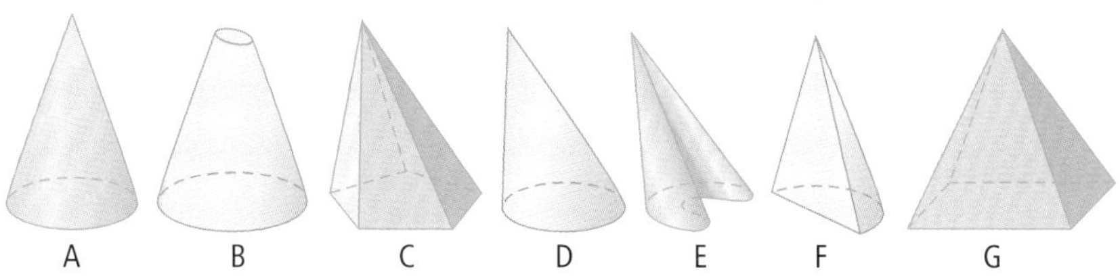

1. a) Examine the definition of a cone. List the three conditions required for a three-dimensional object to be a cone.
   b) Examine each object on page 403. Which objects match the definition of a cone? Explain.
   c) What characteristics of cones are missing from each object that is not a cone?

apex
- the point in a three-dimensional object that is highest from the base

2. a) Use an equilateral triangle and any three congruent triangles from a set of Polydron® pieces to build a tetrahedron. What is another name for this polyhedron?
   b) Could the tetrahedron you built also be called a cone? Explain.

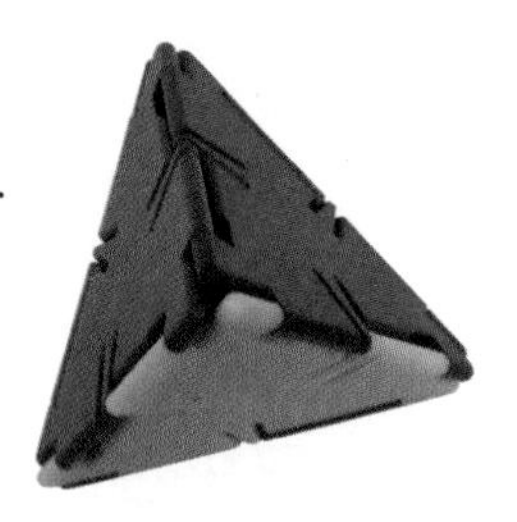

   c) Build a pentahedron using a square and four congruent triangles. Describe its faces.
   d) Look at a Geoblock pentahedron. Describe its faces.
   e) How does a pyramid fit the definition of a cone?

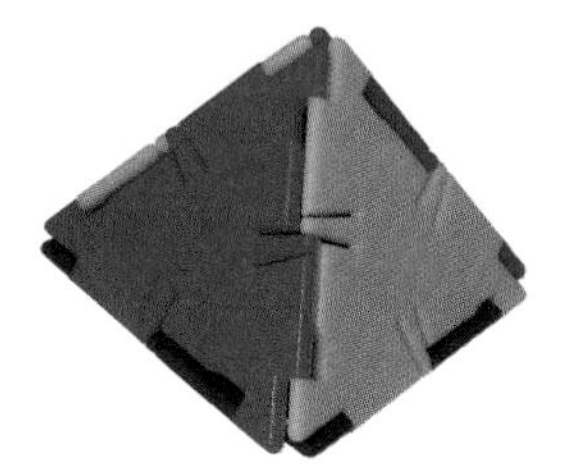

3. Certain cones, such as A and G and the pyramids in question 2 on page 403 can be described as right cones.
   a) How do you think the word *right* applies to these cones?
   b) Look at the polyhedra you built. Imagine a line through the apex of each polyhedron, extending to the middle of the base. What angle is formed by the line at the base?
   c) If the line does not form a right angle, the cone is called **oblique**. Draw an oblique pyramid.
   d) A full description of a cone or pyramid could include *right* or *oblique* as well as the name of its base. For example, the Egyptian pyramids are right square pyramids. Name the Geoblock pentahedron and all the pyramids you built.
   e) How would the faces of a *right* pentagonal pyramid be different from those of an *oblique* pentagonal pyramid?

oblique
- neither parallel nor perpendicular

Communicating Mathematically

Although there are many different types of cones, the term *cone* is most commonly used to refer to right circular cones.

4. a) Copy and complete the table. Use a set of geometric solids to help you. Look for patterns to generalize for an n-gonal pyramid. The bases of these pyramids are all regular polygons.

Base of Pyramid	Number of Faces	Number of Vertices	Number of Edges
triangular			
square			
pentagonal			
hexagonal			
heptagonal			
octagonal			
nonagonal			
decagonal			
20-gonal			
n-gonal			

b) Describe how to use the name of a pyramid to determine the number of faces, vertices, and edges.

c) Among all polyhedra, only pyramids have the same number of vertices as faces. If you are told that a polyhedron has eight faces and eight vertices, what is its name?

5. a) Examine a right square pyramid for symmetry, both reflective and rotational. How many planes of symmetry does the pyramid have? Explain. What order of rotational symmetry does it have? Explain.

b) Examine some other right pyramids. Is Justina correct? Explain. Why is the regular tetrahedron an exception?

The symmetry of any right pyramid is the same as the symmetry of its base.

c) If a square pyramid were oblique, how would these symmetries change? Explain.

d) How would you describe the reflective and rotational symmetries of a right circular cone? Describe how these symmetries would change if the circular cone were oblique.

6. Reflect Describe how your understanding of cones and pyramids has changed.

Part B: Cylinders and Prisms

Which 3-D objects can be classified as cylinders or prisms in mathematics?

Work in pairs.
Some of these three-dimensional objects are **cylinders**.

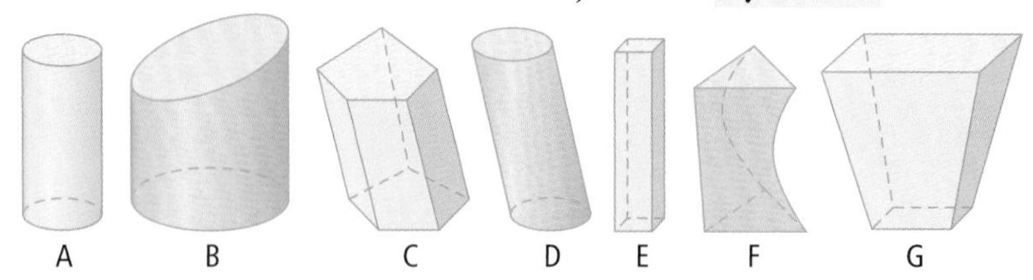

1. **a)** Examine the definition of a cylinder. List the conditions required for a three-dimensional object to be a cylinder.
   b) Examine each object. Which objects match the definition of a cylinder? Explain.
   c) What characteristics of cylinders are missing from each object that is not a cylinder?
   d) Draw other examples of three-dimensional objects that are and are not cylinders. Share your pictures with your classmates.

2. **a)** Examine the cylinders with circular bases. How are they different? What other descriptors could you add to their names to show this difference?
   b) Which objects are prisms? How would you name them to best describe their appearance?

3. **Reflect** How is the relationship between prisms and cylinders like the relationship between pyramids and cones?

Communicate the Key Ideas

1. Explain Ravi's statement.

All pyramids are cones but not all cones are pyramids.

2. Write a similar statement for cylinders and prisms. Explain your statement.

3. Explain the difference between a right square pyramid and an oblique square pyramid.

4. A right circular cylinder is cut in half. How could the cut produce two smaller congruent right circular cylinders? How could the cut produce two congruent three-dimensional objects that are not cylinders?

Communicating Mathematically

Like *cone*, the term *cylinder* is most commonly used to refer to right circular cylinders.

cylinder

- a three-dimensional object with congruent flat bases that are parallel. All corresponding points on the bases are connected by line segments along the surface.
- a right circular cylinder has two parallel circular bases that are perpendicular to the curved surface

Example 1: Confirm That a 3-D Object is a Cone

Are these three-dimensional objects cones? Explain why or why not.

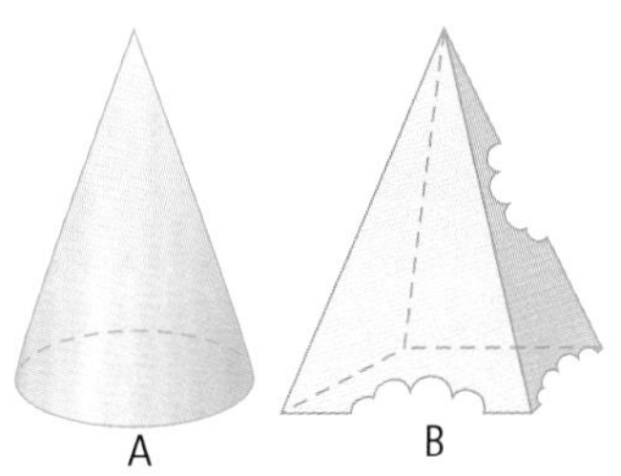

Solution

Object A is a cone because its base is flat, it has an apex, and all line segments that join points of the base to the apex along a surface are straight.

Object B is not a cone. Even though it has an apex, some line segments joining the base to the apex along a surface are not straight.

Example 2: What Shape Am I?

I am a right polyhedron with eight vertices, six faces, and two planes of symmetry. What am I?

Solution

Start with polyhedra that have *eight vertices*.

- A heptagonal pyramid and any of the quadrilateral-based prisms have eight vertices.

The next clue of *six faces* eliminates the heptagonal pyramid which has eight faces.

- All the quadrilateral-based prisms have six faces.

The prism is *right*, so one plane of symmetry will be through its middle, parallel to the bases, and one will be through the bases.

- Most of the quadrilateral-based prisms will be eliminated since their bases have more than one line of symmetry. Isosceles trapezoids have only one line of symmetry.

Therefore, this polyhedron is a right prism with an isosceles trapezoid base.

Check Your Understanding

1. Classify each three-dimensional object a cone or not. Justify your choices.

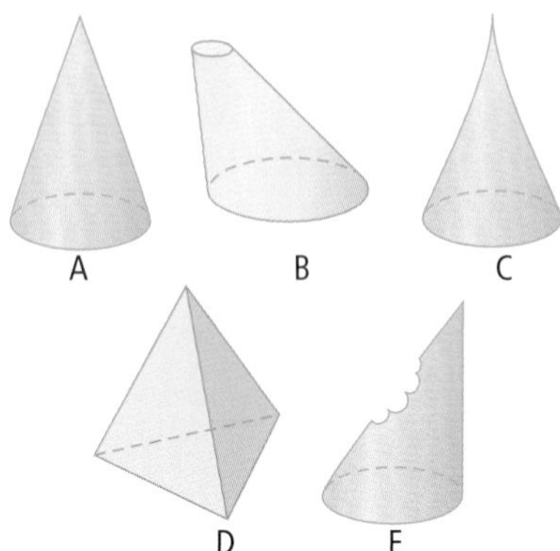

2. a) Are any of the objects in question 1 pyramids? Explain.

   b) Why do you think it is preferable to describe an object as a pyramid, if it is one, rather than a cone?

   c) Why can a right square pyramid also be called a pentahedron? Why can it also be called a cone? Which name do you think best describes the object? Why?

   d) Explain the difference between a right triangular pyramid and an oblique triangular pyramid. Include sketches to illustrate your explanation.

3. a) Classify each three-dimensional object a cylinder or not. Justify your choices.

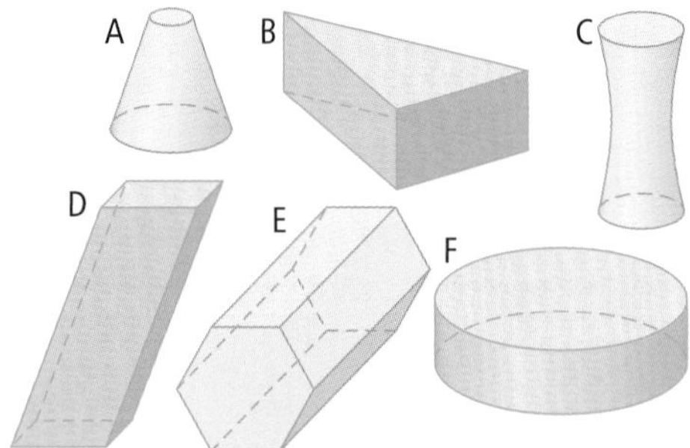

   b) Which of the objects are prisms? Explain.

   c) Why do you think it is preferable to describe an object as a prism, if it is one, rather than a cylinder?

   d) What are the names of the objects? Are there any you cannot name?

   e) What polygons make up the faces of object D? How would these polygons change if the polyhedron became a right square prism?

4. a) What two polyhedra do you see in house A?

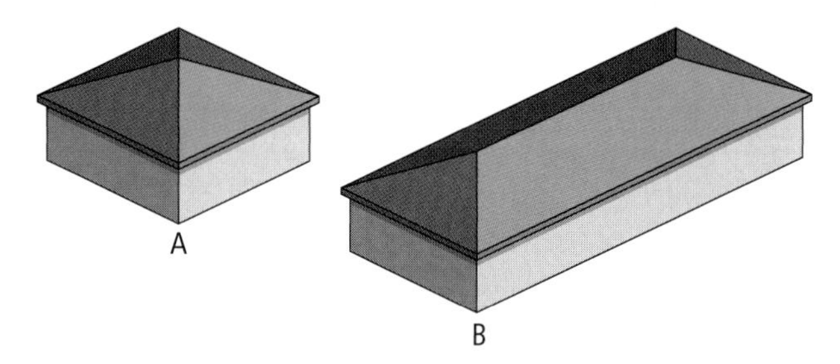

   b) Why is the roof of house B neither a prism nor a pyramid? What name could you give this roof?

   c) What two polyhedra do you see in house B?

   d) Explain the differences in the number and locations of the planes of symmetry of the houses.

   e) Explain the difference in the order of rotational symmetry of the houses.

5. Maxwell has a right nonagonal pyramid.

   a) How can you tell the number of vertices, edges, and faces without seeing his pyramid?

   b) What would you expect to be the shape of nine of its faces? Explain.

   c) How many planes of symmetry does the pyramid have? How do you know?

   d) What order of rotational symmetry does the pyramid have? How do you know?

6. **a)** Copy and complete the table. Use right Geoblock prisms with regular polygon bases to help you.

Base of Prism	Number of Vertices	Number of Edges	Number of Faces	Number of Planes of Symmetry	Order of Rotational Symmetry
triangular					
square					
pentagonal					
hexagonal					
heptagonal					
octagonal					
nonagonal					
decagonal					
20-gonal					
n-gonal					

b) Look at the patterns in the table. Describe how to use the name of a prism to determine the number of vertices, edges, faces, planes of symmetry, and order of rotational symmetry.

c) Do these patterns apply to prisms that are oblique rather than right? Explain.

d) How are the shapes of the faces different in right and oblique versions of these prisms?

e) How would the symmetries of a prism be affected if its bases were not regular?

7. **a)** Use Polydron® pieces to construct examples of right and oblique square and hexagonal prisms. *Hint: Two isosceles triangles or two isosceles right triangles will make parallelograms.*

b) Compare the number and location of the planes of symmetry in right and oblique prisms.

c) Compare the orders of rotational symmetry in right and oblique prisms.

8. What am I?

a) I have nine faces. I also have nine vertices.

b) I have ten vertices. I have fifteen edges. I have six planes of symmetry.

c) I have eight faces. Two of my faces are rectangles and four are parallelograms. I have one plane of symmetry.

9. **a)** What type of three dimensional object is the mailbox? (Ignore the flag.)

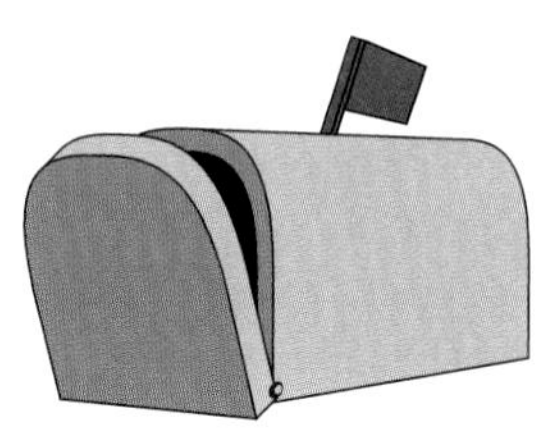

b) Design and draw a different cylindrical mailbox.

Chapter Problem

10. Choose several three-dimensional objects from this section to use in your building. Sketch the building and explain which objects are used in its construction.

Extend

11. **a)** Use eight equilateral triangles and two squares from a set of Polydron® pieces to build a decahedron. *Hint: The squares will not be adjacent faces.*

b) Use two squares and four rectangles to build a hexahedron.

c) How are these polyhedra similar? different?

d) Mathematicians consider this decahedron to be a special polyhedron and have named it an **antiprism**. Why do you think this name was chosen?

e) Build another antiprism using two squares and eight isosceles triangles.

f) Build other examples of antiprisms using two pentagons, two hexagons and equilateral or isosceles triangles or both.

9.4 Draw Polyhedra

Focus on...
- making isometric and orthographic drawings of 3-D shapes
- making and interpreting perspective drawings of prisms

Most of the objects we manufacture are three-dimensional. How are the details of what these objects look like communicated to the people who will build them? Most of these instructions are given in the form of drawings, which are two-dimensional. But how can a three-dimensional object be represented on two-dimensional paper?

Materials
- centimetre cubes or linking cubes
- Geoblocks
- 1-cm square dot paper
- 1-cm isometric dot paper

Discover the Math

isometric drawing
- a diagram of three-dimensional objects in two dimensions

Example:

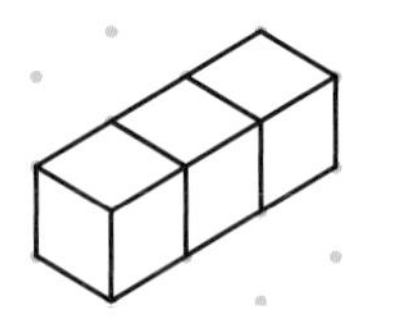

Part A: Isometric and Orthographic Drawings

Isometric and **orthographic** drawings are two unique but different ways to represent a three-dimensional object on a two-dimensional paper. How are they different? What advantages does each have in communicating what an object looks like?

1. a) Use linking cubes to construct a tower that matches this mat plan. The numbers indicate the height of each tower in cubes.

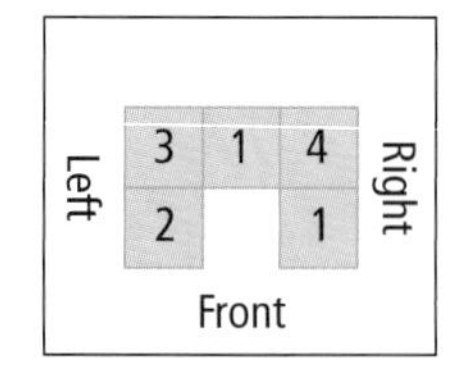

b) On square centimetre dot paper, draw and label the front, back, left, right, and top views of the tower.

c) Which views would you need to accurately construct the tower? Explain why some views would not be necessary.

d) This diagram is an isometric view of your tower. From which direction is it drawn: front left, front right, back left, or back right?

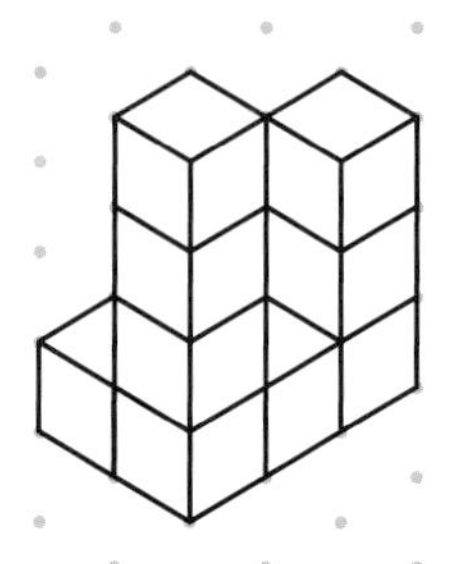

e) Why does this isometric drawing not give enough information for you to construct the tower?

Make another isometric drawing on isometric dot paper that could be used to accurately construct the tower.

2. a) Choose a rectangular Geoblock prism. Then choose a triangular prism that is half the size of the rectangular prism. Measure the dimensions of both prisms. Find a face of the triangular prism that is congruent to a face of the rectangular prism. Place both prisms in front of you, standing on their congruent faces. Use your measurements to make an orthographic drawing of the top, front, and left views of both prisms on 1-cm square dot paper.

b) Are any views of the prisms exactly the same? If so, which ones? Explain why only one view of the triangular prism shows a triangle.

c) Ask three classmates to use your labelled orthographic drawing to select the correct Geoblock prisms. Have your classmates place the prisms in front of them in the same position as you had. Were they able to do it? If not, why?

d) Place the prisms in the same position as in part a). Draw the front right isometric views of both prisms on 1-cm isometric dot paper. Are these views enough for you to recognize the prisms?

e) Which drawing, orthographic or isometric, shows the actual length of the edges of the prisms? Which drawing shows the actual angles on the faces of the prisms? Which drawing shows the parallel relationships among the faces?

3. a) Two isometric views of the same object are shown. Use Geoblocks to build the object.

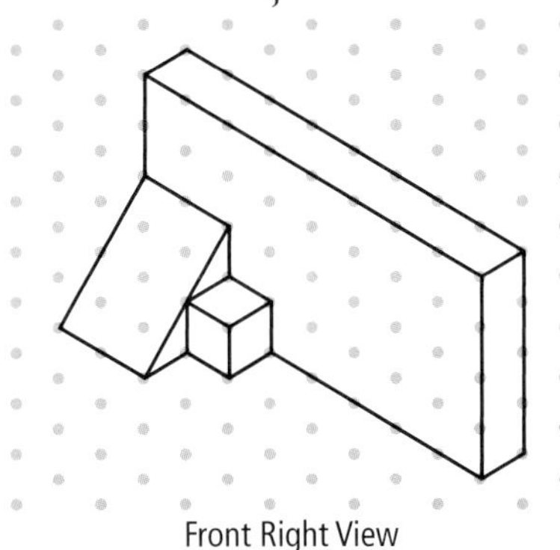

Front Right View

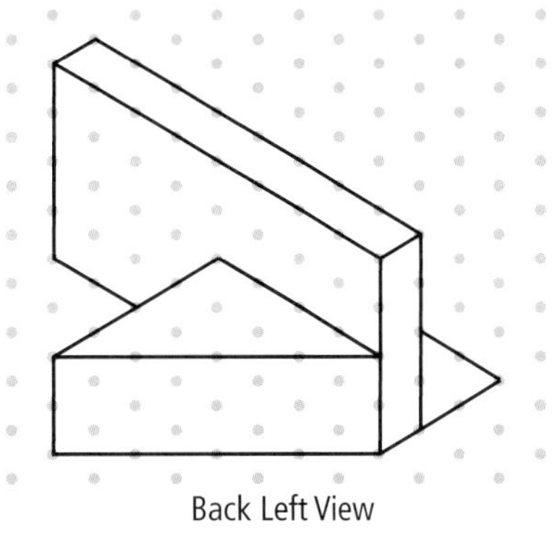

Back Left View

b) Did you need both isometric views to build the object? Explain.

c) Draw the top, front, and left orthographic views of the object.

d) Is there part of the object that is shown in only one orthographic view? Explain.

4. Reflect Suppose you were asked to draw a three-dimensional object using either isometric or orthographic drawings. What are the advantages and disadvantages of using each type of drawing?

orthographic drawing

- a diagram that shows up to six views of an object
- usually front, back, side, and top views are given

Example: For the rectangular prism,

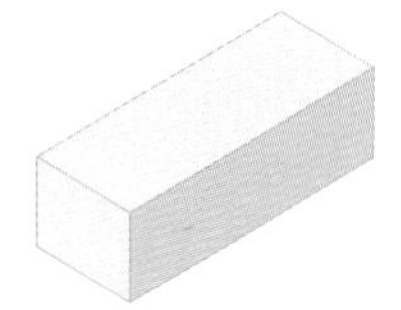

an orthographic drawing is shown.

	back	
left	top	right
	front	

Communicating Mathematically

- *Iso* means equal.
- *Ortho* means perpendicular.

Part B: Perspective Drawings of Right Prisms

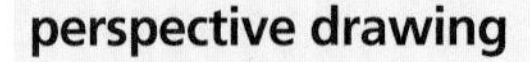

perspective drawing

- a diagram of three-dimensional objects in two dimensions
- parts of the drawing designed to look farther away are smaller

Example:

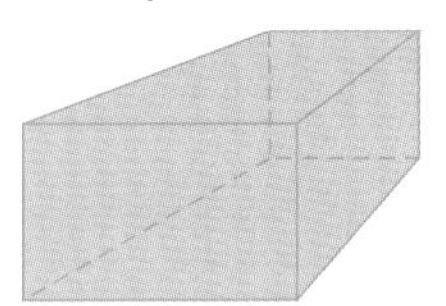

How do you make a perspective drawing? How is this related to a dilatation?

1. a) Near the centre of a sheet of paper mark a point P. On the bottom left of the paper draw any polygon. This will be the base of the prism.

b) Draw a dilatation image of the base with P as the centre and a scale factor of $\frac{1}{2}$.

c) Join the corresponding vertices of the base and its image. Change all unnecessary line segments to dotted lines.

d) In one-point perspective drawings, the point P is called the *vanishing point*. Why is this an appropriate name?

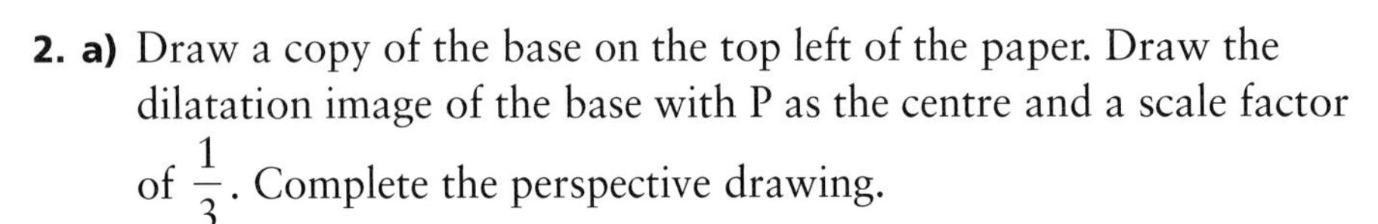

2. a) Draw a copy of the base on the top left of the paper. Draw the dilatation image of the base with P as the centre and a scale factor of $\frac{1}{3}$. Complete the perspective drawing.

b) How does this perspective drawing compare to your original perspective drawing?

3. Reflect If your perspective drawing were based on an actual prism, which side lengths and angles in the drawing would match the prism? Explain.

Strategies

When making a perspective drawing, how far should you extend the object toward the vanishing point to create the illusion of depth? Artists estimate the distance based on what looks right, usually about half the actual length or distance. This can vary depending upon the location and distance of the vanishing point relative to the object.

Example 1: Interpret Orthographic Drawings

Which Geoblocks were used to create the object shown in these orthographic drawings?

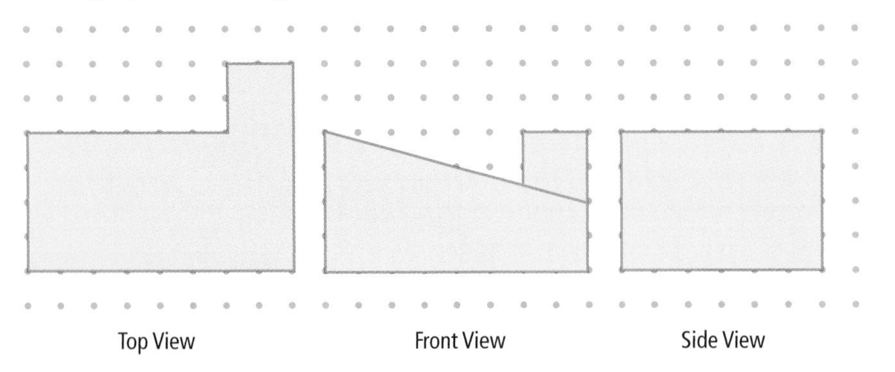

Making Connections

Landscape designers can use the techniques in perspective drawing to trick us into thinking a garden is larger than it really is. They do this by reducing the heights of plants and the widths of paths as you go from the front to the back of a garden.

Solution

Visualize the combination of the three views. While each view gives some information about the object, some views give more.

- The top view shows the rectangular prism and the triangular prism are both 4 cm wide. The object in the back is 2 cm wide, making it a 2 cm by 2 cm by 4 cm square prism.
- The front view shows a rectangular prism that is 8 cm long and 2 cm high with a triangular prism on top that is 8 cm long and 2 cm high. Behind the prisms is an object 2 cm wide and 4 cm high.
- The left view confirms that the top and front views were interpreted correctly.

Example 2: Interpret Isometric Drawings

The diagram is drawn on 1-cm isometric dot paper.

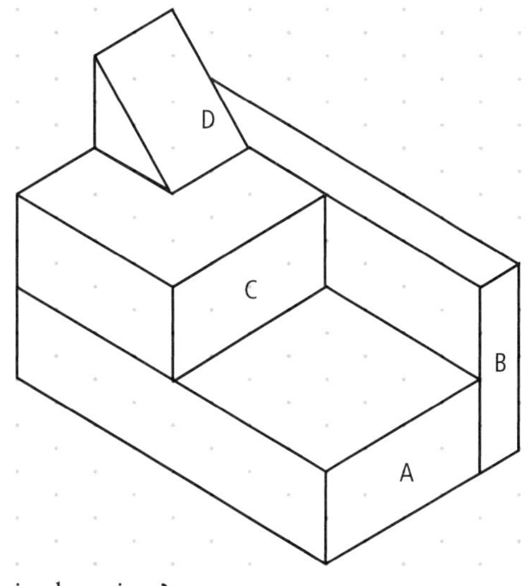

a) What are the measurements of the four objects in the isometric drawing?

b) Are they the only objects in this diagram? Explain.

c) Would a top orthographic view be enough to show that there are only four objects in the diagram? Explain.

d) If there were more objects in the diagram, what could you say about them from the information shown in this isometric drawing?

e) What other isometric view would be needed to show there are only four objects in the diagram?

f) Draw this view.

Solution

a) Object A is a 2 cm by 4 cm by 8 cm rectangular prism.
Object B is a 1 cm by 4 cm by 8 cm rectangular prism.
Object C is a 2 cm by 4 cm by 4 cm square prism.
Object D is a 2 cm by 2 cm by 2 cm triangular prism.

b) Maybe. There could be other objects on the left or at the back that are not shown in this view.

c) Yes. A top orthographic view would show all the objects, including any to the left or the back. However, it would not show the heights of the objects so it would be difficult to tell their exact shape.

Making Connections

Orthographic views are often used in architecture to show floor plans.

d) An object at the back would have to be less than 8 cm long and less than 3 cm high. An object on the left would have to be less than 5 cm long and less than 4 cm high. All objects would have to be narrow, 1 cm or 2 cm depending upon the other dimensions. Otherwise, part of the object would show on this isometric view.

e) A back-left isometric view would show if there were objects at the back or on the left.

f) Use Geoblock prisms to construct the front-right isometric view. Label the sides of the construction. Move around the desk to see the back-left view. It is difficult to represent the triangular prism because only two of its faces show. Part of the largest rectangular prism that is just visible would show entirely if the square prism and the triangular prism were not on top of it.

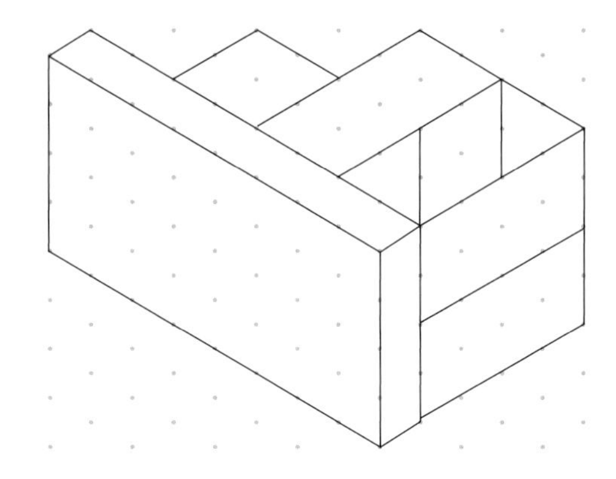

Communicate the Key Ideas

1. List the advantages and disadvantages of using each type of drawing: isometric, orthographic, and perspective.
2. How can a designer be sure that someone will not misinterpret their drawing of an object?
3. Build three towers, 3-cubes, 2-cubes, and 4-cubes high. Place them side by side, left to right, in the order you made them. Make a front-left isometric drawing of the towers and write the steps you follow to complete the drawing. Your steps should be clear enough for a classmate to replicate the drawing without seeing the object.

Check Your Understanding

1. **a)** Choose four different Geoblock prisms and use them to build an object.
   b) On 1-cm isometric dot paper, draw the front-right and back-left isometric views of the object.
   c) Examine the two views. Could a classmate use the drawings to build your object accurately, without worrying about missing prisms? Explain.
   d) On 1-cm square dot paper, draw the top, front, and right orthographic views of the object.
   e) Would back and left views give any more information about the object than the top, front, and right views? Explain.
   f) Could a classmate use the three orthographic views to build your object accurately, without worrying about missing prisms? Explain.

2. These are the front, right and top orthographic views of an object built from Geoblocks.

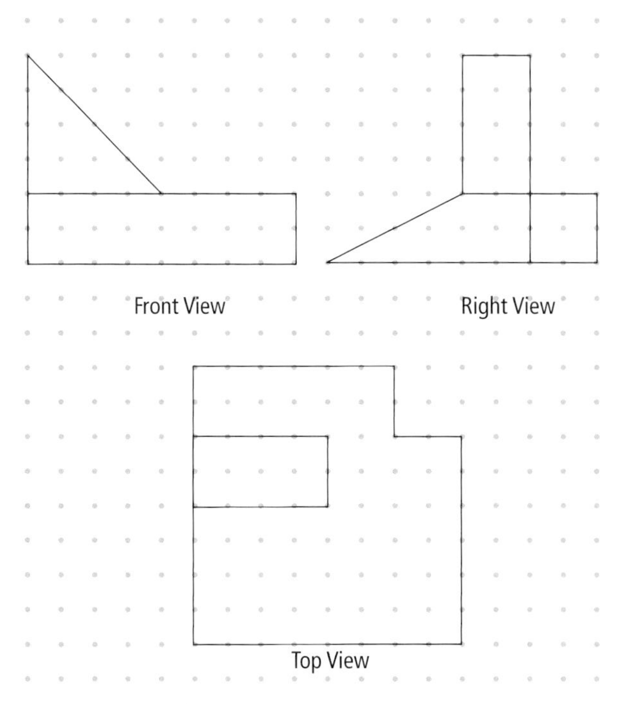

a) Build the object using Geoblocks.
b) Could there be other blocks in the object that are not visible in these views? Explain.
c) Draw a front-right isometric view of the object. Are all the blocks visible in this view? Is it necessary to provide another isometric view to show all the blocks needed to build the object?

3. These are two isometric views of an object built from Geoblocks.

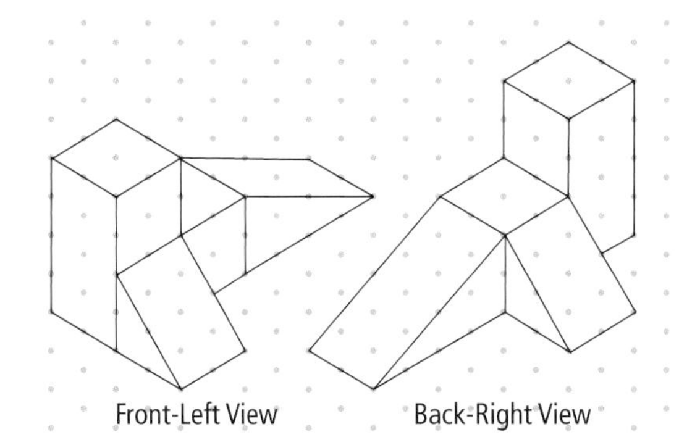

a) Draw top, front, and left orthographic views of this object.
b) Use Geoblock prisms to build the object. Compare the object to your drawings and adjust your drawings as needed.
c) Why are both of the isometric views necessary to build the object?
d) Is it possible to add a small prism to the object that will not show in the isometric views? Explain.
e) Draw the front-right and back-left isometric views of the object. Could these two views have been used instead of the two you were given to create the object? Explain.

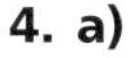

4. **a)**

Isometric drawings are funny because they look as if they contain right angles but when you actually measure the angles, they aren't right angles!

Is Armand correct? What are the measures of the angles in the drawings? What do the faces of a rectangular prism appear as in an isometric drawing? What information about an object cannot be communicated accurately with isometric drawings?
b) Armand's teacher explained that *isometric* comes from the Greek word for "equal measure." Which parts of an isometric drawing of an object have "equal measure" to the object? What can be reliably communicated using isometric drawings?
c) The opposite faces of many 3-D objects are parallel. For example, the faces of a rectangular prism are rectangles. The opposite sides of a rectangle are parallel. Do isometric drawings show these parallel relationships? Explain.

5. Alene made a perspective drawing of a rectangular prism.

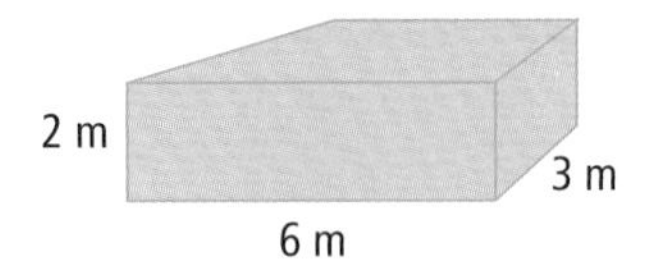

Make isometric and orthographic drawings of the prism.

6. Sue made this perspective drawing of a trapezoidal prism.

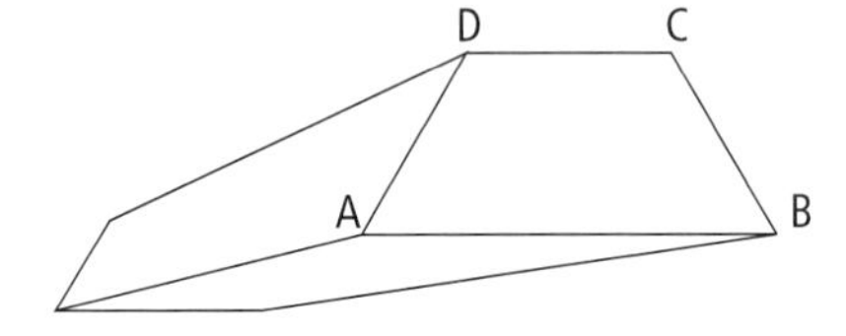

L

Describe how you could copy Sue's drawing by performing a dilatation of trapezoid ABCD.

7. **a)** Make a separate perspective drawing of each object represented here in the isometric drawing. Vary the location of the vanishing point.

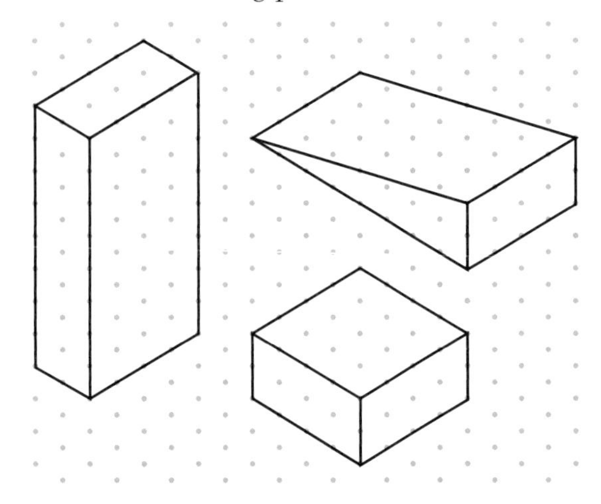

b) For each isometric and perspective drawing, list the parts of the object that could be measured accurately from the drawing.

8. **a)** Find a small box such as one for toothpicks or paper clips. Accurately measure its dimensions to the nearest tenth of a centimetre.

b) On 1-cm isometric dot paper, use the measures to draw one isometric view of the box.

c) On 1-cm square dot paper, draw the top, front, and right orthogonal views of the box.

d) What do the orthographic views show about the faces of the box that is not shown in the isometric drawing?

9. Three orthographic views of an object made from linking cubes are shown on 1-cm dot paper. The drawing does not show the individual cubes.

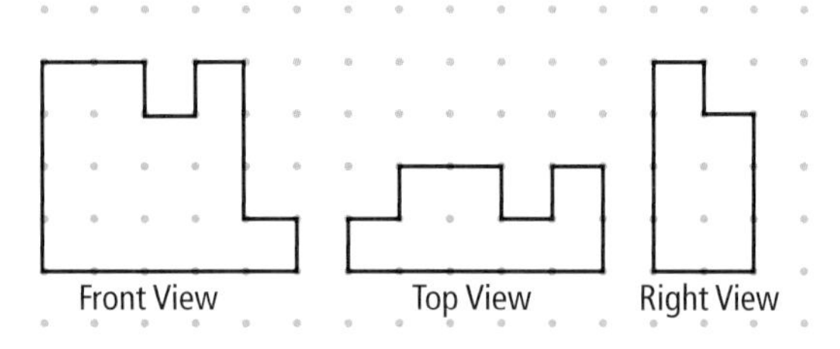

a) Try to visualize this shape using the three views. Predict how many cubes you will need to build the object. Record your prediction.

b) Build the object. Did you have to adjust the number of cubes from your prediction? Explain.

c) What does Justina mean?

This object could be built in different ways because there are places where I'm not sure how many cubes high it is. You can't tell from these orthographic views.

d) If Justina is correct, what is the minimum number of cubes needed to satisfy these orthogonal views? What is the maximum number that could be used?

e) On isometric dot paper, draw two isometric views of the object you built. These two views should leave no doubt as to the number of cubes needed to build the object.

f) What might be a disadvantage of using orthographic drawings? Provide an example.

10. This is a scale drawing of a quilt called *Magic Cubes*.

a) How has the designer achieved the three-dimensional appearance of the quilt?

b) Use triangle dot paper to draw part of the quilt. Show at least eight of the coloured shapes.

c) How has the designer removed the symmetry from the pattern?

Chapter Problem

11. Draw the exterior of the building you are designing using two different methods. Include measurements on your drawings.

Extend

12. a) Build an object using five to ten different Geoblocks.

b) Draw two isometric views and three orthographic views of your object.

c) Which drawings show your object completely, without leaving parts where blocks could be hidden?

d) Trade drawings with a partner. Build the object. Compare your solutions.

Did You Know?

In isometric perspective drawings, the lengths in the drawing of an object match (or are proportional) to the side lengths of the real object.

In linear perspective drawings, the apparent size of an object decreases as the object moves farther away from the observer.

Artists use the concept of a "vanishing point" to achieve the effect of perspective or distance in their art.

To create a cube that is vanishing in the distance,

1. draw a square
2. draw line segments from each vertex of the square to a meet at a point farther away
3. make a smaller copy of the square that fits on the perspective lines
4. erase the perspective lines
5. join corresponding vertices of the squares and erase appropriate lines to create a solid

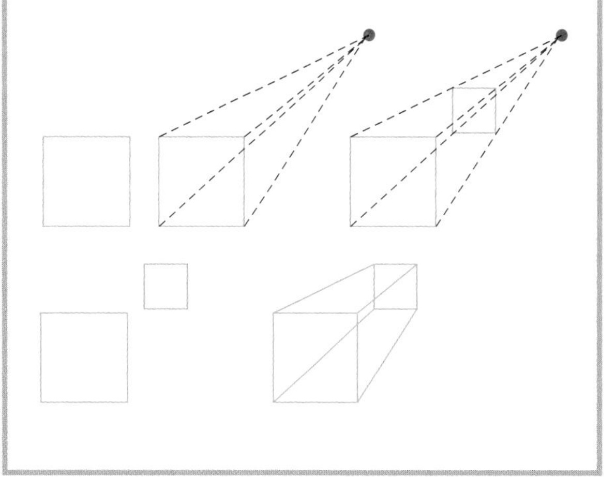

Review

1. a) Does the small triangle appear to be a dilatation image of the $\triangle DEF$? Explain.

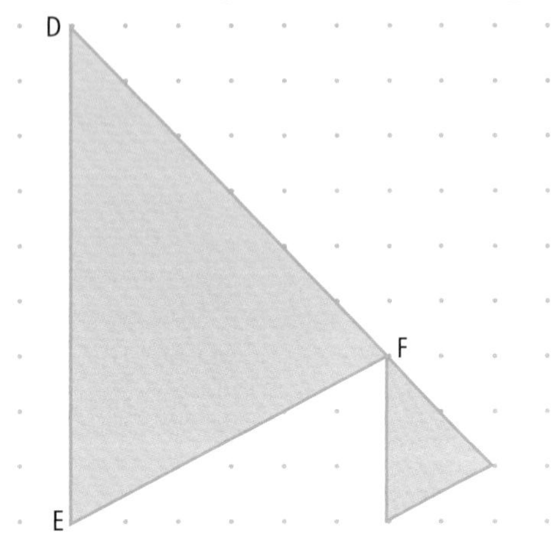

b) Examine the triangles and explain why they are similar.
c) If the small triangle is a dilatation image of $\triangle DEF$, how should you label it?
d) Explain how you can locate the centre of dilatation.
e) What is the scale factor of the dilatation? How do you know?

2. a) Explain the difference between stating that $\triangle ABC \cong \triangle JKL$ and $\triangle ABC \sim \triangle JKL$. Provide a picture to support your explanation.
b) Explain what it means when someone states that the corresponding sides of similar figures are proportional. Give an example to support your explanation.
c) How do the corresponding angles of similar figures compare?
d) Can you claim that two polygons are similar if you only compare the corresponding angles? Explain. Be sure to mention any exceptions.
e) If someone has pictures of two different-sized regular octagons, can you know if these shapes are similar without looking at the pictures? Explain.

3. a) What conditions must be met for a three-dimensional object to be called a cylinder?
b) What conditions must be met for a three-dimensional object to be called a cone?
c) Explain the difference between a right and an oblique cone.

4. Name each three-dimensional object as specifically as possible.

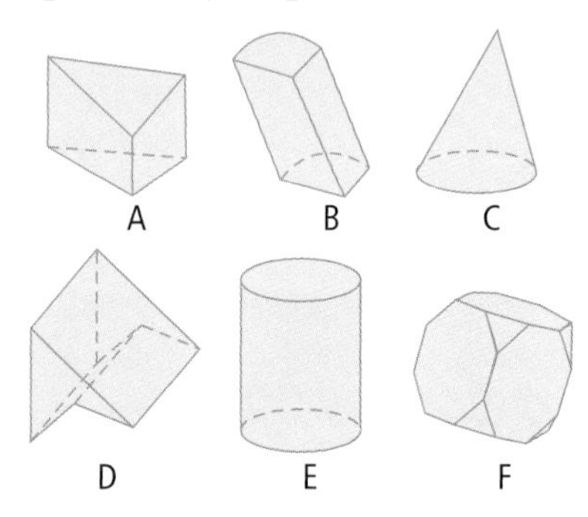

5. Angelica drew a picture of a three-dimensional object on isometric dot paper.

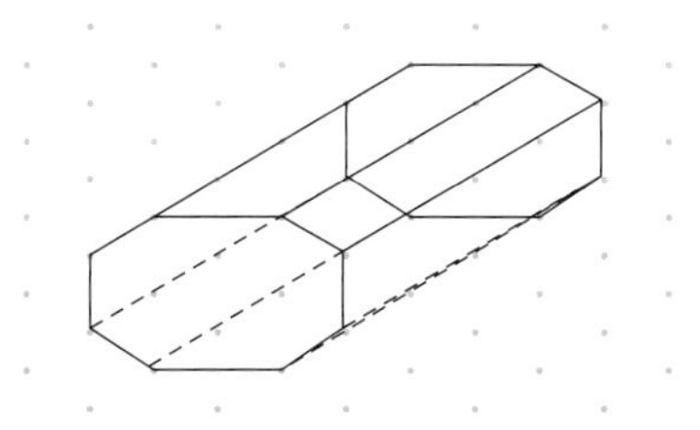

a) Is her drawing a cylinder? Explain.
b) Is her drawing a prism? Explain.
c) Is her drawing an octahedron? Explain.
d) Would Angelica describe her shape as *right* or *oblique*? Explain.
e) Make a perspective drawing of Angelica's object.
f) Describe how the perspective drawing is different from Angelica's isometric drawing.

6. The top view in the orthographic drawings of a pyramid with a regular hexagon base looks like six congruent triangles. Is the pyramid right or oblique? Explain.

7. a) Draw a scalene triangle.
   b) Copy the triangle. Draw and label a dilatation image of the triangle with scale factor of 0.5 and a dilatation centre at one of the vertices.
   c) Repeat part b) twice, using a dilatation centre outside the triangle and inside the triangle.
   d) What does the location of a dilatation image tell you about the location of the dilatation centre?

8. a) Could you describe the boxes as cones, cylinders, pyramids, or prisms? Explain.

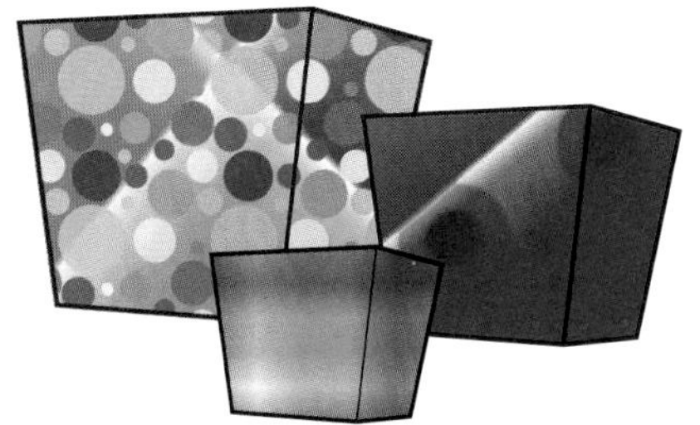

   b) Are the front faces of these boxes congruent or similar? Explain.

9. You want to make a floor plan of your rectangular classroom. You measure the classroom and find it is 10.4 m long and 5.6 m wide.
   a) You need to select a scale so your floor plan will fit on a sheet of paper and will be similar to the actual floor of the classroom. What scale will you use? Explain how you determined the scale.
   b) Make a scale drawing of the classroom floor.
   c) On one of the long walls, 2 m from the corner, is a doorway that is 0.44 m wide. Locate and draw the doorway on your floor plan.
   d) Midway along one of the short walls is a rectangular table that is 1.2 m long and 0.3 m wide. Locate and draw this table on your floor plan.

10. Make a perspective drawing of the object represented by the orthographic drawings.

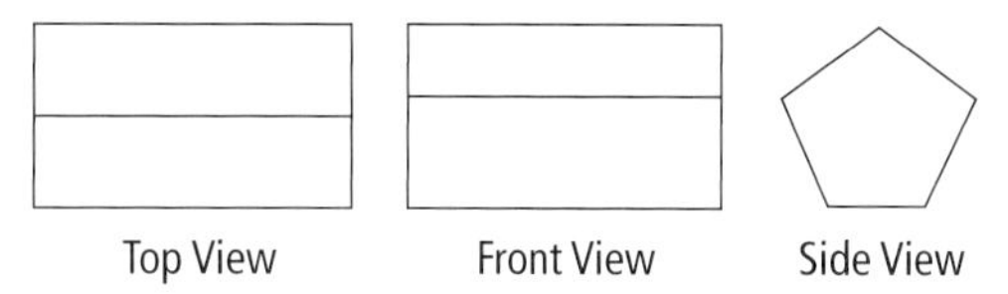

11. Make orthographic drawings of the object represented by the isometric drawing.

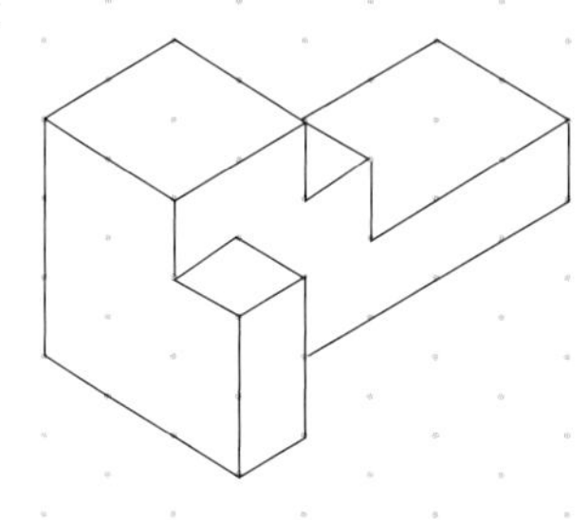

12. a) Mrs. Smith has a large flowerbed in her back garden shaped like a right triangle. The two legs of this triangular bed are 5 m and 12 m. Draw a picture of this triangle using a scale of 1 cm = 1 m.
    b) She decides to create a similar flowerbed with its right angle vertically opposite the right angle in her current flowerbed. She takes an 8-m length of rope with stakes attached to each end and sticks one stake into the vertex of the larger of the non-right angles. She then walks along the 5 m side of her current flowerbed, past the right angle until the rope was tight, and sticks the other stake in the ground. If her new flowerbed is to be similar, how far should she extend the 12-m side?
    c) What is the length of the longest side of her new flowerbed? Show your work.

CHAPTER 9

Practice Test

Selected Response

Select the best answer.

1. Where would be the centre of dilatation that makes the small triangle the image of the large triangle?

   A inside the small triangle
   B outside the large triangle
   C a vertex of one of the triangles
   D outside the small triangle

2. Which figures are always similar to one another regardless of their sizes?

   A equilateral triangles
   B regular hexagons
   C both A and B
   D neither A nor B

3. If the corresponding sides of two similar figures are 12 cm and 16 cm, then which are possible measures of two other corresponding sides?

   A 11 cm and 15 cm
   B 13 cm and 17 cm
   C 8 cm and 12 cm
   D 7.5 cm and 10.0 cm

4. Which is an example of an octahedron?

   A hexagonal prism
   B hexagonal pyramid
   C octagonal prism
   D octagonal pyramid

5. Which cone is guaranteed to have exactly one plane of symmetry?

   A right circular cone
   B oblique circular cone
   C right rectangular cone
   D oblique rectangular cone

6. A right triangular prism will have two planes of symmetry if its bases are which type of triangle?

   A scalene B isosceles
   C equilateral D right

7. The angles and side lengths of all faces of a polyhedron are preserved in which type of drawing?

   A orthographic B isometric
   C perspective D none of the above

Short Response

Provide a complete solution.

8. The Kelly's three rectangular patios measure 2.5 m by 7.5 m, 3 m by 8 m, and 4 m by 12 m.

   a) Are any of the patios similar? Explain.
   b) The Kelly's want a fourth patio similar to the 3 m by 8 m patio, but larger. Give the dimensions of two possible patios.

9. a) What are the three conditions that must be met for a three-dimensional object to be called a cylinder?
   b) What is the mathematical name for a cylindrical tin can?
   c) Describe the cylinders that are also prisms.
   d) Explain the difference between a right and an oblique prism.

10. Name two polyhedra that have

a) ten vertices

b) seven faces

c) eight planes of symmetry

Extended Response

Provide a complete solution.

11. a) Is the small polygon similar to the large polygon? Explain.

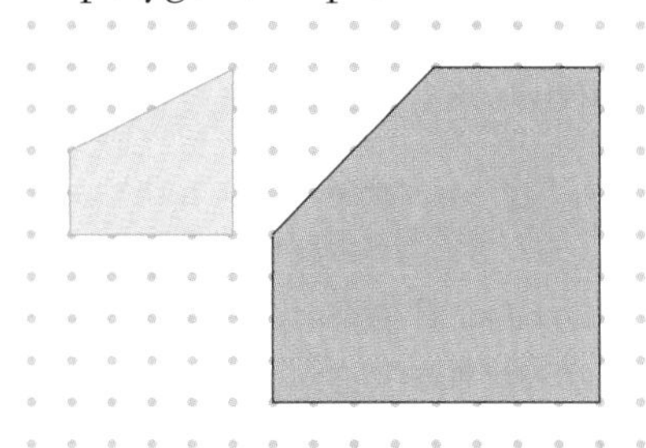

b) Copy the polygons onto square dot paper. If it is not similar, change the small polygon to make it similar to the large polygon.

c) Describe the large polygon as a dilatation image of the small, similar polygon.

d) Use the similar polygons to create a perspective drawing of a prism.

e) What is the name of the prism?

12. A wire is attached to the top of a large pole, to the top of a smaller pole, and staked in the ground as shown. The distance from the stake to the base of the small pole is 3 m and from the stake to the base of the large pole is 11 m.

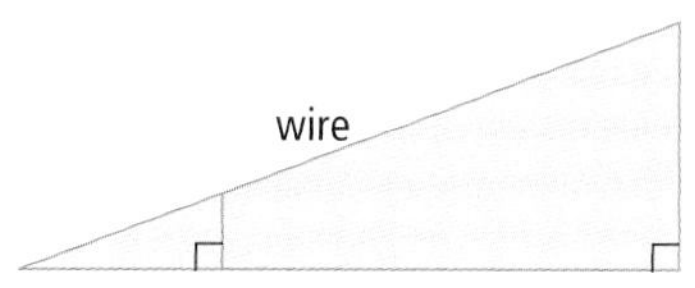

a) Copy the diagram and mark the lengths.

b) If the length of the wire from the stake to the top of the smaller pole is 3.27 m, what is the total length of the wire? Show your calculations.

c) Find the height of the large pole.

13. a) What is the name of the polyhedron in this isometric drawing? Explain how you know.

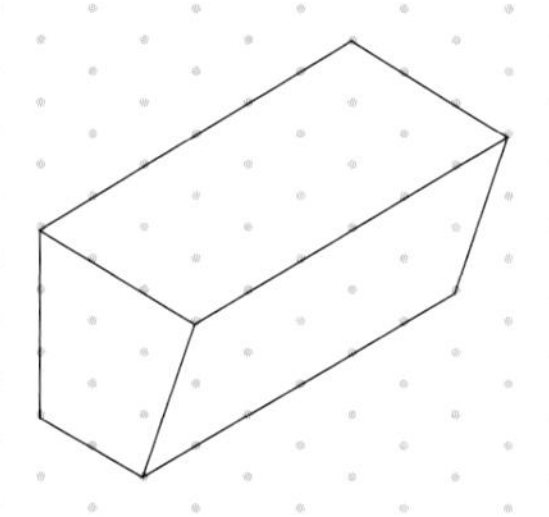

b) Make orthographic drawings of the polyhedron.

c) Explain what the isometric drawing shows about the polyhedron that the orthographic drawings do not.

Chapter Problem Wrap-Up

Complete the plans for your building. The plans should be sufficient for the construction company to understand your design without needing to ask questions. Choose an appropriate scale for your drawings and include the measurements where necessary. Include at least three representations of your building:

- a scale model
- a front isometric view
- a floor plan
- a perspective drawing
- orthographic views

Chapters 7–9 Review

1. Simplify by collecting like terms.

a) $9y + 2 - 4y + 5$
b) $(-w + 7) - (w - 4)$
c) $5w + 2 + 4w^2 - 8w - 3w^2$
d) $(5p^3 + 4p^2) - (3p^3 - p^2)$
e) $-4y^7 - 9z^4 + 4y^7z - 5y^7 + 3z^4$

2. Rent-Us Bikes charges \$5.00 for the bicycle plus \$0.25/h.
We-Rent Bikes charges \$0.75/h.

a) Write an equation for each company's rental costs.
b) Copy and complete the table.

Number of Hours, *h*	Rent-Us Bikes Rental Costs (\$)	We-Rent Bikes Rental Costs (\$)
1		
4		
8		
12		

c) Graph both equations on the same set of axes.
d) What is the point of intersection? What does it mean?

3. Name each 3-D object as specifically as possible.

a) 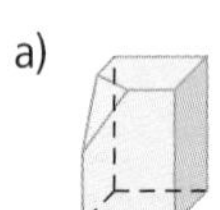
b) 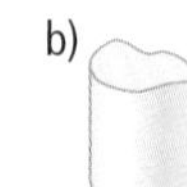
c) 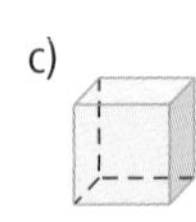
d) 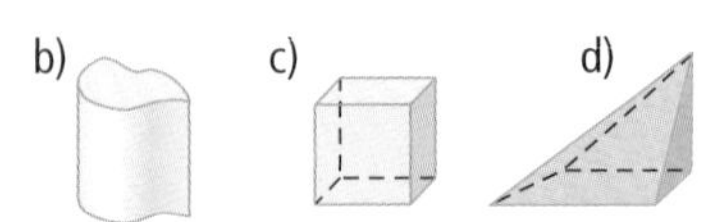

4. Simplify and solve.
$3(2x + 5 - 4x) - 4(3x + 2) = 61$

5. Solve each equation. Verify your solution.

a) $8 - 2y = 6$ **b)** $\frac{1}{3}x - 3 = 1$
c) $\frac{p}{6} + 5 = 9$ **d)** $\frac{w}{30} = \frac{4}{6}$

6. ABCD and EFGH are similar. Determine the measures of ∠A, ∠H, and FG.

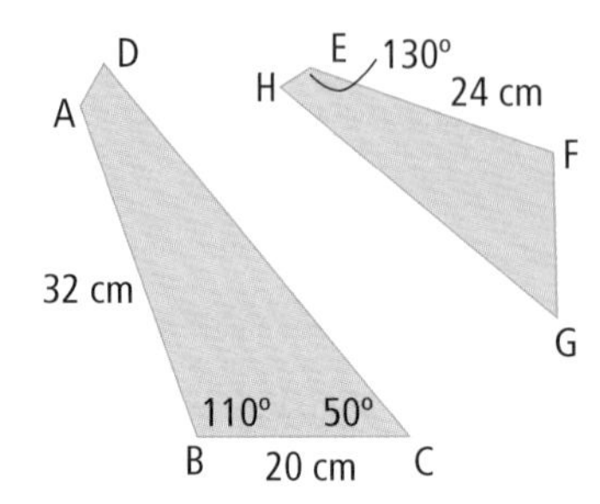

7. a) The dilatation image of a pentagon has an area 100 times greater than the area of the original pentagon. What was the scale factor of the dilatation?
b) The pentagons share a common vertex. What do you know about the dilatation centre?

8. Which expression is equivalent to $-3(x - 4)$? Show your work.

A $-3x + 12$ **B** $3x - 12$
C $3x - 6$ **D** $-2x - 6$

9. The table shows the distance travelled by a person cycling at a constant speed.

Time (s)	Distance (m)
0	0
1	2
2	4
3	6

a) Describe in words how the distance changes as time changes.
b) Is this relation linear or non-linear? Explain how you know.
c) Graph the relation. How far will the person have cycled in 10 s?
d) Write an equation that relates the person's distance and time.
e) Use your equation to find the distance cycled after
- 25 s
- 1 min

f) How long will it take for the person to cycle 1 km at this pace?

10. Justine made a perspective drawing of a triangular prism. Make isometric and orthographic drawings of the prism.

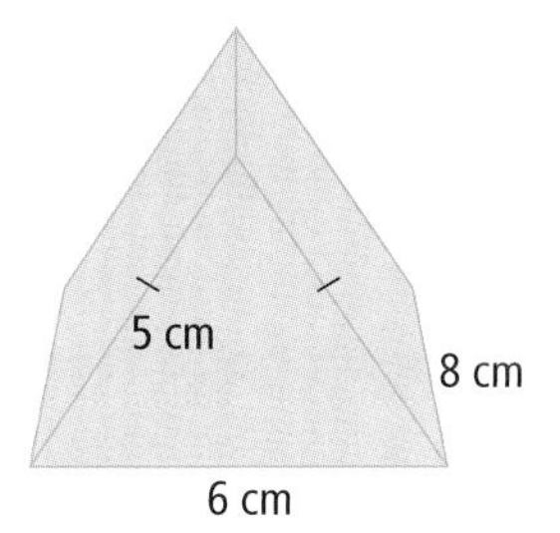

11. Which relation is non-linear? Explain how you know.

A

x	y
0	1
1	2
2	3
3	4

B

x	y
1	0
2	2
3	4
4	6

C

x	y
0	2
1	4
2	8
3	16

D

x	y
2	5
3	7
4	9
5	11

12. a) Graph the relation.
b) Determine the slope of the line. What does the slope mean?

Month, m	Savings, s ($)
0	200
1	170
2	140
3	110
4	80

13. Write an expression for the area of each shaded region.

a)

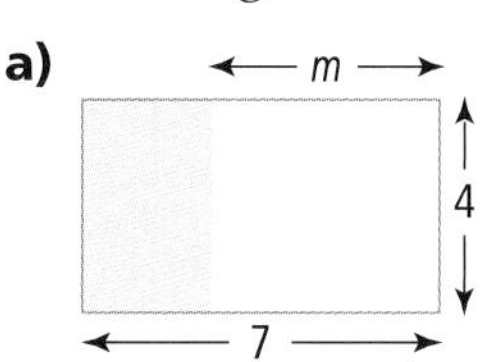

b)

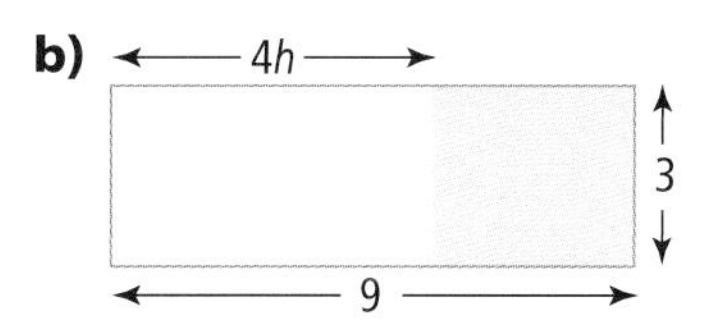

14. The graph shows the distance over time for two sisters who are walking home. Marlene is coming from the grocery store and Ginger is coming from school. They leave for home at the same time.

Marlene and Ginger's Trip Home

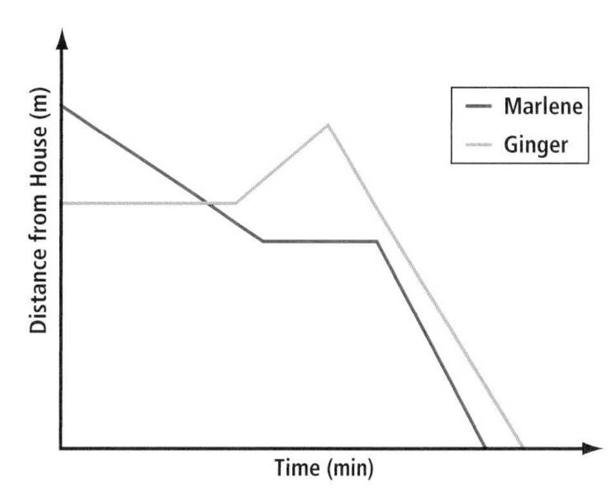

a) Describe each segment of each sister's path towards home. Give some reasons for the shape of each graph.
b) Which is farther from the sisters' home: the grocery store or the school? Explain how you know.
c) Who arrived at home first? Explain how you know.

- describe patterns and generalize the relationships between areas and perimeters of quadrilaterals, and areas and circumferences of circles
- estimate the areas of circles
- develop and use the formula for the area of a circle
- calculate the areas of composite figures
- estimate and calculate volumes and surface areas of right prisms and cylinders
- measure and calculate volumes and surface areas of composite 3-D shapes

Key Words

perimeter
area
volume
composite figure
components
prism
cylinder
base area
lateral face
surface area

CHAPTER 10

Measurement

The Halifax Citadel is a famous Canadian historical landmark, located near downtown Halifax, Nova Scotia. Inside the Citadel, you can take a tour and enjoy a thrilling glimpse into our country's past.

Look at the wall in the photograph. This wall surrounds the entire Citadel. What geometric shapes can you identify? How could you measure the area and perimeter of the Halifax Citadel?

Chapter Problem

The Robinson family is installing an in-ground swimming pool in their backyard. Will there be enough room to build the pool and a pool shed? Where should the pool and the shed be placed? How much water will be needed to fill the pool?

Get Ready

Squares and Square Roots

To square a number means to multiply a number by itself.
For example,

$$5^2 = 5 \times 5$$
$$= 25$$

The square root of a number is the factor that, when multiplied by itself, gives the number.
For example,

$\sqrt{25} = 5$ **The square root of 25 is 5 because 5 × 5 = 25.**

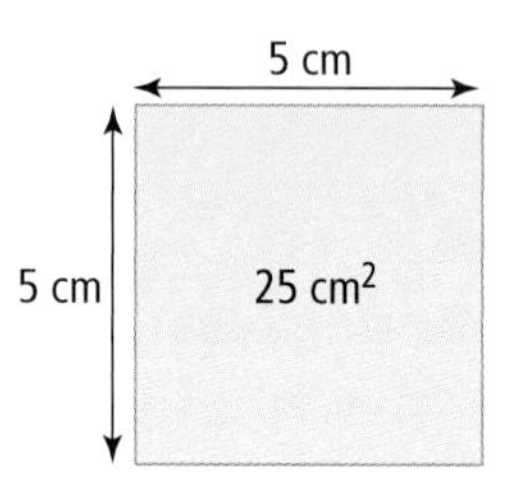

For the diagram shown, the square root of its area (25 cm^2) represents the side length (5 cm) of the square.

1. Square each number.

a) 4 **b)** 15
c) 3.5 **d)** 62.4

2. Evaluate. Use a calculator if necessary and round to the nearest hundredth.

a) $\sqrt{49}$ **b)** $\sqrt{121}$ **c)** $\sqrt{1.44}$
d) $\sqrt{20}$ **e)** $\sqrt{2.5}$ **f)** $\sqrt{12.5}$

3. a) Choose a number and use it as a side length to construct a square on square dot paper. What is the area of the square? Calculate the square root of the area. Describe what you notice.

b) Is this relationship always true? Repeat part a) using two different numbers to help you explain.

c) How are squaring a number and finding the square root of a number related operations?

Area

Area is the space covered by a two-dimensional figure. Area is measured in square units such as square centimetres (cm^2), square metres (m^2), or square kilometres (km^2). The table shows the area formula for some two-dimensional figures.

Figure	Diagram	Area Formula
square	s	$A = s^2$
rectangle	l, w	$A = lw$
parallelogram	height, h; base, b	$A = bh$
triangle	height, h; base, b	$A = \frac{1}{2}bh$

Communicating Mathematically

Variables that are multiplied in formulas are often written without brackets. For example, *lw* means (*l*)(*w*) or *l* × *w*.

You can find the area of a two-dimensional figure by applying an appropriate formula.

For example,

$$A = \frac{1}{2}bh$$
$$= \frac{1}{2}(5.0)(4.6)$$
$$= 2.5 \times 4.6$$
$$= 11.5$$

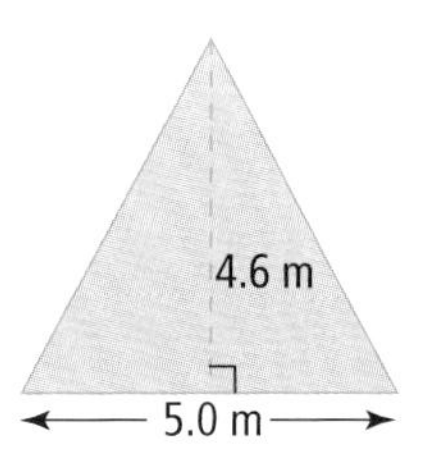

The area of the triangle is 11.5 m^2.

4. Find the area of each figure in square centimetres.

a) **b)**

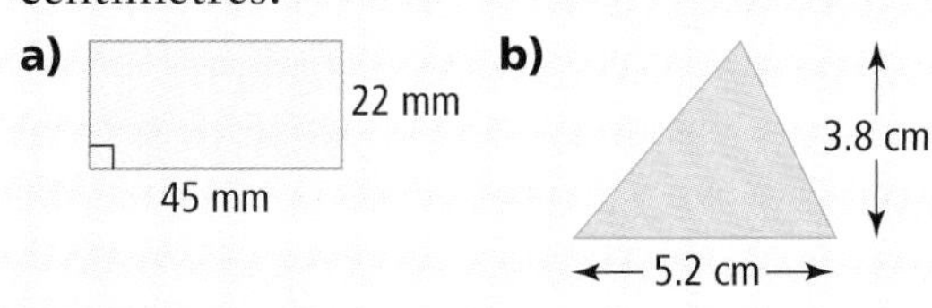

5. Find the area of an isosceles triangle with two sides measuring 6 cm each and a base of 4 cm.

Get Ready

Volume

Volume is the amount of space contained in a three-dimensional object. Volume is measured in cubic units, such as cubic millimetres (mm^3), cubic centimetres (cm^3), and cubic metres (m^3). The volume of a liquid is most often measured in litres, (L) or millilitres (mL).
To calculate the volume, consider the following.

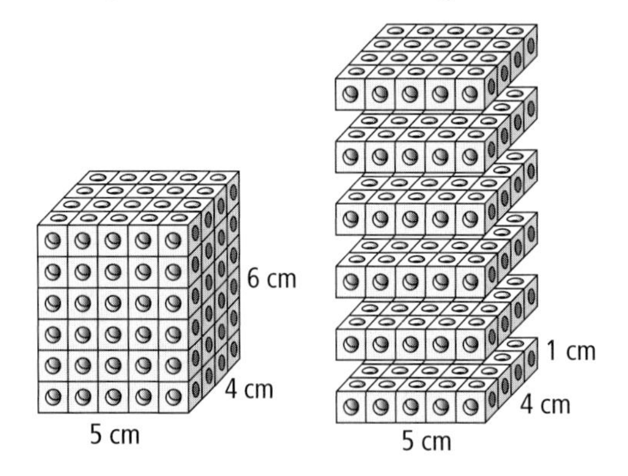

This rectangular prism has a base that is made of 20 cubes.
The base has a surface area of 5 cm × 4 cm = 20 cm^2.
The prism is made of six identical layers. Therefore, the volume of the prism is the area of the base multiplied by the height or
20 cm^2 × 6 cm = 120 cm^3.
120 cm^3 = 120 mL or 0.12 L (1 cm^3 = 1 mL)

6. Find the volume of each rectangular prism.

a)

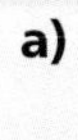
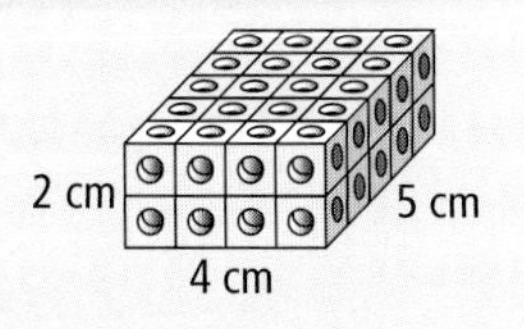

b)

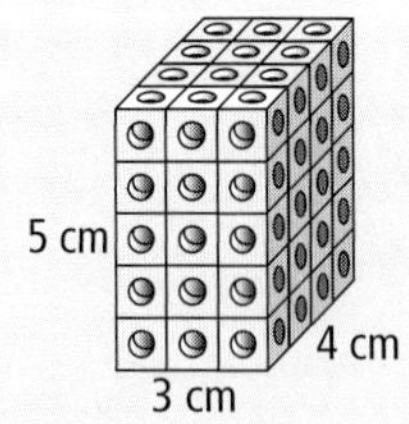

c)

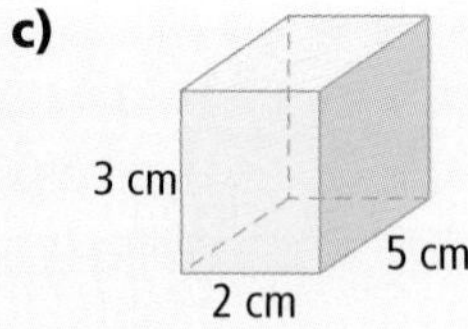

d)

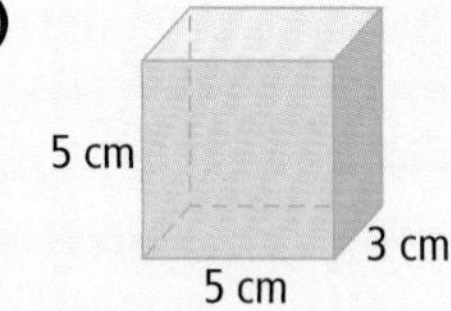

Convert Units of Length

To convert a unit to a smaller unit, multiply by powers of 10.

1 cm = 10 mm

1 m = 100 cm So, 1 m = 100 × 10 = 1000 mm.

1 km = 1000 m So, 1 km = 1000 × 100 = 100 000 cm.

So, 1 km = 100 000 × 10 = 1 000 000 mm.

To convert a unit to a larger unit, divide by powers of 10.

1 mm = 0.1 cm

1 cm = 0.01 m

1 m = 0.001 km

The area of the rectangle can be calculated in three different units.

Area = lw
= 7 cm × 3 cm
= 21 cm^2

3 cm

7 cm

Area = lw
= 70 mm × 30 mm
= 2100 mm^2

Area = lw
= 0.07 m × 0.03 m
= 0.0021 m^2

7. a) Draw a square that measures 1 cm by 1 cm. Imagine the edge of the square is divided into ten 1-mm^2 sections. How many rows of ten 1-mm^2 sections can be marked on the square? How many square millimetres will cover 1 cm^2?

b) On newsprint or a large sheet of paper, sketch a square that is 1 m^2. Imagine a row of 1-cm^2 sections along the top of the square. How many sections are in one row? How many rows can be marked on the square? How many square centimetres are needed to cover 1 m^2?

c) One hectare is equal to the area of a square that is 1 hm by 1 hm (1 hm = 100 m). Use your sketch from part b) to represent 1 km^2. How many hectares will cover 1 km^2?

8. Express each area in question 4 in square millimetres and square metres.

10.1 Area and Perimeter of Quadrilaterals

Focus on...
- describing patterns and generalizing the relationships between areas and perimeters of quadrilaterals

Materials
- geoboards
- toothpicks
- centimetre grid paper

Before solving the problem, ask:
- What tools do I need?
- What measurements should I use?
- How can I determine these measurements?

Discover the Math

Part A: What is the maximum area of a rectangle with a given perimeter?

You are designing a rectangular pen for a dog. Your fence is built using 1-m pieces that join together. What design will give the dog the most room?

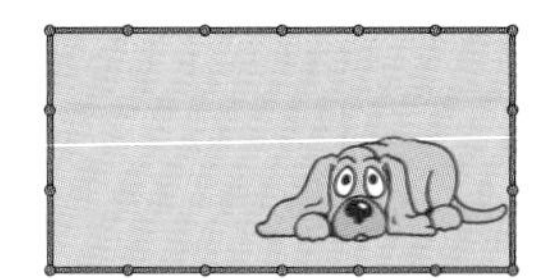

1. Suppose you have 12 m of fencing.

a) On grid paper, draw a scale model of a rectangular pen that has a perimeter of 12 m. Use a scale of 1 cm = 1 m.

b) Calculate the area of the pen.

c) Copy and complete the table.

Drawing of Rectangular Pen	Length (m)	Width (m)	Perimeter (m)	Area (m^2)

2. a) Add a new row to the table. Draw a different pen that has a perimeter of 12 m.

b) Compare your new pen to the first pen. Predict whether the area of the new pen will be greater than, less than, or equal to the area of the first pen.

c) Fill in the table for the new pen. Was your prediction correct? Explain.

3. a) Repeat question 2 until you have drawn all the pens that have a perimeter of 12 m.

b) Which pen has the greatest area?

4. Suppose that you purchase an additional 8 m of fencing.

a) What total length of fencing do you have now?

b) Draw as many pens as you can that have this perimeter.

c) Predict which pen has the greatest area. Explain your reasoning.

d) Measure the length, width, and area of each pen. Was your prediction correct?

e) What relationship between the length and the width of a rectangle gives the maximum area?

Strategies
Find a pattern.

5. Do you think you could create a pen containing a greater area if you could use parts of fence pieces instead of only whole pieces? Draw scale diagrams of the pens with a perimeter of 30 m on grid paper to explore this. Describe your findings.

6. Suppose that you have 16 m of fencing. What should be the dimensions of a pen with this perimeter to give the maximum area? Justify your answer. Draw the pen to test your prediction. Explain your findings.

7. Repeat question 6 for 36 m of fencing.

8. Reflect

a) Do all rectangles with the same perimeter have the same area? Provide an example to support your answer.

b) What type of rectangle will give a maximum area for a given perimeter?

Part B: What is the minimum perimeter required to surround a given rectangular area?

Materials
- square tiles
- linking cubes
- centimetre grid paper

You are designing a rectangular playground for young children. The playground must have an area of 24 m^2. What is the minimum total length of fencing needed to completely enclose the playground?

1. a) Build a rectangle using 24 square tiles or cubes. Make sure you use all the tiles or cubes and that you have no gaps in the rectangle.

b) Copy and complete the table.

Drawing of Playground	Length (m)	Width (m)	Perimeter (m)	Area (m^2)

2. If you make a different rectangle that has the same area, do you think it will have the same perimeter? Explain your thinking.

3. a) Add a new row to the table in question 1. Use the tiles or cubes to build a different rectangle that has an area of 24 m^2.

b) Compare your new rectangle to the first rectangle. Predict whether the perimeter of the new rectangle will be greater than, less than, or equal to the perimeter of the first rectangle.

c) Fill in the table for the new rectangle. Was your prediction correct? Explain.

4. a) Use the tiles or cubes to find as many rectangles as you can that have an area of 24 m^2.

b) Predict which rectangle has the smallest perimeter. Explain your thinking.

c) Measure the length, width, and perimeter of each rectangle. Was your prediction correct?

5. Suppose that you could use fractions or decimals for the length and width of the playground. If the area of the playground is 40 m^2, what is the smallest perimeter possible? Round your answer to one decimal place. Use grid paper to explore this. Explain your findings using words and diagrams.

Strategies
What other methods could you use to explore the perimeters?

6. Reflect

a) Do all rectangles with the same area have the same perimeter? Provide an example to support your answer.

b) What type of rectangle will give a minimum perimeter for a given area?

Communicate the Key Ideas

1. A rectangle has a perimeter of 30 cm. Can you determine the area of the rectangle? Why or why not? Explain using words and diagrams.

2. A square has a perimeter of 36 cm. Can you determine the area of the square? Why or why not? Explain using words and diagrams.

3. These rectangles all have the same area. Which rectangle has the smallest perimeter? Explain how you know.

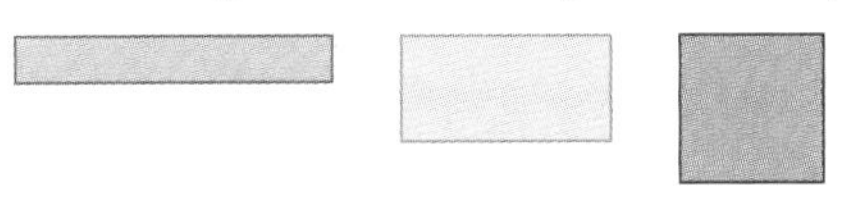

4. A farmer has an enclosed field for her horses, as shown. The farmer wants to increase the grazing area for her horses without adding more fencing. What advice would you give her, and why?

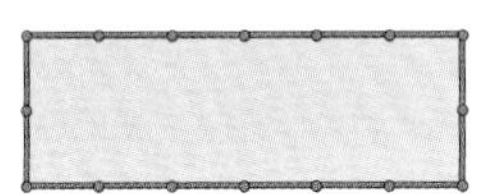

Check Your Understanding

1. Examine how the perimeter of a square changes as the side length increases.
   a) Draw a series of squares having increasing side lengths, as shown.

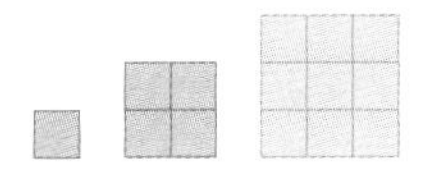

   b) Copy and complete the table.

Side Length (units)	Perimeter (units)
1	4
2	
3	

   c) Look at the pattern of perimeter values. Describe the pattern using words.
   d) Extend the pattern for at least two more squares.
   e) Write an equation that relates side length, s, and perimeter, P.
   f) Graph the relationship. Describe the pattern in the graph.
   g) Does it make sense to join the points on the graph? Explain.

2. Examine how the area of a square changes as the side length increases.
   a) Refer to the squares in question 1, part a). Copy and complete the table.

Side Length (units)	Area (square units)
1	1
2	
3	

   b) Extend the pattern for at least two more squares.
   c) Write an equation that relates side length, l, and area, A.
   d) Graph the data in the table. Describe the pattern in the graph.
   e) Does it make sense to join the points on the graph? Explain.

3. Compare the graphs you drew for questions 1 and 2. Describe how the graphs are
   a) similar
   b) different

4. A rectangular yard has an area of 64 m^2. What is the minimum perimeter of the yard?

5. Maurice's fenced-in garden has an area of 36 m^2, as shown.

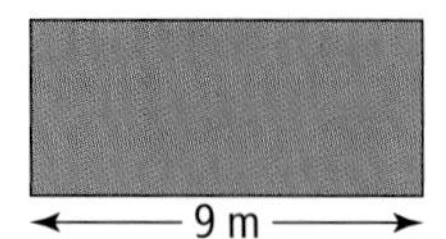

a) What is the width of the garden?

b) Can Maurice decrease the total length of fencing without reducing the area of the garden? If so, by how much? Assume that the garden remains rectangular.

6. A rectangular swimming pool has an area of 60 m^2.

a) What length and width give a minimum perimeter, assuming that these measures are

i) natural numbers?

ii) decimal numbers rounded to the nearest tenth of a metre?

b) Why are the answers to part a) different? Explain using words and diagrams.

7. A rectangle has a perimeter of 18 m.

a) What are the dimensions that will give a maximum area, assuming that the measures are

i) natural numbers?

ii) decimal numbers rounded to the nearest tenth of a metre?

b) Why are the answers to part a) different? Explain using words and diagrams.

8. Describe a situation in which it would be useful to find the dimensions of a rectangle that give a

a) minimum perimeter for a given area

b) maximum area for a given perimeter

9. Jose is fencing a pig-pen behind his shed, which will have an area of 144 m^2. Fencing costs \$12/m.

a) What is the minimum cost, assuming that he completely surrounds the pig-pen with fencing, as shown? Describe any assumptions you must make.

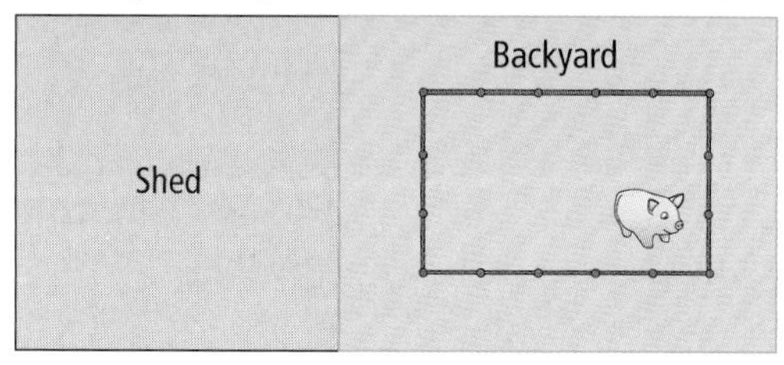

b) Jose thinks he can save money on fencing if he uses the back of his shed as one of the walls, as shown. The shed wall is 16 m long.

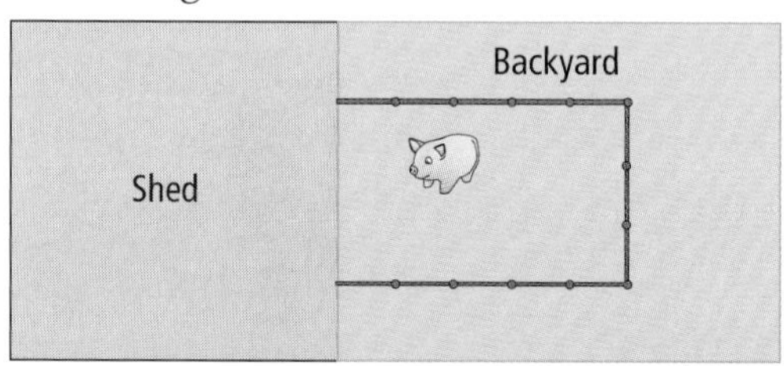

If Jose buys fencing for the other three sides, what is the most he can save? Describe any assumptions you must make.

Chapter Problem

10. The Robinson family's yard is rectangular, as shown.

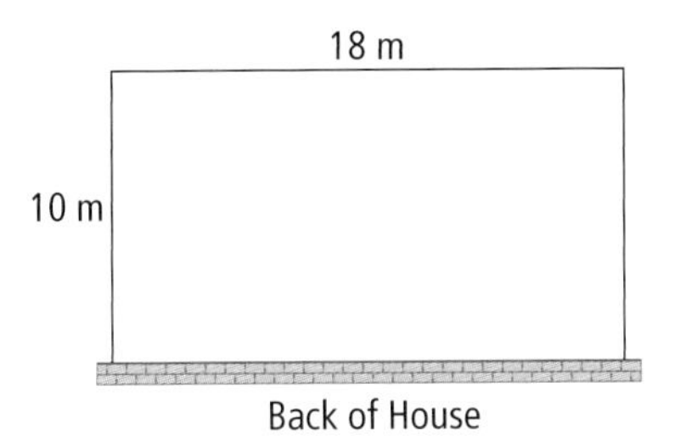

They are planning to build an in-ground pool and a pool shed. The rectangular pool will have a width of 6.0 m, and a length that is double the width. The rectangular shed will be 2.0 m by 3.5 m.

a) Where can the pool and the shed be placed in the yard? If there is more than one arrangement, suggest the best one. Justify your reasoning.

b) The Robinsons want to add a tile trim around the perimeter of the pool's edge. What length of trim will they need?

c) Calculate the area of the yard taken up by the pool and the shed. What percent of the total yard area is this?

Did You Know?

Traditionally, First Nations peoples played lacrosse on fields that were about 183 m wide and between 402 m and 805 m long. The exact size of the field was determined by the number of players and by how much ground they had to play on. Sometimes games had hundreds of players. The field would be several kilometres long and the game would last for days.

Puzzler

The field used in the North American Indigenous Games competition is 55 m wide by 100 m long. A traditional lacrosse field was about 183 m wide and 604 m long. Approximately how many times larger was the traditional field than the field used today?

Extend

11. Repeat question 7 for a rectangle having a minimum area. Discuss any assumptions you must make to answer part a).

10.2 Area and Circumference of Circles

Focus on...

- estimating areas of circles
- developing and using the formula for the area of a circle
- the relationships between areas and circumferences of circles

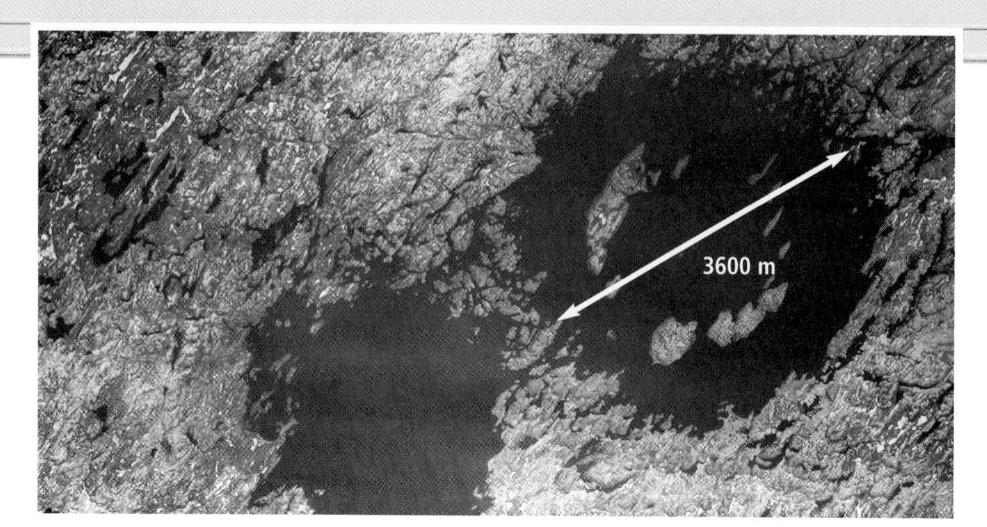

Lac a l'Eau Claire in Québec was formed around 300 million years ago when meteorites crashed to Earth. The photograph shows the two circular craters that form the bed of this 20-m deep lake. Both craters have about the same diameter. About how much land does Lac a l'Eau Claire cover?

Discover the Math

Materials

- centimetre grid paper
- Bullseye or compasses
- scissors

Part A: How do you estimate the area of a circle?

1. a) Use a Bullseye to draw a circle that just fits inside four connected squares of the grid paper, as shown.
b) Measure the radius, r, of the circle.

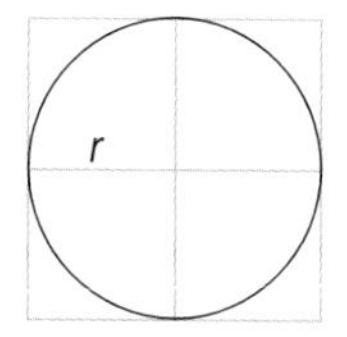

2. a) Lightly shade in one of the squares, as shown.
b) What is the side length of the shaded square?
c) How is this side length related to the radius of the circle?
d) Find the area of the shaded square.

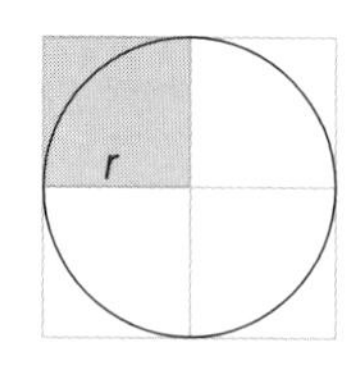

3. a) Explain how you can use your answer to question 2, part d), to find the area of all four squares.
b) Calculate the total area of the four squares.
c) Is the area of the circle greater than, less than, or equal to the area of the four squares? Explain how you know.

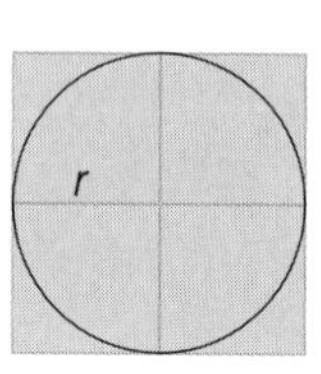

4. **a)** Do you think the area of a circle is greater than, less than, or equal to $3r^2$? Justify your answer.
   b) Carefully cut out three squares from the grid paper and test your prediction. Explain what you notice.

Communicating Mathematically

Read r^2 as "r squared." It means $r \times r$.

5. **Reflect**
   a) Copy and complete the statement.
   The area of a circle is approximately equal to ■ × the square of the radius.
   b) Explain how you chose the value to replace the coloured square in part a).

6. Another method to estimate the area of a circle is to square its diameter. Use this method to find the area of the circle. Is your estimate the same, greater than, or less than that in question 5, part a)?

7. **Reflect** How is the method in question 6 the same as the method used in questions 1 to 4? Explain.

Part B: How do you calculate the area of a circle?

Materials

- Bullseye or compasses
- scissors
- tape

1. **a)** Use a Bullseye to draw a large circle on a blank piece of paper. The radius of the circle should be a whole number of units.
   b) Estimate the area of the circle using the method you learned in Part A.
   c) Calculate the circumference of the circle.

2. **a)** Cut the circle out carefully.
   b) Fold the circle in half. Then fold the half circle in half to form a quarter circle. Finally, fold the quarter circle in half.
   c) Unfold the circle. How many sectors do you have? Alternately colour the sectors using two different colours. Compare your coloured circle with a classmate. Did you both create the same number of sectors?

Strategies
Manipulate a concrete model.

3. **a)** Cut out the sectors and arrange them as shown.
   b) What quadrilateral does this shape resemble? How would you determine the area of the quadrilateral?
   c) Paste the shape into your notebook.

Strategies
Apply a previously learned procedure for finding area.

4. **a)** Is the area of this shape the same as the area of the circle? Explain how you know.
   b) How is the height of this shape related to the circle? Label the height using a variable related to the circle.

c) How is the base of this shape related to the circle? Write an expression you could use to represent the base.

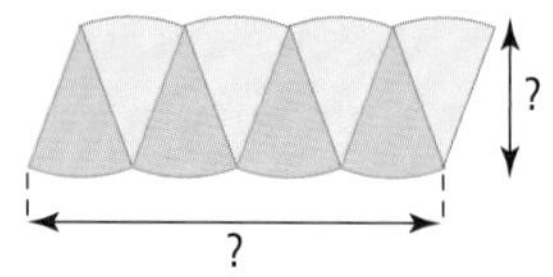

d) If the circumference of a circle can be found using the formula $C = 2\pi r$, how could your expression be represented in another way? Write your expression in simplified form.

e) How can you use these measures to find the area of the shape?

5. **Reflect**

a) Write a formula for the area of a circle A, with radius r.

b) Will your formula give an approximate or an exact answer? Explain.

6. **Reflect** Use your formula to find the areas of the circles in Part A and Part B. How do your calculated answers compare to your estimates?

Example 1: Calculate Area From Radius

Ingrid wants to make a decorative circular patch for her book bag. She draws a circle with a radius of 3 cm.

What is the minimum amount of material that she needs for the patch? Round your answer to the nearest 10 cm^2.

Solution

Apply the area formula.

$A = \pi r^2$

$A = \pi \times (3)^2$ ($3^2 = 3 \times 3$)

$A = \pi \times 9$

$A = 28.2743...$ C π × 3 x^2 = 28.27433388

$A \doteq 30$ **Rounded to the nearest 10 cm^2.**

Ingrid will need at least 30 cm^2 of material.

Technology Tip

- When performing calculations involving π, you can use 3.14 as an approximate value. However, some calculators have a π button, which is more accurate.

Example 2: Calculate Area From Diameter

Lac a l'Eau Claire is formed by two craters that are roughly circular. Each crater has a diameter of about 3600 m.

How much area does each crater cover

a) in square metres, to the nearest million?

b) in square kilometres, to the nearest tenth?

Solution

a) The diameter is 3600 m. The radius is half the diameter.

$r = 3600 \div 2$

$r = 1800$

Apply the area formula.

$A = \pi r^2$

$A = \pi(1800)^2$

$A = \pi \times 3\ 240\ 000$

$A \doteq 10\ 000\ 000$ **Rounded to the nearest million.**

Each crater covers approximately 10 000 000 m^2.

> **Communicating Mathematically**
>
> Read 10 000 000 as "ten million."

b) The radius is 1800 m. Convert the radius to kilometres.

1800 m = 1.8 km

Apply the area formula.

$A = \pi r^2$

$A = \pi \times (1.8)^2$

$A = \pi \times 3.24$

$A \doteq 10.2$

Each crater covers approximately 10.2 km^2.

I am converting from metres to kilometres. I know 1000 m equals 1 km. So, I divide by 1000 because I am changing from a smaller unit to a larger unit.

Example 3: Calculate Radius from Area

Hannah is designing an archery target. She wants the circular target to have an area of 1.5 m^2. What should be the radius of the target, correct to the nearest centimetre?

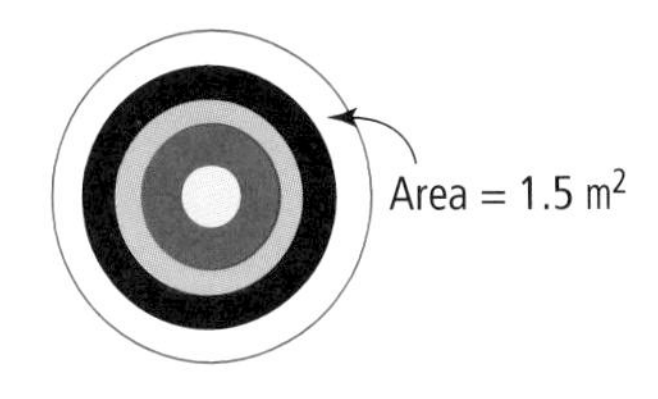

Solution

You can find the radius of a circle if you know its area. Substitute the known area into the formula and solve for the radius.

$A = \pi r^2$

$1.5 = \pi r^2$

$\frac{1.5}{\pi} = \frac{\pi r^2}{\pi}$ **Divide both sides by π.**

$\frac{1.5}{\pi} = r^2$

$0.477 \doteq r^2$ **Calculate the square root of both sides.**

$\sqrt{0.477} \doteq \sqrt{r^2}$

$0.690 \doteq r$

The radius is approximately 0.69 m. To convert this to centimetres, multiply by 100.

$0.69 \times 100 = 69$

Hannah should use a circle having a radius of approximately 69 cm for her target.

Use opposite operations to solve for the variable r.
$1.5 = \pi r^2$
The opposite of multiplying is dividing.
$0.477 \doteq r^2$

The opposite of squaring a value is to calculate its square root.

I know I am converting from a larger unit to a smaller unit, so I multiply.

Communicate the Key Ideas

1. a) What is the expression for the circumference of a circle with radius, r?

b) How is this expression related to the shape made in Discover the Math, Part B?

2.

To find the area of a circle, I measure halfway around the circumference. Then I multiply that value by the radius.

I split the circle into eight congruent sectors. Then I rearrange the sectors to form a parallelogram-like figure. By measuring the base and height of the figure I can estimate the area of the circle.

a) Do you think Linh's method will work? Why or why not?

b) Do you think Pierre's method will work? Why or why not?

c) Are the two methods related? If so, explain how.

3. Explain how to calculate the area of the circle.

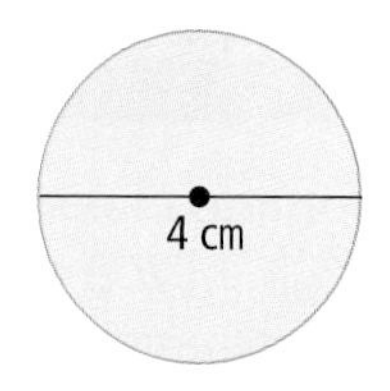

4. What value should result from dividing the area of any circle by the square of its radius? Explain.

5. The lid of a peanut butter jar has an area of 25 cm^2. Explain how to find the

a) radius

b) diameter

c) circumference

Check Your Understanding

1. Choose the best estimate for the area of the circle.

- 5 cm^2
- 10 cm^2
- 25 cm^2
- 50 cm^2

Justify your choice.

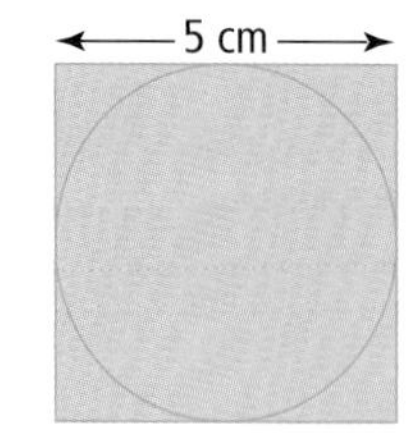

2. Find the area and circumference of each circle. Round the area to the nearest tenth of a square centimetre.

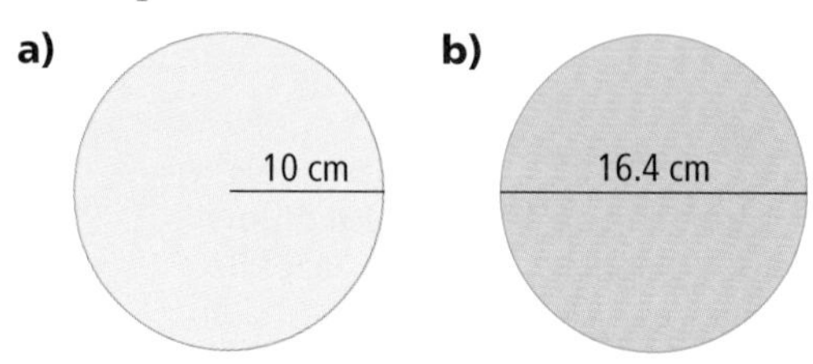

3. A circular skylight has a radius of 0.8 m.
   a) What is the area of the skylight? Round your answer to the nearest tenth of a square metre.
   b) The skylight is surrounded by a 4-cm wide strip of rubber. What length of rubber has been used?

4. A circular porthole has a diameter of 30 cm. What is the area of the porthole? Round your answer to the nearest square centimetre.

5. The circular rug is trimmed with cotton fringe. What length of fringe was used?

6. The Montagnais Crater, located in the Atlantic Ocean south of Nova Scotia, is roughly circular. Its diameter is 45 km. Approximately how much area does the crater cover, in square kilometres?

7. Brent Crater in Algonquin Park, Ontario, is roughly circular. Its diameter is approximately 4100 m. Approximately how much area does the crater cover
   a) in square metres?
   b) in square kilometres?

I know that there are 1000 m in 1 km. How many square metres are in 1 km^2?

Did You Know?

The exposed rock in the Canadian Shield was formed 500 million years ago. It is the oldest surface rock on Earth. This makes Canada one of the best places on Earth to find ancient craters.

8. Mei Yu is creating a label for her jewellery products. The label will be circular and covered in sparkles. The actual size of the label is shown.

One container of sparkles will cover 2000 cm^2. How many labels can she cover with one container? Show two methods for solving this problem.

9. The area of the circle shown below is 12 cm^2. What is the area of the blue square?

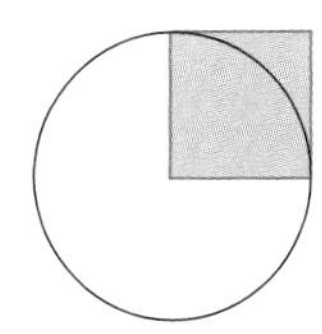

10. Find the radius of the circle, correct to the nearest tenth of a centimetre.

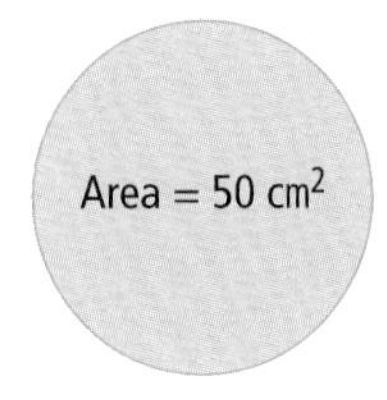

11. A circular bird bath has an area of 2800 cm^2. What is the radius of the bird bath, correct to the nearest centimetre?

12. a) Draw a large circle on centimetre grid paper. Use the grid squares to estimate the area of the circle.

b) Measure the radius of the circle. Calculate the area of the circle, using the area formula.

c) Compare your answers to parts a) and b). How close is your estimate to the actual area?

d) How could you improve your estimate from part a)?

13. A circular skating rink has a diameter of 28 m. How much ice do the skaters have to skate on?

14. This midway ride lies flat when it is not in operation.

Suppose you need to make a rain cover for this ride. How large should the cover be, if there is 0.5 m of overhang on the sides? Use numbers, words, and pictures to explain your answer.

15. Nadia is playing golf. She estimates the diameter of a circular golf green is about 25 m.

What is the area of the green? Round your answer to the nearest square metre.

16. Sandra swims around the edge of a roughly circular lake. The lake has an area of 32 000 m^2. Luke swims across the middle of the lake and back. Estimate who swam farther, and by how much. Do not calculate. Explain how you found your answer and discuss any assumptions you made.

17. If you know the circumference of a circle, can you determine its area? Explain, using words and diagrams. Use an example to support your explanation.

18. Describe a situation in which you might need to find the area of a circle. Choose an example that has not been used in this section. Explain why you would need to find the area. Estimate the area for the situation you have chosen.

19. How are the area and the circumference of a circle related?

a) Draw three circles and measure their **radii**.

b) Copy and complete the table for each circle.

Radius, r (cm)	Circumference, C (cm)	Area, A (cm^2)	$A \div C$

c) Describe any patterns that you notice. How is the value $A \div C$ related to the radius of the circle?

Strategies
Use a table to organize your information.

Communicating Mathematically

The plural of radius is **radii**.

20. Jack is practising for a hockey tournament. His coach tells him to skate with the puck around three circles marked on the ice: first the red circle, then the blue, and then the green. The radius of each circle is 1 m greater than the previous circle.

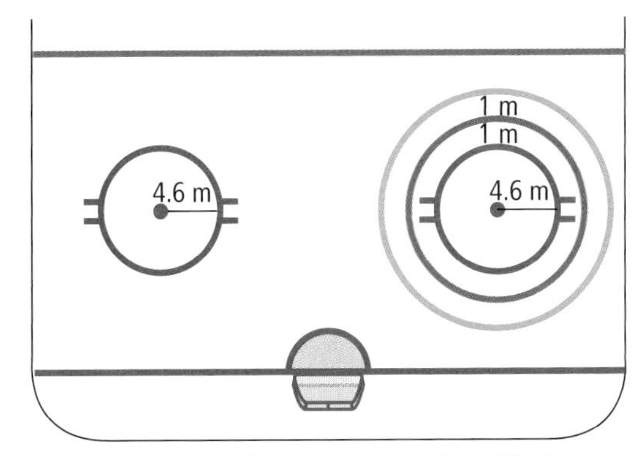

What is the total distance Jack will skate during this practice?

21. A round pizza has an area of 450 cm^2.

a) How long is the edge of the pizza crust?

b) What is the side length of the smallest square box that will hold the pizza?

Extend

22. The distance from the centre of the bulls-eye to the outer ring of a dartboard is 17 cm. Players score points for any dart landing inside the outer ring.

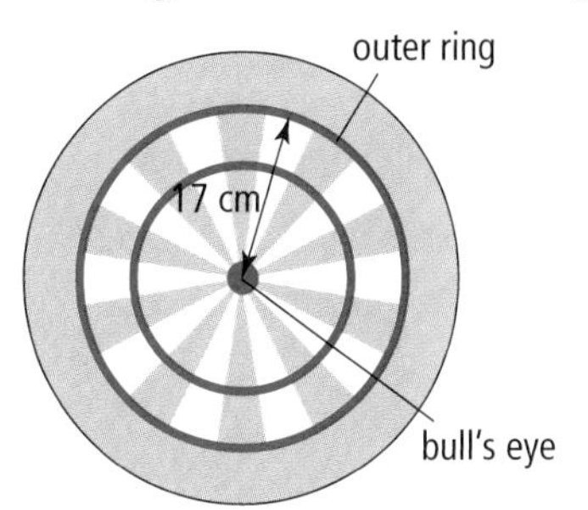

a) What is the area of the scoring region? Round your answer to the nearest square centimetre.

b) In one type of dart game, players must "double-in" to start collecting points. This means that the dart must land within a narrow band called the "double ring."

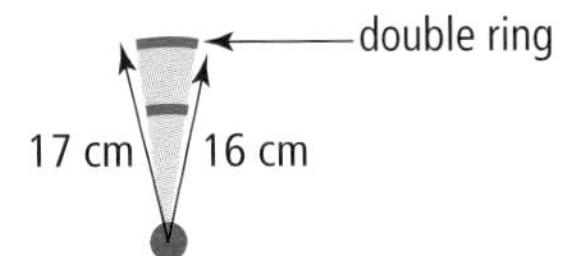

What is the area of the double-ring section?

c) Describe your method for calculating part b). Is there another method you could use? Explain.

10.3 Area of Composite Figures

Focus on...
- describing patterns and generalizing the relationships between areas and perimeter of quadrilaterals
- calculating the areas of composite figures

The photograph is of the Baddeck lighthouse near Kidston Island on the Bras d'Or lakes. How many geometric shapes can you identify in the lighthouse? A figure that consists of more than one simple shape is called a **composite figure**.

composite figure
- a figure made up of two or more simpler shapes, called **components**

Which simple shapes make up each composite figure?

Where have you seen composite figures such as these in everyday life? What other composite figures have you encountered?

Materials
- colour tiles
- centimetre grid paper

Discover the Math

How can I find the area of a composite figure?

Part A: Find the Area Using Rectangles

Larry is redecorating his living room. He has decided to replace his worn out blue carpet with hardwood flooring. How much do you think this will cost?

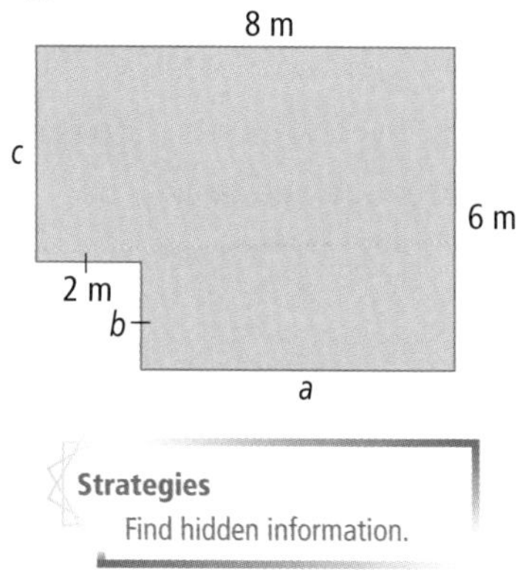

Larry drew a floor plan of his living room with measurements as shown.

Strategies
Find hidden information.

1. Find the missing measures a, b, and c. Show your work.

2. a) Identify two or more simple shapes that combine to form the composite figure. Make a drawing that shows the simple shapes using different colours.

b) Find the total area of Larry's living room floor by adding the areas of the simple shapes. Show how you found your answer using words and a diagram.

3. Repeat question 2 using a different combination of simple shapes.

4. a) Trace an 8×6 rectangle on a piece of grid paper. Find its area.

b) Explain how you can find the area of Larry's living room floor by using subtraction. Illustrate using your diagram.

Hardwood Flooring

Type of Wood	Price ($/m^2)
Oak	40
Maple	30
Birch	25

All prices are subject to Harmonized Sales Tax (HST) of 14%.

5. a) What is the lowest cost for the hardwood flooring needed to cover the entire living room floor

i) not including taxes?

ii) including taxes?

b) If Larry has $1800 to spend on flooring, what types of wood can he choose from? Explain.

Did You Know?

The Gilbert's Cove Lighthouse was built in 1904. While most lighthouse keepers lived away from the lighthouse itself, the keeper of this lighthouse lived in it for over 50 years. This lighthouse is one of three of its kind left in Nova Scotia. The use of the beacon was discontinued in May 1984.

To learn more about the history of the Gilbert's Cove Lighthouse, and many other famous lighthouses in Nova Scotia, go to **www.mcgrawhill.ca/links/math8NS** and follow the links.

Part B: Find the Area Using Circles and Triangles

Larry has installed the new hardwood floor. He is now choosing a throw rug from the two he likes best.

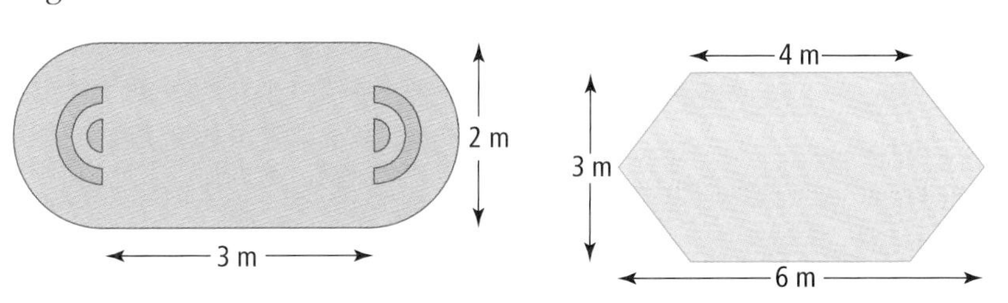

Materials

- scissors
- tape
- BLM 10.3 Rug Areas
- tracing paper

Red Rug

1. Copy the red rug onto tracing paper. Cut the rug into three pieces, as shown.

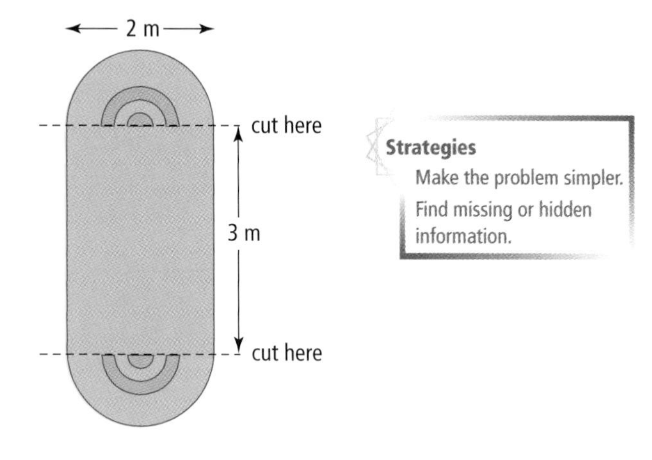
2 m
cut here
3 m
cut here
Strategies
Make the problem simpler.
Find missing or hidden information.

2. **a)** What simple shape is the largest piece? What are its dimensions?
   b) Find the area of the largest piece.

3. **a)** Tape the two remaining pieces together along the cut lines.
   b) What simple shape have you formed? Find the area of this shape.

4. Find the total area of the red rug. Explain your method using words and diagrams.

Green Rug

5. Copy the green rug onto tracing paper.

6. **a)** Show how the green rug can be divided into two or more simple shapes.
   b) Identify the measures needed to calculate the area of each simple shape. Show how you found the measures that are missing from the diagram.

7. Find the total area of the green rug. Explain your method using words and diagrams. Could subtraction be used to find the area of this rug? Explain.

8. Refer to question 7. Which method did you find easiest? Why might you use both methods for finding the area?

9. **Reflect** How can you find the area of a composite figure using a combination of simple shapes? Use an example to help you explain.

Part C: Find the Area of a Trapezoid

1. Draw a trapezoid with the two parallel sides 10 cm and 8 cm long and the height 6 cm. It should not be an isosceles or a right trapezoid.

2. **a)** In a small group, work together to find different ways to split the trapezoid into simple shapes. Choose shapes whose area you can calculate easily, such as triangles, parallelograms, and rectangles.
   b) Use your results to find the area of the trapezoid in as many ways as you can. Share your methods with your class.

3. a) Copy the trapezoid onto tracing paper twice. Label the parallel sides on one copy, as shown.

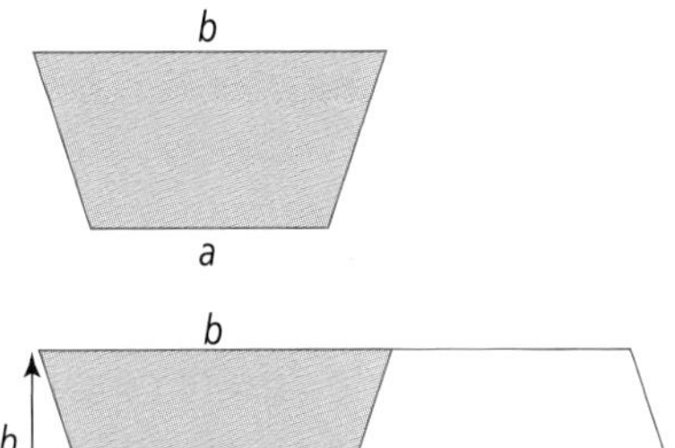

b) Cut out the unlabelled trapezoid. Flip it and rotate it to fit along one of the slanted sides of the labelled trapezoid, as shown. Paste the whole new shape into your notebook. What shape is produced? What algebraic expression could be used to represent the base of this figure?

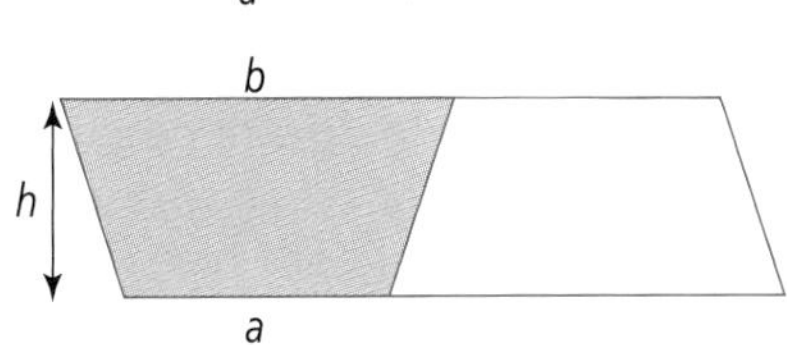

c) Write an expression to determine the area of the new shape using sides a and b, and height h.

d) How could you change your expression in part c) to determine the area of the original trapezoid. Explain how you found your answer.

4. **Reflect**

a) Explain in your own words how you can determine the area of a trapezoid using the two methods, composite figures and transformations, as in questions 2 and 3. Which method is easiest to use? Why?

b) Explain the formula for finding the area of a trapezoid. Where does it come from? What information does it give you? How does it work?

Example 1: Fence a Trapezoidal Garden

a) The garden shown is a right trapezoid. What is the area of the garden?

b) Fencing materials costs \$10/m. What is the total cost to fence in the garden? Assume the fencing materials are sold in whole units. Do not include HST.

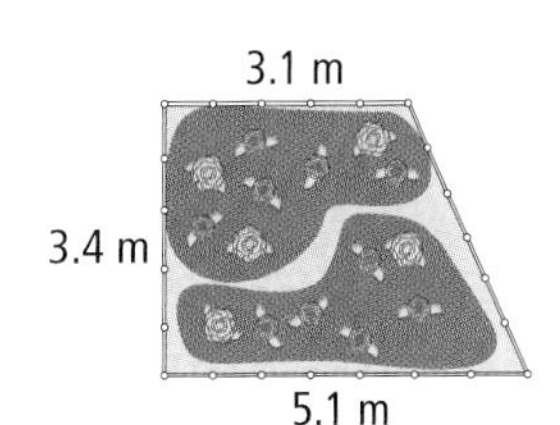

Strategies
Make an assumption.

Solution

a) *Method 1: Use the formula.*

$$A_{\text{trapezoid}} = \frac{1}{2}h(a + b)$$
$$= \frac{1}{2}(3.4)(3.1 + 5.1)$$
$$= 1.7(8.2)$$
$$= 13.94$$

Strategies
Choose a formula.

The area of the garden is 13.94 m^2.

Method 2: Add the areas.

I can divide the right trapezoid into a rectangle and triangle. I can use formulas to find the area of each simple shape.

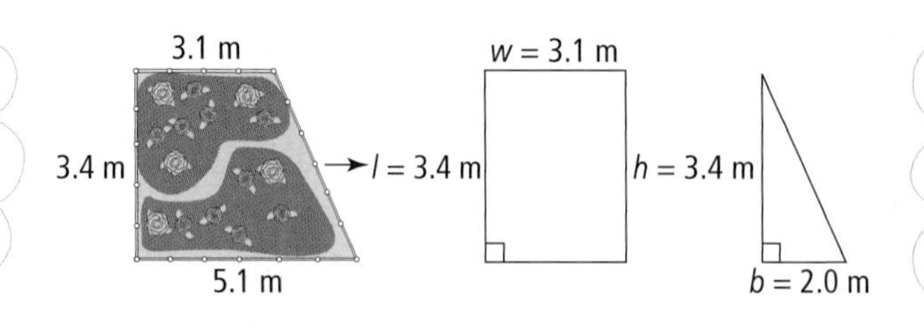

I can subtract the length of the trapezoid base from the top for the length of the triangle base. 5.1 m − 3.1 m = 2.0 m

Area of trapezoid = Area of rectangle + Area of triangle

$$A_{\text{trapezoid}} = lw + \frac{1}{2}bh$$
$$= (3.1 \times 3.4) + \frac{1}{2}(2.0 \times 3.4)$$
$$= 10.54 + 3.4$$
$$= 13.94$$

The area of the garden is 13.94 m^2.

b) To find the cost of the fencing, you need to find the perimeter of the garden. All the side lengths are given except for the slanted side. The diagram from part a), Method 2, shows a right triangle.

Use the Pythagorean relationship to find the length of the slanted side.

$c^2 = h^2 + b^2$
$c^2 = (3.4)^2 + (2.0)^2$
$c^2 = 11.56 + 4.0$
$c^2 = 15.56$
$c = \sqrt{15.56}$
$c \doteq 3.94$

Estimate: $\sqrt{16} = 4$

Use a calculator.

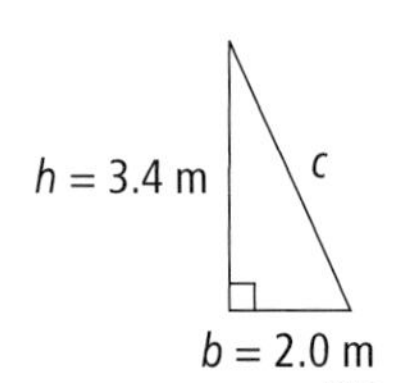

Perimeter of garden $\doteq 3.1 + 3.4 + 5.1 + 3.94$
$\doteq 16$

3.4 + 3.1 = 6.5 and 5.1 + 3.9 = 9.0, so 6.5 + 9.0 = 15.5. Round up to the nearest metre.

The perimeter of the garden is approximately 16 m.

Total cost = Perimeter × Cost of fencing materials
= 16 × 10
= 160

The total cost for fencing materials will be $160.

Example 2: Find the Area of a Complex Composite Figure

Find the area of the wave pool shown. Round your answer to the nearest tenth of a metre.

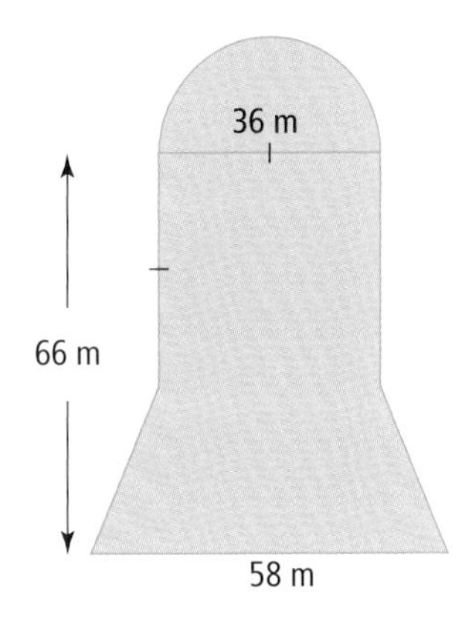

Solution

Divide the composite figure into three components: a semi-circle, a square, and a trapezoid.

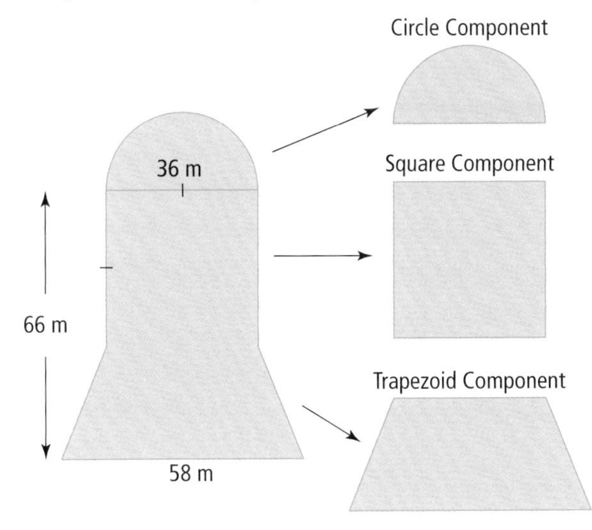

Find the area of each component, and then add the areas.

Circle Component

The circle component is a semi-circle, so you need to find half the area of the circle.

$$
\begin{aligned}
A_{\text{semi-circle}} &= \frac{1}{2}\pi r^2 \\
&= \frac{1}{2}\pi(18)^2 \\
&= 162\pi \\
&\doteq 508.9
\end{aligned}
$$

The diameter of the circle is 36 m, so the radius is half the diameter, or 18 m.

← 36 m →

The area of the circle component is approximately 508.9 m^2.

Square Component

$A = s^2$

$\quad = (36)^2$

$\quad = 1296$

The area of the square component is 1296 m^2.

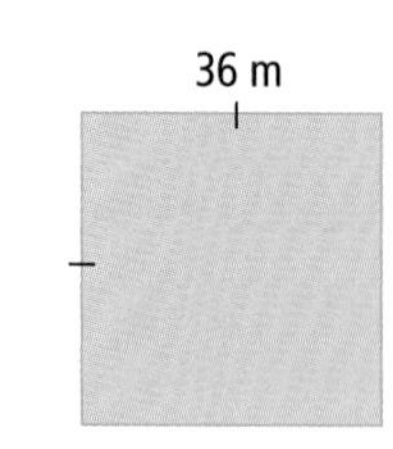
36 m

Area is measured in square units.

Trapezoid Component

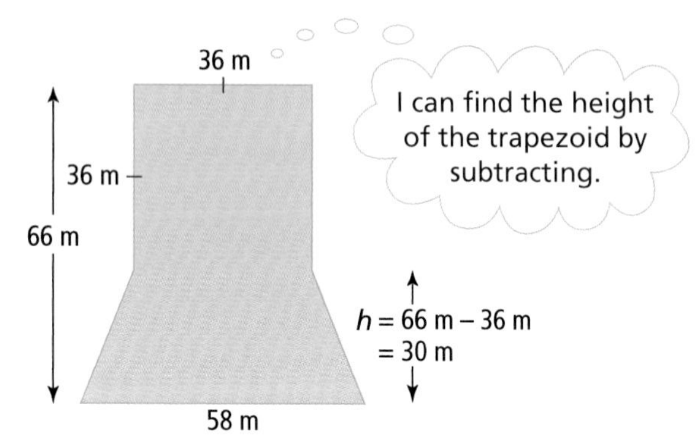
36 m
36 m
66 m
I can find the height of the trapezoid by subtracting.
h = 66 m – 36 m
= 30 m
58 m

Use the formula.

$A = \frac{1}{2}h(a + b)$

$\quad = \frac{1}{2}(30)(36 + 58)$

$\quad = 15(94)$

$\quad = 1410$

a = 36 m

h = 30 m

b = 58 m

The area of the trapezoid component is 1410 m^2.

Composite Figure

Add the areas of the components to find the total area of the composite figure.

Total Area = Area of semi-circle + Area of square + Area of trapezoid

$\quad \doteq 508.9 + 1296 + 1410$

$\quad \doteq 3214.9$

The area of the wave pool is approximately 3214.9 m^2.

Example 3: Subtract to Find Net Area

The red bullseye of the target has a radius of 2 cm. The diameter of the entire target is 10 cm.

Find the area of the blue ring.
Use $\pi = 3.14$

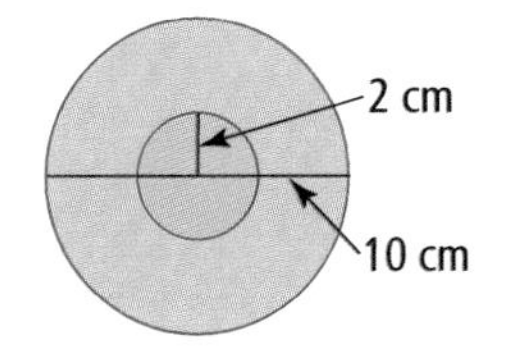

Communicating Mathematically

Net area is the area remaining after part of a two-dimensional shape has been removed.

Solution

To find the area of the blue part of the target:
- Find the total area of the target.
- Find the area of the bullseye.
- Subtract the area of the bullseye from the total area.

The entire target has a diameter of 10 cm. Find its area.

$A = \pi r^2$ **Divide the diameter in half to find the radius: $r = 10 \text{ cm} \div 2 = 5 \text{ cm}$.**

$= (3.14)(5)^2$
$= (3.14)(25)$
$= 78.5$

The area of the entire target is approximately 78.5 cm^2.

The bullseye has a radius of 2 cm. Find its area.

$A = \pi r^2$
$= (3.14)(2)^2$
$= (3.14)(4)$
$= 12.56$

The area of the bullseye is approximately 12.6 cm^2.

Subtract to find the net area of the blue ring.

Net area = Area of entire target − Area of bullseye
= 78.5 − 12.6
= 65.9

The area of the blue ring is 65.9 cm^2.

Communicate the Key Ideas

1. a) What is a composite figure? Draw an example.
b) For the example you chose, identify the simple shapes that combine to form the composite figure.

2. a) Show or describe how each figure can be divided into two or more simple shapes.

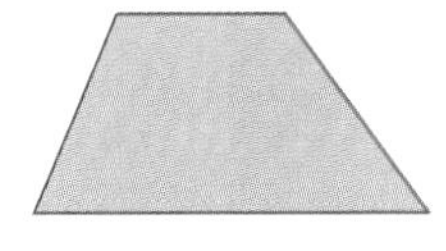

b) Show a different way to divide each figure in part a) into simple shapes.

3. Explain how you can find the measures of x and y in the diagram.

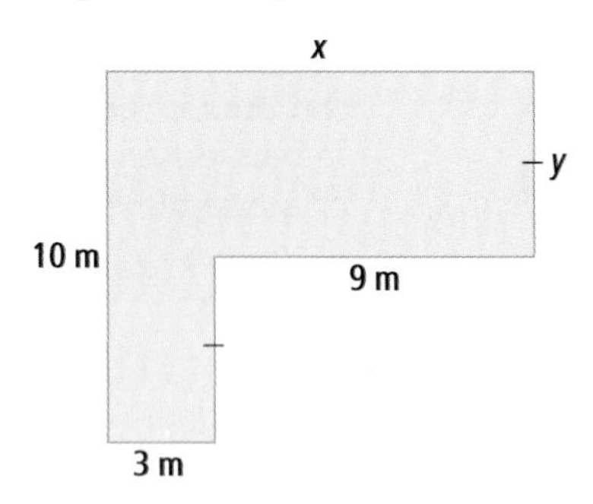

4. Describe the steps to follow for finding the area of the composite figure in question 3.

5. Draw a composite figure that consists of:

a) a rectangle and a triangle
b) a square and a trapezoid
c) part of a circle and another shape
d) three different simple shapes

6. Roger found the area of the trapezoid.

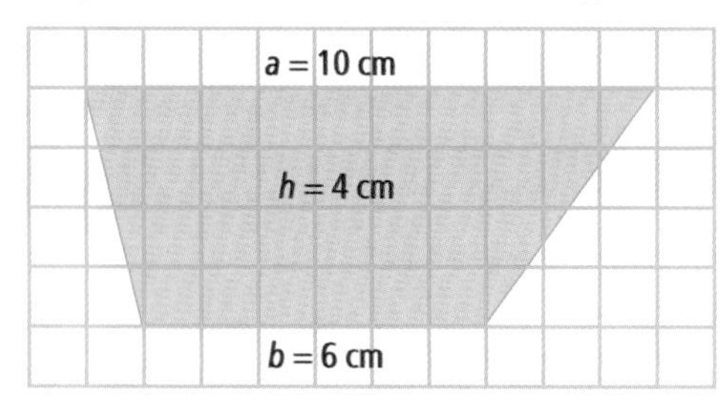

Is he correct? If not, explain where he went wrong in his thinking and correct his work.

$$
\begin{aligned}
A &= \frac{1}{2}h(a)(b) \\
&= \frac{1}{2}(4)(10)(6) \\
&= 2(60) \\
&= 120
\end{aligned}
$$

The area is 120 cm^2.

Check Your Understanding

1. a) Find the area of the swimming pool.

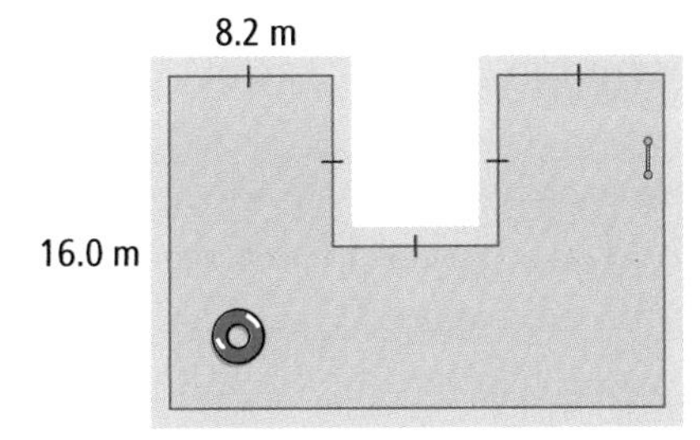

   b) Explain how you solved the problem.
   c) Describe how you can solve this problem using a different method.
   d) Find the perimeter of the pool.

2. Copy this composite figure.

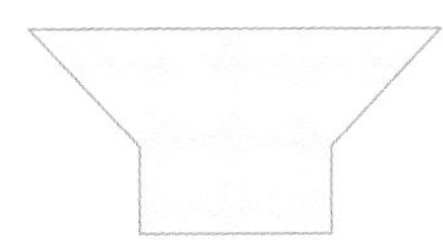

   Show how it can be divided into:
   a) three simple shapes
   b) two different simple shapes

3. Copy this composite figure.

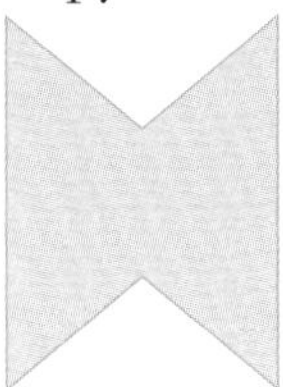

   Show how it can be divided into:
   a) five simple shapes
   b) three simple shapes
   c) two simple shapes

4. Refer to question 3. Which method would you use to find the area of the composite figure? Explain your reasoning.

5. The logo for Hanna's hockey team, the Tough Cats, is shown.

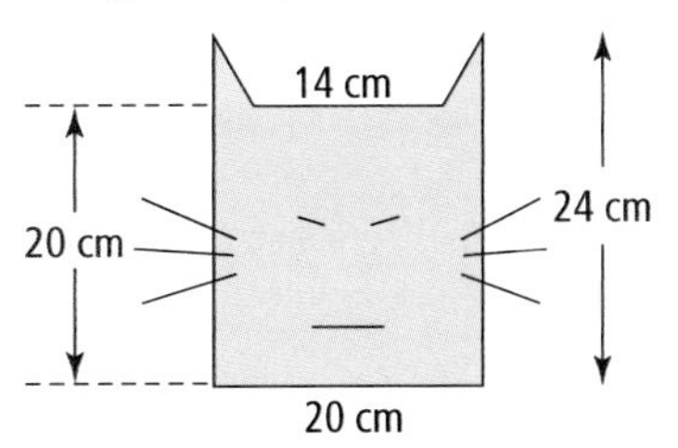

   Find the area of the logo, not including the whiskers. Explain any assumptions you must make.

6. The Tough Cats' archrivals are the Scary Bears, whose logo is shown.

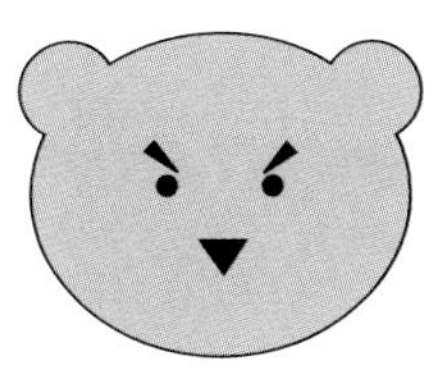

   a) Describe how you could estimate the total area of the Scary Bears' logo.
   b) Take measurements and use your method to estimate the area of the logo. Explain why your estimate may not be entirely accurate.

7. The parking lot of the local shopping mall needs to be coated with sealant.
   a) What is the area of the parking lot? Show how you found your answer.

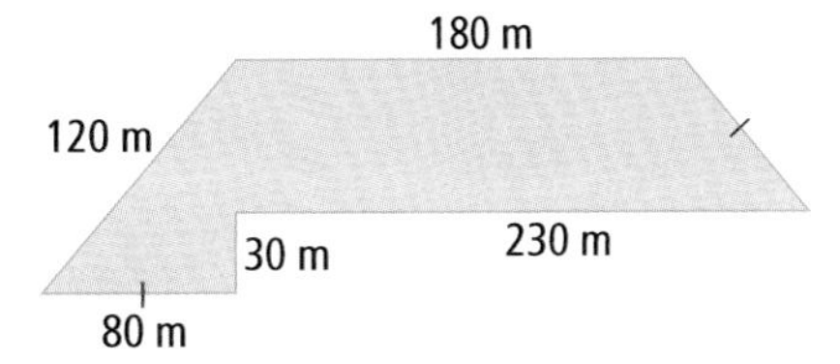

   b) One 19-L pail of sealant covers an approximate area of 2100 m^2. How many pails are needed to apply one coat of sealant to the parking lot?

8. A lighthouse is being repainted. It has four symmetrical white walls. One wall has a red door, as shown. The doorway is 1.2 m wide.

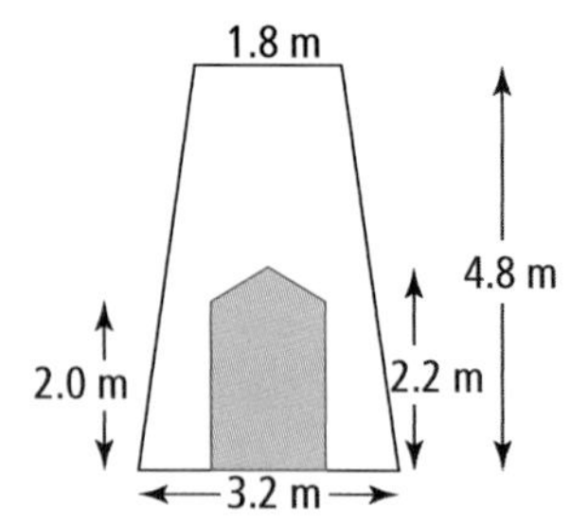

a) Find the total surface area painted if the door is covered with two coats of red paint.

b) Find the total surface area painted, not including the doorway, if the walls are covered with two coats of white paint.

9. The picture frame has a constant thickness of 5 cm.

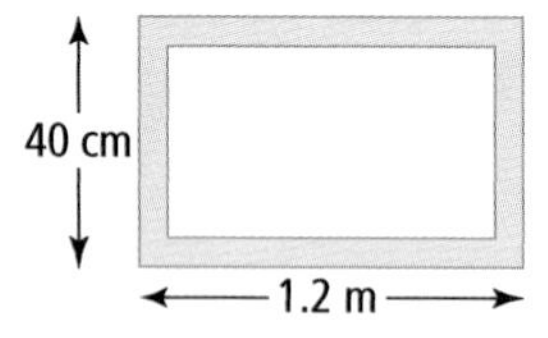

a) What is the length and width of a picture that would just fit in this frame?

b) Find the area inside the frame.

c) Find the total area of the frame itself.

d) Describe another method that could be used to solve part c).

10. A machine part is formed by drilling a hole through a rectangular piece of sheet metal, as shown.

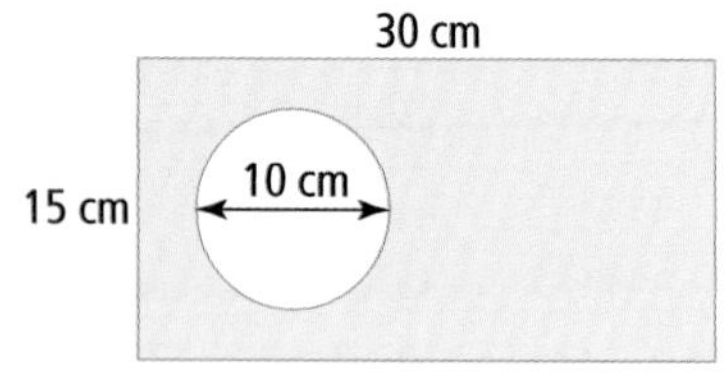

a) What is the perimeter of the rectangle?

b) What is the area of the rectangle before the circle is removed?

c) Both sides of the part will be covered in anti-friction sealant. How many square centimetres of sealant are needed? What assumption did you make?

11. Describe a situation in which you would need to find the area of a composite figure. Describe the composite figure using words and diagrams. Identify the simple shapes that combine to form the composite figure.

12. Hans is building wooden shelf dividers, as shown.

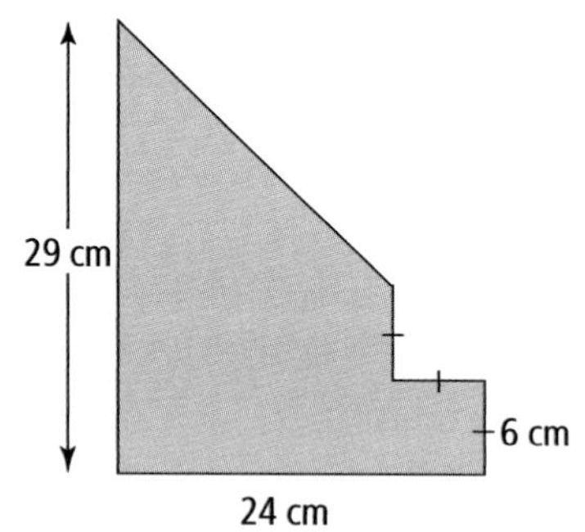

a) Show how you can divide this composite figure into two or more simple shapes.

b) Find the total area of wood needed to build 10 dividers.

c) Describe another method to find the total area of the composite figure.

d) How many dividers could Hans make from one standard sheet of plywood (122 cm by 244 cm)? What assumption did you make?

Chapter Problem

13. The front of the Robinsons' new pool shed is shown. Mrs. Robinson wants to paint it a special colour.

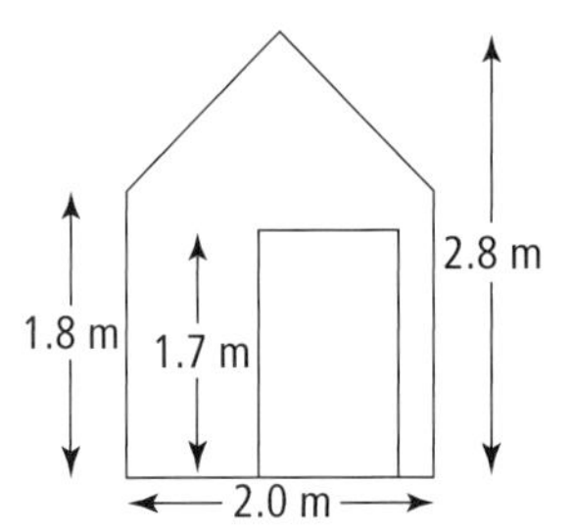

a) Find the area that needs to be painted.

b) The door is 1.0 m wide. How does your answer change if the door is not to be painted?

Extend

14. Look at the overhead photograph of the Halifax Citadel, shown below.

This simplified model shows the approximate shape of the Halifax Citadel grounds.

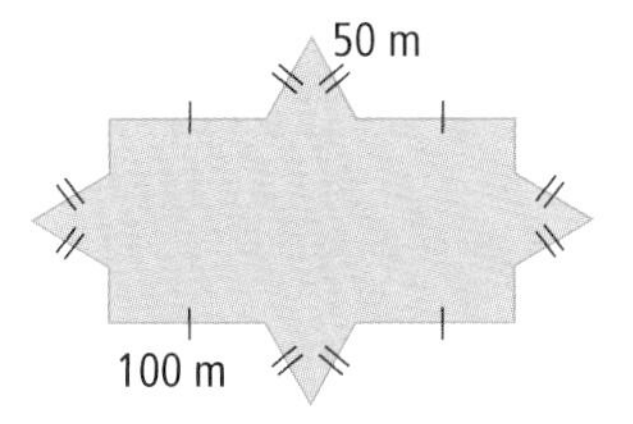

a) Estimate the perimeter of the Halifax Citadel's outer wall. Explain how you found your estimate.

b) Estimate the area of the Halifax Citadel grounds. Explain how you found your estimate.

c) How accurate are your answers? Explain.

Did You Know?

Built between 1828 and 1856, the Halifax Citadel was an important naval defence station. The Citadel is now preserved as a national park.

To learn more about this important part of Canada's heritage go to: www.mcgrawhill.ca/links/math8NS and follow the links.

15. a) Find a logo that uses a composite shape.

b) Sketch the logo and identify the simple shapes that make up the logo.

c) Find the area of the logo.

d) Trade questions with a partner and find the area of your partner's logo. Check your solutions.

10.4 Surface Area of Three-Dimensional Figures

Focus on...
- estimating and calculating the surface areas of right prisms and cylinders
- measuring and calculating the surface areas of 3-D composite figures

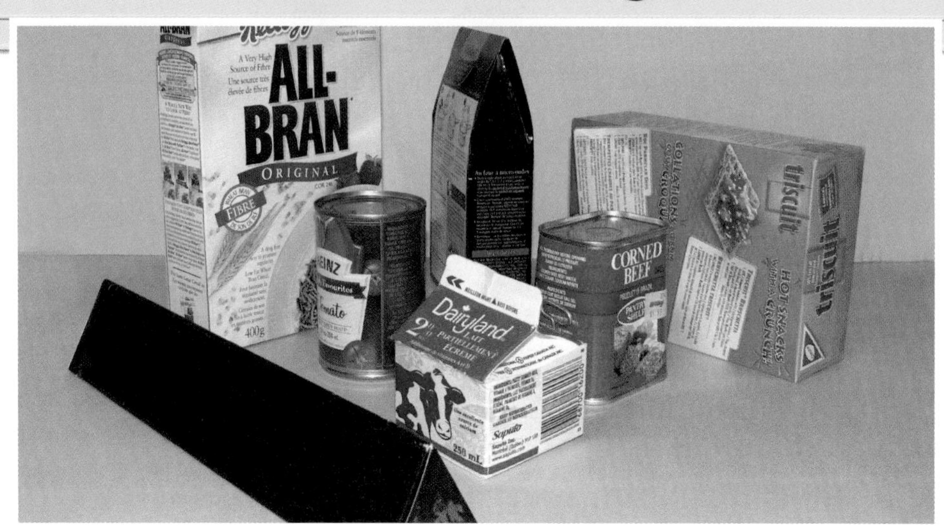

Many packages are in the shape of common geometric solids, such as rectangular prisms and cylinders. Why do you think these figures are chosen? What properties do these figures have that makes them useful?

Manufacturers want packages that:
- hold a maximum amount of material
- are easy to stack on shelves and in boxes
- are large enough for labels to be easily read

How can your understanding of geometry help you determine important measures of packages, such as **surface area**?

surface area
- the total outer area of all the faces of a three-dimensional figure

Discover the Math

Materials
- geoblocks
- centimetre grid paper
- ruler

Part A: How can you find the total surface area of a rectangular prism?

1. Choose a rectangular prism from the set of geoblocks.

a) Look at the prism. Estimate its total surface area, in square centimetres.

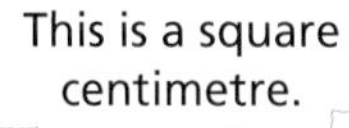

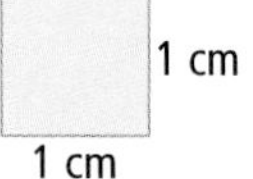

How many of these would it take to cover all faces of the prism?

b) Explain how you found your estimate.

2. a) How many faces does the prism have?

b) Identify the shapes of the faces. Which are congruent faces? Which are not congruent faces?

3. Create a net for the prism by tracing each face of the prism on centimetre grid paper. Make sure all the faces are connected to each other. Label each face with a letter, starting with A.

4. a) Identify pairs of congruent rectangles.

b) Measure the length and width of each different rectangle.

c) Find the area of each different rectangle.

5. a) Add the areas of each face to determine the total surface area.

b) Compare this value with your estimate. Was your estimate very close? Explain.

6. Reflect

a) Write a sequence of steps that you can use to find the surface area of a box that is a rectangular prism.

b) Do you think this method will work for other three-dimensional figures? Explain why or why not.

Nets are useful for finding the surface area of three-dimensional figures. By laying all the faces flat, it is easy to:
- count the faces
- identify the shape of the faces
- recognize congruent faces

Can you identify which three-dimensional figure each net will form?

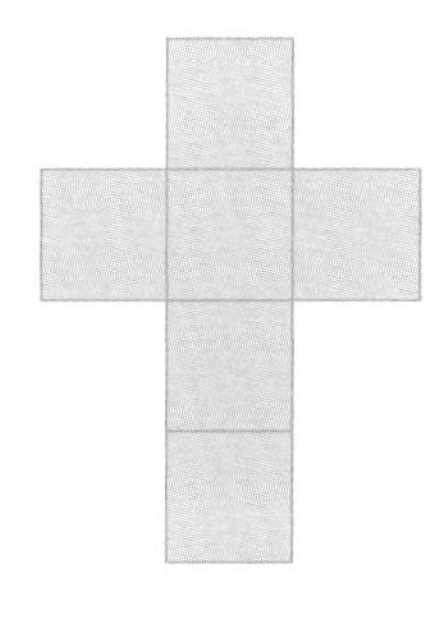

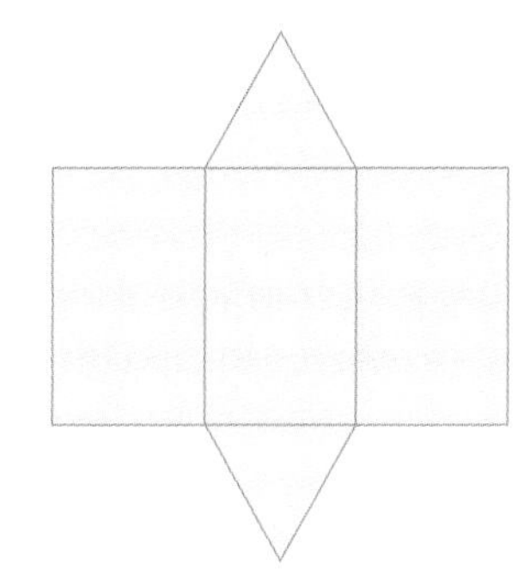

Part B: How can you find the total surface area of a cylinder?

Materials
- toilet paper roll
- scissors
- tape

1. a) Look at the top and bottom of the cylinder. What shapes form the top and bottom?

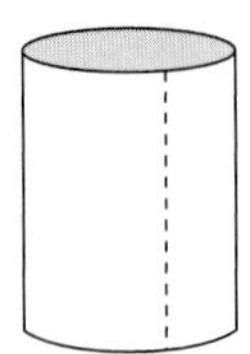

b) Explain how you can find the area of the top and bottom.

2. Look at the **lateral face**.

lateral face
- the surface of a three-dimensional object that is not a base

Example: the curved face of a cylinder

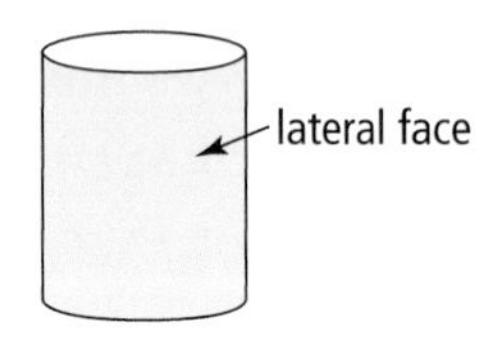
lateral face

a) Predict what two-dimensional shape this will form when it is cut vertically along the dotted line and laid flat.

b) Cut the toilet paper roll along the dotted line. Lay it flat on your desk. What shape is it?

3. a) Draw the three faces of the cylinder on a sheet of paper to form a net.

b) Measure and label the diameters of the circular faces. (To be accurate, fold the circles in half to get the diameter.)

c) Measure and label the length and width of the lateral face.

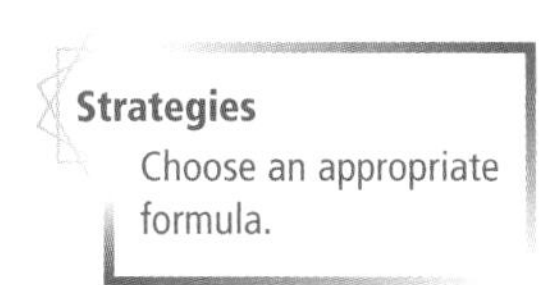
Strategies
Choose an appropriate formula.

4. a) Determine the area of each face of the net.

b) Add the areas of the faces to find the total surface area of the cylinder.

5. Reflect Compare the length and width of the lateral face to the original dimensions of the cylinder.

a) How is the width (in blue) of the rectangle related to the original cylinder?

b) How is the length (in red) of the rectangle related to the original cylinder?

c) What formula could you use to find the length of the rectangle, if you knew the radius of the cylinder?

6. Take a second cylinder. Sketch its net. Measure only the diameter of the circular face and the height of the cylinder. Label your sketch with these measurements. Use this information to find the surface area of the cylinder.

Example 1: Surface Area of a Triangle-Based Prism

How much packaging is needed to make a box for a fruit bar in the shape shown, to the nearest square centimetre?

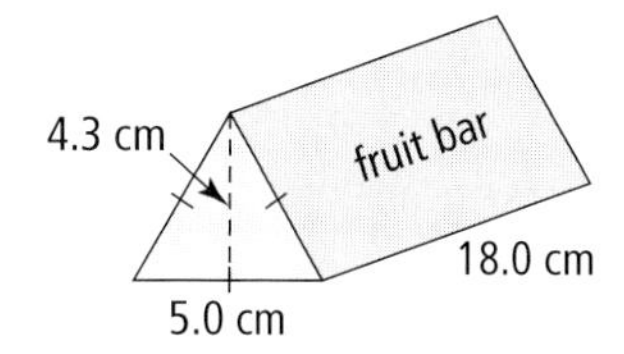

Solution

The fruit bar is in the shape of a triangle-based prism. The base is an equilateral triangle.

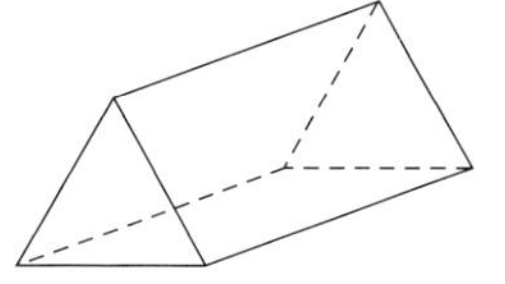

Strategies
Build a model.

Draw a net for the prism.
The prism has five faces:
- 2 congruent equilateral triangles
- 3 congruent rectangles

Find the area of each face.

Triangle

$$A = \frac{1}{2}bh$$
$$= \frac{1}{2}(5.0)(4.3)$$
$$= 2.5 \times 4.3$$
$$= 10.75$$

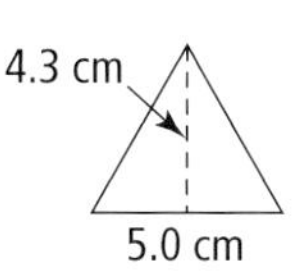

Each triangular face has an area of 10.75 cm^2.

Rectangle

$$A = lw$$
$$= (18.0)(5.0)$$
$$= 90.0$$

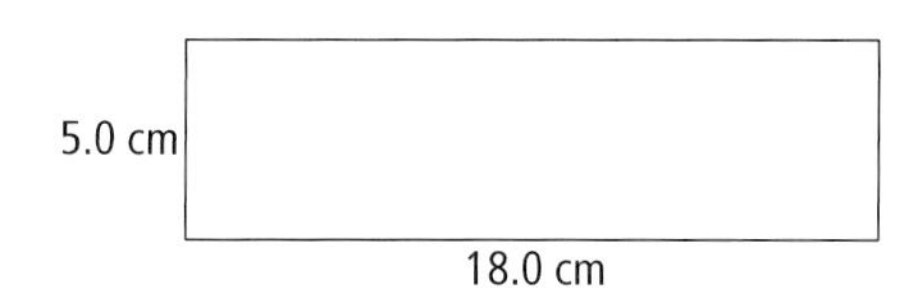

Each rectangular face has an area of 90.0 cm^2.

Add the areas of each face to find the total surface area.

$$\text{S.A.} = 2(\text{area of triangle}) + 3(\text{area of rectangle})$$
$$= 2(10.75) + 3(90.0)$$
$$= 21.50 + 270$$
$$= 291.50$$

The amount of packaging needed to wrap the fruit bar, to the nearest square centimetre, is 292 cm^2.

Example 2: Surface Area of a Composite Three-Dimensional Figure

Risers are wooden boxes that are combined to form elevated platforms for singers in a choir to stand on.

This scale model for a riser platform was built using a R1 (8 cm by 4 cm by 1 cm) geoblock and a S4 (4 cm by 4 cm by 2 cm) geoblock.

The scale for the model is 1 cm = 1 m. What is the total surface area of the actual platform?

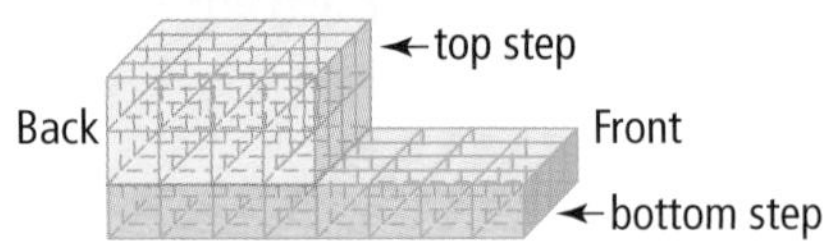

Solution

Examine the structure from each side. Find the surface area of each side.

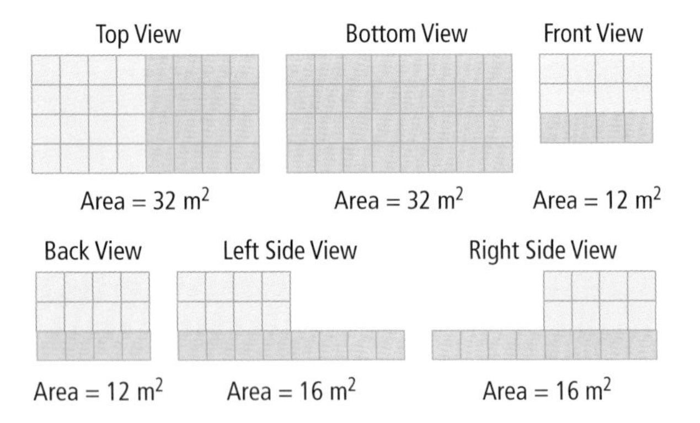

Add the surface areas to find the total surface area.
$32 + 32 + 12 + 12 + 16 + 16 = 120$
The total surface area of the platform is 120 m^2.

Strategies
What other methods could you use to solve this problem?

Communicating Mathematically

The different views represent the orthographic drawings of the platform.

Communicate the Key Ideas

1. a) Consider surface area. Is it a two-dimensional concept, a three-dimensional concept, neither two- nor three-dimensional, or both? Explain.
   b) Describe two activities in which it is necessary to determine surface area.
2. Why is surface area measured in square units?
3. a) What is a net? Is it two-dimensional or three-dimensional?
   b) How is a net related to the solid object it represents?
   c) Describe the advantages of using a net when finding surface area.

4. **a)** Draw a three-dimensional object made from two different geoblocks.
   b) Draw the net for this object.
   c) Trade nets with a partner. Try to guess each other's three-dimensional shape.

Communicating Mathematically

A three-dimensional object whose faces are polygons can also be a called a **polyhedron**.

Here are some examples of polyhedra.

rectangular prism, cube, triangular prism, hexagonal prism, pentagonal prism, triangular pyramid, square-based pyramid

5. **a)** Draw a net for a cylinder. Describe the faces.
   b) Explain how you can find the total surface area of a cylinder.

6. Look at the Robinson family's pool shed.
   a) What measurements would you need to calculate the surface area of the shed?
   b) Describe the steps that you would follow to find the total surface area, including the door.

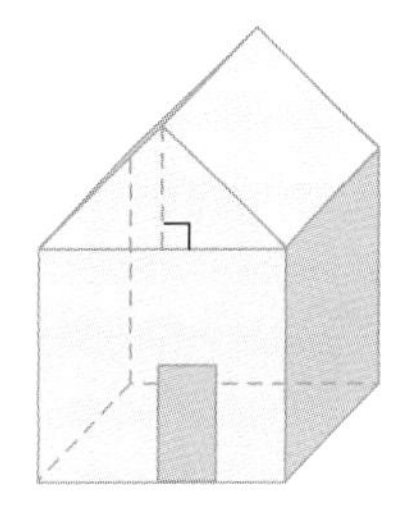

Check Your Understanding

1. A box, in the form of a rectangular prism, measures 25 cm × 35 cm × 50 cm.
   a) Draw a three-dimensional model of the box and label its dimensions.
   b) Draw a net for the box. Identify any pairs of congruent faces.
   c) Label the dimensions of each different face on the net.
   d) Find the surface area of the box.
   e) Make an isometric drawing of the box on isometric dot paper.

2. A cube-shaped jewellery box has a side length of 4.2 cm.

The outside of the box is covered with felt. How much felt would be needed to cover ten boxes?

3. Apple chips come in a container shaped like a pentagonal prism. The entire container is sealed with plastic wrap. How much plastic wrap is needed to cover 12 containers? Show your work.

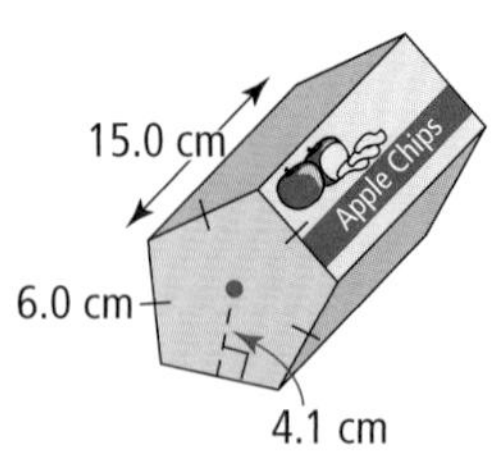

4. The diagram shows the dimensions of a slice of carrot cake. Icing is spread on the top and back of the slice and between the two layers of cake.

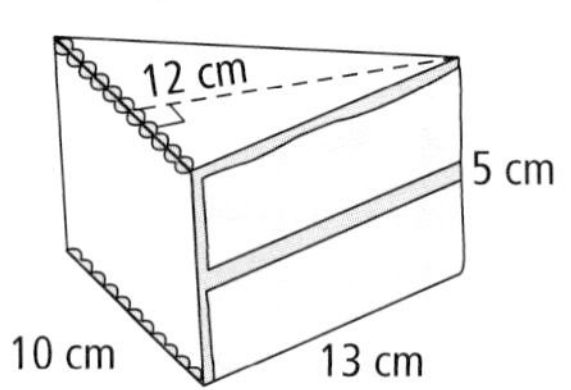

a) What is the surface area of the iced portion of the slice of cake? What assumption did you make?
b) What is the entire surface area of the slice of cake?
c) What percent of the slice of cake is covered with icing?

5. Determine the surface area of the soup can.

6. A box has an open top, as shown.

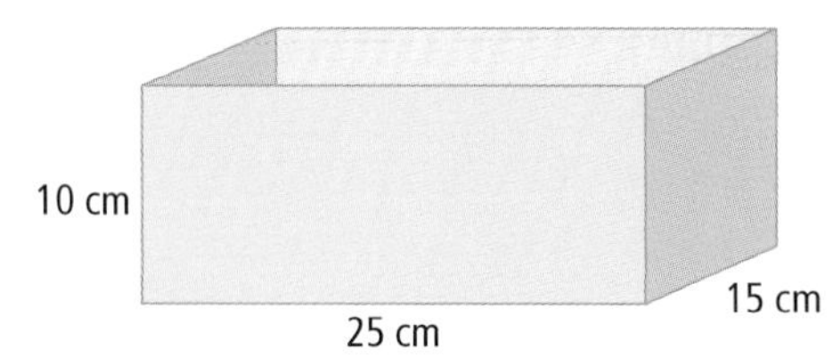

Find the outer surface area of the box.

7. Find the outer surface area of the pail. Note that there is no top.

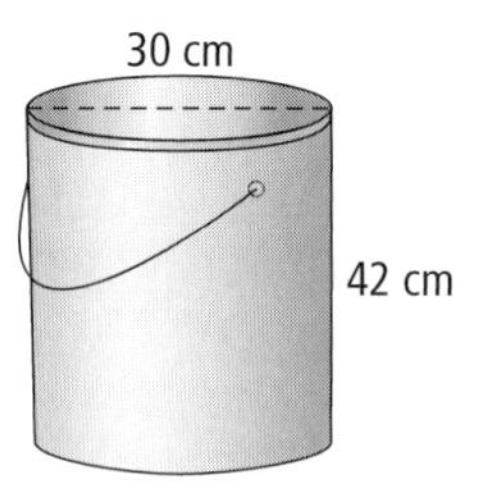

8. Dan sketched the tree house he wants to build.

a) Estimate the surface area of the roof of the tree house, including the gable ends. Give your answer in square metres.
b) What is the actual surface area of the roof? Round your answer to the nearest square metre.
c) What is the percent difference between your estimated and your calculated surface area?

9. a) Build a composite three-dimensional object using geoblocks or linking cubes.
b) Find the surface area of the object.
c) Trade objects with a partner and determine the surface area of each other's object. Compare your answers.

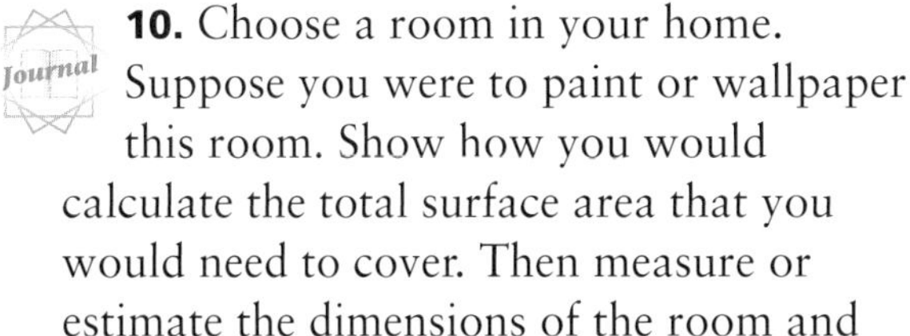

10. Choose a room in your home. Suppose you were to paint or wallpaper this room. Show how you would calculate the total surface area that you would need to cover. Then measure or estimate the dimensions of the room and find the total surface area.

Chapter Problem

11. Mr. Robinson is ordering plywood to build the new shed.

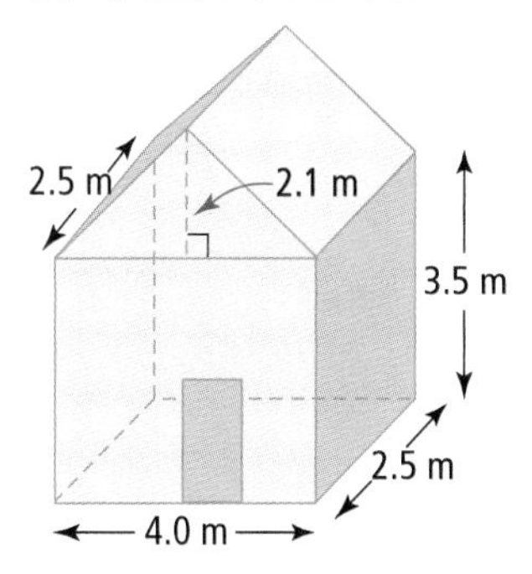

a) Find the total surface area of the shed to the nearest tenth of a square metre.

Strategies
Hint: Use information you have calculated or been given previously.

b) How many standard sheets of plywood (122 cm by 244 cm) will he need?

12. A container is designed to hold three tennis balls, stacked in a row, as shown.

The radius of each tennis ball is 3.2 cm.

a) What are the dimensions of the smallest square-based prism that can hold the tennis balls?

b) What are the dimensions of the smallest cylinder that can hold the tennis balls?

c) Which container uses less material? How much less?

Did You Know?

The word *wigwam* comes from the Mi'kmaq word *wikuom*, a dwelling. Wigwams are cone-shaped structures made of a lattice of spruce poles covered with sheets of birch bark that are layered like giant shingles. Wigwams are designed to be portable. The poles are left standing and the birch bark is taken to the next hunting camp and attached to the poles. The round floor would have a diameter of approximately 4 m. What would be the area of the floor?

Extend

13. The floor of this tent is square. The outside of the tent needs to be waterproofed.

a) Calculate the area to be covered.

b) One can of waterproofing covers 3.5 m^2. How many cans are needed for the tent?

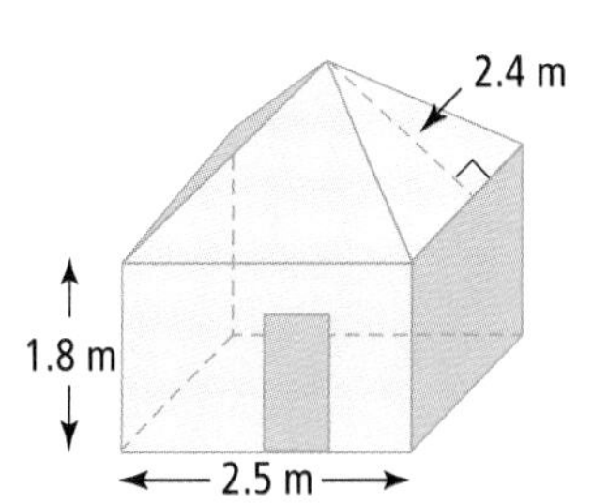

10.5 Volume of Three-Dimensional Figures

Focus on...
- the volumes of right prisms and cylinders
- measuring and calculating the volumes of 3-D composite figures
- solving measurement problems, using appropriate SI units

volume
- amount of space occupied by an object
- measured in cubic units, such as cm^3, m^3, etc.

capacity
- the maximum amount of material or liquid that an object or container can hold
- measured in cubic units or liquid measures

Geometric figures are used in the design of many types of structures, some real and some imagined. Look at the space station in the photograph. Can you see how cylinders are used in the design? What other geometric shapes can you recognize? Why do you think a space station would have such design features?

How much living space do you think the astronauts have in this space station? Questions such as this require finding the **volume** or **capacity** of a three-dimensional object or container.

Discover the Math

Materials
- large, fillable relational geometric solids
- 1000 cm^3 square or rectangular prism
- 1-L graduated cylinder, beaker, or jug
- 10-mL graduated cylinder
- water
- balance
- base-10 blocks

Optional
- fillable base-10 blocks

Part A: What are the SI measurement units for volume and density?

Obtain a 1 L graduated cylinder and a 10 cm by 10 cm square prism.

1. a) Fill the graduated cylinder with 1 L of water.

b) Empty the graduated cylinder into the square prism. What is the volume of the square prism in cubic centimetres?

c) How many millilitres of water does the square prism hold? Write an equality statement that relates the number of cubic centimetres to the number of millilitres the container holds.

d) Use your answer to part c) to relate 1 cm^3 to 1 mL.

2. Density is a rate. It compares the mass of one unit of material to the volume of space it occupies. The mass of water when measured on Earth at 4°C has a special relationship with its volume.

   a) Find the mass of the square prism filled with water from question 1, part b). Find the mass of the empty square prism. What is the mass of the water?
   b) Determine the mass of 1 mL of water. Show your work.
   c) i) What is the relationship between 1 mL of water and 1 g of water?
      ii) What is the relationship between 1 cm^3 of water and 1 g of water? Explain your reasoning.
   d) i) What is the density of water? Express the density in two different ways. Refer to question 1, part c) to help you.
      ii) Write the rates as a proportion. Include the measurement units.

3. a) Express 1 L in cubic centimetres. Explain your reasoning.
   b) Express 1 L in cubic metres. Use a labelled diagram of a large cube to help you. Show your work.

4. What volume, in cubic centimetres, will 1 kg of water have? Explain your reasoning. Use a labelled diagram of a large cube to help you. Show your work.

5. A sugar cube has a volume of approximately 1 cm^3. A stove has a volume of approximately 1 m^3. Use these estimates to determine whether the volume of each object should be measured in cubic centimetres or in cubic metres.

   a) box of paper tissues
   b) washing machine
   c) refrigerator
   d) pack of playing cards
   e) one-car garage

Part B: How can you find the volume of a prism?

Volume of a Rectangular Prism

Materials
- linking cubes
- centimetre grid paper

Work in groups of four. Each group gets 27 linking cubes.

1. Copy the table.

Diagram of Base	Base Area (unit^2)	Number of Layers	Volume (unit^3)

Strategies
Work with models.
Record results in an organized table.
Find patterns.

The black square is an empty space.

2. Calculate the **base area** for the first diagram in the table. Use the base to build the tallest rectangular prism possible using the 27 blocks. Calculate the number of layers and the volume of the prism. Let the volume of one linking cube be 1 unit^3, or 1 cubic unit. Copy and fill in the table.

base area
- area of one of the two congruent faces of a prism

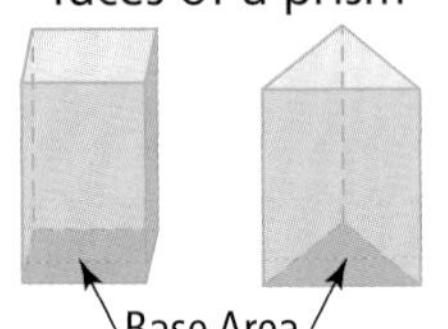

3. Repeat step 2 for the remaining diagrams in the table. You might not be able to use all 27 blocks for some bases.

4. **Reflect** Look at the patterns in your table. Write a rule for finding the volume of a rectangular prism.

Volume of a Triangular Prism

Will this technique work for other types of prisms? Consider this right triangle-based prism made of a cube and half-cubes. The base is the shaded triangle face.

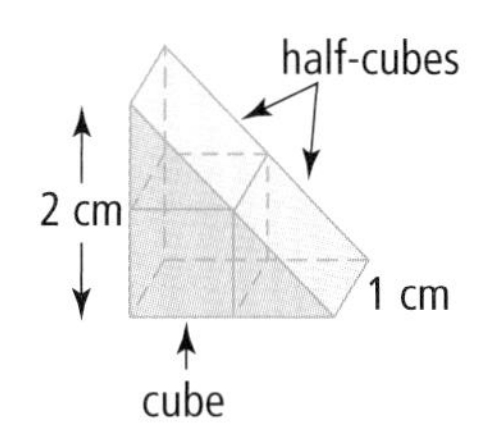

Each cube measures 1 cm × 1 cm × 1 cm. The volume of one cube is 1 cm^3.

1. a) How many whole cubes make up this prism? Explain.

b) What is the volume of the prism?

2. a) What are the base and height measures of the triangular base of the prism?

b) Calculate the area of the base of the prism.

3. How does the volume of the prism change if you add another layer of cubes, as shown?

a) Determine the volume of the new prism.

b) What is the height of the new prism?

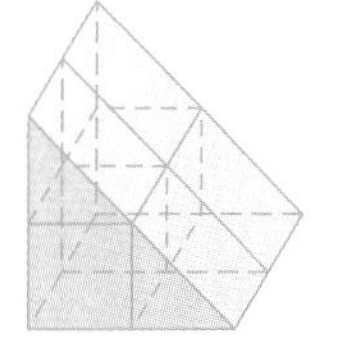

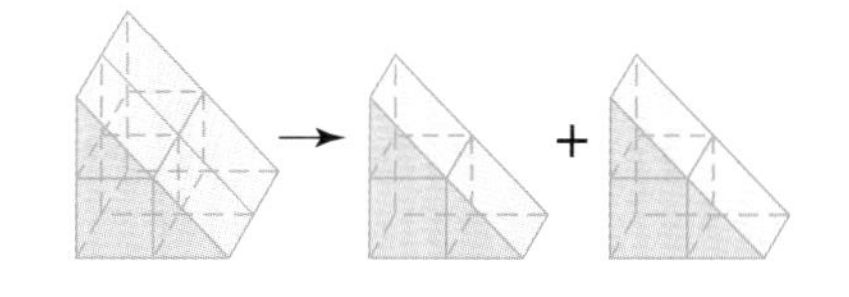

4. Repeat question 3 three more times. Copy and complete the table.

Triangular Prism (sketch)	Base Area (cm^2)	Height (cm)	Volume (cm^3)
half-cubes, 2 cm, 1 cm, cube		1	

Strategies

Build models using Polydron® pieces or draw models on graph paper to help you visualize these prisms.

5. Reflect Look for a pattern in your table. Compare your answers with those of your classmates. Describe how the volume of a triangular prism can be found using the base area and the height.

Part C: How can you estimate the volume of a cylinder?

Materials
- linking cubes
- cylindrical tin cans
- graduated cylinder or beaker
- water

1. a) Select a cylindrical can.

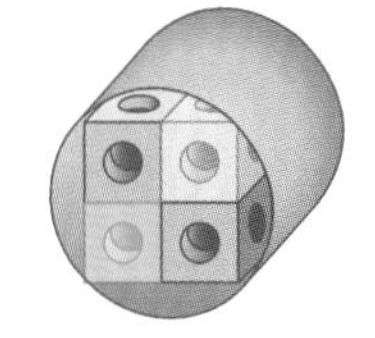

b) Cover the bottom of the can with a single layer of centimetre cubes. Why is it not possible to completely cover the bottom of the can?

c) Estimate the number of cubes needed to completely cover the bottom of the can.

d) Build a column of cubes to the height of the can to estimate how many layers of cubes would be needed to fill the can. Estimate the volume of the can.

e) Trace the bottom of the can on a piece of paper. Mark the centre of the base and draw the radius. How could you find the number of cubes needed to cover this area in a single layer?

f) Measure the height of the can. Calculate the volume of the can.

2. a) Build a frame for a square-based prism that will just contain the cylinder, as shown.

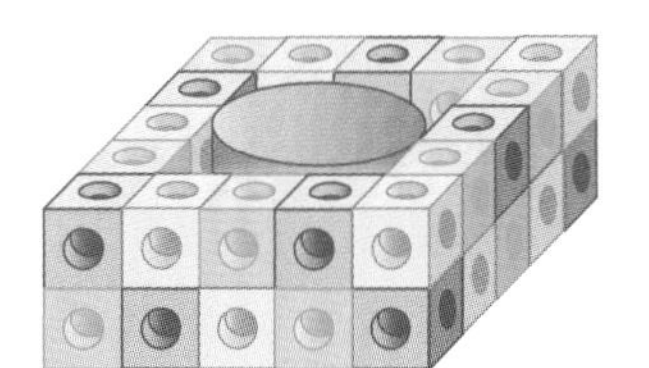

b) Estimate the volume of the cylinder by finding the volume of the square-based prism that would fit inside the frame.

c) Is the volume of the cylinder less than, greater than, or equal to this estimate? Explain your reasoning.

3. Reflect

a) How is a cylinder like a prism?

b) Suggest how you could find the volume of a cylinder if you knew the base area and the height.

Example 1: Volume of a Prism

The base for a garden statue is a trapezoid-based prism. Find the volume of the prism to the nearest tenth of a cubic metre.

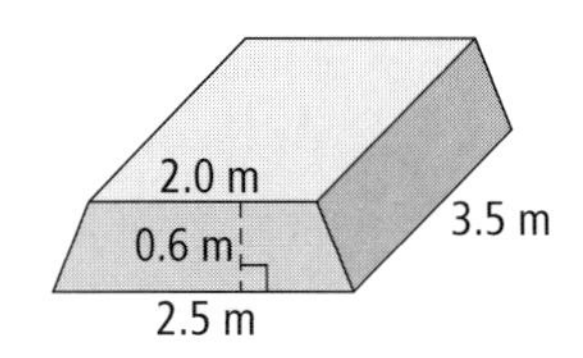

Solution

You can find the volume of any prism by multiplying its base area by its height. First find the base area of the prism.

$$
\begin{aligned}
A_{\text{trapezoid}} &= \frac{1}{2}h(a + b) \\
&= \frac{1}{2}(0.6)(2.0 + 2.5) \\
&= \frac{1}{2}(0.6)(4.5) \\
&= \frac{2.7}{2} \\
&= 1.35
\end{aligned}
$$

The base area of the prism is 1.35 m^2.

Multiply the base area by the height of the prism to find its volume.

$V = (\text{base area}) \times (\text{height})$
$= 1.35 \times 3.5$
$= 4.725$

The volume of the prism is approximately 4.7 m^3.

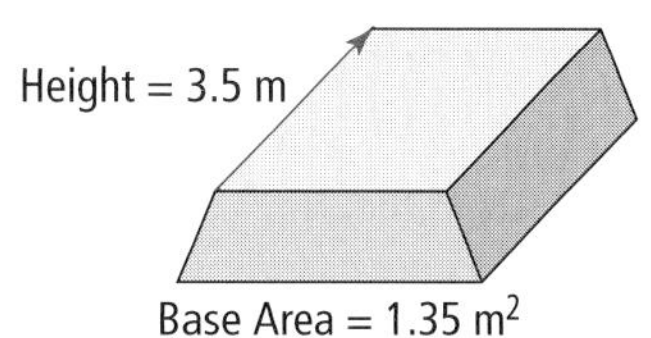

Example 2: Volume of a Cylinder

a) How much sand can the cylindrical container hold, to the nearest cubic centimetre?

b) How much liquid can the cylindrical container hold, to the nearest tenth of a litre?

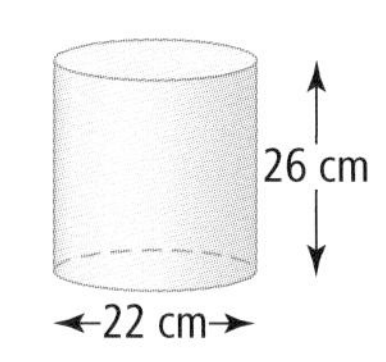

Solution

a) You can find the volume of any cylinder by multiplying its base area by its height. The base of a cylinder is a circle. Find the base area first.

$A = \pi r^2$
$= \pi(11)^2$
$= \pi \times 121$
$= 380.13...$

The base area of the cylinder is approximately 380 cm^2.
Multiply this by the height to find the volume.

$V = (\text{base area}) \times (\text{height})$
$= 380 \times 26$
$= 9880$

The volume of the cylinder is 9880 cm^3. Therefore, it can hold 9880 cm^3 of sand.

> **Strategies**
> Apply the formula for the area of a circle. Divide the diameter in half to find the value of the radius first.

b) To find the amount of liquid the cylinder can hold, in litres, use the relationship:
1 L = 1000 cm^3

Divide the volume to convert cubic centimetres to litres.

$V = 9880 \div 1000$
$= 9.88$ L

The cylinder can hold 9.9 L of liquid.

Example 3: Find the Volume of a Composite Solid

Lucy's paperweight is a solid letter 'L'.

Find the volume of the paperweight two ways.

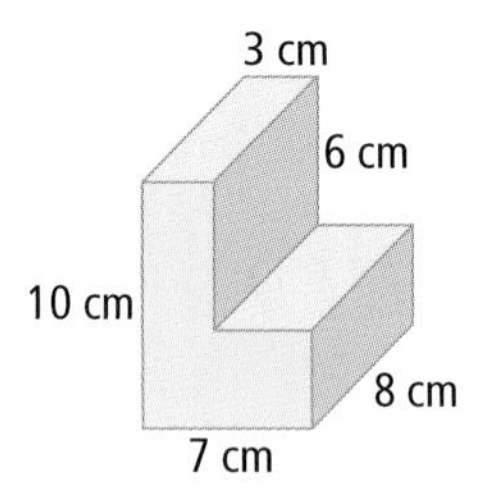

Solution

Method 1: Find the volume by adding.

You can find the volume of the paperweight by dividing the object into two rectangular prisms and adding their separate volumes.

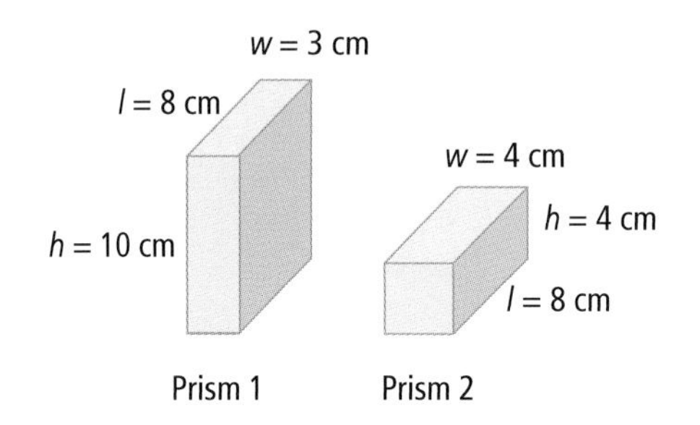

Volume of paperweight = Volume of prism 1 + Volume of prism 2

$V = (\text{base area} \times \text{height}) + (\text{base area} \times \text{height})$

$V = (8 \times 3 \times 10) + (8 \times 4 \times 4)$

$V = 240 + 128$

$V = 368$

The volume of the paperweight is 368 cm^3.

Method 2: Find the volume by subtracting.

Find the volume of a rectangular prism from which the paperweight could be made. Then subtract the volume of a smaller rectangular prism that can be removed to form the L-shape. This will give the **net volume** of the composite shape shown.

net volume

- the volume remaining after part of a solid has been removed

This is the smallest rectangular prism from which the paperweight could be made.

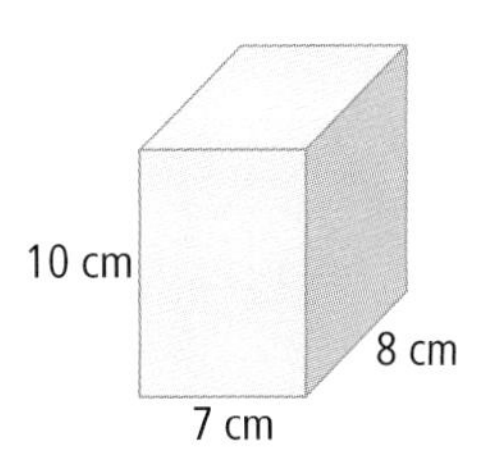

Find the volume of this rectangular prism.

V = base area × height
= (7 × 8) × 10
= 56 × 10
= 560

The volume of this prism is 560 cm^3.

To form the L-shape, this rectangular prism must be removed.

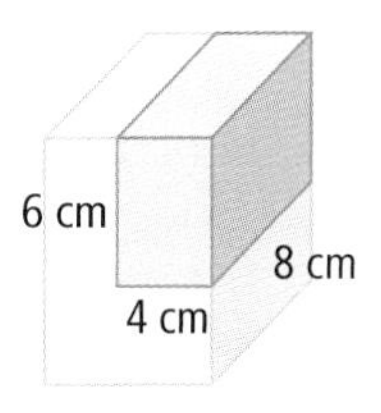

7 cm – 3 cm = 4 cm

Find the volume of this rectangular prism.

V = base area × height
= (8 × 4) × 6
= 32 × 6
= 192

The volume of this prism is 192 cm^3.

Find the net volume of the paperweight by subtracting the volume of the small prism from the volume of the larger prism.

V = 560 cm^3 – 192 cm^3
= 368 cm^3

The volume of the paperweight is 368 cm^3.

Communicating Mathematically

You can think of net volume in the way you think of net profit.
Net profit
= Revenue – Expenses
Net volume
= Total volume of solid
– Volume of solid part removed

Communicate the Key Ideas

1. a) What is volume? Is it a one-, two-, or three-dimensional measure? Explain.

b) What are some typical units of measure for volume?

2. a) What is the base area of a prism? Give two examples to support your explanation.

b) How is the base area important in finding the volume of a prism?

3. a) How are a cylinder and a prism

i) similar?

ii) different?

b) How can you find the volume of a prism and a cylinder?

4. **a)** Explain the difference between surface area and volume.
   b) Describe a real situation in which you would need to find:
      i) surface area **ii)** volume

5. Explain how you can find the volume of each composite solid.

a)

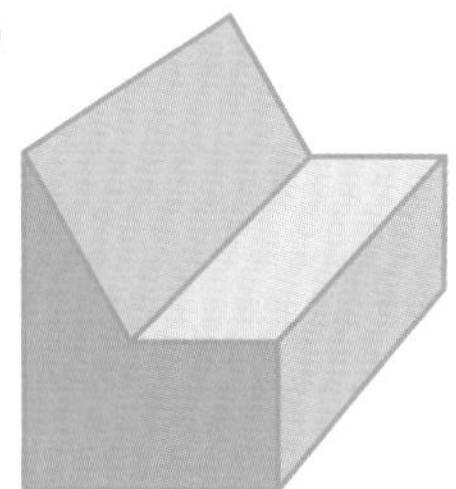

b)

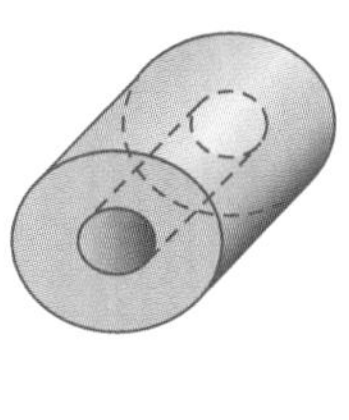

Check Your Understanding

1. Find the volume of each prism.

a)

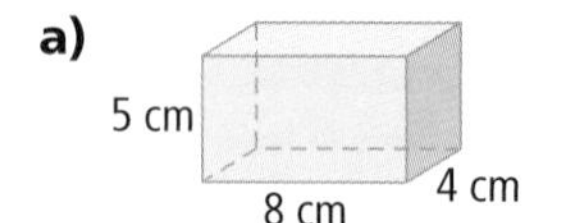

b)

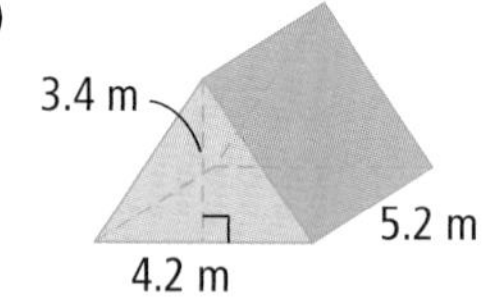

2. Find the volume of each cylinder.

a)

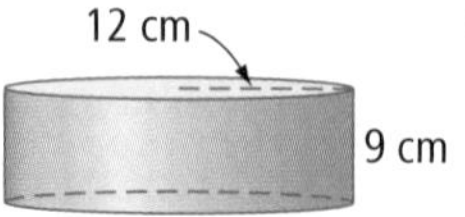

b)

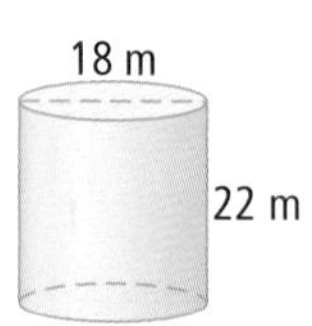

3. A number cube has a side length of 18 mm. Find its volume.

4. A hockey puck has a diameter of 7.6 cm and a height of 2.5 cm. What volume of container would be required to hold five pucks as shown?

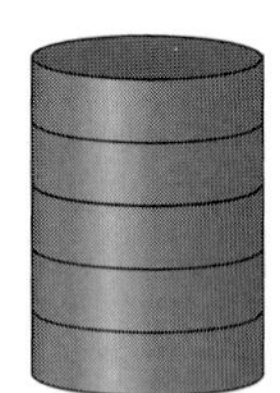

5. Julie and Raul have the same cylindrical pail. Julie records the radius of her pail as 24 cm. Raul records the height as 0.5 m. How many litres of water will one pail hold? *Hint:* Remember to work in common units.

6. How many millilitres of herbal tea will the mug hold?

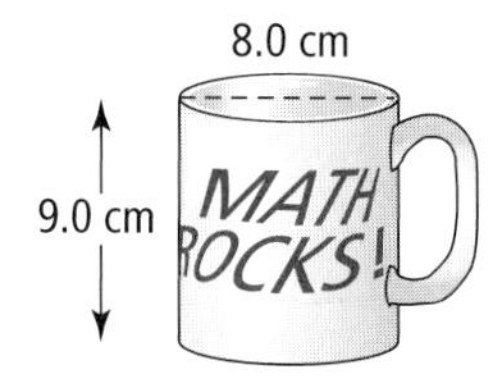

7. Find the volume of the ramp-riser to the nearest cubic metre.

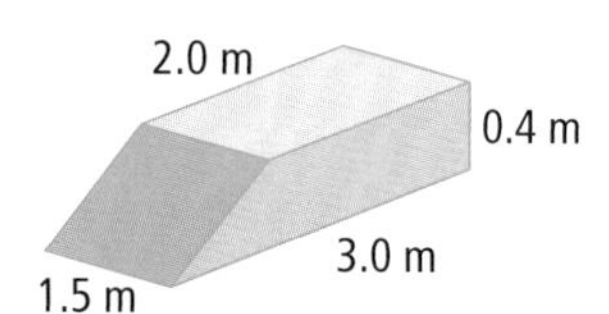

8. During the summer, Minh and Julie plan to sell home made chocolate-covered ice cream "pucks" on vanilla cookie sticks. The square base of the stick has a side length of 1 cm.

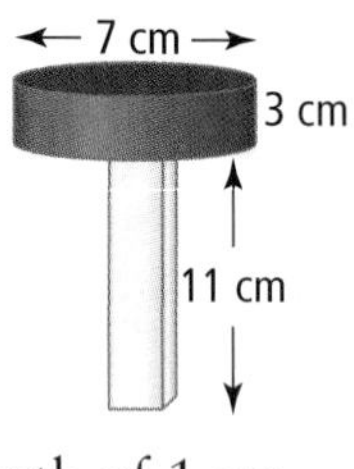

   a) Find the total volume of ice cream needed to make one puck.
   b) Find the volume of one cookie stick.
   c) If the puck is covered in chocolate, how much chocolate is needed for one treat? What assumptions did you make?

9. **a)** Take two identical sheets of paper and roll them to form two different open-ended cylinders.
b) Predict which of these cylinders has the greater volume, if either. Justify your choice.
c) Place the tall cylinder inside the short cylinder on your desk. Carefully fill the tall cylinder with rice or popcorn.
d) What do you think will happen if you lifted the tall cylinder, emptying its contents into the short cylinder? Will the short cylinder be partially filled, completely filled, or over filled? Explain your reasoning, based on your prediction in part b).
e) Carefully remove the tall cylinder and test your prediction. Were you correct? Explain.
f) Measure both cylinders and calculate their volumes. Use your calculations to explain the results of your experiment.

Journal

10. Describe a practical situation in which you would need to find the volume of a composite solid. Sketch the solid. Explain how you could find its volume. Give estimates for the measures that you would need and use these values to estimate the volume. How accurate do you think your estimate is? Explain.

11. A square-based prism has a small square-based prism removed from its centre, as shown. The depth of the composite solid is 10 cm.

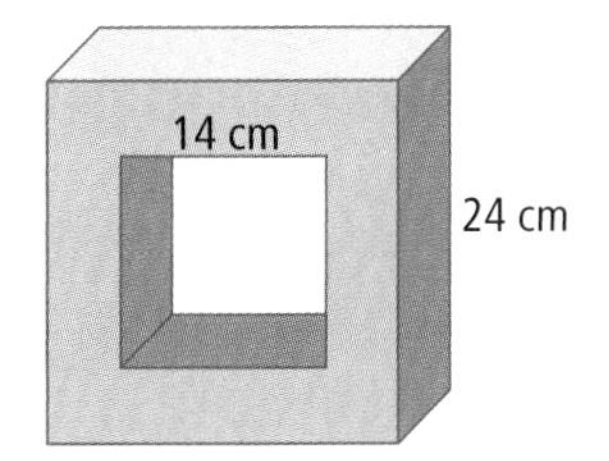

Find the volume of the composite solid.

Chapter Problem

12. The Robinson family's in-ground pool has a deep end and a shallow end, as shown.

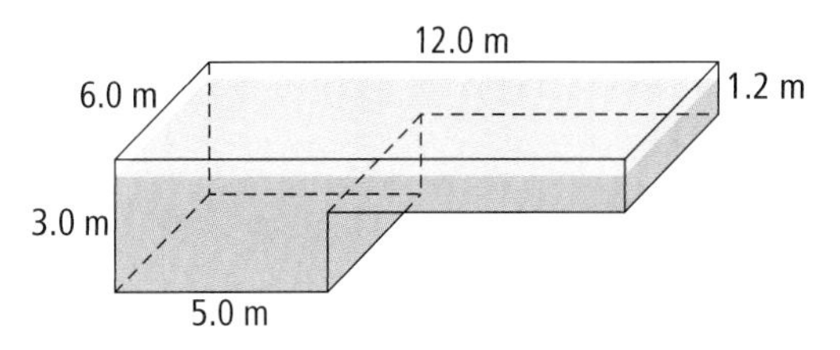

How many litres of water will it take to completely fill the pool?

13. Write a simplified expression for the volume of each solid.

a)

b)

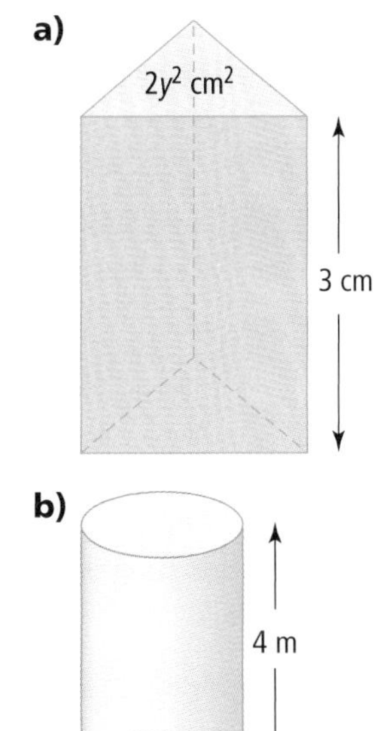

Extend

14. A small plastic octopus is dropped into a cylindrical container that is partly filled with water. The diameter of the base of the container is 8 cm and the height of the water in the container is 11 cm. If the volume of the small octopus is 24 cm^3, by how much will the height of the water in the container be increased?

CHAPTER 10

Review

1. A plate has a diameter of 24 cm. Find the plate's:
   a) circumference
   b) area

2. A coffee cup fits completely on a square coaster, whose side length is 7 cm. Choose the best estimate for the area of its base.
   • 5 cm^2 • 50 cm^2 • 45 cm^2 • 7 m^2
   Justify your choice.

3. Each rectangle has the same perimeter. Which has the greatest area? Explain how you know.

4. a) What simple shapes can be combined to form this composite figure?
   b) Describe how to find the area of this composite figure.
   c) Find the area.

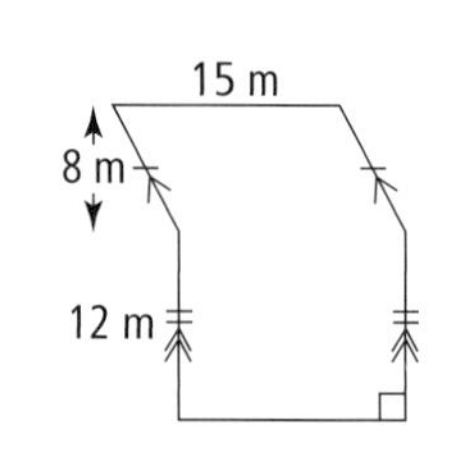

5. Mohan is renovating his bathroom.

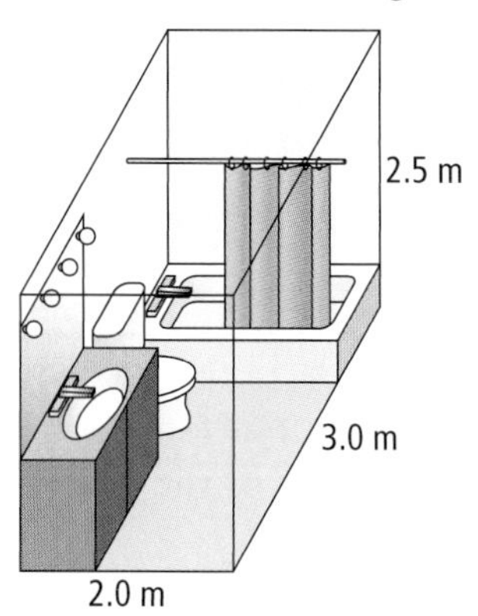

   a) He wants to install an exhaust fan. The manufacturer says the fan is suitable for ventilating spaces between 10 m^3 and 12 m^3. Will the fan be suitable for Mohan's bathroom? Explain.
   b) Mohan would like to paint the walls of his bathroom. A 1-L can of paint will cover 13 m^2 of area. How many cans of paint should Mohan buy? What assumptions did you make?

6. Next, Mohan wants to paint his bedroom, which measures 6.6 m by 6.6 m and has a 3-m high ceiling. Paint costs $25 for a 4-L can and each can of paint covers 50 m^2 of area.
   a) How much will it cost to paint all four walls of the bedroom, if Mohan applies two coats of paint? What assumptions did you make?
   b) Suppose half the wall space in Mohan's bedroom is filled with windows. How does your answer to part a) change?

7. Refer to questions 5 and 6. Mohan has decided to make his bathroom and bedroom larger by removing a large walk-in closet.
   a) What is the new volume of the bathroom if its length and width have both increased by 1 m?
   b) If the length and width of the bedroom have both increased by 0.5 m, how will your answers to question 6 change? How is the volume of the bedroom changed?

8. What is the area of a regular polygon with a perimeter of 48 cm, when the polygon has four sides? six sides? Use grid paper to draw diagrams.

9. Sophie is painting a sign, which is in the shape of a regular octagon. If she makes 20 signs, what total area will need to be painted? What assumptions did you make?

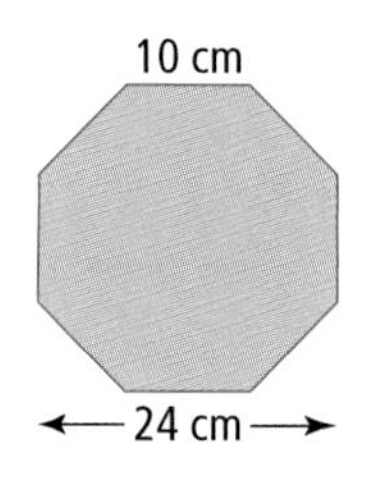

10. Find the area of the composite figure.

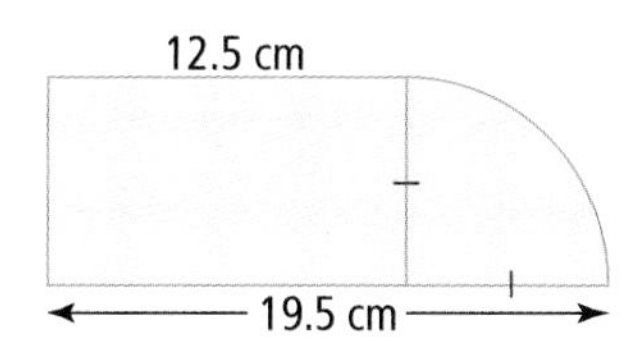

11. A square platform has an area of 36 m^2.

Find the maximum area of a circle that can be painted on the platform.

12. a) Find the total surface area of the monument shown. Do not include the area of the square base.

b) Find the total height of the monument to the nearest tenth of a metre.

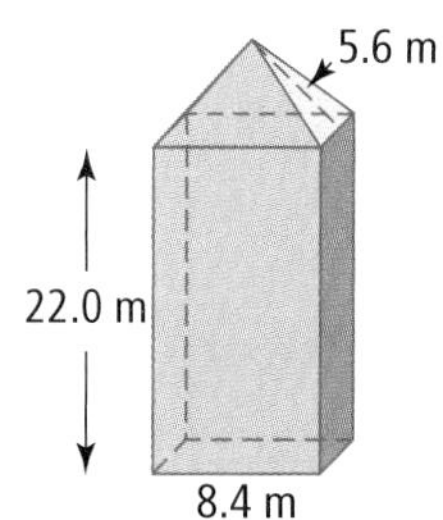

13. Find the volume of this block of cheese.

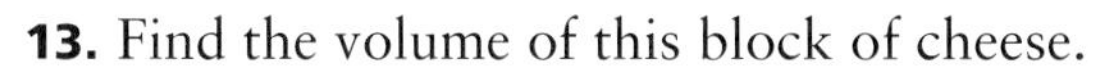

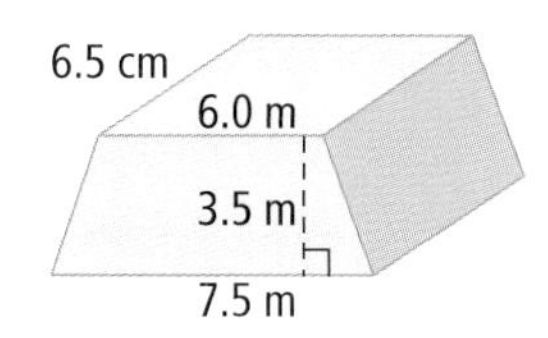

14. a) Find the volume of beans in this can.

b) The interior of the can is covered with a protective coating. What is the total surface area covered?

c) Repeat parts a) and b) for a can of tomatoes that is 3 cm greater in diameter than the can of beans. How do your answers change?

15. a) If you double the height of a square-based prism will you
- double its volume?
- double its surface area?

Record your predictions. Justify your reasoning.

b) Design and carry out an investigation to test your predictions. Use Polydron® pieces, pattern blocks, colour tiles, or other materials of your choice.

c) Write a report of your findings. Include:
- a description of the method you used, using words and diagrams
- an explanation of what you discovered
- a discussion of the correctness of your initial predictions

16. If the perimeter of a rectangular yard is 12 m, what are the possible whole number dimensions of the yard if its area is a maximum? a minimum? Use a table to organize your work.

CHAPTER 10

Practice Test

Selected Response

For questions 1 to 4, select the best answer.

1. Which is the best estimate for the area of one orange circle?

A 30 cm^2 **B** 100 cm^2
C 200 cm^2 **D** 1000 cm^2

2. A circular putting green has an area of 160 m^2. Which is the best estimate for its radius?

A 7 m **B** 14 m
C 25 m **D** 51 m

3. A box in the shape of a rectangular prism measures 14 cm × 18 cm × 30 cm. Its volume is:

A 62 cm^3 **B** 3780 cm^3
C 7560 cm^3 **D** 227 640 cm^3

4. A diagram of Tillie's lawn is shown. The perimeter of her lawn is:

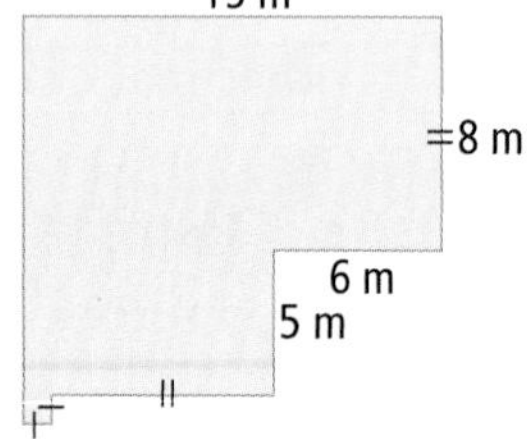

A 46 m **B** 58 m
C 166 m **D** 166 m^2

Short Answer

Provide a complete solution.

5. Refer to question 4. Find the area of Tillie's lawn. Explain how you determined your answer.

6. What is the area of a regular polygon with a perimeter of 36 cm, when the polygon has four sides? six sides? Use grid paper to draw diagrams.

7. a) What is the volume of tuna and liquid in the can?

b) The interior of the can is covered with a protective coating. What is the total surface area covered?

8. Refer to question 7.

a) If the volume of the can stays constant but the radius is changed to 64 mm, what would be the new height?

b) How would the interior surface area of the can change?

9. The red bullseye on the target has a diameter of 10 cm. The white and blue rings each have a width of 5 cm.

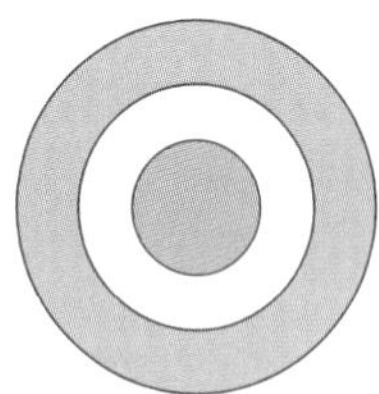

a) What is the area of the red bullseye?

b) What is the area of the blue outer ring?

10. The parallel sides of a trapezoid are 10.4 cm and 6.4 cm long. The height is 2 cm. Draw and label a diagram of the trapezoid. Determine the area of the trapezoid in two different ways.

11. Find the volume of air in this empty storage unit.

1.5 m
2.4 m
Easy-2-Store
2.5 m
2.0 m

Extended Response

Provide a complete solution.

12. A podium is made from two stacked wooden risers.

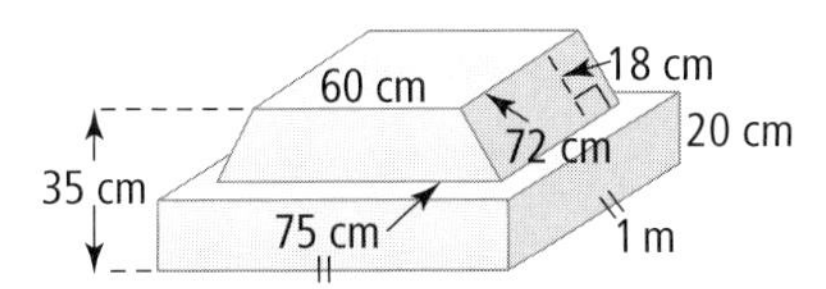

a) Find the total volume of the podium. What assumptions did you make?

b) If you paint only the exposed surfaces of each riser, how much surface area would you cover? What assumptions did you make?

13. A box in the shape of a square-based prism holds two golf balls, as shown. Each golf ball has a diameter of 43 mm.

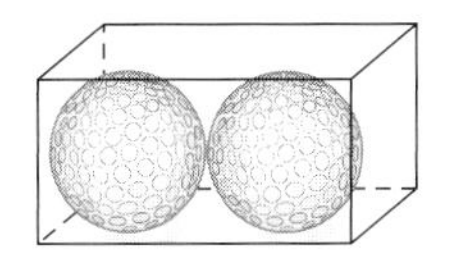

a) What is the volume of the smallest box that can hold the golf balls?

b) What is the surface area of this box?

14. If the perimeter of a rectangular yard is 16 m, what are the possible whole number dimensions of the yard if its area is a maximum? a minimum? Use a table to organize your work.

15. Carlos is renovating his kitchen.

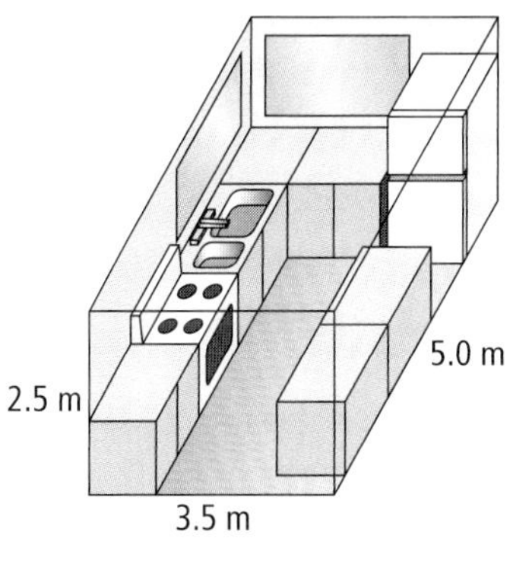

a) He wants to install an exhaust fan that is suitable for ventilating spaces between 40 m^3 and 42 m^3. Will the fan be suitable for Carlos's kitchen? Explain.

b) If Carlos increased both the length and the width of his kitchen by 1 m, how would the volume change?

c) Carlos wants to tile the walls. How much surface area will he cover if 60% of the walls are cupboard space?

Chapter Problem Wrap-Up

a) Design an additional structure for the Robinson family's backyard, such as another shed, a barbecue station, a birdbath, or a change room. The structure must be a composite solid made of at least two different simple components that you can measure easily.

b) Make a scale drawing or a model of your structure. Remember to include the scale.

c) Record the actual measurements of you structure. Calculate two of these measures for your structure: perimeter, area, surface area, or volume. Show your calculations.

d) Make a scale drawing of the backyard that shows where all three structures are located (pool, shed, your structure).

e) Explain why you chose your structure. How does it enhance the Robinson's backyard?

Glossary

A

acute angle An angle whose measure is less than 90°.

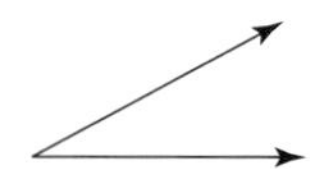

acute triangle A triangle in which each of the three interior angles is less than 90°.

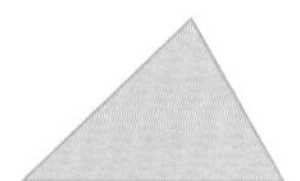

algebraic equation A mathematical statement that says two quantities or expressions are equal. Uses an equal sign.

algebraic expression A mathematical statement that usually includes one or more variables, one or more constants, and at least one mathematical operation (+. −, ×, ÷).

algebraic model A model that represents a relationship by an equation or formula, or represents a pattern of numbers by an algebraic expression.

alternate angles A pair of equal angles formed between two parallel lines and on opposite sides of a transversal.

$c = w$
$d = x$

apex The point in a three-dimensional object that is highest from the base.

approximate To find a number that is close enough to another number to make useful calculations. For example, 3.14 may be used to approximate π.

area The space covered by a two-dimensional figure. Measured in square units.

associative property The sum or product of three or more numbers will remain the same, no matter in what order you do the operations.

$(5 + 4) + 3 = 5 + (4 + 3)$
$(2 \times 5) \times 3 = 2 \times (5 \times 3)$

B

bar graph A graph that uses horizontal or vertical bars to represent data visually. Used to compare categories.

base (exponential form) The number used as a factor for repeated multiplication. In 6^3, 6 is the base.

base area Area of one of the two congruent faces of a prism

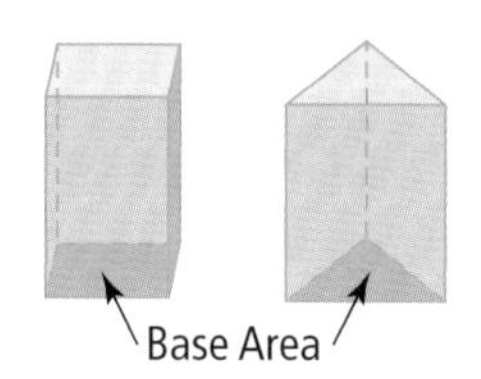

BEDMAS A way of remembering the order of operations.

B Brackets
E Exponents
D, M Division and Multiplication, from *left* to *right*
A, S Addition and Subtraction, from *left* to *right*

bias An emphasis on characteristics that are not typical of an entire population. Certain responses can be encouraged by the wording of a question.

binomial A polynomial with two terms, such as $3x^2 - 2x$.

box-and-whisker plot A diagram that shows the median and range of a numeric data set.

C

capacity The maximum amount of material or liquid that an object or container can hold. Measured in cubic units or liquid measures.

Cartesian grid The two-dimensional or (x, y) plane. Also known as the coordinate grid.

census A survey of an entire population.

central tendency A point or value in a range about which the rest of the data is considered to be balanced. Three measures of central tendency are mean, median, and mode.

circle graph A graph in which a circle is used to represent a whole and is divided into sectors that show how data are divided into parts by percent. Also called a pie chart.

circumference The distance around a circle. Short form is C.

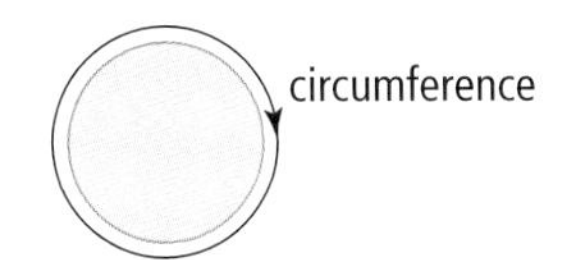

circumscribed circle A circle drawn around a polygon so that the curve passes through every point in the polygon.

co-interior angles A pair of supplementary angles formed between two parallel lines and on the same side of a transversal. Also called interior angles.

a / b
d / c
w / x
z / y

$d + w = 180°$
$c + x = 180°$

common denominator A number that is a multiple of the denominators of two or more fractions.

10 is a common denominator for $\frac{1}{2}$ and $\frac{1}{5}$.

commutative property Two numbers can be added or multiplied in any order and the sum or product will remain the same.

$7 + 4 = 4 + 7$ $3 \times 4 = 4 \times 3$

complementary angles Two angles that add to 90°.

complementary events Two or more distinct outcomes that represent the complete set of outcomes.

composite figure A figure made up of two or more simpler shapes called components.

composite number A number that has factors other than 1 and itself.

8 has four factors: 1, 2, 4, and 8.

compound event An event that contains several events. For example, when choosing a new bicycle, the type of brakes are one event, and the colour is another event.

cone A three-dimensional object with a base that is a flat surface and a point not on the base called an apex. All points on the base are connected to the apex by a line segment. Most cones are right circular cones where the base is cicular and a line drawn through the apex and the centre of the base is perpendicular to the base.

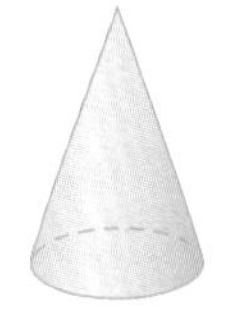

congruent figures Figures that have the same size and shape.

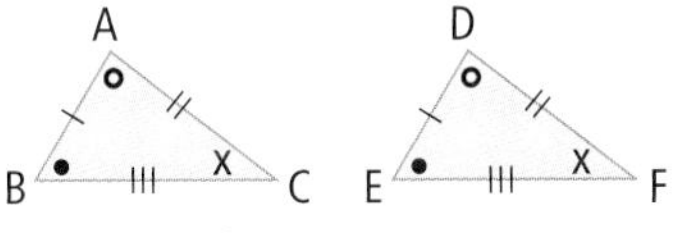

△ABC and △DEF are congruent.

constant 1. A value that does not change.

Represented by a number. In $4x - 1$, 1 is a constant.

2. A fixed rate or quantity. Example: slope represents a constant rate of change.

continuous data A set of data where a variable can be any real number. Represents the measure of a quantity that allows for continuous change, such as speed or temperature.

correlation The measure of how closely the points on a scatterplot fit a line. May be strong or weak, and positive or negative. The degree to which two quantities show a linear relationship.

corresponding angles A pair of angles on the same side of a transversal. If the two lines are parallel, corresponding angles are congruent.

a / b
d / c
w / x
z / y

$a = w$
$b = x$
$c = y$
$d = z$

counter-example An example that illustrates a situation in which a statement does not hold true. For example:

Statement: All perfect squares are even.
Counter-example: 9 is a perfect square that is not even. Therefore, the statement is false.

cube A prism with six congruent square faces.

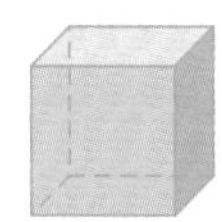

cube (cubic number) The product of three equal factors. Represents the volume of a cube.

$2 \times 2 \times 2 = 2^3$

cube root The factor of a cubed product. 2 is the cube root of 8.

cylinder A three-dimensional object with congruent flat bases that are parallel. All corresponding points on the bases are connected by line segments along the surface. A right circular cylinder has two parallel circular bases that are perpendicular to the curved surface.

D

database An organized collection of information. Often stored electronically.

denominator The number of equal parts in the whole or the group.

$\frac{3}{4}$ has denominator 4.

dependent events Two or more events whose outcomes affect each other.

dependent variable In a relation, the variable whose value is determined by the other variable (independent variable).

You are travelling at 100 km/h. The distance travelled, d, depends on the time, t. In $d = 100t$, d is the dependent variable.

diameter The distance from one side of a circle to the other, passing through the centre. Short form is d.

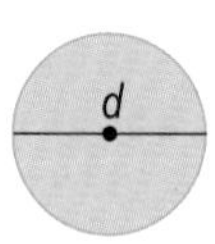

dilatation A transformation that enlarges or reduces a figure by a scale factor.

distributive property Factors can be represented as multiplication expressions. To apply the distributive property, break one of the factors into two parts that are easy to work with. Then multiply each part by the other factor.

$6 \times 12 = 6 \times (10 + 2) = (6 \times 10) + (6 \times 2)$

E

edge Where two faces of a three-dimensional figure meet.

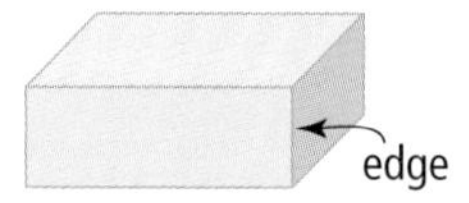

equation A statement formed by two expressions related by an equal sign. $3x + 3 = 2x - 1$ is an equation.

equilateral triangle A triangle with three equal sides and three equal angles.

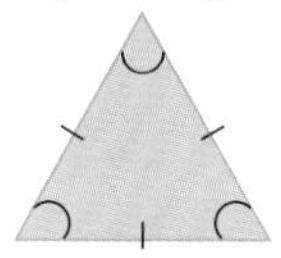

equivalent fractions Two or more fractions that represent the same part of a whole or a group.

$\frac{1}{3}$ and $\frac{2}{6}$ are equivalent fractions.

estimate To calculate approximately the value, amount, or size of something.

evaluate To find the value of an expression or an equation.

event A possible outcome in a probability experiment.

expand To apply the distributive property to an algebraic expression.

expanded form A way of writing a number that shows the value of each digit.

The number 4793 in expanded form is $4000 + 700 + 90 + 3$.

experimental probability A ratio that compares the number of favourable outcomes to the total number of trials in an experiment. The experimental probability may be different from the theoretical probability.

exponent The use of a raised number to denote repeated multiplication of a base. In 3^4, 4 is the exponent.

exponential form A shorter way of writing repeated multiplication.

$3 \times 3 \times 3 \times 3 = 3^4$

expression Numbers and variables, combined by operations. $3x + 2y$ is an expression.

exterior angle Angle formed by the extended side of a polygon and an adjacent side. Appears outside the polygon.

extrapolate To estimate values lying outside the given data. To extrapolate from a graph means to estimate coordinates beyond those that are plotted.

F

face A flat or curved surface of a three-dimensional object.

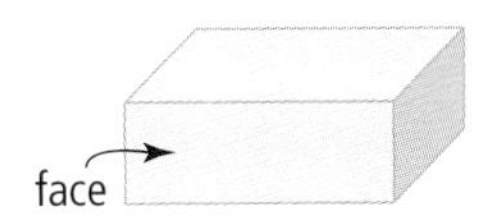

factor tree A diagram used to factor a number into its prime factors.

factors All the numbers that divide evenly into given number with no remainder. The factors of 8 are 1, 2, 4, and 8.

favourable outcome An outcome that produces a desired event. An outcome that counts for the probability being calculated.

first differences The differences between consecutive *y*-values in tables of values with evenly spaced *x*-values.

fraction Any rational number that is not an integer.

fractional percent A percent that contains a part of a percent in fraction form.

Example: $8\frac{2}{3}\%$ and $\frac{1}{4}\%$ are fractional percents.

frequency table A table used to show the number of occurrences in an experiment or survey.

G

greatest common factor (GCF) The greatest number that is a factor of two numbers. Used to write a fraction in lowest terms.

H

Harmonized Sales Tax (HST) Money collected by the federal and provincial government on purchases.

height The perpendicular distance from the base of a polygon to the opposite side or vertex. Short form is *h*.

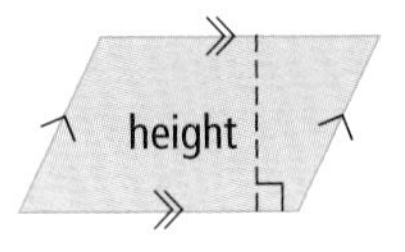

hexagon A polygon with six sides.

hexagonal prism A prism whose bases are congruent hexagons.

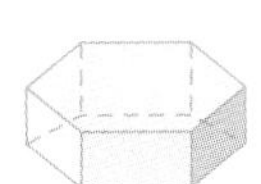

histogram A connected bar graph that shows data organized into intervals.

horizontal change The distance right or left between two points on a graph. On the *x*-*y* plane, the distance between x_1 and x_2.

hypotenuse The longest side of a right triangle. Opposite the right angle.

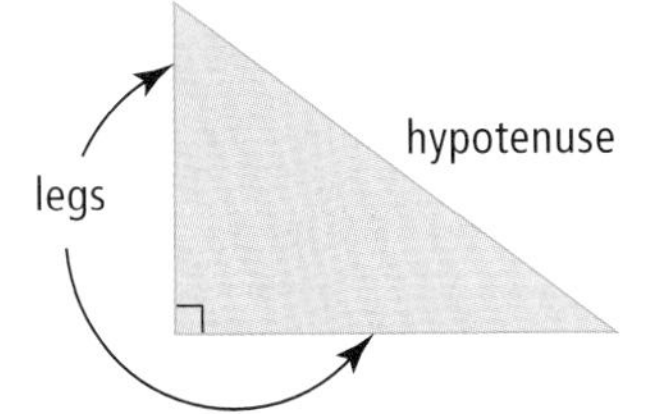

I

image The resulting line or figure after a transformation.

improper fraction A fraction in which the numerator is greater than the denominator, such as $\frac{8}{5}$.

independent events Two or more events whose outcomes do not affect each other.

independent variable In a relation, the variable that determines the value of the other variable.

You are travelling at 100 km/h. The distance travelled, d, depends on the time, t. In $d = 100t$, t is the independent variable.

inequality A mathematical statement relating expressions by using one or more inequality symbols $<$, $>$, $\leq$, or $\geq$. $4 < 5$. $x \geq -3$, and $-2 \leq a < 6$ are inequalities.

inscribed circle A circle drawn inside a polygon so that the curve touches every side of the polygon.

inspection A method of solving equations using mental math.

integer A number in the list ..., −3, −2, −1, 0, +1, +2, +3, ... All natural numbers, their opposites, and zero are integers.

interior angles A pair of supplementary angles formed between two parallel lines and on the same side of a transversal. Also called co-interior angles.

a b d c w x z y

$d + w = 180°$
$c + x = 180°$

interpolate To estimate values lying between given data. To interpolate from a graph means to estimate coordinates of points between those that are plotted.

intersecting lines Lines that cross each other.

intersection point The point on a graph where two lines meet.

inverse operation A mathematical operation that undoes a related operation. + and − are inverse operations. × and ÷ are inverse operations.

irrational numbers Numbers that cannot be expressed as a ratio of two integers.

For example, $\sqrt{2}$.
Decimal numbers that are non terminating and with digits that do not repeat in a fixed pattern.

For example, 3.141 592 653 ... .

irregular polygon A polygon that does not have all sides equal and all angles equal.

isometric drawing A diagram of three-dimensional objects in two dimensions.

isosceles triangle A triangle with exactly two equal sides and two equal angles.

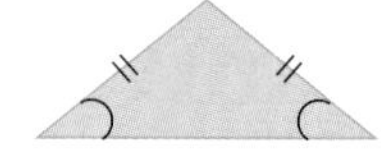

J

justify To provide evidence that supports a statement.

For example:
Statement: $3(5 + 2)$ is equivalent to $3(5) + 3(2)$.
Justification (supporting evidence): because they both are equal to 21.

L

lateral face The surface of a three-dimensional objects that is not a base. Example: the curved face of a cylinder.

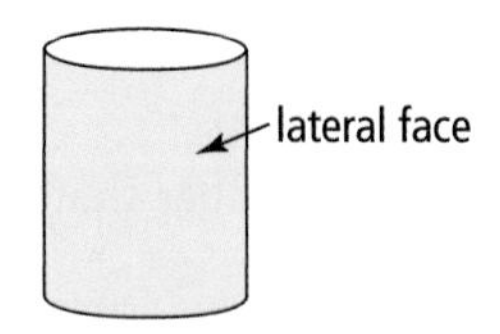

legs (of a right triangle) The two shorter sides of a right triangle. Legs meet at 90°.

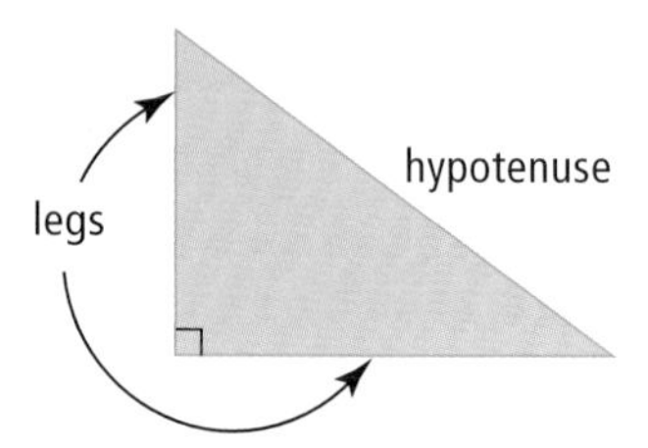

like terms Terms that have the same literal coefficient. $3xy$ and $-5xy$ are like terms but $2x$ and $7x^2$ are not like terms.

line graph A graph in which a line shows in trends in data, or a relationship between the two quantities or variables.

line of best fit The straight line that passes through, or as near as possible to, the points on a scatter plot. The line shows the possible trend or relationship between the two quantities being measured.

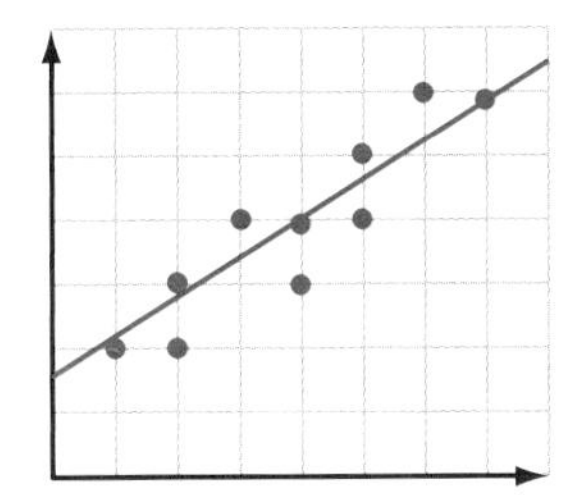

linear relationship As one variable changes by a constant amount, the other variable changes by a constant amount. A graph of the ordered pairs forms a straight line.

literal coefficient The non-numeric factor (or factors) of an algebraic term. For $4x$ the literal coefficient is x, while in $3x^2$ the literal coefficient is x^2.

lower quartile The median value of the first 50% of data points in an ordered set of data.

lowest common denominator (LCD) The lowest common multiple of the denominators of two or more fractions.

The LCD of $\frac{1}{2}$ and $\frac{2}{3}$ is 6.

lowest common multiple (LCM) The lowest multiple that two or more numbers share.

The LCM of 4 and 5 is 20.

lowest terms A way of writing a fraction so that the numerator and denominator have no common factors other than 1.

mapping To associate each point of a geometric shape with a corresponding point in another.

mean The sum of a set of values divided by the number of values in the set.

measure of central tendency A value that represents the centre of a set of data. It can be the mean, median, or mode.

median The middle value when a set of data is arranged in order from least to greatest.

mixed number A number made up of a whole number and a fraction, such as $3\frac{1}{2}$.

mode The most common value in a set of data.

model (noun) A physical model that can be used to represent a situation.

model (verb) To represent the facts and factors of, and the results of, a situation.

monomial A polynomial with one term, such as $3x^2$.

multiple The product of a given number and a natural number. Multiples of 2 are 2, 4, 6, 8, and so on.

N

natural numbers Numbers in the list 1, 2, 3, ... Also called positive integers.

negative exponent An exponent whose value is a negative number.

net A single pattern piece that can be folded to form a three-dimensional figure.

net area The area remaining after part of a two-dimensional figure has been removed.

net volume The volume remaining after part of a solid has been removed.

non-linear relationship As one variable changes by a constant amount, the other variable does not change by a constant amount. A graph of the ordered pairs does not form a straight line.

non-perfect square A number that cannot be written as the square of a whole number.

***n*th term** An item in a sequence or pattern. The variable, n represents the position in the sequence.

number line A line that matches a set of points and a set of numbers one to one.

−3 −2 −1 0 1 2 3

numerator The number of equal parts being considered in the whole or the group.

$\frac{3}{4}$ has numerator 3.

numerical coefficient The numeric factor in an algebraic term. For $4x$, the numerical coefficient is 4, while in $-3x^2$ the numerical coefficient is -3.

oblique Neither parallel nor perpendicular.

obtuse angle An angle that measures more than 90° but less than 180°.

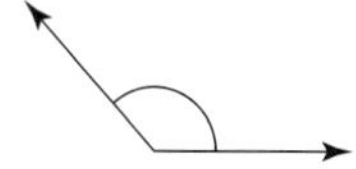

obtuse triangle A triangle containing one obtuse angle.

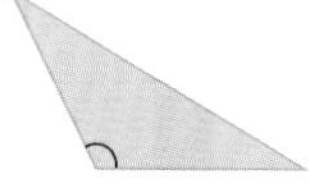

octagon A polygon with eight sides.

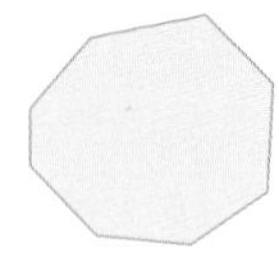

octahedron A polyhedron with eight triangular faces.

opposite angles A pair of equal angles formed by two intersecting lines.

Angles a and c, and b and d are pairs of opposite angles.

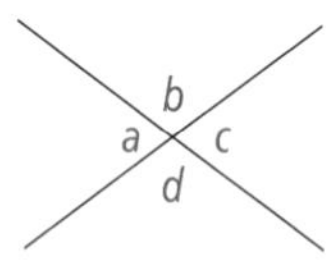

opposite integers Two integers with the same numeral but different signs. +2 and −2 are opposite integers.

order of operations Proper sequence of steps to evaluate an expression. Use BEDMAS to remember.

B Brackets
E Exponents
D } Division and Multiplication,
M } from *left* to *right*
A } Addition and Subtraction,
S } from *left* to *right*

ordered pair A pair of numbers, such as (4, 5), used to locate a point on a coordinate grid.

orientation The way the vertices of a polygon are ordered. If you read a figure as A–B–C in a clockwise direction, and read its image A–C–B in a clockwise direction, then the orientation of the image is the reverse of the object.

origin The point of intersection of the x-axis and the y-axis on a coordinate grid. The point (0, 0).

orthographic drawing A diagram that shows up to six views of an object. Usually front, back, side, and top views are given. Example:

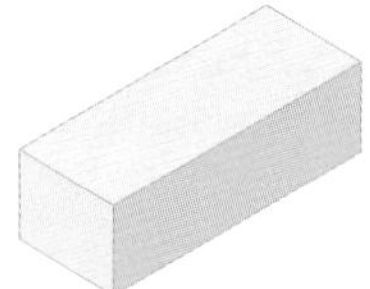

orthographic drawing

	back	
left	top	right
	front	

outcome One possible result of a probability experiment.

outlier An unusual value in a set of data that does not fit the pattern of the other data.

parallel lines Lines in the same plane that do not intersect.

parallelogram A four-sided figure with both pairs of opposite sides parallel.

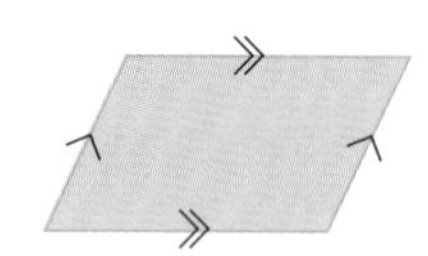

pattern An arrangement of shapes, lines, colours, numbers, symbols, and so on, for which you can predict what comes next.

pentagon A polygon with five sides.

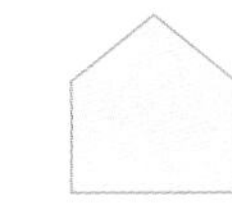

pentagonal prism A prism whose bases are congruent pentagons.

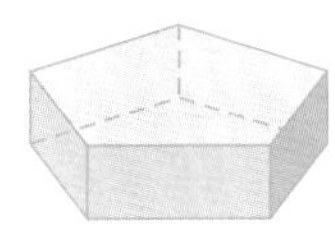

percent Out of 100.

75% means $\frac{75}{100}$ or 0.75.

percent decrease A ratio comparing the change in a quantity compared to the original quantity where the change is less than the original quantity.

percent increase A ratio comparing the change in a quantity compared to the original quantity where the change is greater than the original quantity.

perfect square A number that can can be expressed as the product of two equal integer factors. 16 is a perfect square because $16 = 4 \times 4$.

perimeter The distance around the outside of a two-dimensional figure.

perpendicular lines Lines that intersect at right angles.

perspective drawing A diagram of three-dimensional objects in two dimensions. Parts of the drawing designed to look farther away are smaller. Example:

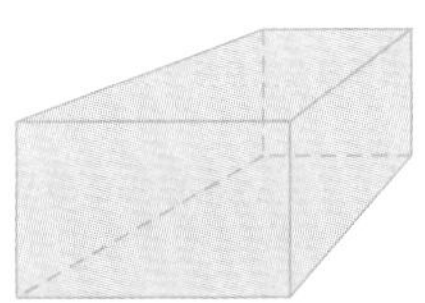

pi The ratio of the circumference of a circle to its diameter. Short form is π.

$\pi = \frac{C}{d}$

pie chart A graph in which a circle is used to represent a whole and is divided into sectors that show how data are divided into parts by percent. Also called a circle graph.

place value The value given to the place in which a digit appears in a number.

In the number 6172, 6 is in the thousands place, 1 is in the hundreds place, 7 is in the tens place, and 2 is in the ones place.

Platonic solids The five regular polyhedrons named after the Greek mathematician Plato. They are the cube, the tetrahedron, the octahedron, the dodecahedron, and the icosahedron.

polygon A two-dimensional closed figure whose sides are line segments.

polyhedron A three-dimensional figure with faces that are polygons. Plural is polyhedra.

polynomial An algebraic expression consisting of one or more terms separated by addition or subtraction signs with at least one variable that is raised to a non-negative whole number power.

population The entire group of people you want to learn about.

power A number in exponential form. Made up of a base and an exponent.

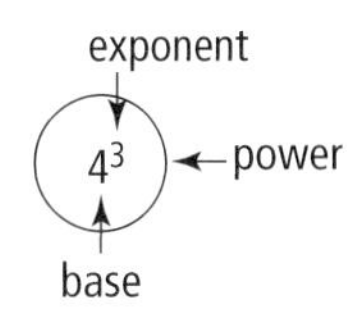

pre-image The original line or figure before a transformation.

prime factors The prime numbers that are multiplied to give a specific product.

The prime factors of 30 are 2, 3, and 5.
$30 = 2 \times 3 \times 5$

prime number A number that has only two different factors, 1 and itself.

prism A three-dimensional object with a parallel base and top that are congruent polygons. A prism is named by the shape of its base, for example, rectangular prism, triangular prism.

probability The chance that something will happen.

product A number resulting from multiplication. Example: 20 is the product of 5 x 4.

proper fraction A fraction in which the denominator is greater than the numerator.

$\frac{5}{8}$ is a proper fraction.

proportion A statement that says two ratios are equal. Can be written in fraction form as $\frac{1}{4} = \frac{4}{16}$ or in ratio form as 1:4 = 4:16. Proportions can be used to determine measurements in scale drawings.

pyramid A polyhedron with one base and the same number of triangular faces as there are sides on the base.

Pythagorean relationship The relationship between the lengths of the sides of a right triangle.

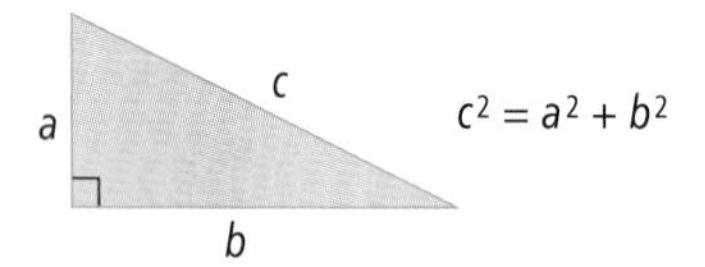

quadrilateral A closed shape with four straight sides.

quotient A number resulting from division.

Example: 7 is the quotient of 21 ÷ 3.

R

radical The square root of a number.

radical sign The square root sign $\sqrt{\ }$.

radicand The number inside the square root sign.

4 is the radicand of $\sqrt{4}$.

radius The distance from the centre of a circle to the outside edge. Short form is *r*. Plural is radii.

random sample A sample in which everyone in a population has an equal chance of being selected.

range The difference between the greatest and least values in a set of data.

rate A comparison of two quantities measured in different units. 120 km in 2 h and $\frac{\$2.97}{\text{kg}}$ are rates.

ratio A comparison of quantities measured in the same units. Can be written in ratio form as 3:4 or in fraction form as $\frac{3}{4}$.

rational numbers Numbers that can be written as the ratio *a*:*b* or fraction $\frac{a}{b}$ of two integers ($b = 0$).

Examples: $-3 = -\frac{3}{1}, 2\frac{3}{4} = \frac{11}{4}$. When *a* is divided by b, the result is a terminating decimal, such as 1.52, or a non-terminating, repeating decimal, such as 0.727 272 … .

reciprocals Two numbers that have a product of 1.

$\frac{3}{4}$ and $\frac{4}{3}$ are reciprocals.

rectangular prism A prism whose bases are congruent rectangles.

rectangular pyramid A pyramid with a rectangular base.

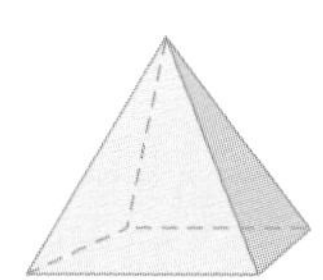

reflection A transformation that reflects a figure over a line. Results in a mirror image of the figure.

regular polygon A polygon with all sides congruent and all angles congruent.

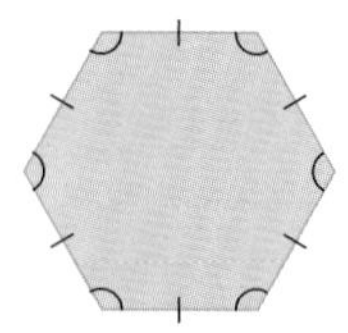

regular polyhedron A polyhedron in which all faces are congruent regular polygons. Also called a Platonic solid.

relation A relationship between two variables.

relationship A pattern formed between two sets of numbers. Can often be seen by plotting ordered pairs on a coordinate grid.

repeating decimal A decimal with a digit or group of digits that repeats forever. Write the repeating digits with a bar or dot:
$0.333... = 0.\overline{3}$ or $0.\dot{3}$

right angle An angle that measures 90°.

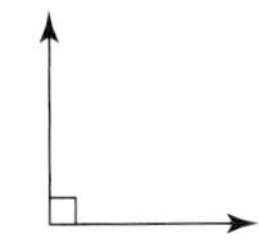

right triangle A triangle containing a 90° angle.

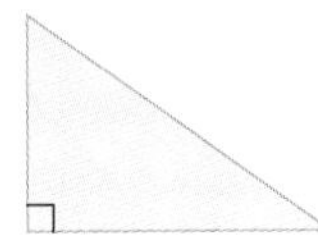

rise The vertical distance between two points on a line or line segment.

rotation A transformation that rotates a figure around a fixed point.

run The horizontal distance between two points on a line or line segment.

sales tax Money collected by a government on purchases. Usually written as a percent.

sample A small group that represents a population.

sample size The size of the data gathered in an experiment. For example, the number of trials conducted or the number of people surveyed.

sample space The set of all possible outcomes of an experiment or random trial.

scalar A real number.

scale The size of a figure or drawing in relation to the original object.

$$\text{Scale} = \frac{\text{diagram measurement}}{\text{actual measurement}}$$

A scale diagram is an accurate drawing that is either an enlargement or a reduction of an actual object.

scale factor A ratio or number that represents the amount by which a figure is enlarged or reduced.

scalene triangle A triangle with no equal sides or angles.

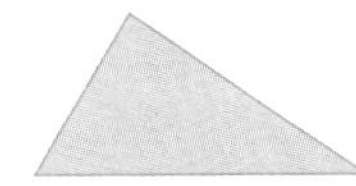

scatterplot A graph of ordered pairs of numeric data. Used to see relationships between two variables or quantities.

scientific notation A method of writing very large and very small numbers as a product of a number greater than or equal to 1 but less than 10 and a power of 10.

$123\ 000 = 1.23 \times 10^5$

$0.000\ 000\ 085 = 8.5 \times 10^{-8}$

sector Part of a circle bounded by two radii and an arc of the circumference.

sequence A pattern of numbers.

similar figures Figures that have the same shape but different size.

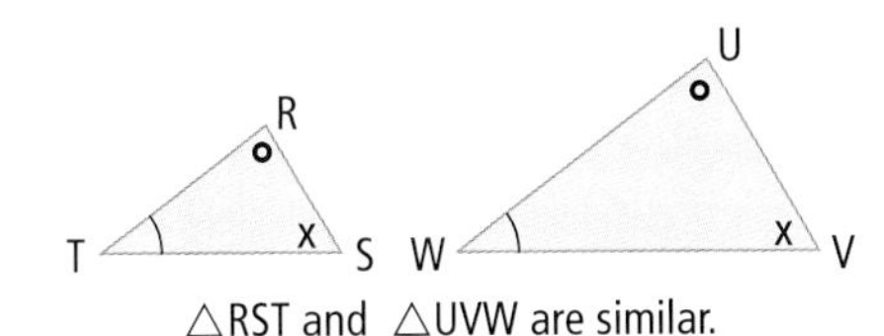

△RST and △UVW are similar.

simplest form A fraction in which the numerator and denominator have no common factors other than 1.

simplify To find an equivalent, simpler, and shorter expression.

Original: $300 - 4x - 100 + 3x$
Simplified: $200 - x$

simulation A probability experiment used to model a real situation.

slope A measure of the steepness of a line. Written as a fraction or a decimal. Represents a constant rate of change

$\text{Slope} = \frac{\text{rise}}{\text{run}}$

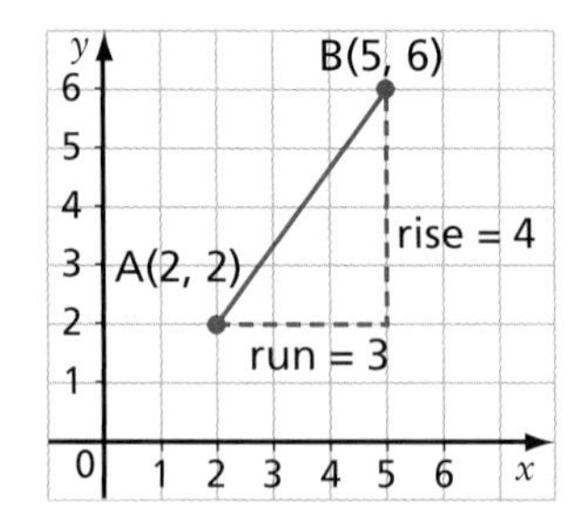

speed A comparison of distance travelled to time taken.

$\text{Speed} = \frac{\text{distance}}{\text{time}}$

sphere A round ball-shaped object. All points on its surface are the same distance from a fixed point called the centre.

spreadsheet A software tool for organizing and displaying numerical data.

square-based pyramid A pyramid with a square base.

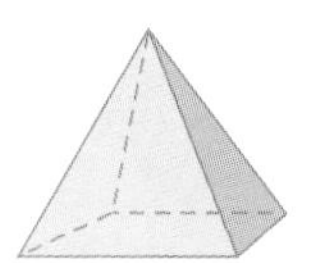

square (number) The product of two equal factors. Represents the area of a square.

$3 \times 3 = 3^2$

square root (of a number) A factor that multiplies by itself to give that number. Symbol is $\sqrt{\ }$.

Since $9 \times 9 = 81$, the square root of 81 is 9.
$\sqrt{81} = 9$

standard form A way of writing a number so that each digit has a place value.

4793 is written in standard form.

statistic A value calculated from a set of data.

stem-and-leaf plot A way of organizing numerical data by representing part of each number as a stem and the other part as a leaf.

straight angle An angle that measures 180°.

supplementary angles Two angles that add to 180°.

surface area The total outer area of all the faces of a three-dimensional figure.

symmetry A balanced arrangement on either side of a centre line or around a point. For example, an equilateral triangle has three lines of reflective symmetry and rotational symmetry of order 3.

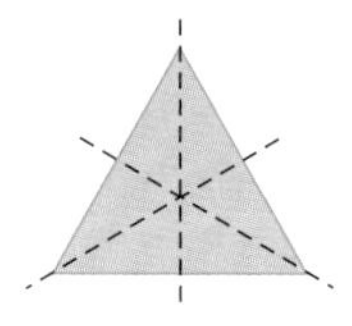

systematic trial A method of solving equations by substituting values for the variable until the correct answer is obtained.

table of values A table listing two sets of numbers that may be related.

tally chart A table used to record experimental results or data. Tally marks are used to count the data.

term 1. An expression formed by the product of numbers and variables. Terms are separated by the mathematical operations + and −.
2. A number or a daigram in a pattern. In the number pattern 20, 15, 10, the third term in 10.

theoretical probability A ratio that compares the number of favourable outcomes to the total number of possible outcomes. To calculate theoretical probability, all outcomes must be equally likely.

transformation A mapping of one geometrical figure to another according to some rule. See dilatation, reflection, rotation, and translation.

translation A transformation that slides a figure along a straight line.

transversal A line that crosses two or more lines.

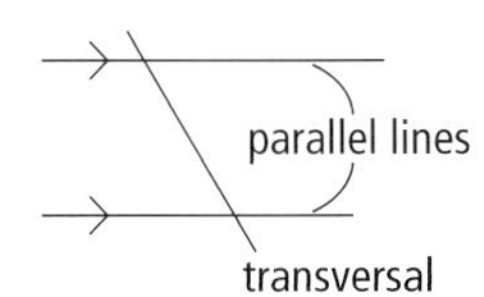

trapezoid A four-sided figure with at least one pair of opposite sides parallel.

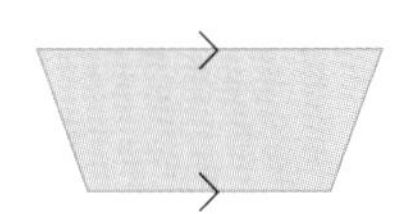

tree diagram A branching diagram used to list all the possible outcomes of a compound event.

trend A pattern found in data.

trial One round of a probability experiment, such as rolling dice once or picking one object from a bag of many objects.

triangular prism A prism whose bases are congruent triangles.

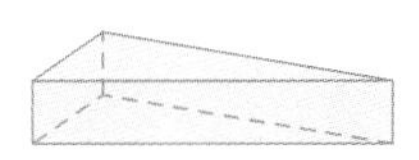

triangular pyramid A pyramid with a triangular base.

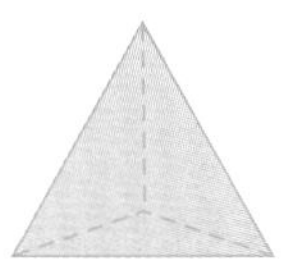

trinomial A polynomial with three terms, such as $y^2 + 3y - 1$.

U

unique (triangle) A triangle that is one-of-a-kind. Any other triangle constructed using the same specific set of instructions will be congruent to the original triangle.

unit price A unit rate that involves prices. $399/kg and $100 per day are unit prices.

unit rate A rate in which the second term is 1.

70 km/h is a unit rate.

upper quartile The median value of the second 50% of data points in an ordered set of data.

V

variable A letter used to represent a value that can change or vary. In $4x - 1$, x is a variable.

vertex A point at which two or more edges of a figure meet. Plural is vertices.

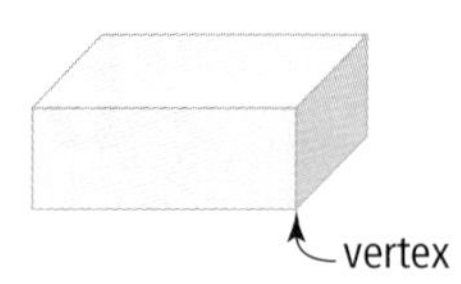

vertical change The distance up or down between the two points on a graph. On the x-y plane, the difference between y_1 and y_2.

volume The amount of space occupied by an object. Measured in cubic units.

W

whole numbers Numbers in the list 0, 1, 2, 3, … .

X

***x*-axis** The horizontal number line on a coordinate grid.

***x*-coordinate** The first number in the ordered pair describing a point on a coordinate grid.

The point P(2, 5) has x-coordinate 2.

***x*-intercept** The x-coordinate of the point where a line or curve crosses the x-axis. At this point $y = 0$.

Y

***y*-axis** The vertical number line on a coordinate grid.

***y*-coordinate** The second number in the ordered pair describing a point on a coordinate grid.

The point P(2, 5) has y-coordinate 5.

***y*-intercept** The y-coordinate of the point where a line or curve crosses the y-axis. At this point $x = 0$.

Z

zero principle The sum of a pair of opposite integers is zero.

Example: $(+7) + (-7) = 0$

Index

S

T

U

Z

Credits

Photo Credits

Contents: iii (left) ©Todd Gipstein/CORBIS; (right) ©Steve Craft/Masterfile; iv (left) ©John & Barbara Gerlach/Visuals Unlimited; (right) Royalty Free/CORBIS; v (left) Gilbert Cove & District Historical Society/Lorraine Lovett; (right) NASA; vi (top left) ©Garry Black/Masterfile; (top right) ©Michael Boys/CORBIS; (middle) PhotoLink/Getty Images; (bottom) ©Dick Hemingway; viii (left) ©Greg Ferens; (right) ©Neil Rabinowitz/CORBIS; p. 4 Roland W. Meisel.

Chapter 1: p. 8-9 Photodisc/Getty Images; p. 9 ©Phil Jason/Stone/Getty Images; p. 14 (top) ©Tony Freeman/Photo Edit. All Rights Reserved; (bottom) First Folio Resource Group, Inc.; p. 18 First Folio Resource Group, Inc.; p. 20 Tess Miller; p. 24 Photodisc Green/Getty Images; p. 25 NSARM, Wallace MacAskill Photographic Collection, #20040012; p. 27 Roland W. Meisel; p. 36 (top) AP Photo/New Haven Colony Historical Society, HO; (bottom) ©Todd Gipstein/CORBIS.

Chapter 2: p. 46-47 ©Garry Black/Masterfile; p. 47 ©Michael Boys/CORBIS; p. 56 ©Myrleen Ferguson Cate/Photo Edit-All Rights Reserved; p. 69 PhotoLink/Getty Images; p. 70 (top) ©Bob Daemmrich/Photo Edit. All Rights Reserved; (bottom) Roland W. Meisel; p. 80 Roland W. Meisel; p. 82 Roland W. Meisel; p. 90 Stockdisc/ Punchstock; p. 91 Photodisc Blue/Getty Images; p. 95 Roland W. Meisel; p. 96 (left) John A. Rizzo/ Getty Images; (right) ©Rebecca Cook/Reuters/ CORBIS; p. 99 Pippin Lee; p. 101 PhotoLink/Getty Images; p. 102 Roland W. Meisel.

Chapter 3: p. 104-105 Quilt courtesy Lee McLean, photography by Michelle Darling; p. 105 Suttles & Seawinds of Nova Scotia Ltd; p. 108 ©Steve Craft/Masterfile; p. 109 First Folio Resource Group, Inc.; p. 114 (top) ©Mikael Karlsson; (bottom left) ©SPL/Photo Researchers, Inc.; p. 120 Photodisc/Getty Images; p. 121 ©Dick Hemingway; p. 131 ©Martin Moos/Lonely Planet Images; p. 138 Nick Beecham.

Chapter 4: p. 144-145 ©David W. Carmichael; p. 150 First Folio Resource Group, Inc.; p. 152 Tom Dart; p. 156 ©Mike Dobel/Masterfile; p. 158 First Folio Resource Group, Inc.; p. 163 First Folio Resource Group, Inc.; p. 165 First Folio Resource Group, Inc.; p. 166 (top left) First Folio Resource Group, Inc.; (top right) Comstock/Punchstock; (bottom) PhotoLink/Getty Images; p. 167 ©Bubbles Photolibrary/Alamy; p. 170 ©Robyn Nelson/Photo Edit-All Rights Reserved; p. 173 ©Brandon Cole Marine Photography/Alamy; p. 175 ©Mike Johnson; p. 176 ©John & Barbara Gerlach/Visuals Unlimited; p. 179 ©Bill Banaszewski/Visuals Unlimited; p. 181 ©Reinhard Dirscherly/Alamy.

Chapter 5: p. 182 (bottom left) CP Photo/Halifax Chronicle Herald/Tim Krochak; p. 183 (bottom left) CP Photo/Paul Chiasson; (right) First Folio Resource Group, Inc.; p. 182-183 (top) Halifax Chronicle Herald; p. 186 First Folio Resource Group, Inc.; p. 188 ©Dennis Mac Donald/Photo Edit-All Rights Reserved; p. 193 First Folio Resource Group, Inc.; p. 195 ©Bettmann/CORBIS; p. 196 First Folio Resource Group, Inc.; p. 200 First Folio Resource Group, Inc.; p. 203 First Folio Resource Group, Inc.; p. 205 Pippin Lee; p. 206 ©Brett Barden; p. 213 ©Dick Hemingway; p. 220 The McGraw Hill Companies, Inc./Ken Cavanagh Photographer; p. 225 (left) Photodisc/Getty Images; (top) Brand X Pictures/Punchstock; (bottom) G.K. & Vikki Hart/ Getty Images; p. 226 Courtesy Nova Scotia Tourism, Culture & Heritage; p. 232 ©Dick Hemingway.

Chapter 6: p. 240-241 ©Digital Art/CORBIS; p. 246 ©Pixal/age Fotostock; p.247 First Folio Resource Group, Inc.; p. 250 Daniel MacDonald; p. 251 ©Scimat/Photo Researchers, Inc.; p. 254 STScI/NASA; (inset) National Science and Technology Council; p. 259 Daniel MacDonald; p. 261 Dynamic Graphics/Jupiter Images; p. 262 ©Ramin Talale/CORBIS; p.268 Pippin Lee; p. 275 Royalty Free/CORBIS; p. 284 (top) ©Biophoto Associates/Photo Researchers, Inc.; (bottom) Hybrid Medical Animation/Photo Researchers, Inc.; p. 285 Peggy Maclean.

Chapter 7: p. 290 ©Neil Rabinowitz/CORBIS; p. 291 ©Dick Killam; p. 296 Royalty Free/CORBIS; p. 305 ©Greg Ferens.

Chapter 8: p. 326-327 ©Greg Ferens; p. 327 ©Greg Ferens; p. 330 ©Dick Hemingway; p. 339 Andy Maclean/Ski Martock; p. 341 Digital Vision/Getty Images; p. 345 ©Michelle Darling; p. 348 ©Greg Ferens; p. 361 ©Chris Smith/Photo Edit-All Rights Reserved; p. 362 Nova Scotia Tourism, Culture & Heritage; p. 368 ©liquidlibrary/PictureQuest; p. 374 Roland W. Meisel.

Chapter 9: p. 382-383 ©Ed Gifford/Masterfile; p. 383 Keith Brofsky/Getty Images; p. 388 ©Denis Pepin/istockphoto; p. 395 courtesy IMAX Corporation; p.403 ©Michelle Darling; p. 404 First Folio Resource Group, Inc.; p. 410 Photodisc/Getty Images; p. 413 (top) Nick Beecham; (bottom) Courtesy Monarch Corporation.

Chapter 10: p. 424-425 ©Parks Canada. All rights reserved. Credit: Ron Garnett, 2007; p. 425 (top) Wolfgang Kaehler/CORBIS; (bottom) ©Tom Rosenthal/Superstock; p.436 NASA/CORBIS; p. 441 ©Michelle Darling; p. 442 (left) ©Paul Hutley, Eye Ubiquitous/CORBIS; (right) Roland W. Meisel; p. 444 ©Buddy Mays/CORBIS; p. 445 Gilbert Cove & District Historical Society/Lorraine Lovett; p. 455 ©Parks Canada. All rights reserved. Credit: Ship to Shore Photography, 2005; p. 456 Roland W. Meisel; p. 460 The McGraw Hill Companies, Inc./Jill Braaten, Photographer; p. 463 NSARM, J.A. Irvine fonds, Album 35, No. 86 (N-5642); p. 464 NASA.

Text Credits

Statistics Canada information is used with the permission of the Ministry of Industry, as Minister responsible for Statistics Canada. Information on the availability of the wide range of data from Statistics Canada can be obtained from Statistics Canda's Regional Offices, its World Wide Web site at http://www.statcan.ca, and its toll-free access number 1-800-263-1136.

Screen Captures

Calculator templates, Texas Instruments Incorporated; *The Geometer's Sketchpad*®, Key Curriculum Press.

Illustration Credits

Ben Hodson; xii, 3, 4, 6, 148, 310, 430
Tina Holdcroft Enterprises Inc.; 7, 92

Technical Art

Tom Dart, Kim Hutchinson, and Adam Wood of First Folio Resource Group, Inc.